Local Exhaust Ventilation

Aerodynamic Processes and Calculations of Dust Emissions

Local Exhaust Ventilation

Aerodynamic Processes and Calculations of Dust Emissions

Ivan Logachev
Konstantin Logachev
Olga Averkova

CRC Press
Taylor & Francis Group
Boca Raton London New York

CRC Press is an imprint of the
Taylor & Francis Group, an **informa** business

CRC Press
Taylor & Francis Group
6000 Broken Sound Parkway NW, Suite 300
Boca Raton, FL 33487-2742

© 2016 by Taylor & Francis Group, LLC
CRC Press is an imprint of Taylor & Francis Group, an Informa business

No claim to original U.S. Government works

Printed on acid-free paper
Version Date: 20150417

International Standard Book Number-13: 978-1-4987-2063-2 (Hardback)

Visit the Taylor & Francis Web site at
http://www.taylorandfrancis.com

and the CRC Press Web site at
http://www.crcpress.com

Contents

SECTION I Aerodynamics of Dust Airflows in the Spectra of Air Exhaust Ducts

SECTION II   *Recirculation of Ejected Air in Chutes*

Preface

This book expounds the theory behind design computations of gas-borne dust flows in local exhaust ventilation systems and provides practical recommendations on energy-efficient containment of gas-borne dust emissions.

The basic approaches to operational energy savings for local exhaust ventilation systems are discussed briefly, including shaping intake openings of open local exhaust devices after determining the boundaries of vortex areas, increasing the working distance of suction openings, inhibiting carryover of dust into aspiration network by promoting rotational aerodynamic fields and installing mechanical screens, minimizing air intake through leaking gaps of aspiration cowls by using mechanical screens and leveraging the jet separation effect, and promoting recirculation in aspiration cowls to suppress ejection of air.

Section I comprises a survey of separated and vortex currents in exhaust duct spectra. Topics covered include determination of vortex field boundaries, development dynamics of vortex flow patterns, and interaction between the exhaust plume and inflow jets. The behavior of individual dust particles and the polydisperse pluralities of dust particles are studied in effective spectra of exhaust ducts. Various classes of problems are considered together with open and closed configurations of local ventilated suction units.

The methods used for surveying currents in local exhaust ventilation systems include those known from the theory of the functions of a complex variable and ideal uncompressible liquid jet theory, boundary integral equations method, discrete vortex methods for stationary and nonstationary problems, vortex method, and numerical method for solving Reynolds-averaged Navier–Stokes equations and continuity equations.

Section II deals with the aerodynamics of loose-matter handling in porous ducts and the identification of regularities in air circulation patterns in bypass ducts. Differential equations for air ejection by a flow of loose matter have been put forward and subsequently solved and studied to demonstrate theoretically the capability for a manifold reduction of aspiration volume by optimizing the parameters of loading facilities. The case of grain handling in bucket elevators is used to illustrate ejection effects of air cross-flow patterns both in adjacent loading and discharge chutes and in the enclosures of the carrying and return runs of a bucket elevator.

The book may be of interest to researchers in fields such as aerodynamics of dust-control ventilation, mechanics of heterogeneous media, and fluid mechanics. It may be of value for process and mechanical engineers and designers involved in ventilation systems design, as well as supervisors of dust-control and ventilation facilities at industrial plants. The book will be useful to young scientists, undergraduate and postgraduate students majoring in the design of heat/gas supply and ventilation systems, occupational safety and health, environmental safety, and fluid mechanics.

Research findings published here would not have been possible without the financial support from the Grants Council of the President of Russia (Project MK-103.2014.1), the Russian Basic Research Foundation (Project No. 14-41-08005r_ofi_m, 14-08-31069mol_a), and the Strategic Development Program of the Belgorod State Technological University (Project No. A-10/12).

Authors

Ivan Nikolayevich Logachev holds a doctorate in engineering and is professor and academician at the Russian Academy of Natural Sciences, Moscow, Russia. He graduated from the Kharkiv Engineering-Building Institute in 1962, with a specialization in "heat and ventilation." For more than 30 years, he worked in the All-Union Scientific Research Institute of Safety and Environment in the Mining and Metallurgical Industry (Krivoy Rog), where he was promoted from research assistant to the head of the laboratory of industrial ventilation. Since 1994, he has been professor at Belgorod State Technological University, named after V. G. Shukhov.

Dr. Nikolayevich is author of more than 300 scientific papers, more than 50 inventions, and several monographs, including "Air Dedusting at Plants of Mining and Concentration Complexes, Moscow, Russia: Nedra, 1972," "Aspiration of Vapor and Dust Mixtures in Dedusting of Process Equipment, Kiev, Ukraine: Naukova Dumka, 1974," "Air Suction and Dedusting in Production of Powders, Moscow, Russia: Metallurgy, 1981," "Aerodynamic Basis of Aspiration, St. Petersburg, FL: Himizdat, 2005," "Dedusting Ventilation, Belgorod, Russia: BSTU, 2010," "Aerodynamics Antidust Ventilation, Saarbruchen, Germany: LAP Lambert, 2012," and "Industrial Air Quality and Ventilation: Controlling Dust Emissions, Boca Raton, FL: CRC Press, 2014."

Konstantin Ivanovich Logachev holds a doctorate in engineering and is professor at Belgorod State Technological University, Belgorod, Russia, where he has been working since 1995. He graduated from the Dnepropetrovsk State University in 1992, with a specialization in "hydroaerodynamics."

Dr. Ivanovich is author of more than 150 scientific publications, including many monographs: "Aerodynamics of Suction Torches, Belgorod, Russia: BSTU, 2000," "Aerodynamic Basis of Aspiration, St. Petersburg, FL: Himizdat, 2005," "Dedusting Ventilation, Belgorod, Russia: BSTU, 2010," "Aerodynamics Antidust Ventilation, Saarbruchen, Germany: LAP Lambert, 2012," and "Industrial Air Quality and Ventilation: Controlling Dust Emissions, Boca Raton, FL: CRC Press, 2014."

Olga Aleksandrovna Averkova received a Candidate of Technical Sciences degree from Belgorod State Technological University, Belgorod, Russia, in 2004 and has been associate professor in that university since then. She is author of more than 100 scientific publications, including some monographs: "Dedusting Ventilation, Belgorod, Russia: BSTU, 2010" and "Aerodynamics Antidust Ventilation, Saarbruchen, Germany: LAP Lambert, 2012."

Section I

Aerodynamics of Dust Airflows in the Spectra of Air Exhaust Ducts

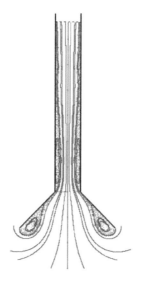

I.1 INTRODUCTION

One of the main causes of occupational diseases in workers is emissions of dust, pollutants, heat, and water vapor that are inherent to production processes in the metal, mining, and chemical industries and building material production and other industries. Levels of harmful contaminants in the atmosphere of industrial premises are usually much higher than in the atmosphere of the adjacent areas. The adverse impact on the workplace affects a lot of people comprising the active working-age part of the population.

It is important to keep up workers' health not only socially, but also economically. Treatment of occupational diseases in workers costs to the state much more than the introduction of ventilation systems. In addition, when penetrating into the moving parts of equipment, dust contributes to their early wear that results in the deterioration of operational performance and output quality.

The most reliable way to control emissions of harmful substances is local exhaust ventilation. Despite the continuous increase in the cost of manufacturing, installation, and operation of local ventilation systems, their efficiency often remains poor. The adequate dimensioning of local suction units will reduce the dust level in the working area below the allowable concentration with the minimum exhaust air volumes, which is directly related to the power consumption of the hood exhaust system. This, in the era of energy crisis, is becoming of particular importance. Therefore, increasing the range of suction flares of local suction units is an important scientific and technical problem, the solution of which will have a significant economic and social benefit: pollutant emissions will be contained with minimum energy consumption.

A right choice of local exhaust design and its location depends not only on production technology but also on the methods used to calculate dust and gas flows near suction inlets. The development of these calculation methods can be classified into three stages:

1. Construction of empirical relationships and absorption spectra
2. Determination of analytical formulas for the simple boundaries of air inflow to a suction unit
3. Numerical simulation of flows at local suction units

The first stage is associated with the research by V.V. Baturin, A.F. Bromley, A.S. Pruzner, Della Valle, Engels, Koop, Willert, and others [1–11].

To calculate the rectangular and circular cross sections of the suction intake, Della Valle proposed an empirical relationship:

$$Q = (10x^2 + A)v_x,$$

where

Q is the volume flow rate
A is the suction inlet area
v_x is the axial air velocity at a point located at distance x from the suction port

For the same inlets with a flange

$$Q = 0.75(10x^2 + A)v_x,$$

where the hole vs. flange areas ratio may range from 1:2 to 1:2.5.

Koop obtained a relationship for a round suction intake

$$v_x = v_o \cdot e^{-3.2 \cdot x/d},$$

where

d is the diameter of the suction inlet
v_o is the air suction velocity

For a rectangular inlet, the following formula is proposed:

$$v_x = v_o \cdot e^{-3.2 \cdot \frac{xU}{4A}},$$

where U is the perimeter.

Engels and Willert obtained a combined relationship for round and rectangular inlets:

$$\begin{cases} \dfrac{v_x}{v_o} = \dfrac{\left(x/r_h\right)^{-1.6}}{1+\left(x/r_h\right)^{-1.6}}, & x/r_h \le 2, \\[4mm] \dfrac{v_x}{v_o} = \dfrac{\left(x/r_h\right)^{-1.7}}{1+\left(x/r_h\right)^{-1.7}}, & x/r_h > 2, \end{cases}$$

where $r_h = A/U$ is the hydraulic radius.

For similar junction pipes with a flange

$$\begin{cases} \dfrac{v_x}{v_o} = \dfrac{1.35\left(x/r_h\right)^{-1.45}}{1+1.35\left(x/r_h\right)^{-1.45}}, & x/r_h \le 2, \\[4mm] \dfrac{v_x}{v_o} = \dfrac{2\left(x/r_h\right)^{-1.9}}{1+2\left(x/r_h\right)^{-1.9}}, & x/r_h > 2. \end{cases}$$

The inlet area relates to the flange area as 1:2.3.

Upon the processing of experimental findings, engineer A.S. Pruzner [4] proposed his empirical formulas. For square and circular inlets for $x/r_h \le 2$

$$\frac{v_x}{0.95 \cdot v_{av} - v_x} = 0.8\left(\frac{x}{r_h}\right)^{-1.4},$$

where v_{av} is the average speed in the suction inlet. For rectangular inlets,

$$\begin{cases} \dfrac{v_x}{v_{av} - v_x} = 0.8\left(\dfrac{x}{r_h}\right)^{-1.4}, & x/r_h \le 2, \\[4mm] \dfrac{v_x}{v_{av} - v_x} = \left(\dfrac{x}{r_h}\right)^{-1.7}, & x/r_h > 2. \end{cases}$$

For circular inlets

$$\frac{v_x}{v_{av}} = \left(\frac{x}{r_h}\right)^{-2}, \quad x/r_h > 2.$$

For square inlets

$$\frac{v_x}{v_{av}} = \frac{4}{\pi}\left(\frac{x}{r_h}\right)^{-2}, \quad x/r_h > 2.$$

M.F. Bromley [5] built an absorption spectrum for a circular inlet with sharp edges based on taking the velocity field experimentally. The absorption spectra for rectangular inlets with sharp edges in various fume-extraction hoods are described in a book by V.V. Baturin [6] and in work [7]. Article [8] studies the absorption spectrum, the scope of what includes a plane.

The experimental data accumulated during the period under review was used 50 years later in works [9,10], which approximated the experimental velocity field as found by M.F. Bromley. A suction spectrum model for the local suction of a stone-cutting machine was built upon the mathematical treatment of the experiment data (findings from 435 points were processed); its sphere of effect including a plane of the stone to be cut, which was accounted for by overlapping the spectrum of the model and its mirror image.

An expression, which is a particular solution of the Laplace equation, can serve as the basis for the experimental data analysis:

$$\Phi = e^{ax+by+cz} \quad \text{for } a^2 + b^2 + c^2 \neq 0,$$

where Φ is the velocity potential that was used in [2].

The above methods can be applied to a narrow class of simple suction inlets that were studied experimentally. Objects more complex than a plane placed in a suction flare cause considerable difficulties in determining the desired velocity field.

Among the experimental methods, we can point the method of electrohydrodynamic analogy (EHDA), which can be named a semianalytic method that allows calculating flat potential flows with any boundaries of the air inflow. This method was applied by Professor I.N. Logachev to construct a velocity field in the suction flares of local suctions of industrial baths [11]. A geometrically similar model of an industrial bath and adjacent enclosing structures (walls, floors, etc.) was plotted on resistance paper. The boundary conditions were implemented in a certain way and the flow velocity $v(x, y)$ was determined at a desired point of the model (x, y):

$$v(x, y) = \frac{\Delta\varphi}{\Delta\varphi_\infty} v_\infty,$$

where
$\Delta\varphi$ is the maximum increment of the potential near a point with coordinates (x, y)
$\Delta\varphi_\infty$ is the maximum increment of the potential near any point of the free-stream flow
v_∞ is the free-stream flow velocity

EHDA was used to obtain adjustment coefficients for the installation of industrial baths in calculating the lateral suction units. The same method was used to find the velocity components at flow around the cylinder.

The second historical stage is connected with such scientists as I.I. Konyshev, G.D. Livshits, I.N. Logachev, G.M. Pozin, V.N. Posokhin, E.V. Sazonov, V.N. Taliev, I.A. Shepelev, and others who calculated the local suction using the methods of conformal mappings, flow superposition, magnetic and vortex analogy, generalized flow superposition method, as well as direct integration of the Laplace equation by the Fourier method [12–51].

The greatest difficulties in the application of the conformal mapping method (see Chapter 1) relate to finding a mapping function, which is not always defined (e.g., for multiply connected domains). However, after this stage has been successfully completed, simple tasks can be solved without using a powerful computer but using a calculator. Unfortunately, the conformal mapping method allows solving only plane problems.

To describe plane potential airflows, a graphical method [12] was used based on the flow superposition method where a complex flow is presented as the sum of the simple ones. In the beginning, we build flow lines of flow components. To this end, we integrate the equation

$$\frac{dx}{v_x} = \frac{dy}{v_y}.$$

We superpose one family of flow lines over another family to obtain a grid in which cell sides represent the velocity vectors in a certain scale. The diagonal of any cell represents the velocity vector value of the resulting flow in the same scale.

The graphic method in [18] is used to calculate the velocity field near a slot-type connection pipe with two mutually perpendicular enclosing surfaces, where the flow was replaced by a system of sinks that are symmetrical mappings of the main sink on the surfaces. Evidently, the graphical method is not as accurate as the conformal mapping method.

The method of sources (sinks) [6,12] was used to calculate, approximately, the velocities at the selected points of both plane and 3D flows. It is based on the fact that at a considerable distance from the local suction, the velocity value varies according to the laws of sink:

$$v = \frac{L}{4\pi r^2},$$

where
L is the exhausted airflow
r is the distance from the suction center to the selected point

For the plane case, the equation for the velocity is as follows

$$v = \frac{L}{2\pi r}.$$

The calculation data obtained by the method of sinks may differ significantly from the experimental data.

The methods of vortex and magnetic analogy [27,52] are used to calculate the suction inlets in an unlimited space. These methods are based on the fact that the air velocity field is identified with the flux density field of a semi-infinite solenoid or semi-infinite *film of vortices*. Using the Biot–Savart law, I.I. Konyshev obtained the formulas to calculate the axial air velocity for the circular, rectangular, triangular, and round variable cross sections of semi-infinite suction connections. The following formula was obtained for a round connection pipe with a radius r with a suction rate v_o at a distance z from the suction opening:

$$v = \frac{v_o}{2}\left(1 - \frac{z}{\sqrt{r^2 + z^2}}\right).$$

For a rectangular pipe with a size of $2a \times 2b$

$$v = \frac{2v_o}{\pi}\arctg\frac{ab}{z\sqrt{a^2 + b^2 + z^2}},$$

and for a regular triangle-shaped pipe with the side b

$$v = \frac{3v_o}{\pi} \operatorname{arctg}\left[\sqrt{3}\,\frac{\sqrt{\frac{1}{3}b^2 + z^2} - z}{\sqrt{\frac{1}{3}b^2 + z^2} + 3z}\right].$$

Works [28–32] used the flow superposition method (see Section 1.4) with suction hole integration of sinks to obtain formulas to calculate the axial velocity at the exhaust openings built into an infinite flat wall. Abroad, the flow superposition method was used to study the velocity field near a rectangular suction inlet [44]. That study did not result in such simple formulas as those obtained by I.A. Shepelev. The sources were integrated by summing up 100 single sinks. A flow confined by walls (one, two, and three mutually perpendicular walls) was studied and described using mirror-image presentation and graphical addition. This method was used to consider a problem in the plane [45] for a single point sink, a point source, and a plane flow.

E.V. Sazonov [50,51] offered new formulas to calculate the extension length of the projecting hood and the required volumes of aspiration to remove escaping gases at loading and unloading of electric furnaces on the basis of formulas by V.N. Posokhin, I.A. Shepelev, M.I. Grimitlin [49], and the methodology [48].

Though the flow superposition method is more than simple, it is difficult to determine how to superpose the elementary flows to get a pattern of the flow of interest, especially when it comes to local suction units in confined conditions. In his work [53], N.Ya. Fabrikant describes the generalized flow superposition method, which enables to solve this problem. Sections 4 and 5 describe the principle of this method. We only note that it consists of solving the Fredholm second-order boundary integral equations. Therefore, abroad, this method was named the boundary integral equation method, or the boundary element method [54–57,59], based on the works by Russian mathematician S.G. Mikhlin [58]. The generalized flow superposition method is a special case of the BIE method, as the latter covers a wider class of problems, including elasticity problems.

The generalized flow superposition method, sometimes referred to as the method of *singularities* in the aerodynamics, was applied by G.D. Livshits and his followers [33–38,128] to calculate circular, square, and ring-type semi-infinite pipes that are loosely spaced. There are some similar studies abroad as well [46,47]. No calculation of flows with complex boundaries was made.

The third historical stage of development of calculation methods for local suction units is connected with an intensive development of computer technology and using IT software for numerical methods of aerohydrodynamics and mathematical analysis. We must mention the contributions made by V.K. Khrushch, N.N. Belyaev, V.G. Shaptala, G.L. Okuneva, V.N. Posokhin, I.L. Gurevich, R.Kh. Akhmadeev, I.I. Konyshev, and other researchers [60–78].

Work [60] investigates a slot-type suction unit from axially symmetric diffusion sources (a ring-type suction unit from a round bath). It provided the solution of the continuity equation for the potential flow in cylindrical coordinate system r, z:

$$\frac{\partial^2 \psi}{\partial z^2} + \frac{\partial^2 \psi}{\partial r^2} - \frac{1}{r}\frac{\partial \psi}{\partial r} = 0,$$

where ψ is the flow function.

This equation was approximated using a five-point stencil and was solved by the Seidel iterative method combined with the overrelaxation method. After that, velocity components were calculated using difference analogues.

I.I. Konyshev proposed a computation algorithm for axially symmetric problems by the numerical integration of the Laplace equation in a spherical coordinate system by the method of finite differences.

An axially symmetric flow at a circular flanged pipe is studied in [61]. The velocity field was also based on the numerical integration of the Laplace equation.

Works [70,71] model flat airflows in a production room when it is ventilated with jets. Here, we have used a vorticity transfer and diffusion equation obtained from the Navier–Stokes equation in the Boussinesq approximation and the Poisson equation for the flow function

$$\frac{\partial \Omega}{\partial t} + \nabla\left(\overline{v}\Omega\right) = \frac{1}{\mathrm{Re}_t} \Delta\Omega,$$

$$\Delta\psi = -\Omega,$$

where $\overline{v}$ is the time-averaged dimensionless air velocity value. The vorticity vector is

$$\Omega = \frac{\partial v_y}{\partial x} - \frac{\partial v_x}{\partial y}.$$

The horizontal and vertical components of velocity are

$$v_x = \frac{\partial \psi}{\partial y}, \quad v_y = -\frac{\partial \psi}{\partial x},$$

where ψ is the flow function. The Reynolds number was defined by the relation

$$\mathrm{Re}_t = 5\frac{v_1 H}{vl}\sqrt[3]{\frac{HD}{l^2}},$$

where
$v_1 = 1$ m/s
H, D are the height and width of the room
l is the basic scale of turbulence

The explicit conservative Arakawa scheme was used to solve the vorticity transfer equation. The successive overrelaxation scheme was used to solve the Poisson equation by the relaxation method.

Simulation of distribution of airflows, temperature, and concentration of impurities in the workplace was performed in [72,75,76] on the basis of the Navier–Stokes equations in the Boussinesq–Oberbeck approximation, equations of heat transfer, and convective diffusion of impurities. The numerical solution of these equations served as a base to study, by the relaxation method, the foundry shake-out department aspiration system that consists of a local exhaust unit and general dilution airflow [73,74].

A software system (SS) *Aspiration* developed by V.K. Khrushch and N.N. Belyaev determines the optimal location of local suction units and aspirated air volumes [77,78] that helps to ensure the greatest concentration of impurities in the exhaust air. The Aspiration SS is based on a numerical integration of the equation

$$\frac{\partial\varphi}{\partial t}+\frac{\partial u\varphi}{\partial x}+\frac{\partial v\varphi}{\partial y}+\frac{\partial(w-w_s)\varphi}{\partial z}+\sigma\varphi$$

$$=\mathrm{div}(\mu\nabla\varphi)+\sum_{i=1}^{N}q_i(t)\delta\big(r-r_i(t)\big)-\sum_{j=1}^{M}\varphi Q_j(t)\delta(r-r_j),$$

where

$q_i(t)$ is the strength of a point source of pollutant emissions

$r_i(x_i, y_i, z_i)$ are the Cartesian coordinates of the specified source

Q_j is the airflow removed by the local suction, which is presented as a 3D point sink in this model

$\delta(r-r_i)$, $\delta(r-r_j)$ are the deltas of the Dirac function

φ is the concentration of pollution

u, v, w are the air velocity projections

w_s is the rate of gravitational sedimentation of contaminants

The most promising method [16,117,126, Chapter 3] currently is the method of discrete vortices [79,112,113,130–141], which is widely used in aerodynamics to calculate bearing surfaces, turbulent jets, and wakes. This method applied to problems of airflows near the suction holes [80,82–88,93,94] enables not only to take into account the formation of transient eddy zones, but also to investigate the interaction of supply air jets and suction flares.

Stochastic and deterministic methods are used to calculate the dynamics of the solids in ventilating air jets in an unlimited and confined space. Considerable computational effort is required to describe the dispersed phase on the basis of the kinetic equation for the one-particle distribution function [142]. The deterministic method based on the discrete vortex method is used to calculate the distribution of dispersed particles in turbulent jets and aircraft trail [116]. Work [104] offers a method to forecast the dispersed composition of the dust particles in the aspirated air at the transshipment of bulk materials on the basis of the method of trajectories and determination of the maximum diameter of dust particles. The value of the maximum diameter was assumed to be the reference point and the particle size distribution was assumed to adhere to the lognormal distribution. Note that the definition of the particle size distribution and concentration of dust particles in the aspirated air is a difficult and ambiguous problem in experimental aerodynamics as well. The experimental results vary in a quite large range and sometimes contradict each other due to the pulsation nature of processes.

1 Determination of the Velocity Field by Conformal Mapping Methods

1.1 UNSEPARATED FLOWS

1.1.1 GENERAL INFORMATION ON THE CONFORMAL MAPPING METHOD

A mapping of the area z on the area ω, as defined by the function $\omega = f(z)$, is called conformal if it preserves the stretches and rotation angles. This means that any infinitesimally small element in z varies the same number of times as it is mapped $\omega = f(z)$ and the linear expansion coefficient is $|f'(z)|$. The curve in the area z is rotated by the same angle $\alpha = \arg f'(z)$ in the same direction when mapped $\omega = f(z)$. Conformal mapping method is used to calculate the potential flows of gas. A flow is called potential if there is a function φ that is called a potential, for which the following equality is true:

$$\vec{v} = \operatorname{grad} \varphi, \tag{1.1}$$

where $\vec{v}$ is the flow velocity.

The horizontal and vertical components of velocity are determined, respectively, by the formulas

$$v_x = \frac{\partial \varphi}{\partial x}, \quad v_y = \frac{\partial \varphi}{\partial y}. \tag{1.2}$$

The continuity equation for a flat potential flow of gas is reduced to the Laplace equation for the potential

$$\frac{\partial v_x}{\partial x} + \frac{\partial v_y}{\partial y} = \frac{\partial^2 \varphi}{\partial x^2} + \frac{\partial^2 \varphi}{\partial y^2} = 0. \tag{1.3}$$

The line on which $\varphi = \mathrm{const}$ is called an equipotential.

In addition to the potential, a flow function ψ is introduced:

$$v_x = \frac{\partial \psi}{\partial y}, \quad v_y = -\frac{\partial \psi}{\partial x}, \tag{1.4}$$

which also satisfies the continuity equation.

The line on which the function ψ is constant is called a flow line. The velocity vector is tangential at each point of the flow line. The equipotentials and flow lines are mutually perpendicular.

The complex flow potential is the function

$$w = \varphi + i\psi. \tag{1.5}$$

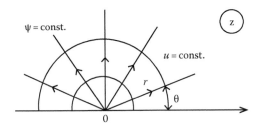

FIGURE 1.1   Point source in the semiplane.

If we know the function w in the flow plane z, then by defining the complex velocity

$$\frac{dw}{dz} = \frac{\partial w}{\partial x}\frac{\partial x}{\partial z} = \frac{\partial \varphi}{\partial x} + i\frac{\partial \psi}{\partial x} = v_x - iv_y, \tag{1.6}$$

we can find the vertical and horizontal components of the velocity at any point in the flow range.

Define, for example, the complex flow potential in the upper semiplane with a point source of a strength Q located in the origin of coordinates (Figure 1.1).

In this case, the flow lines are rays starting at 0 and the equipotentials are semicircumferences. The velocity along the flow line is

$$v_r = \frac{\partial \varphi}{\partial r} = \frac{Q}{\pi r}, \tag{1.7}$$

whence $\varphi = \int \frac{Q}{\pi r} dr = \frac{Q}{\pi}\ln r + C$. Assume that $\varphi = 0$ for $r = 1$, then $C = 0$. In order to determine the flow function, we calculate its partial derivative:

$$\frac{\partial \psi}{\partial y} = v_x = v_r \cos\theta = \frac{Qx}{\pi r^2}. \tag{1.8}$$

For the first quadrant, the angle $\theta = \text{arctg}\frac{y}{x}$ and for the second one, it is $\theta = \pi + \text{arctg}\frac{y}{x}$. Therefore, $\frac{\partial \theta}{\partial y} = \frac{x}{r^2}$. As $\frac{d\psi}{d\theta} = \frac{\partial \psi}{\partial y}\frac{\partial y}{\partial \theta} = \frac{Q}{\pi}$, $\psi = \frac{Q}{\pi}\theta$ (const = 0, as we choose $\psi = 0$ for $\theta = 0$) and the complex potential is $w = \frac{Q}{\pi}\ln r + i\frac{Q}{\pi}\theta = \frac{Q}{\pi}\ln z$. If the point source is at the point z_0, then

$$w = \frac{Q}{\pi}\ln(z - z_0). \tag{1.9}$$

In the case of a point sink

$$w = -\frac{Q}{\pi}\ln(z - z_0). \tag{1.10}$$

The algorithm for calculating the gas velocity field in a given flow plane that is called a physical plane consists of the following steps:

1. Definition of the mapping of the physical flow plane z onto the upper semiplane t (a geometric flow plane, which can be represented by any other area with a known complex potential):

$$z = f(t). \tag{1.11}$$

2. Setting the point z_0, in which it is necessary to determine the flow rate, and calculating the parameter t_0 by Equation 1.11.
3. Calculation of the complex velocity

$$\frac{dw}{dz} = \frac{dw}{dt}\frac{dt}{dz}, \tag{1.12}$$

which is used to determine the coordinates of the velocity vector.

Step 1 is the most demanding one. In addition, mapping Equation 1.11 cannot be defined for multiply connected regions, which are not reduced to simply connected ones, and, therefore, the conformal mapping method cannot be applied for such problems.

An important property of conformal mappings is the principle of symmetry. If a certain line divides the physical flow area into two symmetric subareas, then one of them can be disregarded and the velocity field can be calculated for the remaining subarea. The obtained calculation results of the velocity field will be valid for the symmetric points of the disregarded subarea as well.

Consider the mapping of the upper semiplane onto the interior of a polygon (Figures 1.2 and 1.3), which is often used in engineering practice [89].

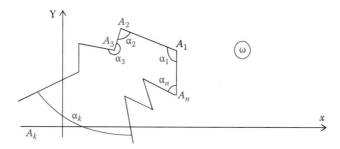

FIGURE 1.2 Polygonal domain.

FIGURE 1.3 Upper semiplane.

The sense of the area traversal is chosen so that the area is always on the left if we move counter-clockwise along the area boundary. The upper semiplane on a polygonal area is mapped using the Schwarz–Christoffel formula:

$$\omega = C_1 \int_{t_k}^{t} (t-t_1)^{\frac{\alpha_1}{\pi}-1}(t-t_2)^{\frac{\alpha_2}{\pi}-1}\cdots(t-t_n)^{\frac{\alpha_n}{\pi}-1}dt + C_2, \tag{1.13}$$

where

$t_1, t_2, \ldots, t_n$ are n independent parameters
$\alpha_1, \alpha_2, \ldots, \alpha_{n-1}$ are $(n-1)$ independent parameters
$\alpha_1 + \alpha_2 + \cdots + \alpha_{n-1} + \alpha_n = (n-2)\pi$
C_1, C_2 are complex constants, and, therefore, four independent parameters

In total, we have $2n + 3$ independent parameters. Note that if t_j is in ∞, then the integral does not contain the term $(t-t_j)^{\frac{\alpha_j}{\pi}-1}$. When mapping by the Schwarz–Christoffel formula, with the preserved sense of traversal, the abscissas in three points in the plane t can be arbitrarily set and the remaining ones must be defined. The reference point t_k is assigned in such a way that the constant C_2 could be easily determined.

1.1.2 Calculation of the Axial Air Velocity at a Freely Spaced Slot-Type Suction

Direct the axis OX along the symmetry axis of the slot-type opening with a width of $2B$ with an air exhaust rate on ∞ equal to 1 (Figure 1.4). Define the air velocity on the axis OX.

Since this area is symmetrical about the horizontal axis, the lower part of the figure can be disregarded (Figure 1.5).

Map the upper semiplane (Figure 1.6) onto the resulting triangular area (Figure 1.5). The points A_1, A_2, A_3 are mapped, respectively, onto t_1, t_2, t_3.

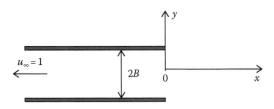

FIGURE 1.4 Freely spaced slot-type suction.

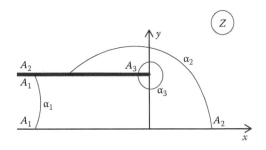

FIGURE 1.5 Physical flow plane.

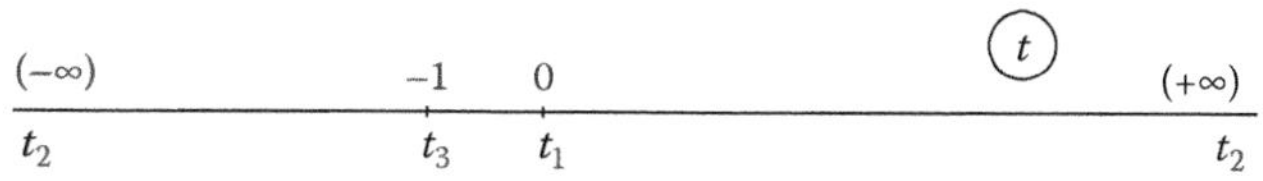

FIGURE 1.6   Geometric flow plane.

The angles $\alpha_1 = 0$, $\alpha_3 = 2\pi$. As $\alpha_1 + \alpha_2 + \alpha_3 = \pi \Rightarrow \alpha_2 = \pi - 2\pi = -\pi$.
The Schwarz–Christoffel integral for this mapping is

$$z = C_1 \int_{t_k}^{t} (t-0)^{0-1}(t+1)^{\frac{2\pi}{\pi}-1} dt + C_2. \tag{1.14}$$

We recommend choosing as a reference point t_k such a point that allows easily determining the constant C_2. In this case, $t_k = t_3 = -1$. Then,

$$z = C_1 \int_{-1}^{t} \frac{t+1}{t} dt + C_2. \tag{1.15}$$

If we assume $t = -1$, then $z = iB$ and $C_2 = iB$. The constant C_1 is determined from the correspondence of the points A_1 and t_1. We use the fact that that in going through the point t_1 along an infinitely small circle with a radius ε from the ray $t_1 t_2$ to the interval $t_1 t_3$, a jump of iB from $A_1 A_2$ to $A_1 A_3$ in the plane z occurs, that is,

$$z(+\varepsilon) - z(-\varepsilon) = 0 - iB = -iB. \tag{1.16}$$

We integrate Equation 1.15 to obtain

$$z = C_1(t+1+\ln t - i\pi) + iB.$$

Let us calculate the values of z at the points ε and $-\varepsilon$ when $\varepsilon \to 0$:

$$z(\varepsilon) = C_1(\varepsilon + 1 + \ln \varepsilon - i\pi) + iB$$

$$z(-\varepsilon) = C_1(-\varepsilon + 1 + \ln \varepsilon + i\pi - i\pi) + iB.$$

Therefore,

$$\lim_{\varepsilon \to 0}(z(\varepsilon) - z(-\varepsilon)) = -i\pi C_1.$$

Use Equation 1.16 to define

$$-iB = -i\pi C_1 \Rightarrow C_1 = \frac{B}{\pi}.$$

Thus, the desired mapping is

$$z = \frac{B}{\pi}(t+1+\ln t). \tag{1.17}$$

TABLE 1.1

Values of the Axial Velocity of Air Inflow to the Slot-Type Suction Unit in an Infinite Space

x/B	0	0.5	1	1.5	2	2.5	3	3.5	4	4.5	5
u	0.78	0.56	0.38	0.27	0.2	0.16	0.13	0.11	0.1	0.09	0.08

On the flow geometric plane (Figure 1.6), we have a point sink at the point t_1. Therefore, the complex potential of this flow is $w = -(Q/\pi)\ln t$. The complex velocity in the physical plane (Figure 1.5) is

$$\frac{dw}{dz} = \frac{dw}{dt}\frac{dt}{dz} = -\frac{1}{t+1}.$$

Thus, the algorithm for calculating the axial velocity at a freely spaced slot-type suction unit consists of four steps (Table 1.1):

1. The initial coordinate $x = 0$ is preset.
2. The formula $x = \dfrac{B}{\pi}(1 + t + \ln t)$ is used to determine the parameter t, for example, by the half-interval method.
3. The formula $u_x = -\dfrac{1}{t+1}$ is used to calculate the axial velocity of the air at x.
4. Perform the step $x = x + \Delta x$ and return to Step 2. Calculate until the predetermined point is reached.

1.1.3 CALCULATION OF THE AXIAL AIR VELOCITY AT SLOT-TYPE SUCTION BUILT IN A FLAT INFINITE WALL

Similar to Section 1.2, we assume that the width of the slot-type opening is $2B$ and the suction velocity $u_\infty = 1$ (Figure 1.7). Define the air inflow velocity on the horizontal axis.

As the flow area is symmetric, we can disregard the lower semiplane (Figure 1.8).

The geometric flow plane (Figure 1.9) is mapped onto a physical one with the Schwarz–Christoffel formula:

$$z = C_1 \int_0^t \frac{\sqrt{t}}{t-1}\,dt + C_2. \tag{1.18}$$

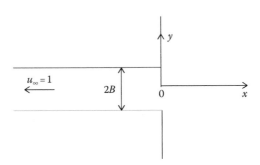

FIGURE 1.7   Slot-type suction built in a flat infinite wall.

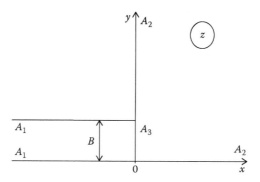

FIGURE 1.8 Physical flow plane.

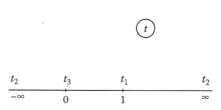

FIGURE 1.9 Geometric flow plane.

The constant $C_2 = iB$, as $z = iB$ for $t = 0$. On integrating expression Equation 1.18 we obtain

$$z = C_1 \left[2\sqrt{t} + \ln\left(\frac{\sqrt{t}-1}{\sqrt{t}+1} \right) - i\pi \right] + iB. \tag{1.19}$$

The constant C_1 is determined from the correspondence of the points t_1 and A_1. When the point t_1 is traversed along the semicircle of the radius $\varepsilon \to 0$ from the ray $t_1 t_2$ to the interval $t_3 t_1$, a jump in the physical plane occurs at $\Delta z = -iB$.

As $\lim_{\tau \to 0} \left[z(1+\varepsilon) - z(1-\varepsilon) \right] = -C_1 \cdot i\pi \Rightarrow C_1 = \dfrac{B}{\pi}$, the desired mapping will be

$$z = \frac{B}{\pi} \left[2\sqrt{t} + \ln\frac{\sqrt{t}-1}{\sqrt{t}+1} \right]. \tag{1.20}$$

The complex flow potential on the geometrical plane is $w = -(Q/\pi)\ln(t-1)$, so the complex velocity is

$$\frac{dw}{dz} = \frac{dw}{dt}\frac{dt}{dz} = -\frac{1}{\sqrt{t}}. \tag{1.21}$$

The calculation algorithm consists of the following steps:

1. The initial coordinate $x = 0$ is preset.
2. The formula $x = \dfrac{B}{\pi}\left(\sqrt{t} + \ln\dfrac{\sqrt{t}-1}{\sqrt{t}+1} \right)$ is used to determine the parameter t.
3. The formula $u_x = -\left(1/\sqrt{t}\right)$ is used to calculate the axial velocity of the air at x.
4. Perform the step $x = x + \Delta x$ and return to Step 2. Calculate until the predetermined point is reached.

TABLE 1.2

Values of the Axial Velocity of Air Inflow to the Slot-Type Suction Unit Built in a Flat Infinite Wall

x/B	0	0.5	1	1.5	2	2.5	3	3.5	4	4.5	5
u	0.83	0.65	0.48	0.37	0.29	0.24	0.2	0.18	0.16	0.14	0.13

The calculation results (Tables 1.1 and 1.2) show that the axial velocity at a slot-type suction unit built into the flat infinite wall is higher than at a freely spaced suction; this can be explained by the influence of the gas inflow boundaries. In the first case, the inflow region is smaller than in the second one. Note that if the air suction rate $u_\infty \neq 1$, then all the results obtained should be multiplied by this value.

1.1.4 CALCULATION OF THE AXIAL AIR VELOCITY AT A SLOT-TYPE SUCTION BELL

Local suction hoods are widely used in containing ventilation. It can be useful to find the optimal length and tilting angle of the hood bell that ensure the maximum range of local suction intake flare. The range of the intake flare refers to the distance from the suction entrance to the point where the axial velocity equals a predetermined value.

Consider the airflow near a slot-type suction with a width $2B$ having a flange (hood) of length l, which is mounted at an angle $\alpha\pi$ ($-1 < \alpha < 1$) to its axis (Figure 1.10). Determine the dependence of the axial velocity $v\xi$ on the angle $\alpha\pi$ and the flange length l.

The Schwarz–Christoffel formula is used to map the upper semiplane z onto the physical plane ω with the point correspondence as shown in Figures 1.10 and 1.11:

$$\omega = \frac{Bm^\alpha}{\pi b} \int_{-1}^{z} \left(\frac{t+1}{t+m}\right)^\alpha \left(\frac{t+b}{t}\right) dt + Bi. \tag{1.22}$$

The complex flow potential in the upper semiplane (the sink is at the point a_1, the source is at a_2)

$$w = -\frac{Q}{\pi} \ln z, \tag{1.23}$$

where $Q = v_\infty B$ is the airflow and the complex velocity is

$$\frac{dw}{dz} = -\frac{Q}{\pi z}. \tag{1.24}$$

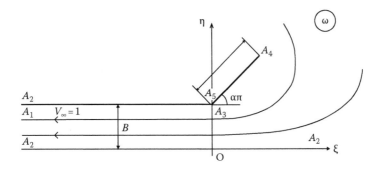

FIGURE 1.10   Air inflow to a freely spaced slot-type flanged pipe.

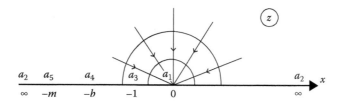

FIGURE 1.11   Sink in the upper semiplane.

The complex velocity in ω can be found using the equation $v(\omega) = \dfrac{dw}{d\omega} = \dfrac{dw}{dz}\dfrac{dz}{d\omega}$ and formula (Equation 1.22):

$$v(\omega) = -\frac{Qb(z+m)^{\alpha}}{Bm^{\alpha}(z+1)^{\alpha}(z+b)}. \tag{1.25}$$

The axial velocity of the slot-type suction (at 0ξ) is determined from the equality

$$-\frac{v_{\xi}}{v_{\infty}} = \frac{b}{m^{\alpha}}\left(\frac{x+m}{x+1}\right)^{\alpha}\frac{1}{x+b}. \tag{1.26}$$

Relations (Equations 1.22 and 1.26) are formally the solution of the problem set.

The unknown parameters m, b in Equation 1.22 are determined based on the correspondence of the points A_4, A_5 and a_4, a_5 (Figures 1.10 and 1.11) by setting the values of the flange length l/B and the angle α:

$$\int\limits_{1}^{m}\left(\frac{t-1}{m-t}\right)dt - b\int\limits_{1}^{m}\left(\frac{t-1}{m-t}\right)^{\alpha}\frac{dt}{t} = 0, \tag{1.27}$$

$$\frac{l\pi b}{Bm^{\alpha}} = \int\limits_{1}^{b}\left(\frac{t-1}{m-t}\right)^{\alpha}\frac{b-t}{t}dt. \tag{1.28}$$

The integrals in Equations 1.27 and 1.28 were calculated by the Gauss quadrature formulas. However, the difficulties in solving this problem extend much further. The integral in Equation 1.22 is convergent only in the sense of the principal value (integral of Cauchy type). It is difficult to find the numerical value of this integral. Thus, we reduce Equation 1.22 to a form more suitable for numerical integration.

Since we are only interested in points on the axis of suction, then $z = x$ in Equation 1.22, in which the integral is transformed as follows:

$$J = \int\limits_{-1}^{x}\left(\frac{t+1}{t+m}\right)^{\alpha}\frac{t+b}{t}dt = bJ_1 + (1-b)J_2,$$

where $J_2 = \displaystyle\int_{-1}^{x}\left(\frac{t+1}{t+m}\right)^{\alpha}$ is a convergent integral for $\alpha > -1$ and the integral $J_1 = \displaystyle\int_{-1}^{x}\frac{(t+1)^{\alpha+1}}{(t+m)^{\alpha}}\frac{dt}{t}$ is

divided into three: $J_1 = \displaystyle\int_{-1}^{-\varepsilon} + \int_{-\varepsilon}^{\varepsilon} + \int_{\varepsilon}^{x} \quad \varepsilon \to 0.$

The integration paths of the first and third integrals are straight-line segments of the real axis and are definite integrals of the real function. The second integral is integrated over the semicircle with a radius $r = \varepsilon$ centered at a_1 (see Figure 1.11):

$$\int_{-\varepsilon}^{\varepsilon} \frac{(t+1)^{\alpha+1}}{(t+m)^{\alpha}} \frac{dt}{t} = \int_{\pi}^{0} \frac{(re^{i\varphi}+1)^{\alpha+1}}{(re^{i\varphi}+m)^{\alpha}} \frac{ire^{i\varphi}}{re^{i\varphi}} d\varphi = \{r = \varepsilon \to 0\} = \frac{-\pi i}{m^{\alpha}}.$$

Using the integration formula by parts, define

$$J_1 = \frac{(1+x)^{\alpha}}{(m+x)^{\alpha}}(1+x)\ln x + \alpha(1-m)\int_{-1}^{x}\left(\frac{1+t}{m+t}\right)^{\alpha}\frac{\ln|t|}{m+t}dt - \int_{-1}^{x}\left(\frac{1+t}{m+t}\right)^{\alpha}\ln|t|dt - \frac{\pi i}{m^{\alpha}}$$

and after substituting the obtained value J_1 in Equation 1.22, we have

$$\omega = \frac{Bm^{\alpha}}{\pi b}\left[b\frac{(1+x)^{\alpha+1}}{(m+x)^{\alpha}} + b\alpha(1-m)I_1 - bI_2 + (1-b)I_3\right], \tag{1.29}$$

where

$$I_1 = \int_{-1}^{x}\left(\frac{1+t}{m+t}\right)^{\alpha}\frac{\ln|t|}{m+t}dt$$

$$I_2 = \int_{-1}^{x}\left(\frac{1+t}{m+t}\right)^{\alpha}\ln|t|dt$$

$$I_3 = \int_{-1}^{x}\left(\frac{t+1}{t+m}\right)^{\alpha}dt$$

The integrals I_1, I_2, I_3 are convergent improper integrals [102] for $\alpha > -1$.

The algorithm for determining the axial velocity consists of four steps:

1. We assign the value of the flange length l/B, flange tilt angle α, and the coordinates of the point on the slot axis ξ/B (and at each time the distance is measured from the flange edge, i.e., from the abscissa of the point A, i.e., for $0.5 < \alpha < 0.5$, the point coordinates are $\xi/B + l/\cos\alpha$).
2. We calculate the parameters m and b from Equations 1.27 and 1.28.
3. Use Equation 1.29 to determine the parameter x.
4. Use Equation 1.26 to determine the axial velocity v_ξ/v_∞.

The calculations have shown that the axial velocity at the suction slot with a flange of a finite length has three extrema (Figure 1.12) when the flange tilt angle changes within $-1 < \alpha < 1$, of which two maximums are at $\alpha = 0.5$ and $\alpha = -0.5$.

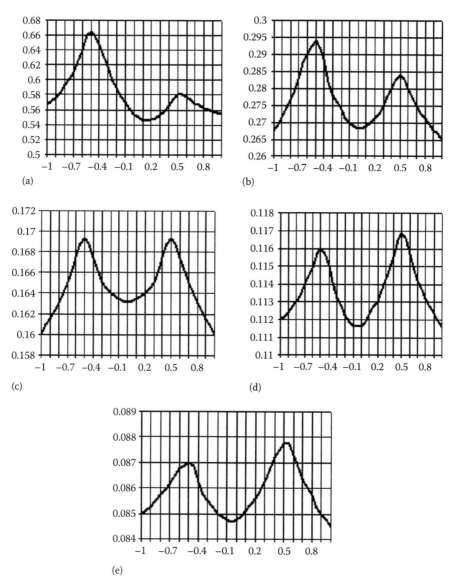

FIGURE 1.12 Change in the axial velocity v_ξ/v_∞ (vertical axis) as a function of the flange tilt angle α (horizontal axis) at the fixed point ξ/B: (a) $\xi/B = 0.5$, (b) $\xi/B = 1.5$, (c) $\xi/B = 2.5$, (d) $\xi/B = 3.5$, and (e) $\xi/B = 4.5$.

In this case, the greatest value of the axial velocity is found at a distance $0 < \xi/B < 2.5$ from the entry section for $\alpha = -0.5$ and in $\xi/B > 2.5$ for $\alpha = 0.5$.

The range of the suction flare increases with the increase in flange length (Figure 1.13). In the area $\xi/B < 3$, the axial velocity for $l/B = 3$ almost reaches its limit (corresponding to $l/B \to \infty$).

The results obtained confirm the well-known fact that the flow boundaries play the dominant role. In our case, the boundaries are the flange and walls of the suction channel. The role of the former is essential for the flow near the suction section, but the suction channel walls have a greater influence at a considerable distance. The regular change in the axial velocity follows the law of the line sink in the latter case.

Despite that the maximum axial velocity in $\xi/B < 2.5$ is observed when $\alpha = -0.5$; it is not advisable to install the flange at a negative angle due to an increase in power consumption resulting from an increase in the velocity of air entry into the suction channel, which would be inevitable in this case.

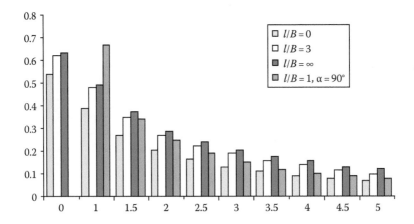

FIGURE 1.13 Change in the axial velocity v_ξ/v_∞ (vertical axis) as a function of the distance from the suction entrance ξ/B (horizontal axis).

Thus, we plotted the dependence of the axial velocity of the open-type slotted suction on the length and tilt angle of its hood (flange). The least axial velocity attenuation is achieved when the flange is installed at the right angle to the axis of the slot whose optimal length is three gauges (a gauge is a half-width of the slot).

1.2   SEPARATED FLOWS

We will assume the following classical assumptions used to calculate the sucked airflows [16,89]: the liquid is weightless and incompressible, no vortex is present, and the flow is steady. The normal component of velocity on the solid walls and unknown free flow lines (resulting from the separation of the flow) is equal to zero. In addition, the module of velocity u_0 is constant on free flow lines. In the described approach, the free flow line is the boundary between the vortex region, occurring at the walls, and a potential flow that will be studied. It follows that a flow is determined if the velocity potential $w(z)$ is known where $z = x + iy$ is a physical plane of the airflow.

The Joukowski function is as follows:

$$\omega = -\ln\frac{dw}{u_o dz} = -\ln\frac{u_x - iu_y}{u_0} = -\ln\frac{u(\cos\theta - i\sin\theta)}{u_0} = -\ln\left(\frac{u}{u_0}e^{-i\theta}\right) = -\ln\frac{u}{u_0} + i\theta, \qquad (1.30)$$

where θ is the angle of the velocity vector $\vec{u}$ to the positive direction of the axis OX.

The plane of the Joukowski function ω must be mapped on the parametric plane t where the complex potential of the flow is known, and then this parametric plane t must be linked with the physical plane z. If the mapping $\omega = \omega(t)$ is known, then we use the expression $\omega(t) = -\ln(dw/u_o dz)$ to determine that

$$z = \frac{1}{u_0}\int e^{\omega(t)}\frac{dw}{dt}\,dt. \qquad (1.31)$$

Use this formula to find the desired complex velocity:

$$\frac{dw}{dz} = \frac{dw}{dt}\frac{dt}{dz} = u_0 e^{-\omega(t)}. \qquad (1.32)$$

1.2.1 CALCULATION OF THE AIRFLOW AT THE SLOT-TYPE SUCTION BUILT IN A FLAT INFINITE WALL

Let the air from the half space $x > 0$ go into the slotted opening with a width $2B$. The flow is separated from the edges of the suction unit. The problem consists of determining the width of the potential flow 2δ for $x \to -\infty$ and the air velocity on the axis OX (Figure 1.14).

As the flow range is symmetric, we can disregard the lower semiplane (Figure 1.15).

On the line A_1A_2, the velocity varies from 0 to u_0 and the angle $\theta = -\pi/2$ is constant. The real part of the Joukowski function Re ω decreases here from ∞ to 0, the imaginary part is Im $\omega = -\pi/2$. On A_2A_3, the velocity is constant and equal to u_0 and the velocity angle θ varies from $-\pi/2$ to $-\pi$, that is, Re $\omega = 0$ and Im ω decreases from $-\pi/2$ to $-\pi$. On the line A_3A_1, the velocity decreases from u_0 down to 0 (Re ω increases from 0 up to ∞) and the angle $\theta = $ Im $\omega = -\pi$ is constant. Thus, the area of the Joukowski function is a semiband (Figure 1.16).

Map the area of the Joukowski function on the parametric plane t (Figure 1.16) using the Schwarz–Christoffel formula:

$$\omega = C_1 \int_{-1}^{t} \frac{dt}{\sqrt{t+1}\sqrt{t}} + C_2, \tag{1.33}$$

where $C_2 = -\pi/2 \cdot i$, as for $t = -1 \Rightarrow \omega = -\pi/2 \cdot i$.

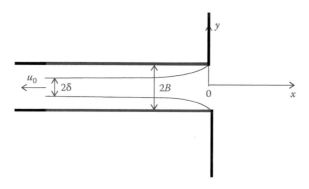

FIGURE 1.14 Separated airflow at the slot-type suction built in a flat infinite wall.

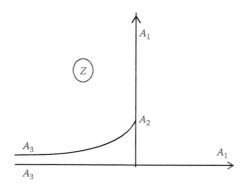

FIGURE 1.15 Physical flow plane.

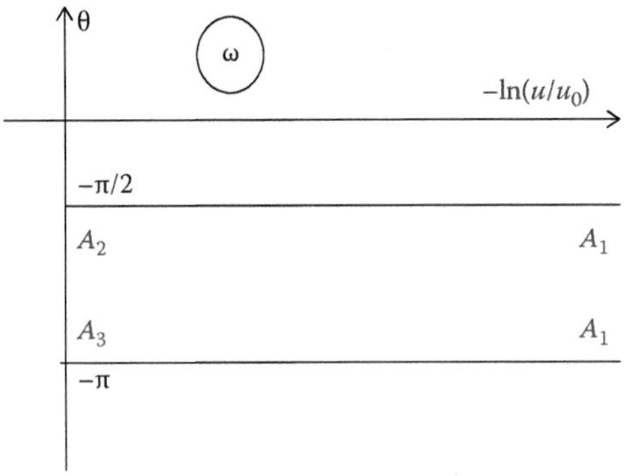

FIGURE 1.16   Plane of the Joukowski function.

On integrating, we obtain $\omega = C_1\left(2\ln\left(\sqrt{t+1}+\sqrt{t}\right)-i\pi\right)-\pi/2\cdot i$. The constant C_1 is determined from the correspondence of the point A_3. For $t = 0$, the function is $\omega = -i\pi$; therefore, $C_1 = 1/2$ and

$$\omega = \ln\left(\sqrt{t+1}+\sqrt{t}\right)-i\pi. \tag{1.34}$$

Using Equation 1.31, we will find a connection between the physical plane z and the parametric one t:

$$z = \frac{1}{u_0}C_1\int_{-1}^{t} e^{\ln\left(\sqrt{t+1}+\sqrt{t}\right)-i\pi}\frac{dw}{dt}dt + C_2. \tag{1.35}$$

Since a point sink is in the plane t at A_3, then the complex velocity is $\dfrac{dw}{dt} = -\dfrac{Q}{\pi t} = -\dfrac{u_0\delta}{\pi t}$, so, after simple transformations of Equation 1.35 we obtain

$$z = C_1\frac{\delta}{\pi}\int_{-1}^{t}\left(\frac{\sqrt{t+1}}{t}+\frac{1}{\sqrt{t}}\right)dt + C_2, \tag{1.36}$$

where $C_2 = iB$, as with $t = -1 \Rightarrow z = iB$.
    On integrating, we obtain

$$z = C_1\frac{\delta}{\pi}\left[2\left(\sqrt{t+1}+\sqrt{t}\right)+\ln\frac{\sqrt{t+1}-1}{\sqrt{t+1}+1}-i(\pi+2)\right]+iB. \tag{1.37}$$

The constant C_1 is determined from the correspondence of the point A_3 in the planes t and z. For $t = \varepsilon \rightarrow 0$, the function is $z = -\infty$ (Figures 1.15 and 1.17). Then, substituting in Equation 1.37 we obtain

$$-\infty = C_1\frac{\delta}{\pi}\left[2-\infty-i(\pi+2)\right]+iB. \tag{1.38}$$

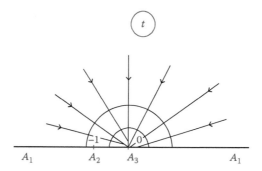

FIGURE 1.17 Parametric plane.

We equate the imaginary parts and find $C_1 = \dfrac{B\pi}{(\pi + 2)\delta}$. We have to now determine the half-width of the jet at infinity δ. For $t = -\varepsilon \to 0$ the function is $z = -\infty + i\delta$ (Figures 1.15 and 1.17). From Equation 1.37, we obtain

$$-\infty + i\delta = C_1 \frac{\delta}{\pi}\left[2 - \infty + i\pi - i(\pi + 2)\right] + iB. \tag{1.39}$$

From the equality of the imaginary parts, find $\delta = B\pi/(\pi + 2C_1)$ using the expression for C_1, then, finally define $C_1 = 1$, $\delta = B(\pi/\pi + 2)$, and substituting in Equation 1.37, we obtain

$$z = \frac{B}{\pi + 2}\left[2\left(\sqrt{t+1} + \sqrt{t}\right) + \ln\frac{\sqrt{t+1} - 1}{\sqrt{t+1} + 1}\right]. \tag{1.40}$$

The complex velocity in the plane z is

$$\frac{dw}{dz} = \frac{dw}{dt}\frac{dt}{dz} = -\frac{u_0}{\left(\sqrt{t+1} + \sqrt{t}\right)}. \tag{1.41}$$

The algorithm to calculate the axial velocity at a slot-type suction built into a flat infinite wall consists of the following steps:

1. The initial coordinate $x = 0$ is preset.
2. The formula $x = \dfrac{B}{\pi + 2}\left[2\left(\sqrt{t+1} + \sqrt{t}\right) + \ln\left(\dfrac{\sqrt{t+1} - 1}{\sqrt{t+1} + 1}\right)\right]$ is used to determine the parameter t.
3. The formula $u_x = -\dfrac{u_0}{\sqrt{t+1} + \sqrt{t}}$ is used to calculate the axial velocity of the air at x.
4. Perform the step $x = x + \Delta x$ and return to Step 2 until the set point is reached.

If the air suction rate is set as $u_0 = 1$, the axial velocity will take the values shown in Table 1.3.

TABLE 1.3
Values of the Axial Velocity of Separated Air Inflow to a Slot-Type Suction Unit Built in a Flat Infinite Wall

x/B	0	0.5	1	1.5	2	2.5	3	3.5	4	4.5	5
u	0.65	0.45	0.31	0.23	0.18	0.15	0.13	0.11	0.1	0.09	0.08

1.2.2 Calculation of the Airflow at the Slot-Type Suction in Infinite Space

Assume that the air from infinite space flows into a suction slot with a width $2B$ (Figure 1.18). The width of the jet at infinity is 2δ. Define the velocity value on the axis OX.

Due to the symmetry of the flow area (Figure 1.18), we disregard its lower part (Figure 1.19).

On the ray A_1A_2, the velocity varies from 0 to u_0 and the real part of the Joukowski function Re ω decreases from ∞ down to 0. The direction of the velocity does not change, the angle to the axis OX $\theta = 0$, and, respectively, Im $\omega = 0$. On the line A_2A_3, the velocity equals u_0 and Re ω and the imaginary part of the Joukowski function Im ω varies from 0 to $-\pi$. On the line A_3A_1, Re ω increases from 0 to ∞ and Im $\omega = -\pi$. Thus, the area of the Joukowski function is a semiband (Figure 1.20).

Map the area ω onto the parametric plane t (see Figure 1.17) using the Schwarz–Christoffel formula:

$$\omega = C_1 \int_0^t \frac{dt}{\sqrt{t}\sqrt{t+1}} + C_2. \tag{1.42}$$

The constant C_1 is determined from the correspondence $t = 0 \Rightarrow \omega = -i\pi \Rightarrow C_1 = -i\pi$. Integrate Equation 1.42 using $t = -1 \Rightarrow \omega = 0$, we define $C_1 = 1$. Thus,

$$\omega = 2\ln\left(\sqrt{t} + \sqrt{t+1}\right) - i\pi. \tag{1.43}$$

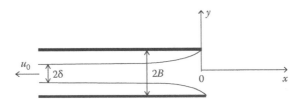

FIGURE 1.18   Jet separation at the freely spaced slot-type suction.

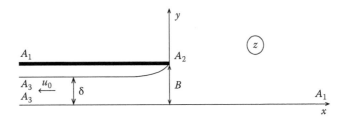

FIGURE 1.19   Physical flow plane.

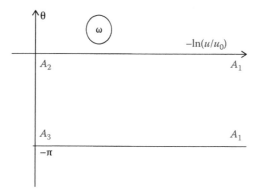

FIGURE 1.20   Area of the Joukowski function.

Using Equation 1.31, we find a connection between the physical flow plane and the parametric one:

$$z = C_1 \frac{\delta}{\pi} \int_{-1}^{t} \frac{\left(\sqrt{t} + \sqrt{t+1}\right)^2}{t} dt + C_2, \tag{1.44}$$

where $C_1 = iB$, as for $t = -1 \Rightarrow z = iB$.

After integrating Equation 1.44 and using the correspondences $t = +\varepsilon \to 0 \Rightarrow z = -\infty$ and $t = -\varepsilon \to 0 \Rightarrow z = -\infty + i\delta$, we obtain $C_1 = 1$, $\delta = B/2$, so

$$z = \frac{B}{2\pi} \left[\left(\sqrt{t} + \sqrt{t+1}\right)^2 + 1 + \ln t + 2\ln\left(\sqrt{t} + \sqrt{t+1}\right) \right]. \tag{1.45}$$

The complex velocity is

$$\frac{dw}{dz} = -\frac{u_0}{\left(\sqrt{t} + \sqrt{t+1}\right)^2}. \tag{1.46}$$

Then, the algorithm for calculating the axial velocity at a freely spaced slot-type suction unit consists of four steps:

1. The initial coordinate $x = 0$ is preset.
2. The formula $x = \dfrac{B}{2\pi}\left[\left(\sqrt{t} + \sqrt{t+1}\right)^2 + 1 + \ln t + 2\ln\left(\sqrt{t} + \sqrt{t+1}\right)\right]$ is used to determine the parameter t.
3. The formula $u_x = -\dfrac{u_0}{\left(\sqrt{t} + \sqrt{t+1}\right)^2}$ is used to calculate the axial velocity of the air at x.
4. Perform the step $x = x + \Delta x$ and return to Step 2. Calculate until the predetermined point is reached.

The calculation results for $u_0 = 1$ (Tables 1.3 and 1.4) show that a suction unit built in a flat wall is more efficient just similar to the unseparated flow model.

TABLE 1.4

Values of the Axial Velocity of Separated Air Inflow to a Slot-Type Suction Unit in an Infinite Space

x/B	0	0.5	1	1.5	2	2.5	3	3.5	4	4.5	5
u	0.64	0.39	0.23	0.16	0.11	0.09	0.07	0.06	0.05	0.04	0.039

1.3 CALCULATION OF THE FLOW AT THE ENTRANCE OF THE SLOT-TYPE SUCTION BELL

This chapter presents some results of a study of a flow at a slot-type *long* suction bell, that is, having two successive vortex zones (Figure 1.21). A series of articles [124] contain detailed descriptions of the development of design ratios using the complex variable theory.

1.3.1 DESIGN RATIOS

The system of equations required for the calculation is as follows:

$$
\left\{
\begin{aligned}
& \frac{\beta}{K\pi} = \frac{(n-h)(h-m)}{\sqrt{(h+1)(h-e)(h-d)h(1-h)}}, \quad K = \frac{\sqrt{2(1-e)(1-d)}(1-h)}{(1-n)(1-m)}, \\[2ex]
& \beta = K \int_{-1}^{e} \frac{(t-n)(t-m)}{(t+1)(e-t)(d-t)|t|} \frac{dt}{(t-h)(t-1)}, \\[2ex]
& \ln \frac{v_1}{v_2} = -K \int_{e}^{d} \frac{(t-n)(t-m)}{(t+1)(t-e)(d-t)|t|} \frac{dt}{(t-h)(t-1)}, \\[2ex]
& l = \frac{L}{\pi v_2} \int_{0}^{h} e^{-\operatorname{Re}\chi(t)} \frac{dt}{1-t^2}, \\[2ex]
& \operatorname{Re}\chi(t) = K \int_{0}^{\zeta} \frac{(t-n)(t-m)}{\sqrt{(t+1)(t-e)(t-d)t}} \frac{dt}{(t-h)(t-1)} + \ln \frac{v_1}{v_2}, \\[2ex]
& \frac{L}{\pi v_2} \operatorname{Im} \int_{0}^{e} \frac{e^{-\chi(t)}}{t^2-1} dt = -l \sin\beta, \quad \frac{L}{\pi v_2} \operatorname{Re} \int_{0}^{e} \frac{e^{-\chi(t)}}{t^2-1} dt = -l \cos\beta, \\[2ex]
& B = \frac{L}{2v_1} \left(\frac{1-h}{h}\right)^{-\beta/\pi} e^{-\chi_1(1)} \left[\frac{1}{2} + \frac{\beta}{\pi}\frac{1}{1-h} + \chi'(1)\right],
\end{aligned}
\right.
$$

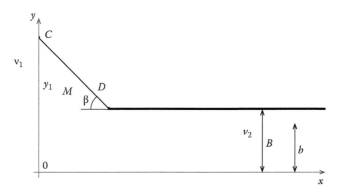

FIGURE 1.21 Slot-type suction bell with two vortex zones.

where

h, d, m, n, e are the parameters of mapping

$$\chi_1(1) = \int_0^1 \left[Kf(t) + \frac{\beta}{\pi f(h)} \frac{f(t)-f(h)}{t-h} + \frac{1}{f(1)} \frac{f(t)-f(1)}{t-1} \right] dt$$

$$\chi'(1) = Kf(t) + \frac{\beta}{\pi f(h)} \frac{f(1)-f(h)}{t-h} + \frac{f'(1)}{f(1)}$$

$$f(t) = \frac{1}{\sqrt{t(t-e)(t-d)(t+1)}}$$

$$f'(1) = -\frac{1}{2} \frac{2(1-d)(2-e)+(1-e)(3-d)}{\left[2(1-e)(1-d)\right]^{3/2}}$$

This system of equations can be reduced to the solution of two equations for two unknowns e, d:

$$\begin{cases} f_1(e, d) = \sin\beta - (\text{int}\,1 \cdot \sin\beta + \text{int}\,2)/\text{int}\,3 = 0, \\ f_2(e, d) = \cos\beta - (\text{int}\,1 \cdot \cos\beta + \text{int}\,4)/\text{int}\,3 = 0, \end{cases} \tag{1.47}$$

where

$$\text{int}\,1 = \int_e^d \exp\left[K \int_e^t f(\tau)d\tau \right] \frac{dt}{1-t^2}$$

$$\text{int}\,2 = \exp(S_2) \cdot \int_d^0 \sin\left(\beta - K \cdot \int_d^t f(\tau)d\tau \right) \frac{dt}{1-t^2}$$

$$S_2 = K \int_e^d f(t)dt$$

$$\text{int}\,3 = \exp(S_2) \int_0^h \exp\left[-K \int_0^t f(\tau)d\tau \right] \frac{dt}{1-t^2}.$$

$$\text{int}\,4 = \exp(S_2) \cdot \int_d^0 \cos\left(\beta - K \cdot \int_d^t f(\tau)d\tau \right) \frac{dt}{1-t^2}$$

$$K = \frac{\beta + B_1 \int_{-1}^{e} \frac{dt}{(t-h)(t-1)r(t)} - A_1 \int_{-1}^{e} \frac{tdt}{(t-h)(t-1)r(t)}}{\int_{-1}^{e} \frac{dt}{r(t)}} \tag{1.48}$$

$$m = \frac{A}{2} - \sqrt{\frac{A^2}{4} + B}$$

$$n = \frac{A}{2} + \sqrt{\frac{A^2}{4} + B}$$

$$r(t) = \sqrt{\left| t(t+1)(t-e)(t-d) \right|}$$

$$f(t) = \frac{(t-n)(t-m)}{(t-h)(t-1)r(t)}$$

$$A = 1 + h - \frac{A_1}{K}$$

$$B = \frac{B_1}{K} - h$$

$$B_1 = \frac{\beta}{\pi} r(h) + h \cdot r(1)$$

$$A_1 = \frac{\beta}{\pi} r(h) + r(1)$$

The integrals in Equation 1.47 generally have singularities of the order $(x-a)^{1/2}$ (i.e., the denominator of the integrand goes to zero for $x = a$), and, thus, they are convergent improper integrals.

Integrals of the type $\int_{a}^{b} \frac{f(y)}{\sqrt{b-y}} dy$, $\int_{a}^{b} \frac{f(y)}{\sqrt{(y-a)(b-y)}} dy$ are calculated using the Lobatto quadrature formulas [123]:

$$\int_{a}^{b} \frac{f(y)}{\sqrt{b-y}} dy = \sqrt{b-a} \sum_{i=1}^{n/2} 2A_i f(y_i), \tag{1.49}$$

where
$y_i = a + (b-a)(1 - x_i^2)$, x_i are the nodes of the Gauss quadrature (positive terms)
n is their number (even-numbered)
A_i is the weighting factors (corresponding to positive nodes x_i)

$$\int_a^b \frac{f(y)dy}{\sqrt{(y-a)(b-y)}} = \frac{\pi}{n}\sum_{i=1}^n f\left(\frac{b+a}{2} + \frac{b-a}{2}\cdot\cos\frac{(2i-1)\pi}{2n}\right). \tag{1.50}$$

Integrals without singularities were calculated using the Gauss quadratures

$$\int_a^b f(y)dy = \frac{b-a}{2}\sum_{i=1}^n A_i f(y_i), \tag{1.51}$$

where

$y_i = (a+b)/2 + (b-a)/2\cdot x_i$, x_i are the nodes of the Gauss quadrature

n is *their* number

A_i *are* the weighting factors

In our calculations, the number of nodes n is 96 in the quadrature formulas of Equations 1.49 and 1.51 (and the interval of integration was sometimes divided into 10–500 parts to improve the accuracy of calculations) and is 40 in quadrature formula of Equation 1.50.

The parameter K is calculated by the formula

$$K = \frac{\beta + B_1\cdot S_1 - A_1\cdot S_2}{S_3}, \tag{1.52}$$

where the integrals S_1, S_2, S_3 are defined as follows:

$$S_1 = \int_{-1}^e \frac{dt}{(t-h)(t-1)\sqrt{t(t+1)(t-e)(t-d)}}$$

$$= \int_{-1}^e \frac{1}{(t-h)(t-1)\sqrt{(t(t-d)(}} \frac{dt}{\sqrt{(e-t)(t+1)}} = \frac{\pi}{n}\sum_{i=1}^n \frac{1}{(t_i-h)(t_i-1)\sqrt{t_i(t_i-d)}},$$

where

$$t_i = \frac{e-1}{2} + \frac{e+1}{2}\cos\frac{(2i-1)\pi}{2n}$$

$$S_2 = \int_{-1}^e \frac{tdt}{(t-h)(t-1)\sqrt{\left|t(t+1)(t-e)(t-d)\right|}} = \frac{\pi}{n}\sum_{i=1}^n \frac{t_i}{(t_i-h)(t_i-1)\sqrt{\left|t_i(t_i-d)\right|}}$$

$$S_3 = \int_{-1}^e \frac{dt}{\sqrt{\left|t(t+1)(t-e)(t-d)\right|}} = \frac{\pi}{n}\sum_{i=1}^n \frac{1}{\sqrt{\left|t_i(t_i-d)\right|}}$$

Consider the calculation of the integrals in Equation 1.47.

The inner integral is

$$\int_e^t f(\tau)d\tau = \int_e^t \frac{(\tau-n)(\tau-m)d\tau}{(\tau-h)(\tau-1)\sqrt{\left|\tau(\tau+1)(\tau-e)(\tau-d)\right|}} = \{\text{particularity, if }\tau = e\}$$

$$= \int_e^t \frac{g(\tau)}{\sqrt{\tau-e}}d\tau = \int_{2e-t}^e \frac{g(2e-\tau)}{\sqrt{e-\tau}}d\tau = \sqrt{t-e}\sum_{j=1}^{n/2} 2A_j g(2e-\tau_j),$$

where $\tau_j = (2e-t) + (t-e)(1-x_j^2)$.

The outer integral is

$$\text{int}\,1 = \frac{d-e}{2}\sum_{i=1}^{n} A_i \frac{\exp\left[K\sqrt{t_i-e}\sum_{j=1}^{n/2}2A_jg(2e-\tau_j)\right]}{1-t_i^2},$$

where

$$t_i = \frac{d+e}{2} + \frac{d-e}{2}x_i$$

$$\tau_j = (2e-t_i)+(t_i-e)(1-x_j^2)$$

$$S_2 = K\int_e^d \frac{(t-n)(t-m)dt}{(t-h)(t-1)\sqrt{|t(t+1)|}\sqrt{(t-e)(d-t)}} = \frac{\pi}{n}K\sum_{i=1}^{n}\frac{(t_i-n)(t_i-m)}{(t_i-h)(t_i-1)\sqrt{|t_i(t_i+1)|}}$$

$$t_i = \frac{d+e}{2} + \frac{d-e}{2}\cos\frac{(2i-1)\pi}{2n}$$

The integral is

$$\text{int}\,2 = \exp(S_2)\cdot\int_d^0 \sin\left(\beta - K\cdot\int_d^t f(\tau)d\tau\right)\frac{dt}{1-t^2}.$$

The inner integral is

$$\int_d^t f(\tau)d\tau = \{\text{see Section 1.3.1}\} = \sqrt{t-d}\sum_{j=1}^{n/2}2A_jg(2d-\tau_j),$$

where $\tau_j = (2d-t)+(t-d)(1-x_j^2)$.

The outer integral is

$$\text{int}\,2 = \exp(S_2)\cdot\frac{(-d)}{2}\sum_{i=1}^{n}\frac{A_i}{1-t_i^2}\cdot\sin\left(\beta - K\sqrt{t_i-d}\sum_{j=1}^{n/2}2A_jg(2d-\tau_j)\right),$$

where

$$t_i = \frac{d}{2}(1-x_i)$$

$$\tau_j = (2d-t_i)+(t_i-d)(1-x_j^2)$$

The integral int 4 is calculated in the same way.
The integral is

$$\text{int}\,3 = \exp(S_2)\int_0^h \exp\left[-K\int_0^t f(\tau)d\tau\right]\frac{dt}{1-t^2}.$$

The inner integral has a singularity for $\tau = 0$ and is calculated as follows:

$$\int_0^t f(\tau)d\tau = \int_0^t \frac{(\tau-n)(\tau-m)d\tau}{(\tau-h)(\tau-1)\sqrt{|\tau(\tau+1)(\tau-e)(\tau-d)|}} = \int_0^t \frac{g(\tau)}{\sqrt{\tau}}d\tau = \int_{-t}^0 \frac{g(-\tau)}{\sqrt{0-\tau}}d\tau = \sqrt{t}\sum_{j=1}^{n/2}2A_j g(-\tau_j),$$

where $\tau_j = -t \cdot x_j^2$.

The required integral is determined by the formula

$$\text{int }3 = \exp(S_2)\cdot\frac{h}{2}\sum_{i=1}^{n}\frac{A_i}{1-t_i^2}\cdot\exp\left(-K\sqrt{t_i}\sum_{j=1}^{n/2}2A_j g(-\tau_j)\right),$$

where

$$t_i = \frac{h(x_i+1)}{2}$$

$$\tau_j = -t_i \cdot x_j^2$$

System of Equation 1.47 is nonlinear and can be solved using the half-interval method. For ease of computation, we write the system of equations to be solved as follows:

$$\begin{cases} f_1(e,d)-f_2(e,d)=0, \\ f_2(e,d)=0. \end{cases}$$

The algorithm for determining the roots is as follows:

1. Assign the interval at which we find the root of e: $[niz_e, ver_e]$.
2. Calculate the mean $e_c = (niz_e + ver_e)/2$.
3. Separate the interval $[niz, ver]$ at which the root d of $f_1(niz_e, d) - f_2(niz_e, d) = 0$ is located: To this end, make a loop on d and perform it until it satisfies the inequality

$$[f_1(niz_e, d)-f_2(niz_e, d)]\cdot f_1(niz_e, d+\Delta d)-f_2(niz_e, d+\Delta d)] < 0.$$

   Then, $niz = d$, $ver = d + \Delta d$, if d becomes positive (otherwise, there are no roots as the problem provides that $d > 0$).
4. Revision of the value d, for which $f_1(niz_e, d) - f_2(niz_e, d) = 0$. To do this, make a loop that is executed as long as $|f_1(niz_e, d) - f_2(niz_e, d)| > \varepsilon$ (ε is the set accuracy). Calculate $d = (niz + ver)/2$. Let us denote $d_1 = f_1(e, niz)$, $d_2 = f_2(e, niz)$, $d_3 = f_1(e, d)$, $d_4 = f_2(e, d)$. If $(d_1 - d_2)$ $(d_3 - d_4) < 0$, then equate $ver = d$, otherwise $niz = d$.
5. Calculate the value $e_1 = f_2(niz_e, d)$.
6. Take the root of d, for which $f_1(e_c, d) - f_2(e_c, d) = 0$. Perform this as in Steps 3 and 4, where niz_e is used instead e_c.
7. If $e_1\cdot f_2(e_c, d) < 0$, then $ver_e = e_c$, otherwise $niz_e = e_c$ and go to Step 2. This iterative process is terminated when the inequality $|niz_e - ver_e| < \varepsilon$ is achieved.

If at the first iteration $e_1\cdot f_2(e_c, d) > 0$, then there are no roots and computing will end.

In our calculations, the computational error $\varepsilon = 0.00001$.

Calculate the velocity on the flow axis.

For this, we first calculate the parameter x_0 required to calculate the velocity,

$$x_0 = -\frac{2}{\pi v_1} \int_0^\infty \exp\{K \cdot G_1(t)\} \cdot \sin\left[\pi + \beta + K \cdot G_2(t)\right]\frac{dt}{1+t^2}$$

$$= \left\{ t = -\frac{x-1}{x}, \quad x = \frac{1}{1+t}, \quad dt = -\frac{dx}{x^2} \right\}$$

$$= -\frac{2}{\pi v_1} \int_0^1 \exp\left\{K \cdot G_1\left(\frac{1-x}{x}\right)\right\} \cdot \sin\left[\pi + \beta + K \cdot G_2\left(\frac{1-x}{x}\right)\right]\frac{dt}{x^2 + (x-1)^2},$$

where

$$G_1\left(\frac{1-x}{x}\right) = \int_0^{(1-x)x} \frac{L_1(\tau)}{\sqrt{\tau}} \sin L_2(\tau)d\tau = \{t = -\tau, \, d\tau = -dt\} = \int_{(x-1)x}^0 \frac{L_1(t)}{\sqrt{0-t}} \sin L_2(t)dt.$$

The last integral is calculated using quadrature formula of Equation 1.49. The following denotations are introduced here:

$$L_1(t) = \sqrt{\frac{(t^2 + m^2)(t^2 + n^2)}{(t^2 + h^2)(t^2 + 1)\sqrt{(t^2 + e^2)(t^2 + d^2)(t^2 + 1)}}},$$

$$L_2(t) = \left[\text{arc cos}\frac{-n}{\sqrt{t^2 + n^2}} + \text{arc cos}\frac{-m}{\sqrt{t^2 + m^2}} - \text{arc cos}\frac{-h}{\sqrt{t^2 + h^2}} - \text{arc cos}\frac{-1}{\sqrt{t^2 + 1}} \right]$$

$$- \frac{1}{2}\left[\frac{\pi}{2} + \text{arc cos}\frac{-d}{\sqrt{t^2 + d^2}} + \text{arc cos}\frac{-e}{\sqrt{t^2 + e^2}} + \text{arc cos}\frac{1}{\sqrt{t^2 + 1}} \right].$$

Introduce the denotations

$$P_1(\xi) = \exp\left(-K \int_{-1}^{-\infty} f(t)dt + K \int_\xi^\infty f(t)dt \right), \quad 1 < \xi < \infty,$$

$$P_2(\xi) = \exp\left(-K \int_{-1}^{-\infty} f(t)dt - K \int_{-\infty}^\xi f(t)dt \right), \quad -\infty < \xi < -1.$$

We show that $P_1(\xi)$ and $P_2(\xi)$ are equal. Actually, the integrals

$$\int_{\xi}^{\infty} f(t)dt = \left\{\tau = \frac{1}{t}, dt = -\frac{d\tau}{\tau^2}\right\} = -\int_{1/\xi}^{0} f\left(\frac{1}{\tau}\right)\frac{d\tau}{\tau^2},$$

$$\int_{-\infty}^{\xi} f(t)dt = \left\{\tau = \frac{1}{t}, dt = -\frac{d\tau}{\tau^2}\right\} = \int_{1/\xi}^{0} f\left(\frac{1}{\tau}\right)\frac{d\tau}{\tau^2}.$$

Thus,

$$\int_{\xi}^{\infty} f(t)dt = -\int_{-\infty}^{\xi} f(t)dt.$$

It follows $P_1(\xi) = P_2(\xi) = P(\xi)$.

In the same manner, we can show that

$$\int_{\xi}^{\infty} P(t)\frac{dt}{t^2-1} = -\int_{-\infty}^{\xi} P(t)\frac{dt}{t^2-1}.$$

Thus, the required correspondence of the points for $|\xi| > 1$ will be determined by the relation

$$x(\xi) = x_0 - \frac{2}{\pi v_2}\int_{\xi}^{\infty} P(\tau)\frac{d\tau}{\tau^2-1},$$

where the integral can be represented for computational convenience as

$$\int_{\xi}^{\infty} P(\tau)d\tau = \left\{\tau = \frac{1}{t}\right\} = \int_{0}^{\frac{1}{\xi}} P\left(\frac{1}{t}\right)\frac{dt}{1-t^2}.$$

The integrand is

$$P\left(\frac{1}{t}\right) = \exp(-K\cdot S_1)\cdot\exp\left(K\cdot\int_{1/t}^{\infty} f(\tau)d\tau\right), \quad S_1 = \int_{-1}^{-\infty} f(t)dt.$$

To avoid infinitely large limits of integration, transform the integrals as follows:

$$\int_{1/t}^{\infty} f(\tau)d\tau = \int_{0}^{t} f\left(\frac{1}{\tau}\right)\frac{d\tau}{\tau^2} = \int_{0}^{t} \frac{(1-\tau n)(1-\tau m)d\tau}{(1-h\tau)(1-\tau)\sqrt{|(1+\tau)(1-e\tau)(1-d\tau)|}} = S_2(t).$$

Given the above, we finally obtain

$$x(\xi) = x_0 - \frac{2e^{-KS_1}}{\pi v_2} \int\limits_0^{1/\xi} \exp\left(K \cdot S_2(t)\right) \frac{dt}{t^2 - 1} \quad \text{for } |\xi| > 1.$$

The velocity on the flow axis is calculated from the formula

$$v_x = \frac{v_2}{P(\xi)}, \quad |\xi| > 1,$$

where

$$P(\xi) = \exp\left[-K \cdot \{S_1 + S_2(\xi)\}\right], \quad S_2(\xi) = \int\limits_{1/\xi}^0 f\left(\frac{1}{\tau}\right)\frac{d\tau}{\tau^2}.$$

Thus,

$$v_x = v_2 \cdot \exp\left[K \cdot \{S_1 + S_2(\xi)\}\right], \quad |\xi| > 1.$$

The positions of the points on the first free stream line (CMD) are determined from

$$x = -\frac{2}{\pi v_1} \int\limits_0^\xi \frac{J_1(t)dt}{t^2 - 1}, \quad y = -\frac{2}{\pi v_1} \int\limits_0^\xi \frac{J_2(t)dt}{t^2 - 1} + 1 + l \cdot \sin\beta.$$

The following denotations are introduced here:

$$J_1(t) = \cos\left[K\int\limits_0^t f(\tau)d\tau - \beta\right], \quad J_2(t) = \sin\left[K\int\limits_0^t f(\tau)d\tau - \beta\right].$$

The integrals were calculated using the Gauss quadratures.

1.3.2 CALCULATION RESULTS

Some results of the calculation that were not included in articles [124] are shown in Figures 1.22 through 1.33.

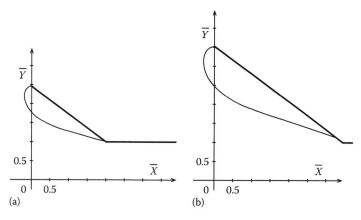

FIGURE 1.22 Form of the first free flow line for a bell expansion angle of 36: (a) $e = -0.99919$; $d = -0.99906$; $m = -0.9946$; $n = 0.462$; $h = 0.405$; $l = 2.474$; $v_1 = 0.818$; $v_2 = 1.209$; and (b) $e = -0.99985$; $d = -0.99935$; $m = -0.9941$; $n = 0.497$; $h = 0.44$; $l = 4.279$; $v_1 = 0.570$; $v_2 = 1.250$.

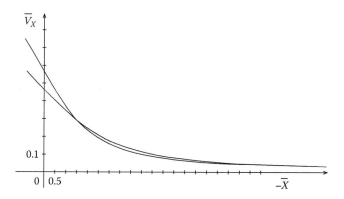

FIGURE 1.23 Change in the dimensionless axial velocity at a distance from the inlet for the cases (a) and (b) in Figure 1.22 (from top downward, respectively).

1.4 SIMULATION OF FLOW SEPARATION AT THE INLET OF A PROJECTING FLAT SUCTION CHANNEL

This chapter focuses on the numerical and experimental study of a separated flow at the inlet of a slot-type hooded suction inlet.

1.4.1 DETERMINATION OF DESIGN RATIOS

Assume that an infinitely wide horizontal pipe with a width $2B$ is projected to a distance S from the vertical wall. The pipe axis is directed along the axis OX of the physical plane of the complex variable $z = x + iy$. We will consider the upper part of the flow in the semiplane $y \geq 0$. We will use the upper semiplane of the complex variable $t = x_1 + iy_1$ as a parametric one. The correspondence of the points in these ranges, as well as the ranges of the unknown quantities of the Joukowski function $\omega = \ln(u_\infty/u) + i\theta$ and the complex potential function $w = \varphi + i\psi$, are shown in Figure 1.34. Here, u_∞ is the velocity value on the *free flow line* CD; u is also in the arbitrary points of the area under review in the quadrilateral $ADCBA$ of the physical area $\text{Im}(z) \geq 0$.

The quadrilateral has two vertexes at infinity: the point A, at which a line source with a strength q is placed, and the point D, where a sink with the same strength is placed. Thus, the range of

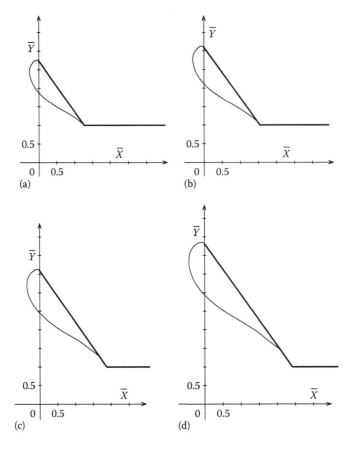

FIGURE 1.24 Form of the first free flow line for a bell expansion angle of 54°: (a) $e = -0.995$; $d = -0.993$; $m = -0.975$; $n = 0.382$; $h = 0.3$; $l = 2.143$; $v_1 = 0.704$; $v_2 = 1.341$; (b) $e = -0.997$; $d = -0.994$; $m = -0.975$; $n = 0.393$; $h = 0.31$; $l = 2.596$; $v_1 = 0.618$; $v_2 = 1.358$; (c) $e = -0.998$; $d = -0.994$; $m = -0.975$; $n = 0.403$; $h = 0.32$; $l = 3.214$; $v_1 = 0.531$; $v_2 = 1.371$; and (d) $e = -0.999$; $d = -0.995$; $m = -0.975$; $n = 0.413$; $h = 0.33$; $l = 4.108$; $v_1 = 0.442$; $v_2 = 1.378$.

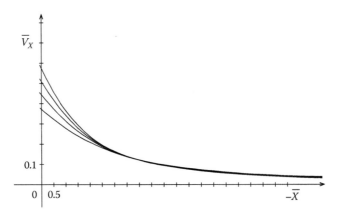

FIGURE 1.25 Change in the dimensionless axial velocity at a distance from the inlet for the cases (a), (b), (c), and (d) in Figure 1.24 (from top downward, respectively).

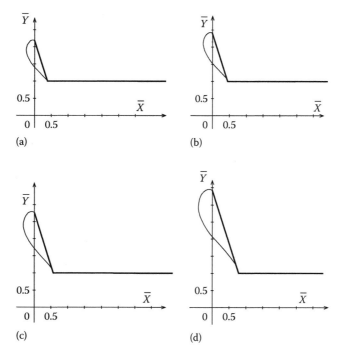

FIGURE 1.26 Form of the first free stream line with a bell expansion angle of 72°: (a) $e = -0.97$; $d = -0.968$; $m = -0.936$; $n = 0.301$; $h = 0.202$; $l = 1.235$; $v_1 = 0.835$; $v_2 = 1.435$; (b) $e = -0.975$; $d = -0.971$; $m = -0.936$; $n = 0.310$; $h = 0.21$; $l = 1.470$; $v_1 = 0.747$; $v_2 = 1.446$; (c) $e = -0.983$; $d = -0.975$; $m = -0.938$; $n = 0.321$; $h = 0.22$; $l = 1.854$; $v_1 = 0.64$; $v_2 = 1.461$; and (d) $e = -0.990$; $d = -0.979$; $m = -0.939$; $n = 0.334$; $h = 0.232$; $l = 2.533$; $v_1 = 0.513$; $v_2 = 1.476$.

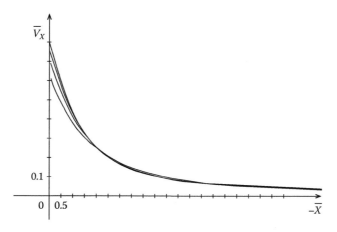

FIGURE 1.27 Change in the dimensionless axial velocity at a distance from the inlet for the cases (a), (b), (c), and (d) in Figure 1.26 (from top downward, respectively).

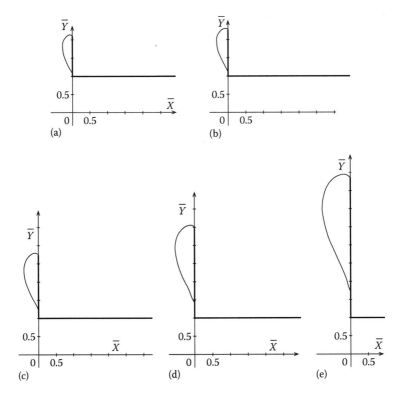

FIGURE 1.28 Form of the first free flow line for a bell expansion angle of 90°: (a) $e = -0.934$; $d = -0.926$; $m = -0.871$; $n = 0.255$; $h = 0.145$; $l = 1.096$; $v_1 = 0.771$; $v_2 = 1.546$; (b) $e = -0.944$; $d = -0.932$; $m = -0.874$; $n = 0.261$; $h = 0.15$; $l = 1.27$; $v_1 = 0.700$; $v_2 = 1.554$; (c) $e = -0.962$; $d = -0.942$; $m = -0.880$; $n = 0.274$; $h = 0.16$; $l = 1.738$; $v_1 = 0.564$; $v_2 = 1.564$; (d) $e = -0.976$; $d = -0.949$; $m = -0.856$; $n = 0.286$; $h = 0.17$; $l = 2.483$; $v_1 = 0.434$; $v_2 = 1.563$; and (e) $e = -0.988$; $d = -0.954$; $m = -0.890$; $n = 0.298$; $h = 0.18$; $l = 3.843$; $v_1 = 0.307$; $v_2 = 1.548$.

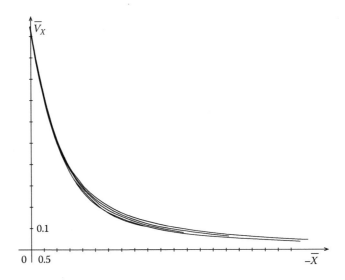

FIGURE 1.29 Change in the dimensionless axial velocity at a distance from the inlet for the cases (a), (b), (c), (d), and (e) in Figure 1.28 (from top downward, respectively).

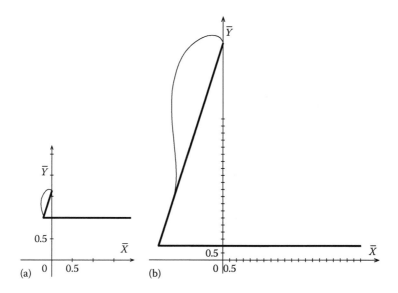

FIGURE 1.30   Form of the first free flow line with a bell expansion angle of 108°: (a) $e = -0.834$; $d = -0.828$; $m = -0.763$; $n = 0.200$; $h = 0.090$; $l = 0.66$; $v_1 = 0.917$; $v_2 = 1.636$; and (b) $e = -0.995$; $d = -0.919$; $m = -0.829$; $n = 0.271$; $h = 0.14$; $l = 15.079$; $v_1 = 0.075$; $v_2 = 1.209$.

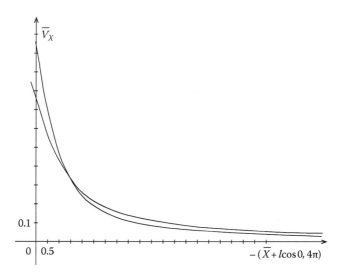

FIGURE 1.31   Change in the dimensionless axial velocity at a distance from the inlet for the cases (a) and (b) in Figure 1.30 (the upper and lower line, respectively).

the complex potential is a band bounded by the flow lines $\psi = 0$, $\psi = q$, and equipotential lines at infinity $\varphi = \pm\infty$, and the range of the Joukowski function is a half-band with a cut along the ray $MA(\theta = -\pi/2)$ bounded by horizontal lines $\theta = 0$ and $\theta = -\pi$ (θ is the angle between the positive axis OX and the direction of the velocity vector $\vec{u}$).

Find a conformal mapping of the upper semiplane $\text{Im}(t) > 0$ onto the interior of the range ω (the interior of the pentagon $ADCBMA$ with two vertexes at infinity).

We use the Schwarz–Christoffel integral, after simple transformations, taking into account the transition conditions of the singular points B and B on semicircles $t = b + \varepsilon_b e^{i\alpha}$ and $t = 1 + \varepsilon_\alpha e^{i\alpha}$

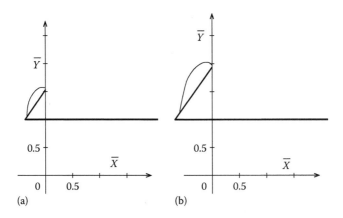

FIGURE 1.32   Form of the first free flow line for a bell expansion angle of 126°: (a) $e = -0.760$; $d = -0.739$; $m = -0.653$; $n = 0.171$; $h = 0.06$; $l = 0.653$; $v_1 = 0.790$; $v_2 = 1.714$; and (b) $e = -0.843$; $d = -0.790$; $m = -0.689$; $n = 0.189$; $h = 0.07$; $l = 1.174$; $v_1 = 0.534$; $v_2 = 1.674$.

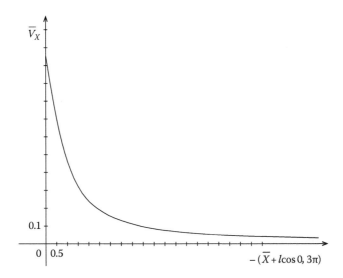

FIGURE 1.33   Change in the dimensionless axial velocity at a distance from the inlet for the case (b) in Figure 1.32.

$(\varepsilon_b \to 0, \varepsilon_a \to 0, \alpha = \pi...0)$, as well as taking into account the correspondence of the points C and M, we find the desired Joukowski function

$$\omega = \frac{1}{2}\ln\left(\frac{\sqrt{t}+\sqrt{b}}{\sqrt{t}-\sqrt{b}}\cdot\frac{\sqrt{t}+1}{\sqrt{t}-1}\right) = \ln\frac{\sqrt{t}+\sqrt{b}}{\sqrt{t}-b}\cdot\frac{\sqrt{t}+1}{\sqrt{t}-1} \qquad (1.53)$$

and define the parameter

$$\mu = \ln\frac{1+b^{1/4}}{1-b^{1/4}}. \qquad (1.54)$$

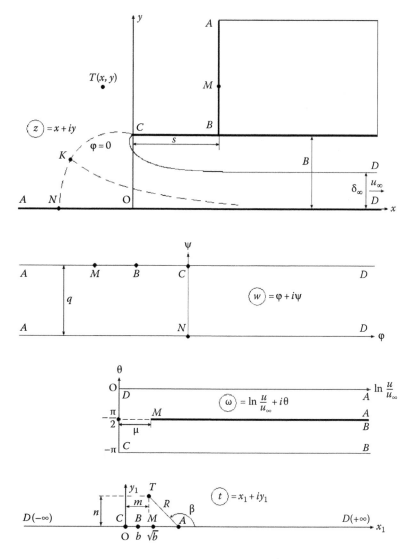

FIGURE 1.34 To the determination of the separated flow near a projecting suction slot.

Using the same techniques, find the complex potential function

$$w = \frac{q}{\pi} \ln(t-1),$$
(1.55)

describing the velocity field in the semiplane $\text{Im}(t) > 0$ that is induced by a line source at the point A.
Bearing in mind that

$$z = \frac{1}{u_\infty} \int_{t_k}^{t} e^{\omega(t)} \cdot \frac{dw}{dt} \, dt + z_k$$
(1.56)

and

$$v = \frac{dw}{dz} = u_\infty e^{-\omega(t)}$$
(1.57)

and if we use the functions found ω and w, and take as a reference point $C(t_k = 0; z_k = B \cdot i)$, we obtain a parametric solution of the problem:

$$z = \frac{\delta_\infty}{\pi} \int_0^T \frac{\sqrt{t}+\sqrt{b}}{\sqrt{t}-b} \cdot \frac{\sqrt{t}+1}{(t-1)^{3/2}} dt + i, \tag{1.58}$$

$$v \equiv u_x - iu_y = \frac{\sqrt{T-b}}{\sqrt{T}+\sqrt{b}} \cdot \frac{\sqrt{T-1}}{\sqrt{T}+1}, \tag{1.59}$$

enabling us to build a hydrodynamic grid ($\psi = 0...1 = \text{const}$; $\varphi = -\infty... + \infty = \text{const}$) and the velocity field

$$u_x = \text{Re}(v); \quad u_y = -\text{Im}(v). \tag{1.60}$$

Hereinafter, the linear dimensions are related to the half-height of the slot B and the velocities to the velocity u_∞; δ_∞ is the dimensionless half-height of the jet for $t \to \infty$ (at the point D) and $T = m + n \cdot i$ is an arbitrary point of the upper semiplane $\text{Im}(t) > 0$ and a point of physical half-plane $\text{Im}(z) > 0$ corresponding to it owing to Equation 1.58, in which we define the velocity vector projection $\vec{u}$.

The maximum speed equal to Equation 1.61 (due to the fact that at this point $T \equiv \sqrt{b}, b = 0...1$) is seen at the point M on the ray BA:

$$u_M = u_y = -\frac{1-b^{1/4}}{1+b^{1/4}}. \tag{1.61}$$

It is easy to determine the position of this point in the physical plane:

$$x_M = S = \frac{\delta_\infty}{\pi} \int_0^{b-\varepsilon} \frac{\sqrt{t}+\sqrt{b}}{\sqrt{b}-t} \cdot \frac{1+\sqrt{t}}{(1-t)^{3/2}} dt, \tag{1.62}$$

$$y_M = 1 + \frac{\delta_\infty}{\pi} \int_{b+\varepsilon}^{\sqrt{b}} \frac{\sqrt{t}+\sqrt{b}}{\sqrt{b}-t} \cdot \frac{1+\sqrt{t}}{(1-t)^{3/2}} dt, \quad \varepsilon \to 0. \tag{1.63}$$

We find the coordinates of the flow line CD to determine the half-height of the jet at infinity δ_∞. Given that the points of this line correspond to the points on the negative semiaxis $OX_1 (-\infty < t < 0)$ based on Equation 1.58, we can write the following system of equations:

$$x_{CD} = \frac{\delta_\infty}{\pi} \int_0^\eta \frac{(v-\sqrt{b})dv}{\sqrt{b}+v(1+v)^{1.5}}, \tag{1.64}$$

$$y_{CD} = 1 - (1+\sqrt{b})\frac{\delta_\infty}{\pi} \int_0^\eta \frac{\sqrt{v}dv}{\sqrt{b}+v(1+v)^{1.5}}, \quad 0 \le \eta < +\infty, \tag{1.65}$$

describing the coordinates of the line *CD* in the upper semiplane of the physical space. Given that $y_{DC} \to \delta_\infty$ for $\eta \to \infty$, we have $\delta_\infty = 1 - \dfrac{\delta_\infty}{\pi} E(b)$ whence it follows

$$\delta_\infty = \frac{\pi}{\pi + E(b)}, \tag{1.66}$$

where $E(b)$ is the number depending on the parameter b:

$$E(b) = \left(1 + \sqrt{b}\right) \int_0^\infty \frac{\sqrt{v}\,dv}{\sqrt{b+v}\left(1+v\right)^{1.5}}. \tag{1.67}$$

Taking into account the result obtained based on Equation 1.62, we can write the following relationship:

$$S = \frac{1}{\pi + E(b)} \int_0^{b-\varepsilon} \frac{\sqrt{t} + \sqrt{b}}{\sqrt{b-t}} \cdot \frac{1+\sqrt{t}}{\left(1-t\right)^{1.5}}\,dt, \tag{1.68}$$

connecting the length of the projection (hood) with the parameter b.
  In particular, for $b = 0$ we have

$$E(0) = \int_0^\infty \frac{dv}{\left(1+v\right)^{1.5}} = 2, \quad \delta_\infty = \frac{\pi}{\pi+2} \approx 0.611, \quad S = 0, \quad \mu = 0, \tag{1.69}$$

that is, the case of a separated airflow at the slot-type suction built into a flat unlimited wall. In this case, the relations to determine the velocity field as well as the coordinates of the free flow line *CD* are much simplified:

$$x + y \cdot i = \frac{1}{\pi+2} \int_0^T \frac{\sqrt{t}+1}{\left(t-1\right)^{1.5}}\,dt + i = \frac{2}{\pi+2}\left[\ln\left(\sqrt{T-1}+\sqrt{T}\right) - \frac{1+\sqrt{T}}{\sqrt{T-1}}\right], \tag{1.70}$$

$$u_x - iu_y = \frac{\pi+2}{\pi} \cdot \frac{\sqrt{T-1}}{\sqrt{T+1}}, \tag{1.71}$$

$$x_{CD} = \frac{2}{\pi+2}\left[\ln\left(\sqrt{\eta+1}+\sqrt{\eta}\right) - \frac{\sqrt{\eta}}{\sqrt{\eta+1}}\right], \tag{1.72}$$

$$y_{CD} = \frac{2}{\pi+2}\left[\frac{\pi}{2} + \frac{1}{\sqrt{\eta+1}}\right]; \quad \eta = 0\ldots\infty. \tag{1.73}$$

For $b = 1$, we have another classic case of the separated airflow at the slot-type suction in infinite space:

$$E(1) = 2\int_0^\infty \frac{\sqrt{v}}{\left(1+v\right)^2}\,dv = \pi, \quad \delta_\infty = 0.5, \tag{1.74}$$

$$S = \frac{1}{2\pi} \int_0^1 \frac{\sqrt{t}+1}{(1-t)^2} dt \to \infty, \quad \mu \to \infty,$$

$$x + y \cdot i = \frac{1}{2\pi} \int_0^T \frac{dt}{(\sqrt{t}-1)^2} + i = \frac{1}{\pi} \left[\ln\left(\sqrt{T}-1\right) - \frac{\sqrt{T}}{\sqrt{T}-1} \right], \tag{1.75}$$

$$u_x - iu_y = \frac{1}{2} \cdot \frac{\sqrt{T}-1}{\sqrt{T}+1}, \tag{1.76}$$

$$x_{CD} = \frac{1}{\pi} \left(\ln\sqrt{1+\eta} - \frac{\eta}{\eta+1} \right), \tag{1.77}$$

$$y_{CD} = \frac{1}{\pi} \left(\pi - \operatorname{arctg}\sqrt{\eta} + \frac{\sqrt{\eta}}{\eta+1} \right), \quad \eta = 0\ldots\infty. \tag{1.78}$$

For the general case of ($b = 0\ldots1$; $S = 0\ldots\infty$) based on the numerical solutions of Equations 1.67, 1.68, and 1.63 and calculations by Equations 1.54, 1.61, and 1.66, the parameters of the problem are given in Table 1.5.

There can be some difficulties in constructing a hydrodynamic grid (a family of orthogonal curves $\varphi = \text{const}$; $\psi = \text{const}$) and determining the velocity field that are related to finding the integral Equation 1.58 that is not expressed in terms of elementary functions (it can only be reduced to the sum of elliptic integrals).

In this case, we can use numerical methods, assuming that the constants δ_∞ and b are known for a given S (e.g., they can be taken from Table 1.5 or from the solution of Equation 1.68 and δ_∞ can be calculated by the formula in Equation 1.66).

After the integrand is divided by the real and imaginary parts, the integral Equation 1.58 is reduced to a sum of contour integrals. In this case, we use the analytic property of the integrand enabling the arbitrary choice of the integration path between the points $C(0, 0)$ and $T(m; n)$. The best option is to integrate first in a circumferential direction (centered at $t = 1$) with a radius $r = 1$ from the point C to the intersection with the ray AT and then along this ray from the intersection point to the preset point T.

However, it is simpler to use a universal mathematical environment package Maple-9 that immediately gives a result of the numerical integration for the given T, b, δ_∞ as a complex number. This allows avoiding complicated transformations of the integrand in separating it into the real part and imaginary one.

As an example, Figure 1.35 shows the flow pattern for $S = 0$ and $S = \infty$ built in Maple-9. We can see that the maximum air velocity is in the area of the inertial flow separation (point C) and along the *free* flow line CD (here, there are the highest frequency of equipotentials $\varphi = \text{const}$).

The longer the length of the projection the greater the speed of the jet separation is and the smaller the width of the jet is (Figure 1.36). The rate of the separation curve (flow line CD) also varies depending on the value S/B (Figure 1.37), and a significant change occurs in the area $S/B = 0\ldots0.5$.

TABLE 1.5

Main Parameters of the Separated Flow at a Projecting Suction Slot

b	$E(b)$	$S(b)$	$\delta_\infty(b)$	$W(b)$	$\mu(b)$	u_M	y_M
0.00	2.000	0.0000	0.6110	1.637	0.0000	−1.0000	1.000
0.02	2.240	0.0151	0.5838	1.713	0.4717	−0.4534	1.068
0.04	2.325	0.0317	0.5747	1.740	0.5928	−0.3820	1.112
0.06	2.386	0.0496	0.5684	1.759	06830	−0.3379	1.153
0.08	2.435	0.0689	0.5633	1.775	0.7589	−0.3056	1.193
0.10	2.477	0.0894	0.5591	1.789	0.8263	−0.2801	1.232
0.12	2.514	0.1114	0.5555	1.800	0.8881	−0.2590	1.272
0.14	2.547	0.1348	0.5522	1.811	0.9460	−0.2409	1.313
0.16	2.577	0.1597	0.5493	1.820	1.001	−0.2251	1.354
0.18	2.605	0.1863	0.5467	1.829	1.054	−0.2111	1.397
0.20	2.630	0.2146	0.5443	1.837	1.105	−0.1985	1.441
0.22	2.654	0.2448	0.5420	1.845	1.155	−0.1870	1.486
0.24	2.677	0.2770	0.5399	1.852	1.204	−0.1765	1.534
0.26	2.698	0.3114	0.5380	1.859	1.252	−0.1668	1.583
0.28	2.718	0.3481	0.5361	1.865	1.300	−0.1578	1.635
0.30	2.738	0.3874	0.5344	1.871	1.347	−0.1494	1.689
0.32	2.756	0.4294	0.5327	1.877	1.395	−0.1415	1.746
0.34	2.773	0.4744	0.5311	1.883	1.442	−0.1340	1.806
0.36	2.790	0.5228	0.5296	1.888	1.490	−0.1270	1.870
0.38	2.806	0.5747	0.5282	1.893	1.538	−0.1204	1.937
0.40	2.822	0.6307	0.5268	1.898	1.586	−0.1140	2.008
0.42	2.837	0.6910	0.5255	1.903	1.635	−0.1080	2.084
0.44	2.852	0.7562	0.5242	1.908	1.684	−0.1023	2.165
0.46	2.866	0.8269	0.5230	1.912	1.735	−0.09676	2.252
0.48	2.879	0.9036	0.5218	1.917	1.786	−0.09149	2.345
0.50	2.893	0.9872	0.5206	1.921	1.838	−0.08643	2.446
0.52	2.905	1.078	0.5195	1.925	1.892	−0.08156	2.554
0.54	2.918	1.178	0.5185	1.929	1.947	−0.07687	2.672
0.56	2.930	1.288	0.5174	1.933	2.003	−0.07235	2.800
0.58	2.942	1.409	0.5164	1.936	2.061	−0.06799	2.940
0.60	2.954	1.544	0.5154	1.940	2.121	−0.06377	3.093
0.62	2.965	1.693	0.5145	1.944	2.184	−0.05968	3.263
0.64	2.976	1.860	0.5135	1.947	2.248	−0.05573	3.451
0.66	2.987	2.048	0.5126	1.951	2.316	−0.05189	3.661
0.68	2.997	2.261	0.5117	1.954	2.387	−0.04817	3.897
0.70	3.008	2.504	0.5109	1.957	2.461	−0.04455	4.164
0.72	3.018	2.784	0.5100	1.961	2.540	−0.04104	4.469
0.74	3.028	3.108	0.5092	1.964	2.624	−0.03762	4.820
0.76	3.037	3.489	0.5084	1.967	2.713	−0.03429	5.230
0.78	3.047	3.942	0.5076	1.970	2.810	−0.03105	5.713
0.80	3.056	4.489	0.5069	1.973	2.914	−0.02789	6.294
0.82	3.066	5.161	0.5061	1.976	3.028	−0.02480	7.002
0.84	3.075	6.008	0.5054	1.979	3.155	−0.02179	7.888
0.86	3.083	7.098	0.5047	1.982	3.297	−0.01885	9.026
0.88	3.092	8.563	0.5040	1.984	3.459	−0.01598	10.54
0.90	3.101	10.62	0.5033	1.987	3.650	−0.01317	12.66

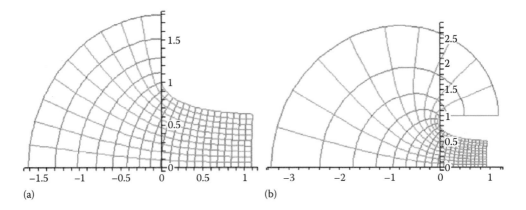

FIGURE 1.35   Flow lines and equipotential lines at the entrance to the suction inlet: (a) built into a flat wall and (b) hooded.

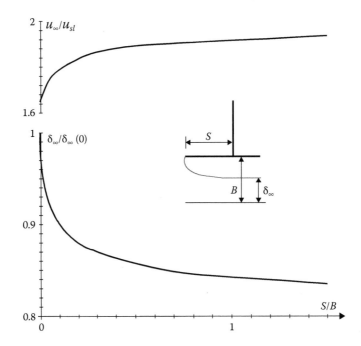

FIGURE 1.36   Change in the flow separation velocity and the jet width at infinity, depending on the length of projection S ($\delta_\infty(0) = \delta_\infty$ for $S \to 0$, u_{sl} is the average air velocity in the slot).

Offsets of the paths toward the flow in the area $S/B = 0.5...\infty$ are close enough. We can see that when $S/B = 0.4...0.5$ the jet width is virtually very similar to the limit width (δ_∞) (more exactly, a significant change in the separation velocity (u_∞) and the jet width δ is seen only in the area $S/B < 0.4...0.5$). It is a known fact that the efflux theory takes a section located at a distance B from the inlet as a typical section.

1.4.2   EXPERIMENTAL STUDY

Experimental studies of the velocity field in the vicinity of the inlet were performed in the pilot plant (Figure 1.38), whose working part was the channel formed by two vertical planes

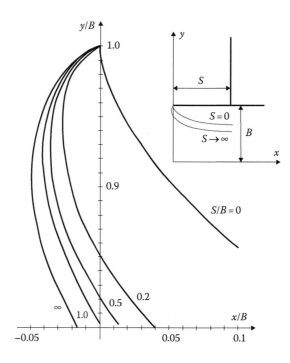

FIGURE 1.37 Theoretical curves of stream flow separation near the exhaust hole with a change in the projection length.

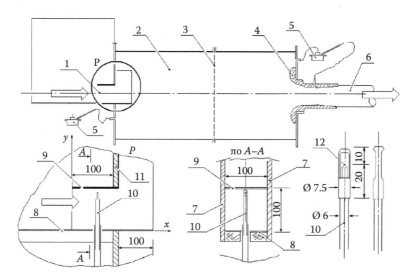

FIGURE 1.38 Pilot plant for the study of the velocity field near the suction slot: 1, suction inlet; 2, plenum (500 × 500 × 1100 mm); 3, partition; 4, measuring collector (∅ = 112 mm); 5, MCM-2400 micromanometers; 6, air duct (∅ = 125 mm); 7, vertical planes; 8, horizontal strip (100 × 600 mm; δ = 0.55 mm), 9, projection (hood 100 × 100 mm; δ = 0.55 mm), 10, testo 425 anemometer probe (∅ = 6/7.5 mm); 11, plenum vertical wall; 12, velocity sensor.

500 × 500 mm (made of Plexiglas, 8 mm thick) and two horizontal strips of galvanized iron sheet (0.55 mm thick). The distances both between the planes and between the strips were 100 mm. The total length of the lower strip was equal to 600 mm (and the strip entered inside the adjacent plenum chamber by 100 mm). The outer 100 mm long strip was attached to the vertical wall of the chamber and to the vertical planes and formed a protrusion of the channel (with a cross

section of 100 × 100 mm). The vertical planes were extended inside the plenum chamber at a distance of 100 mm (and the lower channel wall was extended too). Thus, the working part of the plant was as close as possible to the model of plane problems of the flow near an exhaust opening.

The velocity field measurements were carried out in a vertical plane passing through the axis of symmetry of channel with a testo-435 hot-wire anemometer (with an error of ±(0.03 + 0.05u), m/s). The velocity was automatically averaged over the measurement time interval $\Delta t = 20-25$ s (in this case, about 50 velocity measurements were made at a predetermined point on an automatic basis). A collector and MMN-2400 micromanometer were used to determine the airflow from a sealed plenum chamber sucked with two series-connected fans (Vents VKM 150 and Systemair EX-18 4c). The plenum was equipped with a partition of filter cloth to avoid deformation of the flow entering the collector.

The typical velocity was determined from the flow equation

$$q = u_{sl}^* B^* = u_\infty^* \cdot \delta_\infty^*; \quad u_\infty^* = \frac{u_{sl}^*}{\delta_\infty},$$

where
u_{sl}^* is the average velocity in the slot
q is the specific airflow defined on the basis of the differential static pressure measured by a micromanometer in the collector Δp (Pa)

Here (with the slot length of 0.1 m), these values were determined by the formulas

$$q = \frac{S_k}{0.1} \cdot \sqrt{\frac{2\Delta p}{(1+\zeta_k)\rho}} = 0.0507 \sqrt{\frac{\Delta p}{\rho}}; \quad \frac{m^3}{s \cdot m}; \quad u_{sl}^* = 0.507 \sqrt{\frac{\Delta p}{\rho}}; \quad m/s,$$

where
S_k is the area of the collector metering section (here, $S_k = 0.112^2 \cdot \pi/4 = 0.00985$ m^2)
ρ is the air density, kg/m^3
$\zeta_k = 0.073$ is the local drag factor coefficient of the collector

The accuracy of measured longitudinal components in the vertical sections of the slot was checked by graphical integration and comparison of the average velocity u_{sl}^* with the calculations derived from the given formula. The error did not exceed 2%–3%.

1.4.3 CALCULATION RESULTS AND DISCUSSION

Four vertical sections of the jet in the suction slot (located at $x/B = 0.05, 0.3, 0.55,$ and 0.8 from the input section, respectively) and three horizontal sections were selected as typical sections: under ($y/B = 0.8$), at a level ($y/B = 1$) of and above ($y/B = 1.2$) the hood. With a measurement pitch of $0.1B$, the horizontal component of the velocity vector u_x was measured in the first case and the vertical component u_y in the second case. The results are shown in Figures 1.39 through 1.41. The solid lines in the figures show the calculation results in Maple-9 using Equations 1.58 and 1.59 for $S/B = 1.0$ ($b = 0.50286, \delta_\infty = 0.520474$).

For comparison, we also show the velocity profiles obtained by the discrete vortex methods (DVMs) and by solving the time-averaged Reynolds-averaged Navier–Stokes equation (RANS) in

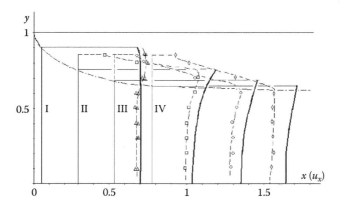

FIGURE 1.39   Change in the longitudinal component of the velocity throughout the height of a flat channel for $S = 0$ (the dash–dotted line is the flow line $\psi = q$; in orthographic representations, solid lines represent theoretical values and dotted ones represent experimental values; I, II, III, and IV are velocity profile curves in the cross sections.

the Fluent software. As can be seen from these results, the theoretical description of the velocity field with models of separated flows presents a practically sufficient accurate description of the behavior of the velocity components, except for the areas near the jet separation and on its free boundary (CD). Here, we have a fully developed turbulence and due to this, the potentiality of the flow seems to be violated. For example, in the vertical sections of the channel near the line CD (Figures 1.39 and 1.40), a well-pronounced boundary mixing layer is seen with a sharp change in the horizontal component of velocity and a significant deviation of the experimental values from the theoretical ones with an increase in the distance of the metering sections from the air inlet into the channel. The theoretical value u_x exceeds the empirical one due to the fact that the true thickness is higher than the theoretical one δ. The dead zone (between CB and CD) is filled with a moving stream, although its velocities are low. Naturally, the velocity within the theoretical jet separation will be lower in this case.

As to the qualitative description, the experimental data are in good agreement with the theoretical values. The longitudinal velocities increase in each section toward the boundary CD, and the peak area (similar to the line CD) becomes more distant from the hood. The dead zone is filled with a flow whose velocity is significantly lower than the velocity within the range of the jet (between the lines CD and AD).

$$x = 0.05; \quad x = 0.3; \quad x = 0.55; \quad x = 0.8.$$

In horizontal sections, the greatest deviation from the theoretical values is also observed near the separation point, although the qualitative character of the change in the vertical component of velocity is in good agreement with the experimental values: the highest value u_y both in the experimental studies and in the theoretical ones (by the CM method and DVM) is seen near the point C (Figure 1.41). The more distant it is from the point C, the closer the value of the measured u_y is to the theoretically calculated one. At a distance $0.5...0.8B$, the deviations will not interrupt the measurement errors. Note that the RANS method gives values close to 0 in the range C.

Thus, the model of potential separated flows can be deemed to adequately describe the nature of the flows almost in the whole area of the suction flare. Some convention in the existence of a *break* of flows on the free boundary CD does not prevent the widespread use of jet contraction (namely δ_∞) in the practical determination of pressure losses. The main cause of energy losses during the inlet into the inlets is undoubtedly the gating effect of the flow that is separated from the walls adjacent to this inlet. The longer the length of these walls (hoods and projections), the

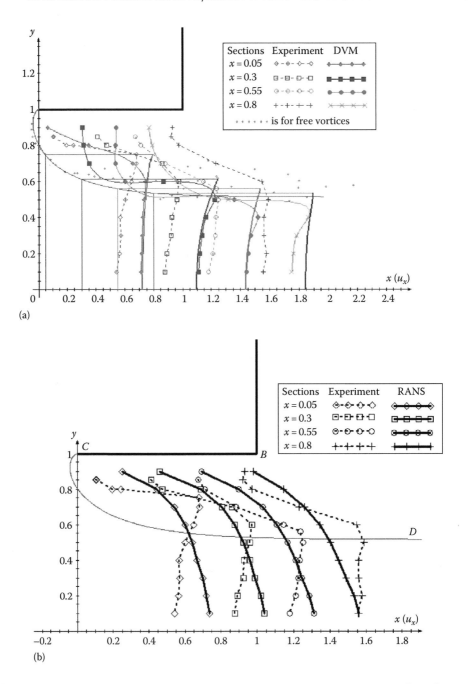

FIGURE 1.40 Measurement of the longitudinal air velocity through the entire height of the flat channel equipped with a hood of a unit length: (a) calculations on DVM, CM method and experiment and (b) calculations on RANS and experiment.

greater is the separation velocity, the higher is the strength of the flow into the inlet, and, consequently, the higher are the pressure losses.

We show this by two specific examples: air inlet into a flat pipe and its outflow from a flat pipe through the diaphragm. These examples have been chosen due to the availability of experimental data, obtained by Idelchik [129].

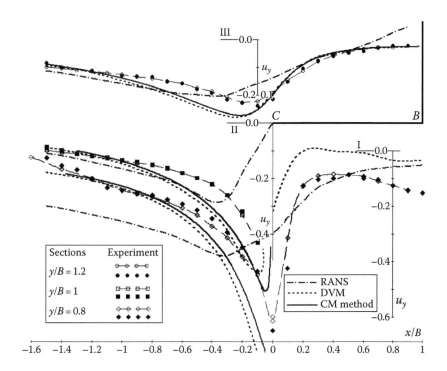

FIGURE 1.41 Change in the vertical velocity component near the inlet of the hooded suction slot (*CB*) of a unit length; in orthographic representations, solid lines are theoretical values; circles, diamonds, squares are experimental values; I, II, and III are the velocity profiles in the cross sections *y/B* = 0.8; 1; 1.2, respectively.

We will use the difference in velocities on a free surface (separation velocity) and the average velocity in the slotted opening to describe the value of inertia

$$\Delta u = \frac{u_\infty^* - u_{sl}^*}{u_{sl}^*} \equiv w(b) - 1; \quad w(b) = \frac{u_\infty^*}{u_{sl}^*} = \frac{1}{\delta_\infty(b)}.$$

We use the data from Table 1.5 to determine the theoretical coefficient of jet contraction $\delta_\infty = \delta_\infty^*/B^*$ (the values δ_∞^*, B^* are dimensional, which is shown by the superscript *) for an inlet into a flat pipe and we use the results of theoretical studies by Zhukovsky (1897) and von Mises (1917) described in [89,143] to assess the degree of flow contraction in the outflow through the diaphragm:

$$\delta_\infty \equiv \frac{\delta_\infty^*}{B^*} = \frac{\pi}{\pi + 2\dfrac{2\alpha}{tg2\alpha}}; \quad n \equiv \frac{B^*}{L^*} = \frac{tg\alpha}{\delta_\infty}.$$

Whence

$$S \equiv \frac{S^*}{B^*} = \frac{1}{n} - 1 = \frac{\delta_\infty}{tg\alpha} - 1,$$

where α is the parameter varying within $0 \le \alpha \le \pi/4$. The latter can be eliminated, allowing for a direct connection between δ_∞ and n (or S) in the form of a transcendental equation:

$$\delta_\infty = \frac{\pi}{\pi + 2 \cdot A}, \tag{1.79}$$

where $A = A_n$ is the function of δ_∞ and n:

$$A_n = \left(\frac{1}{n\delta_\infty} - n\delta_\infty \right) \operatorname{arctg}(n\delta_\infty),$$

or of δ_∞ and S:

$$A = A_S = \left(\frac{S+1}{\delta_\infty} - \frac{\delta_\infty}{S+1} \right) \operatorname{arctg}\left(\frac{\delta_\infty}{S+1} \right) \qquad (1.80)$$

It is easy to see that for $S = 0$ ($n = 1$), the coefficient of contraction is $\delta_\infty = 1$ (no contraction occurs) and for $S \to \infty$ ($n = 0$) the function $A = 1$, $\delta_\infty = 0.611$ (the outflow through holes in an infinite wall).

As can be seen from Figures 1.42 and 1.43, the relative separation velocity Δu is increasing with an increase in the length of a straight wall before the outflow of the jet, and the nature of this increase corresponds with the change in the local drag coefficient.

Thus, the model of potential separated flows adequately describes most of the real suction flare of a flat slot. Significant deviations of the theoretical velocity field from the experimental one are only seen in the jet separation area and on its free boundary. A developed boundary layer with a great transverse gradient of the longitudinal velocities is seen along the free boundary of the jet in the channel.

The degree of the cross-section jet contraction in a channel is determined by the inertia of airflow flowing along flat surfaces at the inlet into the suction slot. The bigger the acceleration path of this flow, the higher the rate of its separation is and the more pronounced is the gating effect at the air inlet into the inlet. The behavior of the relative velocity of jet separation is almost identical to the experimental law of change in the local drag coefficient depending on the length of the projections adjacent to the inlet.

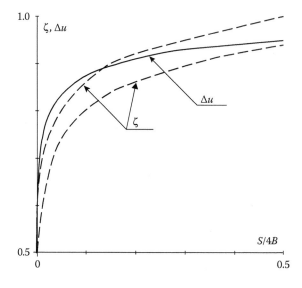

FIGURE 1.42 Change in the local drag coefficient at the inlet into the flat pipe (ζ) and the jet separation velocity (Δu) with increase in the projection length ($S/4B$): the solid line is based on theoretical data (Table 1.5) and dashed ones on experimental data by Idelchik (Curve 1 for the relative wall thickness of $\delta/D_g = 0$, Curve 2 for $\delta/D_g = 0.004$).

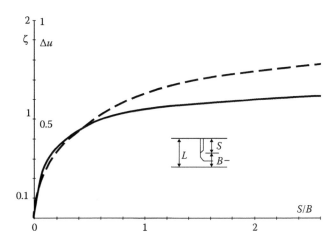

FIGURE 1.43 Change in the local drag coefficient at the inlet into the flat pipe (ζ) and the jet separation velocity (Δu) with increase in the projection length (S/B): the solid curve is based on Equations 1.79 and 1.80 and the dotted one represents the experimental data by Idelchik.

1.5 MATHEMATICAL SIMULATION OF INFLUENCE OF A SCREEN ON THE AERODYNAMIC DRAG VALUE OF A SUCTION SLOT

Applied problems of industrial aerodynamics most often focus on reducing the drag of a suction slot [129,144–146]. But increasing this drag is equally important. First of all, it relates to the practical reducing of the intake of the gaseous medium through openings when there are no options of sealing left. One of the technical means to reduce unwanted air inflows is to use special screens that increase the drag and reduce the flow coefficients of air entering through minimized openings. These may be benefitial, say, for specialists engaged in confinement of harmful emissions by local exhaust ventilation [147], as well as in various heat and power plants.

The chapter focuses on the determination of flow characteristics near a suction slot built into a flat infinite wall, depending on a screen mounted perpendicular to its axis.

1.5.1 DETERMINATION OF DESIGN RATIOS

Consider a potential separated airflow at the inlet into a flat slot, in front of which two screens are mounted perpendicular to its axis (Figure 1.44): one is impermeable (AA') and the other has an opening HH'.

Due to the symmetry, we will consider the flow in the upper semiplane of the physical plane $z = x + iy$. This figure also represents the correspondence of the flow boundaries to the range of the complex potential $w = \varphi + i\psi$, which is the band $0 \le \psi \le q$ with a cut MNA ($\psi = kq$, $-\infty < \varphi < -\xi$), the Joukowski function $\omega = \ln\dfrac{u_0}{u} > 0$ whose range is a half-band $\left(\ln\dfrac{u_0}{u} > 0; \ -\dfrac{\pi}{2} \le \theta \le 0 \right)$, and the upper accessorial semiplane of the complex variable $t_1 = x_1 + iy_1$.

Perform a conformal mapping of the upper semiplane Im(t_1) > 0 onto the interior of polygonal ranges of the complex potential w and the Joukowski function ω.

Use the Schwarz–Christoffel integral, considering the taken correspondence of the points, and determine the constants of the integral with the transition of singular points along the semicircles with an infinitely small radius to obtain the following formula:

$$w = (1-k)\frac{q}{\pi}\ln\left(\frac{t_1 - m}{m}\right) + k\frac{q}{\pi}\ln\left(t_1 - 1\right), \tag{1.81}$$

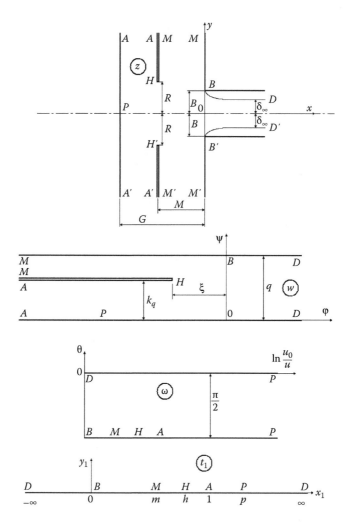

FIGURE 1.44 To the determination of the parameters of the separated flow at a plane slot with a screen.

$$\omega = -\frac{1}{2}\ln\left(\frac{\sqrt{t_1}-\sqrt{p}}{\sqrt{t_1}+\sqrt{p}}\right) = \ln\left(\frac{\sqrt{t_1}+\sqrt{p}}{\sqrt{t_1}-\sqrt{p}}\right), \tag{1.82}$$

$$z = i + (1-k)\frac{\delta_\infty}{\pi}N_m + k\frac{\delta_\infty}{\pi}N_1, \quad k = \frac{1-h}{1-m}, \tag{1.83}$$

$$v \equiv u_x - iu_y = u_0 e^{-\omega(t_1)} = u_0\frac{\sqrt{t-p}}{\sqrt{t}+\sqrt{p}}, \tag{1.84}$$

where t is the point of the upper semiplane $\operatorname{Im}(t_1) > 0$, which corresponds to a given point $t_z(x, y)$ of the physical plane z; N_m, N_1 are the integrals:

$$N_m = \int_0^t \frac{\sqrt{t_1}+\sqrt{p}}{\sqrt{t_1}-\sqrt{p}}\frac{dt_1}{t_1-m}, \tag{1.85}$$

$$N_1 = \int_0^t \frac{\sqrt{t_1} + \sqrt{p}}{\sqrt{t_1 - p}} \frac{dt_1}{t_1 - 1}, \tag{1.86}$$

which can be simply expressed in terms of elementary functions. After the integration variable is replaced according to the formula

$$v = \frac{\sqrt{t_1} + \sqrt{p}}{\sqrt{t_1 - p}} \Rightarrow t_1 = p \left(\frac{v^2 + 1}{v^2 - 1} \right)^2 \tag{1.87}$$

the integrand of the integral N_m is reduced to a rational function

$$\Pi_m = -\frac{8p}{p - m} \cdot \frac{v^2 (v^2 + 1)}{(v^2 - 1) \left[(v/n)^2 + 1 \right] \left[(vn)^2 + 1 \right]}, \tag{1.88}$$

which can undergo some simple transformations, given the fact that

$$p = m \left(\frac{1 + n^2}{1 - n^2} \right); \quad \frac{8p}{p - m} = 2 \left(\frac{1 + n^2}{n} \right)^2, \tag{1.89}$$

and is expressed as follows:

$$\Pi_m = -2 \left(\frac{2}{v^2 - 1} + \frac{1}{(v/n)^2 + 1} + \frac{1}{(vn)^2 + 1} \right). \tag{1.90}$$

Then the integral N_m can be expressed as the integral of the sum of partial fractions:

$$N_m = -2 \int_{-i}^{v} \left[\left(\frac{1}{v - 1} - \frac{1}{v + 1} \right) + \frac{1}{2i} \left(\frac{1}{v/n - i} - \frac{1}{v/n + i} \right) + \frac{1}{2i} \left(\frac{1}{vn - i} - \frac{1}{vn + i} \right) \right] dv. \tag{1.91}$$

The final result for the integrals of Equations 1.85 and 1.86 is

$$N_m = F_m(t) - F_m(0); \quad N_1 = N_m \big|_{m=1} = F_1(t) - F_1(0), \tag{1.92}$$

$$F_m(t) = \ln \left(\frac{v + 1}{v - 1} \right)^2 + \frac{n}{i} \ln \frac{v/n + i}{v/n - i} + \frac{1}{in} \ln \frac{vn + i}{vn - i}, \tag{1.93}$$

$$F_m(0) = \frac{\pi}{n} + i \left[\pi + \left(n + \frac{1}{n} \right) \ln \left(\frac{1 + n}{1 - n} \right) \right], \tag{1.94}$$

$$F_1(t) = F_m(t) \big|_{n = n_1}; \quad F_1(0) = F_m(0) \big|_{n = n_1}, \tag{1.95}$$

where for the simplicity of formulation we use the denotation

$$v = \frac{\sqrt{t} + \sqrt{p}}{\sqrt{t - p}}, \tag{1.96}$$

$$n = \frac{\sqrt{p - m}}{\sqrt{p} + \sqrt{m}}, \tag{1.97}$$

$$n_1 = \frac{\sqrt{p - 1}}{\sqrt{p} + 1}. \tag{1.98}$$

To identify the physical meaning of the parameters n and n_1 we will use the relation of Equation 1.84 and obvious expressions for the airflow:

$$u_M M = (1 - k)q = (1 - k)u_0 \delta_\infty, \tag{1.99}$$

$$u_A(G - M) = kq = ku_0 \delta_\infty, \tag{1.100}$$

where u_M, u_A are the velocities, respectively, at the points M (for $t = m$) and A (for $t = 1$), m/s; u_0 is the velocity on the boundary line of the flow (BD), m/s.

After the velocities u_M, u_A are determined by Equation 1.84, we obtain the following relations:

$$\bar{u}_M \equiv \frac{u_M}{u_0} = \frac{\sqrt{p - m}}{\sqrt{p} + \sqrt{m}} \equiv n, \tag{1.101}$$

$$\bar{u}_A \equiv \frac{u_A}{u_0} = \frac{\sqrt{p - 1}}{\sqrt{p} + 1} \equiv n_1 \tag{1.102}$$

that explain the physical meaning of the denotations from Equations 1.97 and 1.98.

We will substitute these expressions for the velocities $\bar{u}_M$ and $\bar{u}_A$ in Equations 1.99 and 1.100 and using Equation 1.83, we will find the following relations:

$$\frac{u_A}{u_M} = \frac{\sqrt{p - 1}}{\sqrt{p} + 1} \frac{\sqrt{p} + \sqrt{m}}{\sqrt{p - m}} = \frac{h - m}{1 - h}\left(\frac{G}{M} - 1\right), \tag{1.103}$$

$$\delta_\infty = \frac{\sqrt{p - 1}}{\sqrt{p} + 1}(G - M) + \frac{\sqrt{p - m}}{\sqrt{p} + \sqrt{m}} M \tag{1.104}$$

that relate the parameters of the problem (p, m, h) to the geometrical dimensions of the physical domain (δ_∞, G, M).

Note that for $h \to m$ based on Equation 1.83, we have $k \to 1$, that is, the internal screen HM merges with the vertical wall BM ($M \to 0$). In this case, air inflows to the opening through a strip with a width G and thus

$$\frac{\delta_\infty}{G} = \frac{\sqrt{p - 1}}{\sqrt{p} + 1}, \tag{1.105}$$

whence we see that when the impermeable screen is moved to infinity ($G \to \infty$), the parameter is $p \to 1$.

To close the system of Equations 1.103 and 1.104, which include three unknown parameters p, m, and h besides the two known geometrical dimensions G and M, as well as an unknown half-height of the jet at infinity δ_∞, we will consider the correspondence of the flow boundaries in the semiplanes z and t_1.

To determine the value δ_∞, we consider the free line BD, to which the points of the negative axis $OX_1(t_1 = x_1 \le 0)$ correspond.

In this area for $p > 1;\ 0 > m > 1;\ t = 0 \dots -\infty$ we obtain

$$N_m = A_m + iB_m; \quad N_1 = A_1 + iB_1 \tag{1.106}$$

where

$$A_m = \ln \frac{1+\mu_1}{1-\mu_1} + n\left(\text{arctg}\, \frac{n-\mu_2}{\mu_1} + \text{arctg}\, \frac{n+\mu_2}{\mu_1} \right) + \frac{1}{n}\left(-\pi + \text{arctg}\, \frac{n^{-1}-\mu_2}{\mu_1} + \text{arctg}\, \frac{n^{-1}+\mu_2}{\mu_1} \right), \tag{1.107}$$

$$B_m = \pi - 2\left(\text{arctg}\, \frac{\mu_2}{1+\mu_1} + \text{arctg}\, \frac{\mu_2}{1-\mu_1} \right) + \left(n + \frac{1}{n} \right)\left(\ln \frac{1-n}{1+n} - \ln \sqrt{\frac{W-\mu_2}{W+\mu_2}} \right), \tag{1.108}$$

$$\mu_1 = \frac{\sqrt{-t}}{\sqrt{p-t}}; \quad \mu_2 = \frac{\sqrt{p}}{\sqrt{p-t}}; \quad W = \frac{1+n^2}{2n}; \quad W_1 = \frac{1+n_1^2}{2n_1}. \tag{1.109}$$

We will define the value of the integral N_1, its real and imaginary parts by the substitution into Equations 1.107 through 1.109 $m = 1$ $(n = n_1)$:

$$N_1 = N_m\Big|_{\substack{m=1 \\ (n=n_1)}}; \quad A_1 = A_m\Big|_{\substack{m=1 \\ (n=n_1)}}; \quad B_1 = B_m\Big|_{\substack{m=1 \\ (n=n_1)}}. \tag{1.110}$$

Then, the parametric equation of the free flow line BD on the basis of Equation 1.83 is as follows:

$$x_{BD} = \text{Re}(z_{BD}) = (1-k)\frac{\delta_\infty}{\pi} A_m + k\frac{\delta_\infty}{\pi} A_1, \tag{1.111}$$

$$y_{BD} = \text{Im}(z_{BD}) = 1 + (1-k)\frac{\delta_\infty}{\pi} B_m + k\frac{\delta_\infty}{\pi} B_1, \tag{1.112}$$

and based on the condition $y_{BD} \to \delta_\infty$ for $t \to -\infty$ $(\mu_1 = 1;\ \mu_2 = 0)$ we find

$$\delta_\infty = \frac{\pi}{\pi + B}, \tag{1.113}$$

where

$$B = (1-k)\left(n + \frac{1}{n} \right)\ln \frac{1+n}{1-n} + k\left(n_1 + \frac{1}{n_1} \right)\ln \frac{1+n_1}{1-n_1} \tag{1.114}$$

or

$$B = (1-k)\frac{\sqrt{p}}{\sqrt{p-m}}\ln\frac{\sqrt{p}+\sqrt{p-m}}{\sqrt{p}-\sqrt{p-m}} + k\frac{\sqrt{p}}{\sqrt{p-1}}\ln\frac{\sqrt{p}+\sqrt{p-1}}{\sqrt{p}-\sqrt{p-1}}. \tag{1.115}$$

In particular, for $k = 0$ letting sequentially first $m \to 1$, then $p \to 1$, $n_1 \to 0$, we obtain

$$B = 2; \quad \delta_\infty = \frac{\pi}{\pi+2} \approx 0.611,$$

that is, the case of a flat slope in a vertical wall (without any screen). We obtain the same result for $k = 1$ (letting $p \to 1$).

Before we examine the correspondence of the boundaries in the area $t \geq 0$, we consider the behavior of the integrals N_m, N_1 at the points M and A. Making the transition of points along the arcs of the circles $t = m + \varepsilon_m e^{i\varphi}$; $t = 1 + \varepsilon_a e^{i\varphi}$, $\varphi = \pi...0$ with infinitely small radii ($\varepsilon_m \to 0$, $\varepsilon_a \to 0$), we obtain from Equations 1.83, 1.85, and 1.86 the following expressions for the increment of the function Δz at these points

$$\Delta z_M = (1-k)\frac{\delta_\infty}{\pi}\left(-\pi\frac{\sqrt{m}+\sqrt{p}}{\sqrt{p-m}}\right), \tag{1.116}$$

$$\Delta z_A = k\frac{\delta_\infty}{\pi}\left(-\pi\frac{1+\sqrt{p}}{\sqrt{p-1}}\right). \tag{1.117}$$

Otherwise, the analysis, Figure 1.44, shows that at such a transition $\Delta z_M = -M$, $\Delta z_A = -(G-M)$ and thus

$$\left.\begin{array}{l} (1-k)\delta_\infty\dfrac{\sqrt{m}+\sqrt{p}}{\sqrt{p-m}} = (1-k)\dfrac{\delta_\infty}{n} = M, \\[4mm] k\delta_\infty\dfrac{1+\sqrt{p}}{\sqrt{p-1}} = k\dfrac{\delta_\infty}{n_1} = G - M. \end{array}\right\} \tag{1.118}$$

or

$$(1-k)\frac{\delta_\infty}{n} + k\frac{\delta_\infty}{n} = G. \tag{1.119}$$

Whence, taking into account Equation 1.83, we obtain the following important relation between the parameters of the problem and the geometrical dimensions of the physical flow range:

$$\frac{1-h}{h-m}\cdot\frac{1+\sqrt{p}}{\sqrt{p-1}}\cdot\frac{\sqrt{p-m}}{\sqrt{m}+\sqrt{p}} = \frac{G}{M} - 1. \tag{1.120}$$

For $m \to 1$ we obtain

$$\frac{1-h}{h-m} = \frac{G}{M} - 1, \tag{1.121}$$

which is identical to the case of the equality $u_A = u_M$ in Equation 1.103.

Let us now consider the positive real semiaxis of the complex plane t_1. Assume that $t = x_1 > 0$. Now, it is obvious that $t > p(x_1 > p)$ ν is the real positive value, while $t < p(x_1 < p)$ is purely imaginary:

$$\nu = -i\mu \quad \text{at } x_1 < p, \tag{1.122}$$

$$\nu = \frac{\sqrt{t} + \sqrt{p}}{\sqrt{t - p}} \quad \text{at } x_1 > p, \tag{1.123}$$

where

$$\mu = \frac{\sqrt{t} + \sqrt{p}}{\sqrt{p - t}}. \tag{1.124}$$

After we substitute Equations 1.122 and 1.123 in Equations 1.92 and 1.95, we obtain the following value of the integral N_m on the boundary $BMHAP$:

1. On the interval $BM \left(1 < \mu < \dfrac{1}{n} \right)$:

$$N_m = 4i \arctan\frac{1}{\mu} - in\ln\frac{\mu/n - 1}{\mu/n + 1} - \frac{i}{n}\ln\frac{1 - \mu n}{\mu n + 1} - i\left(\pi + \left(n + \frac{1}{n} \right)\ln\frac{1 + n}{1 - n} \right), \tag{1.125}$$

2. On the intervals MH, HA, AP $\left(\dfrac{1}{n} < \mu < \infty \right)$:

$$N_m = 4i \arctan\frac{1}{\mu} - in\ln\frac{\mu/n - 1}{\mu/n + 1} - \frac{i}{n}\ln\frac{\mu n - 1}{\mu n + 1} - \frac{\pi}{n} - i\left(\pi + \left(n + \frac{1}{n} \right)\ln\frac{1 + n}{1 - n} \right). \tag{1.126}$$

The second integral on this boundary has a similar form:

1. On the interval $1 < \mu < \dfrac{1}{n_1}$:

$$N_1 = 4i \arctan\frac{1}{\mu} - in_1\ln\frac{\mu/n_1 - 1}{\mu/n_1 + 1} - \frac{i}{n_1}\ln\frac{1 - \mu n_1}{\mu n_1 + 1} - i\left(\pi + \left(n_1 + \frac{1}{n_1} \right)\ln\frac{1 + n_1}{1 - n_1} \right), \tag{1.127}$$

2. On the interval $\dfrac{1}{n_1} < \mu < \infty$:

$$N_1 = 4i \arctan\frac{1}{\mu} - in_1\ln\frac{\mu/n_1 - 1}{\mu/n_1 + 1} - \frac{i}{n_1}\ln\frac{\mu n_1 - 1}{\mu n_1 + 1} - \frac{\pi}{n_1} - i\left(\pi + \left(n_1 + \frac{1}{n_1} \right)\ln\frac{1 + n_1}{1 - n_1} \right) \tag{1.128}$$

On the ray PD $(t > p, \nu > 0)$, these integrals are, respectively, equal to

$$N_m = \ln\left(\frac{\nu + 1}{\nu - 1} \right)^2 + 2n \arctan\frac{n}{\nu} + \frac{2}{n}\arctan\frac{1}{\nu n} - \frac{\pi}{n} - i\left(\pi + \left(n + \frac{1}{n} \right)\ln\frac{1 + n}{1 - n} \right), \tag{1.129}$$

$$N_1 = \ln\left(\frac{\nu + 1}{\nu - 1} \right)^2 + 2n_1 \arctan\frac{n_1}{\nu} + \frac{2}{n_1}\arctan\frac{1}{\nu n_1} - \frac{\pi}{n_1} - i\left(\pi + \left(n_1 + \frac{1}{n_1} \right)\ln\frac{1 + n_1}{1 - n_1} \right). \tag{1.130}$$

Using these relations, we can check whether the points of the axis OX_1 correspond to the flow boundaries of the physical range since the point B ($t = 0$; $\mu = 1$) corresponds to the point $z_B = i$ (since for $\mu = 1$ we have $N_m = 0$; $N_1 = 0$ from Equations 1.125 and 1.127). The point from Equation 1.131 corresponds to the point M^- on the interval BM for $t = m - \varepsilon$, $\varepsilon \to 0$, $\mu = \dfrac{1}{n} - \varepsilon_1$, $\varepsilon_1 \to 0$:

$$z_{M^-} = i + (1-k)\frac{\delta_\infty}{\pi} N_m\Big|_{t=m-\varepsilon} + k\frac{\delta_\infty}{\pi} N_1\Big|_{t=m-\varepsilon} = iI_{M^-}, \tag{1.131}$$

where I_{M^-} is the positive infinity, which occurs in the case of the limiting value of the third summand of Equation 1.125:

$$\lim_{\substack{\varepsilon_1 \to 0, \\ n \neq 0}} \left(-\frac{1}{n}\ln\frac{1-(1/n-\varepsilon_1)n}{2} \right) = +\infty.$$

The below point corresponds to the other *part* of this point M^+ on the interval MH for $t = m + \varepsilon$; $\varepsilon \to 0$; $\mu = \dfrac{1}{n} + \varepsilon_1$; $\varepsilon_1 \to 0$:

$$z_{M^+} = i + (1-k)\frac{\delta_\infty}{\pi} N_m\Big|_{t=m+\varepsilon} + k\frac{\delta_\infty}{\pi} N_1\Big|_{t=m+\varepsilon} = iI_{M^+} - (1-k)\frac{\delta_\infty}{n}$$

or, taking into account Equation 1.118, we obtain

$$z_{M^+} = iI_{M^+} - M, \quad I_{M^+} \to +\infty. \tag{1.132}$$

Here, the nature of *infinity* is the same as that for the point M^-.

The point from Equation 1.133 corresponds to the point H on the ray MH for $t = h$, $m < h < 1\left(\dfrac{1}{n} < \mu < \dfrac{1}{n_1} \right)$:

$$z_H = i + (1-k)\frac{\delta_\infty}{\pi} N_m(H) + k\frac{\delta_\infty}{\pi} N_1(H), \tag{1.133}$$

where

$$H = \mu_h = \frac{\sqrt{h} + \sqrt{p}}{\sqrt{p-h}} \tag{1.134}$$

$$N_m = 4i\,\text{arctg}\frac{1}{H} - in\ln\frac{H/n-1}{H/n+1} - \frac{i}{n}\ln\frac{Hn-1}{Hn+1} - i\left(\pi + \left(n + \frac{1}{n}\right)\ln\frac{1+n}{1-n}\right) - \frac{\pi}{n} \tag{1.135}$$

$$N_1 = 4i\,\text{arctg}\frac{1}{H} - in_1\ln\frac{H/n_1-1}{H/n_1+1} - \frac{i}{n_1}\ln\frac{1-Hn_1}{1+Hn_1} - i\left(\pi + \left(n_1 + \frac{1}{n_1}\right)\ln\frac{1+n_1}{1-n_1}\right) \tag{1.136}$$

Otherwise, as is evident from Figure 1.44,

$$z_H = -M + iR \tag{1.137}$$

and, therefore, we can write the equation

$$R = \frac{\delta_\infty}{\pi} \left\{ 4 \arctan \frac{1}{H} - (1-k)\left(n \ln \frac{H/n+1}{H/n-1} + \frac{1}{n} \ln \frac{Hn+1}{Hn-1} \right) \right.$$

$$\left. + k\left(n_1 \ln \frac{H/n_1+1}{H/n_1-1} + \frac{1}{n_1} \ln \frac{1+Hn_1}{1-Hn_1} \right) \right\} \quad (1.138)$$

that relates the parameter h to R.*

The point from Equation 1.139 corresponds to the point A^- on the ray HA for $t = 1-\varepsilon, \mu = \frac{1}{n_1} - \varepsilon_1, \varepsilon \to 0; \varepsilon_1 \to 0$:

$$z_{A^-} = i + (1-k)\frac{\delta_\infty}{\pi} N_m\left(\frac{1}{n_1} - \varepsilon_1 \right) + k\frac{\delta_\infty}{\pi} N_1\left(\frac{1}{n_1} - \varepsilon_1 \right) = -M + I_A \cdot i, \quad (1.139)$$

where I_A is the positive infinity, which occurs in the case of the limiting value of the third summand of Equation 1.127:

$$\lim_{\substack{\varepsilon_1 \to 0, \\ n_1 \neq 0}} \left(-\frac{1}{n_1} \ln \frac{1 - (1/n_1 - \varepsilon_1)n_1}{2} \right) = +\infty.$$

Similarly, the point from Equation 1.140 corresponds to the point A^+ on the ray AP for $t = 1 + \varepsilon$; $\varepsilon \to 0; \mu = \frac{1}{n_1} + \varepsilon_1; \varepsilon_1 \to 0$:

$$z_{A^+} = i + (1-k)\frac{\delta_\infty}{\pi} N_m\left(\frac{1}{n_1} + \varepsilon \right) + k\frac{\delta_\infty}{\pi} N_1\left(\frac{1}{n_1} + \varepsilon \right) = iI_{A^+} - (1-k)\frac{\delta_\infty}{\pi}\frac{\pi}{n} - k\frac{\delta_\infty}{\pi}\frac{\pi}{n_1}, \quad (1.140)$$

or, taking into account Equation 1.118, we obtain

$$z_{A^+} = -G + iI_{A^+},$$

where I_{A^+} is a positive infinity, which occurs in the case of the limiting value of the third summand of Equation 1.128:

$$\lim_{\substack{\varepsilon_1 \to 0, \\ n_1 \neq 0}} \left(-\frac{1}{n_1} \ln \frac{(1/n_1 + \varepsilon_1)n_1 - 1}{2} \right) = +\infty.$$

The point in Equation 1.141 corresponds to the point P^- on the ray AP ($t = p - \varepsilon$; $\mu = +\infty$):

$$z_P = i + (1-k)\frac{\delta_\infty}{\pi} N_m(+\infty) + k\frac{\delta_\infty}{\pi} N_1(+\infty) \quad (1.141)$$

or, considering that due to Equations 1.126 and 1.128,

$$N_m(+\infty) = -\frac{\pi}{n} - i\left(\pi + \left(n + \frac{1}{n} \right) \ln \frac{1+n}{1-n} \right); \quad (1.142)$$

---

* It is assumed that the separation point of the flow line on the screen coincides with its bottom point H. Possible cases in which those points do not coincide are shown in problems of jet flow around finite length sheets [33].

$$N_1(+\infty) = -\frac{\pi}{n_1} - i\left(\pi + \left(n_1 + \frac{1}{n_1}\right)\ln\frac{1+n_1}{1-n_1}\right), \tag{1.143}$$

and taking into account the relations, Equations 1.114 and 1.118

$$z_P = -G + i\left(1 - \delta_\infty + \frac{\delta_\infty}{\pi}(-B)\right) = -G. \tag{1.144}$$

We will obtain a similar result with Equations 1.129 and 1.130 for the point P^+ on the ray PD for $t = p + \varepsilon$; $\mu = +\infty$. The point D on the ray PD for $t \to \infty$; $\nu = 1 + \varepsilon$, given that

$$N_m(1+\varepsilon) = \infty - i\left(\pi + \left(n + \frac{1}{n}\right)\ln\frac{1+n}{1-n}\right), \tag{1.145}$$

$$N_1(1+\varepsilon) = \infty - i\left(\pi + \left(n_1 + \frac{1}{n_1}\right)\ln\frac{1+n_1}{1-n_1}\right), \tag{1.146}$$

we have

$$z_D = \infty - i \cdot 0. \tag{1.147}$$

For an arbitrary point belonging to this ray we can write

$$z_{PD} = -G + \frac{\delta_\infty}{\pi}\ln\left(\frac{\nu+1}{\nu-1}\right)^2 + (1-k)\frac{\delta_\infty}{\pi}\left(2n \operatorname{arctg}\frac{n}{\nu} + \frac{2}{n}\operatorname{arctg}\frac{1}{\nu n}\right)$$

$$+ k\frac{\delta_\infty}{\pi}\left(2n_1 \operatorname{arctg}\frac{n_1}{\nu} + \frac{2}{n_1}\operatorname{arctg}\frac{1}{\nu n_1}\right). \tag{1.148}$$

Resulting ratios (Equations 1.83, 1.97, 1.98, 1.113, 1.114, 1.118, and 1.138) enable to solve the direct problem. From the known geometrical dimensions of the physical flow range G, M, and R, using the solution of the system of two equations, Equation 1.138 and the equality*

$$M \cdot n + (G - M)n_1 = \frac{\pi}{\pi + B}, \tag{1.149}$$

we find the unknown parameters m and p, and, consequently, the auxiliary variables n and n_1 (according to Equations 1.97 and 1.98), and then we determine the value k by the formula

$$k = \frac{(G-M)n_1}{(G-M)n_1 + Mn}, \tag{1.150}$$

obtained from Equation 1.118; the value B is calculated by formula Equation 1.114 and then an important parameter δ_∞ by Equation 1.113. Note that Equations 1.138 and 1.149 are reduced to equations that depend also on m and p by substituting Equations 1.150, 1.114, 1.113, 1.134, and

$$h = 1 - k(1 - m), \tag{1.151}$$

derived from Equation 1.83.

---

* Equation 1.149 is based on Equations 1.113 and 1.118.

After the parameters of the problem (m, p, and k) have been determined, it is easy to build a hydrodynamic grid of a potential flow and the velocity field. For example, the flow function is as follows:

$$\psi = \ln(w) - \text{const}$$

or taking into account Equation 1.81

$$\frac{\psi}{q} = \frac{1-k}{\pi} \text{Im}\left[\ln\frac{t_1 - m}{m}\right] + \frac{k}{\pi} \text{Im}\left[\ln(t_1 - 1)\right] - \text{const.} \tag{1.152}$$

The flow lines are built as follows. The value ψ/q and one of the coordinates, x_1 or y_1 the point $t = x_1 + iy_1$ is set. From Equation 1.152, we determine the second coordinate of this point belonging to a given flow line.

1.5.2 Calculation Results and Discussion

Figures 1.45 and 1.46 show the flow lines for two cases: for a slot equipped with an impermeable screen (Figure 1.45) and the case of two screens (Figure 1.46), and the screen with an opening equal to the height of the slot is located at a distance $M = 0.25\,B$ from the inlet section.

In the latter case, we can see a part of the flow separated by the screen with a pronounced vertical direction and an increased velocity, which enables a gating effect on the remaining flow. This contracts the flow to the axis OX resulting in a decrease of the jet height in the slot at infinity δ_∞ compared to the height for $G \to \infty$ and $G \to \infty$ ($\delta_\infty = 0.611$ in the case of a slot in the wall without any screen). If we analyze the degree of contraction (Figure 1.47) depending on the distance from a holed screen, we see that the greatest effect is achieved for $M/B = 0.2 \div 0.4$.

When the hole is decreased (Figure 1.48), the degree of jet contraction also increases (and the flow of the main stream is reduced).

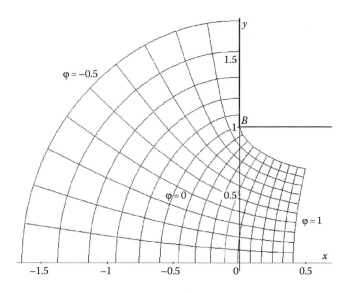

FIGURE 1.45 Flow lines of the suction flare of the slot equipped with an impermeable screen at a distance ($G = 10B$).

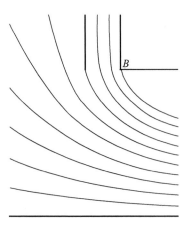

FIGURE 1.46 Flow lines of a suction flare of the slot equipped with a screen having a central opening of a height equal to the height of the slot (for $M = 0.25$; $G = 10$; $R = 1$).

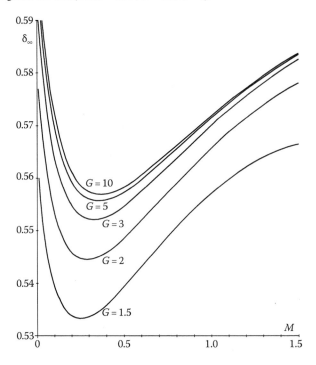

FIGURE 1.47 Change in the jet thickness of the separated flow δ_∞ at a distance from the screen (with a central opening of a height $R = 1$) and at a distance M from the inlet section of the jet.

To identify the role of the inner screen, we consider the special case of an airflow in a slot with one impermeable screen. By the passage to the limit $m \to 1$ ($h \to 1$), we obtain $k = 0$ and the following design relations:

$$w = \frac{q}{\pi} \ln(t_1 - 1); \quad z = i + \frac{\delta_\infty}{\pi} N_m,$$

$$n = n_1 = \frac{\sqrt{p-1}}{\sqrt{p+1}}; \quad M = G = \frac{\delta_\infty}{n_1}, \tag{1.153}$$

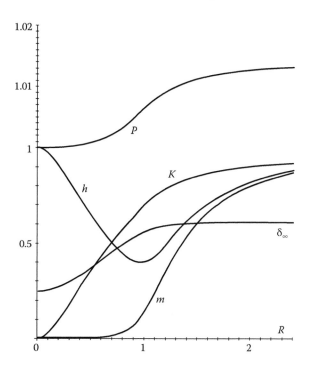

FIGURE 1.48   Change in the parameters of the problem with an increase in the size of the hole of the inner screen (for $M = 0.25$; $G = 10$).

$$\delta_\infty = \frac{\pi}{\pi + B}, \quad B = \left(n_1 + \frac{1}{n_1} \right) \ln \frac{1 + n_1}{1 - n_1}, \tag{1.154}$$

$$N_m = N_1 = \ln \left(\frac{v+1}{v-1} \right)^2 + \frac{n_1}{i} \ln \left(\frac{v/n_1 + i}{v/n_1 - i} \right) + \frac{1}{in_1} \ln \left(\frac{vn_1 + i}{vn_1 - i} \right) - \frac{\pi}{n_1} - i(\pi + B). \tag{1.155}$$

Thus, there is one unknown parameter p (or n_1) left, which defines the hydrodynamic picture of the separated flow. To determine it, it is necessary to solve transcendental equation, Equation 1.153, taking into account Equation 1.154. The results of the computational solution of this equation for given G (or M) are shown in Table 1.6. As can be seen from these data, the jet thickness at infinity varies considerably, especially if the impermeable screen is located in close vicinity (when $G < 1$). When the distance to the screen in this range decreases, the flow resistance significantly decreases δ_∞ and dramatically increases (Figure 1.49) and the nature of this growth is in good agreement with the increase in the flow kinetism at its separation from the vertical wall (at the point B)

$$\Delta u^2 = \left(\frac{1}{\delta_\infty} - 1 \right)^2, \tag{1.156}$$

and with the experimental data obtained by Idelchik [145] for an oblique elbow piece (both for $L = \infty$ and for $L = 4G$ where L is the size of the rectangular elbows perpendicular to the plane of symmetry) and for entry into a rectangular pipe with an impermeable screen. When the distance to the screen increases the jet height increases (Figure 1.50), and the smaller the G, the faster this height attains its limit at infinity $-\delta_\infty$. Thus, if $G = 0.2$, then this height $\delta = 1.1\delta_\infty$ is observed as close as $x_{1.1} = 0.3$ and if $G = 5$, then at the distance $x_{1.1} = 0.6$.

TABLE 1.6

Parameters of the Problem

G	p	B	δ_∞	Δu	G	p	B	δ_∞	Δu
0.20	75900	12.62	0.1993	4.018	2	1.408	2.236	0.5842	0.7117
0.25	3800	9.630	0.2460	3.065	2.5	1.253	2.154	0.5933	0.6855
0.30	589.3	7.771	0.2879	2.474	3	1.173	2.108	0.5985	0.6710
0.35	171.6	6.548	0.3242	2.084	3.5	1.126	2.080	0.6017	0.6620
0.40	72.19	5.698	0.3554	1.814	4	1.095	2.061	0.6038	0.6561
0.45	38.08	5.080	0.3821	1.617	4.5	1.075	2.049	0.6053	0.6521
0.50	23.28	4.612	0.4052	1.468	5	1.061	2.039	0.6064	0.6492
0.55	15.75	4.248	0.4251	1.352	5.5	1.050	2.033	0.6072	0.6470
0.60	11.47	3.957	0.4426	1.260	6	1.042	2.027	0.6078	0.6454
0.65	8.813	3.721	0.4578	1.184	6.5	1.036	2.023	0.6082	0.6441
0.70	7.062	3.525	0.4712	1.122	7	1.031	2.020	0.6086	0.6431
0.75	5.847	3.362	0.4831	1.070	7.5	1.027	2.018	0.6089	0.6422
0.80	4.970	3.223	0.4936	1.026	8	1.023	2.015	0.6092	0.6416
0.85	4.316	3.105	0.5030	0.9882	8.5	1.021	2.014	0.6094	0.6410
0.90	3.815	3.002	0.5113	0.9557	9	1.019	2.012	0.6096	0.6405
0.95	3.423	2.914	0.5188	0.9275	9.5	1.017	2.011	0.6097	0.6401
1.00	3.109	2.836	0.5255	0.9028	10	1.015	2.010	0.6098	0.6398
1.05	2.854	2.768	0.5316	0.8811	10.5	1.014	2.009	0.6099	0.6395
1.10	2.644	2.708	0.5371	0.8619	11	1.012	2.008	0.6100	0.6392
1.15	2.469	2.654	0.5421	0.8448	11.5	1.011	2.008	0.6101	0.6390
1.20	2.321	2.606	0.5466	0.8296	12	1.010	2.007	0.6102	0.6388
1.25	2.195	2.564	0.5507	0.8160	12.5	1.010	2.006	0.6103	0.6386
1.30	2.087	2.525	0.5544	0.8037	13	1.009	2.006	0.6103	0.6385
1.35	1.993	2.490	0.5578	0.7927	13.5	1.008	2.005	0.6104	0.6384
1.40	1.911	2.459	0.5610	0.7827	14	1.008	2.005	0.6104	0.6382
1.45	1.839	2.430	0.5638	0.7736	14.5	1.007	2.005	0.6105	0.6381
1.50	1.776	2.404	0.5665	0.7653	15	1.007	2.004	0.6105	0.6380

To determine the influence of the inner screen on the jet contraction value, we study the change in the relative height of the jet

$$k_\delta = \frac{\delta_\infty}{\delta_\infty^G},$$

where

δ_∞^G is the jet height in the slot with only an impermeable screen placed at a distance G

δ_∞ is the jet height in the slot with two screens mounted G and M, respectively

Similarly, the relative coefficient of kinetism is as follows:

$$k_{\Delta u^\alpha} = \frac{\left(\dfrac{1}{\delta_\infty} - 1\right)^2}{\left(\dfrac{1}{\delta_\infty^G} - 1\right)^2}. \tag{1.157}$$

The behavior graphs of these relative values depending on G and M, shown in Figures 1.51 and 1.52, show the presence of clear extrema in the area $M/B = 0.2...0.4$, and the larger the distance to the

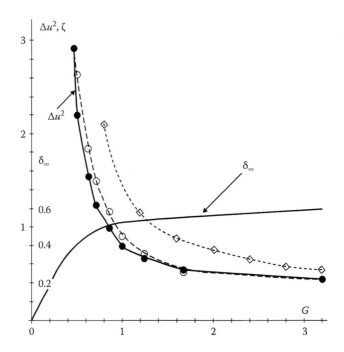

FIGURE 1.49 Change in the jet thickness δ_∞ and Δu depending on the distance to the impermeable screen G (solid lines show the graphs of functions (Equations 1.154 and 1.156), dashed lines show the experimental data obtained by Idelchik for the local drag coefficient of the elbow piece for $L = \infty(\bullet)$ and for $L = 4G(\circ)$ and for the inlet into a rectangular pipe with an impermeable screen ($\Diamond$)).

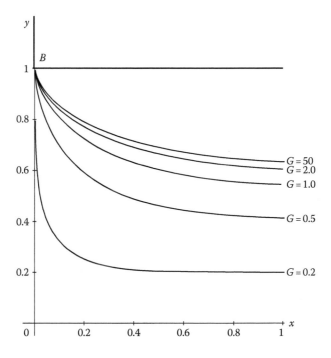

FIGURE 1.50 Change in the free flow line depending on the distance to the impermeable screen G.

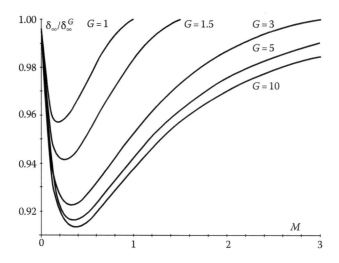

FIGURE 1.51   Change in the relative jet height in the slot depending on the distance of the screen with a center hole $R/B = 1$.

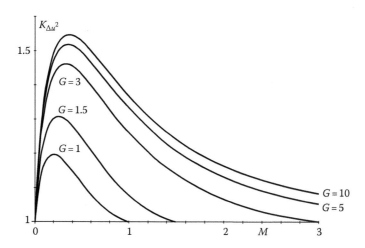

FIGURE 1.52   Change in the relative coefficient of kinetism of the jet in the slot depending on the distance of the screen with a center hole of $R/B = 1$.

impermeable screen ($G \geq 10$) is, the larger is the value of these extrema. Thus, the minimum value of the jet contraction height with an inner screen installed was decreased from $0.957 \, \delta_\infty^G$ (for $G = 1$) to $0.913 \, \delta_\infty^G$ for ($G = 10$), that is, decreased by more than 10% and the maximum rate of kinetism $k_{\Delta u^\alpha}$ was increased by 28.3% in that case.

1.5.3   CONCLUSIONS

1. The equipment of the suction slot with a screen with a hole of the same height results in an increase in the jet contraction, the maximum value of which is achieved when the screen is installed at a distance of 0.2…0.4 of the slot half-height.
2. The jet contraction in the slot is significantly increased when an impermeable screen is installed closer, which increases the flow kinetics and, as a result, causes an increase in the local drag coefficient when the flow enters the slot.

1.6 SIMULATION OF AIR JET FLOW AT THE INLET INTO A HOODED FLAT CHANNEL EQUIPPED WITH AN IMPERMEABLE SCREEN

The ideal fluid theory, in particular the method by Zhukovsky, was used in the study of separated flows at the inlet into flat channels [148,149] due to the fact that it allows determining not only the kinematic characteristic of the flow but also the deterministic boundary of the flow. For example, the jet thickness, which is used to estimate the air drag at the inlet into holes. An increased drag of the channel entrance caused by jet contraction reduces unwanted air inflowing through the holes and the working openings of environmental and technological equipment covers.

The chapter focuses on the description of the separated flow in the slot-type suction channel, which extends beyond the flat wall and the range of which comprises an impermeable screen, and on determination of the effect caused by the distance to this screen and by the degree of the channel extension (hood length) on the jet thickness at infinity (δ).

Note that the achievement of this goal using the method by Zhukovsky does not enable construction of a *pure* analytical solution. This can be achieved only for certain problems with a simple geometry of area boundaries. In this case, the problem is reduced to solving a system of nonlinear equations that contain improper integrals of Cauchy type. This chapter begins with building the said system of equations and extracting the singularity of integrals (in this case they are reduced to a combination of elliptical integrals of different types, which reduces the calculation time), and then implements the solution of the system in the universal mathematical environment *Maple* and offers a numerical determination of the problem parameters, analysis of the dependencies obtained, and their comparison with experimental data.

1.6.1 CONSTRUCTION OF DESIGN RATIOS

Assume that the airflow is directed along the axis OX of the complex variable plane $z = x + iy$. And the axis OX is the axis of symmetry of the flow, and, therefore, we consider the flow range in the upper semiplane $\text{Im } z > 0$ (Figure 1.53). Thus, the flat potential airflow is restricted by the screen PA at a distance G from the entry section of the channel, the vertical wall BA, the hood CB with a length S, and the free flow line CD, which is the boundary of the jet and the ray PD of the axis OX. The half-width of the channel B will be used as a unit of length (i.e., $B = 1$).

The variation range of the complex potential $w = \varphi + i\psi$ (where φ is the velocity potential and ψ is the flow function) is a strip $0 \leq \psi \leq q$ with a width of q (q is the airflow, m^2/s). The range of the Joukowski function $\omega = \ln(u_0/u) + i\theta$ (where u is the flow velocity, m/s; u_0 is the flow velocity at the jet boundary CD, m/s; θ is the angle between the positive axis OX and the direction of the velocity wind $\vec{u}$) is a half-strip with a cut along the ray $MB(\theta = -\pi/2)$ bounded by the rays $CB(\theta = -\pi)$ $DP(\theta = 0)$ and the interval $CD\left(\ln\left(u_0/u\right) = 0; 0 \geq \theta \geq -\pi\right)$.

Find a conformal mapping of the auxiliary half-plane $\text{Im } t > 0$ onto the interior of the ranges of the functions w and ω. As the boundaries of these areas are straight, we can use the Schwarz–Christoffel integral, which maps the upper semiplane onto the interior of a polygonal area.

Thus, the formula for the complex potential in the assumed correspondence of the points is

$$w = C_1 \int_0^t \frac{dt}{t-1} + i = \frac{q}{\pi} \ln\left(t-1\right). \tag{1.158}$$

The constant C_1 is found by the transition of the singular point A ($t = 1$) along a semicircle $t - 1 = \varepsilon e^{i\alpha}$, $\pi \leq \alpha \leq 0$ with an infinitely small radius $\varepsilon \to 0$:

$$\Delta w_A = C_1 \int_{1-\varepsilon}^{1+\varepsilon} \frac{dt}{t-1} = C_1 \int_\pi^0 \frac{i\varepsilon e^{i\alpha}}{\varepsilon e^{i\alpha}} \cdot d\alpha = -iC_1\pi. \tag{1.159}$$

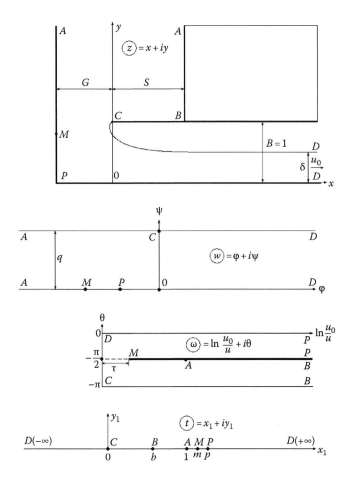

FIGURE 1.53 To the determination of the flow pattern of the separated flow in a flat channel with an impermeable screen.

Alternatively, the increment at the point A (at the transition from the interval CA on AD) is $\Delta w_A = -iq$. Therefore, $C_1 = q/\pi$.

For the Joukowski function

$$\omega = C_2 \int_0^t \frac{t-m}{\sqrt{t}\,(t-b)(t-p)}\,dt - \pi i = C_2 \left[\frac{m-b}{p-b} \int_0^t \frac{dt}{\sqrt{t}\,(t-b)} + \frac{p-m}{p-b} \int_0^t \frac{dt}{\sqrt{t}\,(t-p)} \right] - \pi i. \quad (1.160)$$

To determine the constant C_2 requires finding the increment of the function ω in the singular points B ($t = b$) and P ($t = p$) by transition along the semicircles $t - b = \varepsilon_b e^{i\alpha}$ and $t - p = \varepsilon_p e^{i\alpha}$ (for $\pi \leq \alpha \leq 0$) with infinitely small radii $\varepsilon_b \to 0$ and $\varepsilon_p \to 0$:

$$\Delta\omega_B = C_2 \frac{m-b}{p-b} \cdot \int_{b-\varepsilon_b}^{b+\varepsilon_b} \frac{dt}{\sqrt{t}\,(t-b)} = \frac{C_2}{\sqrt{b}} \cdot \frac{m-b}{p-b}(-\pi i),$$

$$\Delta\omega_p = \frac{C_2}{\sqrt{p}} \cdot \frac{p-m}{p-b}(-\pi i). \quad (1.161)$$

Compare them with the obvious increments in the case of the transition from the ray CB to the ray BM $\left(\Delta\omega_b = -\dfrac{\pi}{2}i - (-\pi i) = \dfrac{\pi}{2}i \right)$ and from the ray MP to the ray DP $\left(\Delta\omega_p = 0 - \left(-\dfrac{\pi}{2}i\right) = \dfrac{\pi}{2}i \right)$. We obtain

$$C_2 \frac{m-b}{p-b} = -\frac{\sqrt{b}}{2}; \quad C_2 \frac{p-m}{p-b} = -\frac{\sqrt{p}}{2}. \tag{1.162}$$

Whence

$$m = \sqrt{bp}; \tag{1.163}$$

$$\omega = -\frac{\sqrt{b}}{2} \int_0^t \frac{dt}{\sqrt{t}\,(t-b)} - \frac{\sqrt{p}}{2} \int_0^t \frac{dt}{\sqrt{t}\,(t-p)} - \pi i. \tag{1.164}$$

After some simple transformations, given the fact that

$$\int_0^t \frac{dt}{\sqrt{t}\,(t-a)} = \frac{1}{\sqrt{a}} \left(\ln \frac{\sqrt{t} - \sqrt{a}}{\sqrt{t} + \sqrt{a}} - \pi i \right), \quad \text{at } a > 0, \tag{1.165}$$

we obtain the explicit form of the Joukowski function in the parametric semiplane $\operatorname{Im} t > 0$:

$$\omega = \ln \left(\sqrt{\frac{\sqrt{t} + \sqrt{b}}{\sqrt{t} - \sqrt{b}}} \cdot \sqrt{\frac{\sqrt{t} + \sqrt{p}}{\sqrt{t} - \sqrt{p}}} \right) = \ln \left(\frac{\sqrt{t} + \sqrt{b}}{\sqrt{t} - b} \cdot \frac{\sqrt{t} + \sqrt{p}}{\sqrt{t} - p} \right). \tag{1.166}$$

Taking into account (Equation 1.163) we find

$$\tau = \frac{1}{2} \ln \left(\frac{\sqrt{p} + 1}{\sqrt{p} - 1} \cdot \frac{1 + \sqrt{b}}{1 - \sqrt{b}} \right). \tag{1.167}$$

To relate it to the physical area z we have

$$z = \frac{\delta}{q} \int_0^t e^{\omega} \frac{dw}{dt}\, dt = i + \frac{\delta}{\pi} A; \quad A = \int_0^t \frac{\sqrt{t} + \sqrt{b}}{\sqrt{t} - b} \frac{\sqrt{t} + \sqrt{p}}{\sqrt{t} - p} \frac{dt}{t - 1}. \tag{1.168}$$

Now define the jet thickness δ at the point D $(t = -\infty)$. To do so, find the value of the integral A in the range $t = x_1$, $0 \leq x < -\infty$. After the integrand is divided by the real and imaginary parts, we can write a general expression for the coordinates of the free flow line CD:

$$x_{CD} = \frac{\delta}{\pi} A_1; \quad y_{CD} = 1 + \frac{\delta}{\pi} \left(\sqrt{p} + \sqrt{b} \right) \cdot B_1, \tag{1.169}$$

where

$$A_1 = \int_0^{x_1} \frac{x_1 + m}{\sqrt{b - x_1}\sqrt{p - x_1}} \frac{dx_1}{1 - x_1}; \quad B_1 = \int_0^{x_1} \frac{\sqrt{-x_1}}{\sqrt{b - x_1}\sqrt{p - x_1}} \frac{dx_1}{1 - x_1}. \tag{1.170}$$

The first integral is expressed in terms of elementary functions:

$$A_1 = 2\ln\frac{\sqrt{b-x_1}+\sqrt{p-x_1}}{\sqrt{b}+\sqrt{p}} + \frac{2\left(1+\sqrt{p\cdot b}\right)}{\sqrt{p-1}\sqrt{1-b}}\left[\operatorname{arctg}\frac{\sqrt{p-x_1}\sqrt{1-b}}{\sqrt{b-x_1}\sqrt{p-1}} - \operatorname{arctg}\frac{\sqrt{p}\sqrt{1-b}}{\sqrt{b}\sqrt{p-1}}\right], \quad (1.171)$$

and the second one in terms of elliptic integrals:

$$B_1 = -\frac{2b}{\sqrt{p}(1-b)}\left[\operatorname{Elliptic}\ Pi\left(\sqrt{\frac{-x_1}{b-x_1}}, 1-b, \sqrt{\frac{p-b}{b}}\right) - \operatorname{Elliptic}\ F\left(\sqrt{\frac{-x_1}{b-x_1}}, \sqrt{\frac{p-b}{b}}\right)\right]. \quad (1.172)$$

Here, the elliptic integrals are notated according to the universal mathematical environment *Maple*:

$$\operatorname{Elliptic}\ Pi(z, v, k) = \int_0^z \frac{dt}{\left(1-vt^2\right)\sqrt{1-t^2}\sqrt{1-k^2t^2}}, \quad (1.173)$$

$$\operatorname{Elliptic}\ F(z, k) = \int_0^z \frac{d\alpha_1}{\sqrt{1-\alpha_1^2}\sqrt{1-k^2\alpha_1^2}}. \quad (1.174)$$

Taking into account that $y_{CD} \to \delta$ for $x_1 \to -\infty$, we obtain the following equation to determine the jet thickness at infinity (as a function of the parameters b and p):

$$\delta = 1 + \frac{\delta}{\pi}\left(\sqrt{p}+\sqrt{b}\right)\cdot B_1\big|_{x_1=-\infty} \quad (1.175)$$

or

$$\delta = \frac{\pi}{\pi+B}, \quad (1.176)$$

where

$$B = 2\frac{\sqrt{p}+\sqrt{b}}{\sqrt{p}}\cdot\frac{b}{1-b}\left[\operatorname{Elliptic}\ Pi\left(1-b, \sqrt{\frac{p-b}{p}}\right) - \operatorname{Elliptic}\ K\left(\sqrt{\frac{p-b}{p}}\right)\right], \quad (1.177)$$

$$\operatorname{Elliptic}\ Pi\left(1-b, \sqrt{\frac{p-b}{p}}\right) = \operatorname{Elliptic}\ Pi\left(1, 1-b, \sqrt{\frac{p-b}{p}}\right),$$

$$\operatorname{Elliptic}\ K\left(\sqrt{\frac{p-b}{p}}\right) = \operatorname{Elliptic}\ F\left(1, \sqrt{\frac{p-b}{p}}\right).$$

From Equation 1.168, write the equation relating the hood length S with the same parameters p and b:

$$S = \frac{\delta}{\pi} \int_0^b \frac{\sqrt{t} + \sqrt{b}}{\sqrt{b-t}\sqrt{p-t}} \cdot \frac{dt}{1-t}$$

or, given that $0 < t \leq b$,

$$S = 2\frac{\delta}{\pi} \cdot \left[-\ln\frac{\sqrt{p-b}}{\sqrt{p}-\sqrt{b}} + \frac{1+\sqrt{p \cdot b}}{\sqrt{p-1}\sqrt{1-b}} \cdot \operatorname{arctg}\frac{\sqrt{b}\sqrt{p-1}}{\sqrt{p}\sqrt{1-b}} - \left(1 + \sqrt{\frac{b}{p}}\right) \cdot \right.$$

$$\left. \left(\text{Elliptic } K\left(\sqrt{\frac{b}{p}}\right) - \text{Elliptic } Pi\left(b, \sqrt{\frac{b}{p}}\right)\right)\right]. \tag{1.178}$$

The second equation defining the direct problem (the parameters p and b, and consequently δ are determined using preset S and G) can be written on the basis of the obvious equation for the airflow:

$$\delta \cdot u_0 = u_A (G + S), \tag{1.179}$$

where u_A is the air velocity at the point A ($t = 1$), m/s.

The air velocity at an arbitrary point $z(t)$ is determined by the equation

$$\frac{dw}{dz} \equiv u_x - iu_y = \frac{dw}{dt} \bigg/ \frac{dz}{dt} = u_0 \frac{\sqrt{t-b}\sqrt{t-p}}{\left(\sqrt{t}+\sqrt{b}\right)\left(\sqrt{t}+\sqrt{b}\right)}. \tag{1.180}$$

We use this equation to find the absolute value of the velocity at the point A:

$$u_A = u_0 \frac{\sqrt{1-b}\sqrt{p-1}}{\left(1+\sqrt{b}\right)\left(1+\sqrt{p}\right)}. \tag{1.181}$$

Then Equation 1.179 is as follows:

$$G = \delta \frac{1+\sqrt{b}}{\sqrt{1-b}} \cdot \frac{1+\sqrt{p}}{\sqrt{p-1}} - S. \tag{1.182}$$

After we have solved the system of two nonlinear equations, Equations 1.178 and 1.182, with Equations 1.176 and 1.177, we can find the accessory parameters of the problem b and p and the jet thickness δ. Table 1.7 shows these parameters for some geometric dimensions of the flow.

As we know the parameters of the problem and δ, we can find the velocity field from Equations 1.180 and 1.168 and construct a hydrodynamic flow pattern (a family of equipotential lines and flow lines) taking into account (Equations 1.158 and 1.168). To do this, it is necessary to find the value of the integral A for an arbitrary point of the upper semiplane Im $t > 0$. Unfortunately, this integral can only be expressed in terms of elementary functions in a special case: $S = 0$ (for $b = 0$), that is, when the inlet section of a flat channel is located in the wall (as discussed in the previous sections).

TABLE 1.7

Parameters of the Separated Flow in a Hooded Flat Channel with an Impermeable Screen (For S = 0.1)

G	b	p	δ	b	p	δ
$S = 0.1$				$S = 1$		
0.5	0.215945	133.8063	0.332612	0.844875	592.6906	0.295440
1.0	0.147974	5.427048	0.463379	0.712623	10.47901	0.422440
1.5	0.129457	2.380932	0.507312	0.642020	3.643530	0.464163
2.0	0.121676	1.689067	0.526700	0.601091	2.298623	0.483388
2.5	0.117706	1.418292	0.536772	0.575628	1.794975	0.494030
3.0	0.115421	1.282736	0.542620	0.558831	1.545575	0.500600
3.5	0.113990	1.204600	0.546298	0.547207	1.401412	0.504963
4.0	0.113037	1.155236	0.548755	0.538846	1.309520	0.508017
4.5	0.112371	1.121966	0.550474	0.532637	1.246870	0.510243
5.0	0.111888	1.098436	0.551723	0.527903	1.202007	0.511917
$S = 0.5$				$S = 5$		
0.5	0.667823	433.2963	0.302275	0.988035	740.5073	0.290827
1.0	0.507886	8.733717	0.431985	0.966650	12.62019	0.413708
1.5	0.443875	3.187741	0.475319	0.946293	4.319172	0.451920
2.0	0.411927	2.069036	0.495131	0.928711	2.692852	0.468897
2.5	0.393847	1.648102	0.505870	0.914008	2.078646	0.478246
3.0	0.382684	1.440196	0.512345	0.901805	1.769754	0.484108
3.5	0.375330	1.320693	0.516550	0.891653	1.587649	0.488109
4.0	0.370239	1.245044	0.519434	0.883155	1.468978	0.491007
4.5	0.366572	1.193842	0.521498	0.875987	1.386162	0.493197
5.0	0.363846	1.157443	0.523026	0.869895	1.325435	0.494906

In general, this integral can be expressed as the sum of elliptic integrals. In particular, for Im $t > 0$ for $0 < b < 1 < m < p$ we have

$$A = \int_0^t \frac{\sqrt{t}+\sqrt{b}}{\sqrt{t-b}} \cdot \frac{\sqrt{t}+\sqrt{p}}{\sqrt{t-p}} \cdot \frac{dt}{t-1} = a_1 + a_2 + a_3,$$

where

$$a_1 = \int_0^t \frac{dt}{\sqrt{t-b}\sqrt{t-p}} = \frac{\sqrt{(t-p)(t-b)}}{\sqrt{t-p}\sqrt{t-b}} \ln\left[-\frac{1}{2}\left(b+p-2\left(t+\sqrt{(b-t)(p-t)}\right)\right)\right]$$
$$+ \ln\left[-\frac{1}{2}\left(b+p-2\sqrt{bp}\right)\right],$$

$$a_2 = \left(1+\sqrt{bp}\right)\int_0^t \frac{dt}{\sqrt{t-b}\sqrt{t-p}(t-1)} = \frac{\left(1+\sqrt{bp}\right)i}{\sqrt{p-1}\sqrt{1-b}} \cdot \left\{\frac{\sqrt{t-b}\sqrt{t-p}}{\sqrt{(t-b)(t-p)}}\right.$$
$$\cdot \ln\frac{(t-p)(1-b)+(b-t)(p-1)+2\cdot i\sqrt{1-b}\sqrt{p-1}\sqrt{(t-p)(t-b)}}{t-1}$$
$$+ \ln\left[-b(p-1)+p(1-b)-2i\sqrt{bp}\sqrt{p-1}\sqrt{1-b}\right]\right\},$$

$$a_3 = \left(\sqrt{p} + \sqrt{b}\right) \int_0^t \frac{\sqrt{t}\,dt}{\sqrt{t-b}\sqrt{t-p}\,(t-1)} = \left(\sqrt{p} + \sqrt{b}\right)\left(P_3(t) - P_3(0)\right),$$

$$P_3(t) = \frac{2}{\sqrt{p-b}} \text{Elliptic } F\left(\sqrt{\frac{b-t}{b}}, -i\sqrt{\frac{b}{p-b}}\right)$$

$$- \frac{2}{\sqrt{p-b}\,(1-b)} \text{Elliptic } Pi\left(\sqrt{\frac{b-t}{b}}, \frac{b}{b-1}, -i\sqrt{\frac{b}{p-b}}\right),$$

$$P_3(0) = \frac{2}{\sqrt{p-b}} \text{Elliptic } F\left(1, -i\sqrt{\frac{b}{p-b}}\right) - \frac{2}{\sqrt{p-b}\,(1-b)} \text{Elliptic } Pi\left(1, \frac{b}{b-1}, -i\sqrt{\frac{b}{p-b}}\right)$$

$$= \frac{2}{\sqrt{p}} \text{Elliptic } K\left(\sqrt{\frac{b}{p}}\right) - \frac{2}{\sqrt{p}} \text{Elliptic } Pi\left(b, \sqrt{\frac{b}{p}}\right).$$

These formulas help construct a hydrodynamic grid easily. This requires only using Equation 1.158 to find the value of point

$$t = 1 + e^{\pi\bar{\varphi}} \cdot e^{i\pi\bar{\psi}} \tag{1.183}$$

for a given flow function $\bar{\psi} = \psi/q$ and the value of the potential $\bar{\varphi} = \varphi/q$.

After substituting t in Equation 1.168 with 1.183, we define the coordinates of the point in the physical area in which $\varphi = \bar{\varphi} \cdot q$ and $\psi = \bar{\psi} \cdot q$. To do this, we must first enter the parameters of the problems b and p and the jet thickness δ.

1.6.2 CALCULATION RESULTS AND DISCUSSION

As an example, Figure 1.54 shows the pattern of flow for $G = 0.5$ and $S = 1.0$ ($b = 0.844875$; $p = 592.690614$; and $\delta = 0.295440$) built using *Maple* according to the aforementioned algorithm.

Here, a change pitch for the pattern parameters is assumed for the flow line $\Delta\bar{\psi} = 0.1$ and the equipotential lines $\Delta\bar{\varphi} = 0.2$, and to eliminate the singular points B ($t = b$) and P ($t = p$), the boundary flow lines are taken with displacements at 10^{-6}; that is, the flow line $\bar{\psi} = 10^{-6}$ is taken instead of the flow line $\bar{\psi} = 0$ and $\bar{\psi} = 0.999999$ instead of $\bar{\psi} = 1$.

As can be seen from the figure, the jet contraction in a flat channel is much higher than the contraction due to a jet separation from the hood (remember that for $S \to \infty$ and $G \to \infty$ the jet thickness at infinity is $\delta = 0.5$).

Consider charts on Figures 1.55 and 1.56 to identify the effect of the hood with a length S and the distance G of the impermeable screen on the jet contraction. The jet thickness at infinity δ changes dramatically in the ranges $0 < G < 1$ and $0 < S < 0.5$. Notably, the effect of the screen located at a small distance from the suction port (for $G < 1$) is more pronounced. If compared to the infinite distance ($G \to \infty$), the jet thickness can be decreased (for $S = 0.5$) by 1.75 for $G = 0.5$, while increasing the hood length up to $S = 0.5$ decreases (for $G = 5$) the jet thickness by only 16% if compared with the case without a hood ($S = 0$). However, if no impermeable screen can be installed near the opening, the effect of the hood is obvious: thus, when $G = 5$ and there is no hood ($S = 0$) the jet thickness is $\delta = 0.6064$ (for $G = \infty$, $\delta = 0.611$), that is, it is only reduced by 0.76%, while when the channel is equipped with a hood of a length $S = 0.5$, the jet thickness decreases down to $\delta = 0.523$ or by 16.8% under the same conditions.

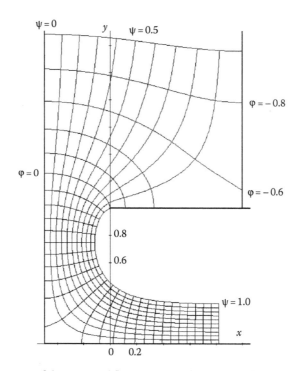

FIGURE 1.54 Flow pattern of the separated flow at the inlet into a projecting channel with an impermeable screen (for $G = 0.5$ and $S = 1.0$).

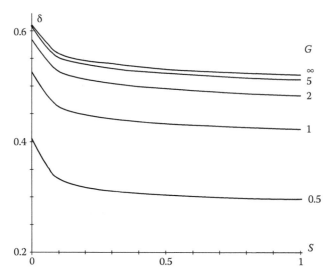

FIGURE 1.55 Change in the jet thickness in a flat channel depending on the hood length.

The nature of the jet separation from the hood (Figure 1.57) also depends on the closer distance of the screen. The smaller G is the greater is the migration of jet toward the screen; however, for $G < 0.5$ this migration begins to decrease (Figure 1.58).

The coordinates of the jet maximum migration toward the flow (point m) can be easily found by testing parametrical equation of jet boundaries (Equation 1.169) for extrema:

$$\frac{dx_{CD}}{dy_{CD}} = \frac{dx_{CD}}{dt} \Big/ \frac{dy_{CD}}{dt} = \frac{1}{\sqrt{p} + \sqrt{b}} \cdot \frac{t + \sqrt{bp}}{\sqrt{-t}} = 0, \tag{1.184}$$

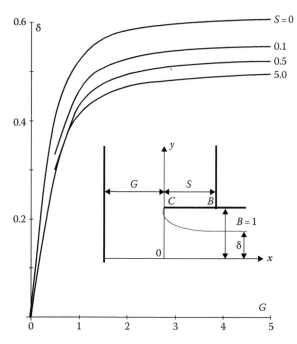

FIGURE 1.56 Change in the jet thickness in a flat channel with an impermeable screen located at a distance from the entry section.

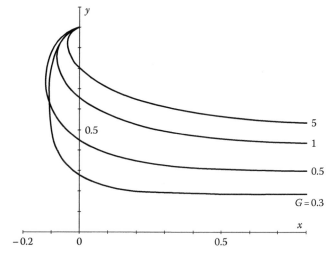

FIGURE 1.57 Change in the jet boundaries in a channel depending on the distance to the screen G (for $S = 1$).

$$x_m = x\big|_{t=-\sqrt{bp}}\,; \quad y_m = y\big|_{t=-\sqrt{bp}}\,. \tag{1.185}$$

The jet migration toward the screen and its contraction to the axis results in an increase in jet inertia, the value of which is determined by the velocity u_0 or the factor

$$K_u = \left(\frac{u_0 - u_{av}}{u_{av}}\right)^2 = \left(\frac{1}{\delta} - 1\right)^2; \quad u_{av} = \frac{q}{B}, \tag{1.186}$$

where u_{av} is the average flow velocity in the channel, m/s.

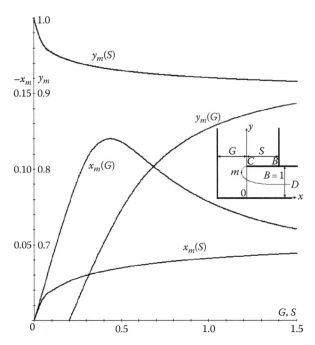

FIGURE 1.58   Change in the coordinates of the point of the jet maximum migration toward the screen depending on its distance G (for $S = 1$) and the channel extension value S (for $G = 5$).

This ratio is well correlated to a local drag coefficient relating to the average velocity u_{av} both for free air entry into the extended channel and for entrance into a screened channel without a hood. Apparently, this can be explained by the fact that the factor of inertia in the form of Equation 1.186 is numerically equal to the local drag coefficient with a sharp narrowing of the channel [143,150], where the process of aerodynamic drag is essential for this class of air entry into flat channels.

For this projecting channel with an impenetrable screen, we can compare the change in the coefficient K_u (Figure 1.59) with the experimental data used by the Central Institute of Aerohydrodynamics as documented by Idelchik [129] and Khanzhonkov [151].

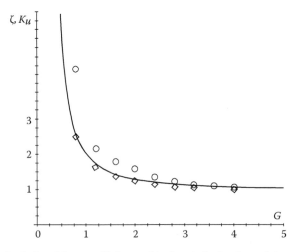

FIGURE 1.59   Change in the local drag coefficient and jet inertia in the channel depending on the distance of the screen from the entry section for $S = 5$ (a solid line means a theoretical change in inertia, diamonds stand for the experimental data obtained by Idelchik for a screened entrance into a flat channel, circles stand for the experimental data obtained by Khanzhonkov for an entrance into a circular shaft with a flat screen).

As can be seen from the graphs, the data of the screened entry into a flat channel match mostly with the experimental data. Some distinction can be seen in the data of the entrance into a circular tube with a flat screen of a finite size (the screen diameter in the experiments made by Khanzhonkov was equal to twice the diameter of the inflow chamber). This can be explained by the difference between the axially symmetric flow and the flat one and that with a small-sized screen. Although the general nature of the drop in the local drag coefficient in the case of a distant screen is satisfactorily in line with the change in the jet inertia in this case as well.

1.6.3 CONCLUSIONS

Where the flat channel is equipped with a hood, a countermigration of the jet occurs and a marked reduction of its thickness in the channel is seen. Moreover, the degree of approximation of the screen begins to have a significant effect when the distance to the entry section of the channel is less than the half-height of the channel and the degree of channel extension becomes crucial when the hood length is equal to half that height. Thus, the maximum (minimum) thickness of the jet for $S = 0.5$ and $G = 0.5$ decreases by 1.75 times compared to the case with no screen and for $G = 5$ and $S = 0.5$ by 16.8% compared to the case with no hood ($S = 0$).

The maximum jet migration along the axis OX occurs for the value $t = \sqrt{bp}$. Moreover, the migration value (x_m) increases steadily with the increase in the hood length; when the screen is installed closer to the channel inlet, the migration initially increases and then decreases. Thus, if there is a hood of the unit length, the maximum value is reached at $G \approx 0.5$ and is $x_m \approx -0.12$.

The behavior of the jet inertia factor in a screened hooded channel is in satisfactory agreements with the change in the experimental local drag coefficient at the air inlet into a flat channel and a circular delivery pipe depending on the distance of the flat screen.

1.7 LAWS OF SEPARATED FLOWS AT THE INLET INTO A PROJECTING CHANNEL WITH SCREENS

1.7.1 DETERMINATION OF DESIGN RATIOS

As it was discussed in the previous chapter, screening greatly affects the jet contraction value in the channel and its aerodynamic drag. This chapter considers a more general problem of separated flow in a horizontal channel, the entry section of which extends from the vertical wall at a distance S (Figure 1.60). In addition, the flow is limited by the other vertical wall (impermeable screen) located at the distance G from the entry section of the flat channel with a height $2B$ and divided at the distance M by a vertical board (a screen with a central opening of a height $2R$) into two areas.

As the suction flare is symmetric, we will consider the flow in the upper semiplane Im $z > 0$ of the complex variable $z = x + iy$, the real axis of which coincides with the symmetry axis of the problem. Assume that the flow line $\psi = 0$ passes through the rays A_2P and PD_2 and the flow line $\psi = q$ passes along the ray M_1B, the interval BC, and the free flow line CD_1. Assume that the zero equipotential ($\varphi = 0$) passes through the jet separation point C. In this case, the range of the complex potential w is a band with a width q with the ray M_2HA_1 ($\psi = kq$, $-\infty < \varphi < \xi$), and the range of the Joukowski function $\omega = \ln \dfrac{u_0}{u} + i\theta$ (where u_0 is the velocity at the free line CD_1, u is the velocity at any arbitrary point, and θ is the angle of the velocity vector $\vec{u}$ inclination to the positive axis OX) is the half-band $\left(\ln \dfrac{u_0}{u} > 0, -\pi < \theta < 0 \right)$ with the ray $B_2FP_1 \left(\ln \dfrac{u_0}{u} > \zeta; \theta = -\dfrac{\pi}{2} \right)$.

Taking into account the change limits of these areas are straight, we take the upper semiplane of the complex variable $t = x_1 + iy_1$ as an auxiliary one and use the Schwarz–Christoffel integral [100] to perform a conformal mapping of this semiplane onto the interior of the said polygonal ranges. When we pass *singular points* M and A in the range w, as well as the points P and B in the range ω

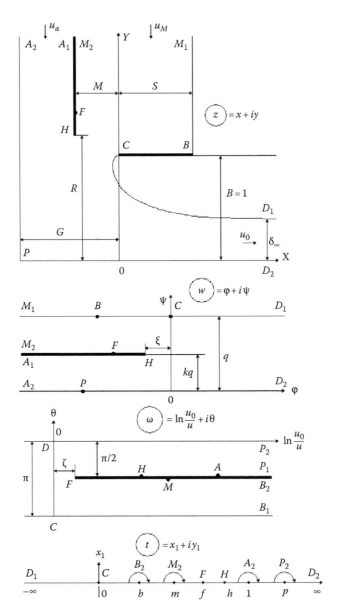

FIGURE 1.60 To the determination of the orthogonal grid and the velocity fields of the suction flare of a screened projecting channel.

along the semicircles with infinitely small radii, we define the constants of this integral and the relations between the parameters of the problem from the correspondence of the boundaries.

Obtain for the Joukowski function

$$\omega = \frac{1}{2}\ln\frac{\sqrt{t}+\sqrt{b}}{\sqrt{t}-\sqrt{b}} + \frac{1}{2}\ln\frac{\sqrt{t}+\sqrt{p}}{\sqrt{t}-\sqrt{p}} \tag{1.187}$$

or

$$e^{\omega} = \frac{\sqrt{t}+\sqrt{b}}{\sqrt{t}-\sqrt{b}} \cdot \frac{\sqrt{t}+\sqrt{p}}{\sqrt{t}-\sqrt{p}}. \tag{1.188}$$

In addition, based on an analysis of transition of the *discontinuous* points P and B of the range ω, we have

$$\frac{\sqrt{b}}{\sqrt{p}} = \frac{f - b}{p - f},$$

whence

$$f = \sqrt{b \cdot p}, \tag{1.189}$$

and from the correspondence of the point with the maximum velocity $F\left(\zeta - i\frac{\pi}{2}\right)$ in the range ω and the point $F(f + i \cdot 0)$ of the complex variable t, we find

$$\zeta = \ln \frac{\sqrt{f} + \sqrt{b}}{\sqrt{f} - \sqrt{b}} \cdot \frac{\sqrt{f} + \sqrt{p}}{\sqrt{p} - f}$$

or taking into account (Equation 1.189)

$$\zeta = \ln \frac{p^{1/4} + b^{1/4}}{p^{1/4} - b^{1/4}}. \tag{1.190}$$

For the complex potential, we obtain

$$w = (1 - k) \cdot \frac{q}{\pi} \cdot \ln\left(\frac{t - m}{m}\right) + k \cdot \frac{q}{\pi} \cdot \ln(t - 1).$$

Based on an analysis of the transition of the singular points M and A in the range w, we find

$$k = \frac{1 - h}{1 - m}. \tag{1.191}$$

The connection of the points of the auxiliary plane t and the points of the physical area are defined using the Joukowski function

$$z = \frac{1}{u_0} \int_0^t e^{\omega(t)} \frac{dw}{dt} dt + i,$$

which is as follows:

$$z = i + (1 - k)\frac{\delta_\infty}{\pi} \cdot A + k \frac{\delta_\infty}{\pi} \cdot B, \tag{1.192}$$

where the integrals of the problem equal, respectively,

$$A = \int_0^t f_0 \frac{dt}{t - m}; \quad B = \int_0^t f_0 \frac{dt}{t - 1}; \quad f_0 = e^\omega. \tag{1.193}$$

The integrals of Equation 1.193 are expressed in terms of elementary functions only in a special case (e.g., for $S = 0$, $b = 0$, a case of no projection when the entry section of the channel is in the plane of the vertical wall, as was considered in previous chapters). In the general case, the integrals A and B can be represented as the sums of four integrals, two of which are expressed in terms of elementary functions and the other two in terms of elliptic integrals:

$$A = a_1 + \left(m + \sqrt{b}\sqrt{p}\right) \cdot a_2 + \left(\sqrt{p} + \sqrt{b}\right) \cdot \left(a_3 + m a_4\right),$$

$$B = a_1 + \left(1 + \sqrt{b}\sqrt{p}\right) \cdot b_2 + \left(\sqrt{p} + \sqrt{b}\right) \cdot \left(a_3 + b_4\right),$$

where

$$a_1 = \int_0^t f_1 dt; \quad a_2 = \int_0^t f_1 \frac{dt}{t-m}; \quad a_3 = \int_0^t f_1 \frac{dt}{\sqrt{t}}; \quad a_4 = \int_0^t f_1 \frac{dt}{\sqrt{t}\left(t-m\right)}$$

$$b_2 = \int_0^t f_1 \frac{dt}{t-1}; \quad b_4 = \int_0^t f_1 \frac{dt}{\sqrt{t}\left(t-1\right)}; \quad f_1 = \frac{1}{\sqrt{t-b}\sqrt{t-p}}$$

We substitute the agreed denotations in Equation 1.192 to obtain the following calculated ratio:

$$z = i + \frac{\delta_\infty}{\pi} \left\{ a_1 + \left(\sqrt{p} + \sqrt{b}\right) a_3 + (1-k)\left[\left(m + \sqrt{b}\sqrt{p}\right) a_2 + m\left(\sqrt{p} + \sqrt{b}\right) a_4\right] \right.$$

$$\left. + k\left[\left(1 + \sqrt{b}\sqrt{p}\right) b_2 + \left(\sqrt{p} + \sqrt{b}\right) b_4\right] \right\}, \tag{1.194}$$

which relates the points of the physical area z to the points of the upper semiplane Im $t > 0$.

Notice that relation in Equation 1.194 makes it easier to find the real and imaginary part of the point z. If we know the parameters of the problem b, m, h, and p and preset the point $t = x_1 + iy_1$, we can perform numerical integration and find the correspondence of the points using the universal mathematical environment *Maple* without solving the integrals and dividing them by A and B. However, this solution method for Equation 1.192 (or Equation 1.194) is hardly suitable for the determination of the four geometric dimensions S, M, R, and G of the physical area. More often, it is necessary to solve a direct problem: determine the parameters of the problem b, m, h, and p using the given geometric dimensions of the physical area.

Determine the explicit form of the integrals a_1, a_2, b_2, a_3, a_4, and b_4 to create the equations of the direct problem. Express the first three integrals in the upper semiplane Im $t > 0$ along the real axis for $0 < b < m < h < 1 < p$ in terms of elementary functions and the last three in terms of elliptic functions.

Thus, we have on the negative real axis (for $-\infty < t < 0$)

$$f_1 = -\frac{1}{\sqrt{b-t}\sqrt{p-t}},$$

$$a_1 = \ln\left(\frac{\sqrt{p} - \sqrt{b}}{\sqrt{p-t} - \sqrt{b-t}}\right)^2, \tag{1.195}$$

$$a_2 = \frac{2}{\sqrt{p-m}\sqrt{m-b}}\left[\operatorname{arctg}\left(\sqrt{\frac{b}{p}}\sqrt{\frac{p-m}{m-b}}\right) - \operatorname{arctg}\left(\sqrt{\frac{b-t}{p-t}}\sqrt{\frac{p-m}{m-b}}\right)\right], \tag{1.196}$$

$$b_2 = a_2\big|_{m=1} = \frac{2}{\sqrt{p-1}\sqrt{1-b}}\left[\operatorname{arctg}\left(\sqrt{\frac{b}{p}}\sqrt{\frac{p-1}{1-b}}\right) - \operatorname{arctg}\left(\sqrt{\frac{b-t}{p-t}}\sqrt{\frac{p-1}{1-b}}\right)\right]. \tag{1.197}$$

These integrals are real unlike the integrals a_3, a_4, b_4 that are purely imaginary in the case of negative values t.

After the change of variable in integration $t = p - \dfrac{p}{t_1^2}$, we obtain

$$a_3 = \frac{2i}{\sqrt{p}}\int_0^{z_1}\frac{dt_1}{\sqrt{1-t_1^2}\sqrt{1-k_1^2\cdot t_1^2}} = 2\frac{i}{\sqrt{p}}\left[\text{Elliptic } F\left(z_1, k_1\right) - \text{Elliptic } K\left(k_1\right)\right], \tag{1.198}$$

$$a_4 = \frac{2i}{\sqrt{p}\left(p-m\right)}\left[\int_1^{z_1}\frac{dt_1}{\sqrt{1-t_1^2}\sqrt{1-k_1^2t_1^2}} - \int_1^{z_1}\frac{dt_1}{\left(1-vt_1^2\right)\sqrt{1-t_1^2}\sqrt{1-k_1^2t_1^2}}\right]$$

$$= \frac{2i}{\sqrt{p}\left(p-m\right)}\left[\text{Elliptic } F\left(z_1, k_1\right) - \text{Elliptic } K\left(k_1\right) + \text{Elliptic } P_i\left(v, k_1\right) - \text{Elliptic } P_i\left(z_1, v, k_1\right)\right], \tag{1.199}$$

$$b_4 = a_4\big|_{m=1} = \frac{2i}{\sqrt{p}\left(p-1\right)}\left[\text{Elliptic } F\left(z_1, k_1\right) - \text{Elliptic } K\left(k_1\right)\right.$$

$$\left. + \text{Elliptic } P_i\left(v_1, k_1\right) - \text{Elliptic } P_i\left(z_1, v_1, k_1\right)\right], \tag{1.200}$$

where for simplicity of record we assume

$$k_1 = \sqrt{1-\frac{b}{p}}; \quad v = 1-\frac{m}{p}; \quad v_1 = 1-\frac{1}{p}; \quad z_1 = \sqrt{\frac{p}{p-t}}. \tag{1.201}$$

Elliptic integrals are written in the transcription of *Maple*, namely, the complete and reduced integrals of the first kind,

$$\text{Elliptic } K\left(k_1\right) = \int_0^1\frac{dt_1}{\sqrt{1-t_1^2}\sqrt{1-k_1^2t_1^2}}, \quad \text{Elliptic } F\left(z_1, k_1\right) = \int_0^{z_1}\frac{dt_1}{\sqrt{1-t_1^2}\sqrt{1-k_1^2t_1^2}},$$

and the complete and reduced integrals of the third kind,

$$\text{Elliptic } P_i\left(v, k_1\right) = \int_0^1\frac{dt_1}{\left(1-vt_1^2\right)\sqrt{1-t_1^2}\sqrt{1-k_1t_1^2}},$$

$$\text{Elliptic } P_i\left(z_1, v, k_1\right) = \int_0^{z_1}\frac{dt_1}{\left(1-vt_1^2\right)\sqrt{1-t_1^2}\sqrt{1-k_1^2t_1^2}}.$$

On dividing Equation 1.194 by imaginary and real parts, we obtain the following parametric equation:

$$x_{CD} = \text{Re}\, z_{CD} = \frac{\delta_\infty}{\pi}\left[a_1 + (1-k)\left(m+\sqrt{b}\right)a_2 + k\left(1+\sqrt{b}\right)b_2\right],$$

$$y_{CD} = \text{Im}\, z_{CD} = 1 + \frac{\delta_\infty}{\pi}\left(\sqrt{p}+\sqrt{b}\right)\left[a_3 + (1-k)ma_4 + kb_4\right],$$

(1.202)

that describes the free flow line CD_1 taking into account the system of Equations 1.195, 1.196, 1.197, 1.198, 1.199, and 1.200, omitting the imaginary unit i in the last equations.

Note that at the point D_1 of the flow line (for $t \to -\infty$) we have

$$a_{1D_1} = \lim_{t\to-\infty} a_1 = \infty; \quad a_{2D_1} \neq \infty; \quad b_{2D_1} \neq \infty,$$

$$a_{3D_1} = \lim_{t\to-\infty} a_3 = -i\frac{2}{\sqrt{p}}\,\text{Elliptic}\,K(k_1),$$

$$a_{4D_1} = \lim_{t\to-\infty} a_4 = -i\frac{2}{\sqrt{p(p-m)}}\left[\text{Elliptic}\,K(k_1)-\text{Elliptic}\,P_i(v,k_1)\right],$$

$$b_{4D_1} = \lim_{t\to-\infty} b_4 = -i\frac{2}{\sqrt{p(p-1)}}\left[\text{Elliptic}\,K(k_1)-\text{Elliptic}\,P_i(v_1,k_1)\right],$$

so

$$x_{D_1} = \infty : y_{D_1} = \delta_\infty = 1 - \frac{\delta_\infty}{\pi}B_1,$$

(1.203)

where

$$B_1 = 2\left(1+\sqrt{\frac{b}{p}}\right)\cdot\left[\left(1+\frac{(1-k)m}{p-m}+\frac{k}{p-1}\right)\text{Elliptic}\,K(k_1)\right.$$

$$\left. -\frac{(1-k)m}{p-m}\text{Elliptic}\,P_i(v,k_1)-\frac{k}{p-1}\text{Elliptic}\,P_i(v_1,k_1)\right].$$

(1.204)

From Equation 1.203 we find an important relation to determine the half-width of the jet at infinity as a function of the parameters b, m, h, and p:

$$\delta_\infty = \frac{\pi}{\pi + B_1}.$$

(1.205)

For $k = 1$; $S \neq 0$ the function B_1 is much simpler:

$$B_1 = 2\left(1+\sqrt{\frac{b}{p}}\right)\cdot\left[\frac{p}{p-1}\text{Elliptic}\,K(k_1)-\frac{1}{p-1}\text{Elliptic}\,P_i(v_1,k_1)\right]$$

and, as can be seen, it is numerically equal to the function B (Equation 1.177) obtained for the case with no screen.

For another special case—when the suction channel is flush with the vertical wall and there is a board in front of the suction port ($k \neq 1$), this function is

$$B_1 = 2\left(1+\sqrt{\varepsilon}\right)\left[\text{Elliptic } K\left(\sqrt{1-\varepsilon}\right)+\frac{(1-k)m}{p-m}\left(\text{Elliptic } K\left(\sqrt{1-\varepsilon}\right)-\text{Elliptic } P_i\left(v,\sqrt{1-\varepsilon}\right)\right)\right.$$

$$\left.+\frac{k}{p-1}\left(\text{Elliptic } K\left(\sqrt{1-\varepsilon}\right)-\text{Elliptic } P_i\left(v_1,\sqrt{1-\varepsilon}\right)\right)\right],$$

where ε is the small value ($\varepsilon \ll 1$).

Already for B_1 the function $\varepsilon = 10^{-12}$ is numerically equal to B computed with the same k, m and p by the formula given as Equation 1.114 obtained for the condition $S = 0$.

For the interval CB (for $0 \leq t \leq b$), the integrals a_1, a_2, and b_2 are still real and numerically defined by Equations 1.195, 1.196, and 1.197. The integrals a_3, a_4, b_4 in this range of the parameter t are also real and, when replaced with the variable $t = bt_1^2$, are expressed in terms of the following elliptic integrals:

$$a_3 = -\frac{2}{\sqrt{p}}\text{Elliptic } F\left(\sqrt{\frac{t}{b}},\sqrt{\frac{b}{p}}\right), \tag{1.206}$$

$$a_4 = \frac{2}{m\sqrt{p}}\text{Elliptic } P_i\left(\sqrt{\frac{t}{b}},\frac{b}{m},\sqrt{\frac{b}{p}}\right), \tag{1.207}$$

$$b_4 = \frac{2}{\sqrt{p}}\text{Elliptic } P_i\left(\sqrt{\frac{t}{b}},b,\sqrt{\frac{b}{p}}\right). \tag{1.208}$$

Therefore, from the correspondence of the points B of the planes z and t on the basis of Equation 1.194 we can write

$$S = 2\frac{\delta_\infty}{\pi}\left\{\ln\frac{\sqrt{p}-\sqrt{b}}{\sqrt{p-b}}+(1-k)\frac{m+\sqrt{bp}}{\sqrt{m-b}\sqrt{p-m}}\text{arctg}\left(\sqrt{\frac{b}{p}}\frac{\sqrt{p-m}}{\sqrt{m-b}}\right)\right.$$

$$+\frac{k\left(1+\sqrt{bp}\right)}{\sqrt{1-b}\sqrt{p-1}}\text{arctg}\left(\sqrt{\frac{b}{p}}\frac{\sqrt{p-1}}{\sqrt{1-b}}\right)+\left(1+\sqrt{\frac{b}{p}}\right)\left[(1-k)\text{Elliptic } P_i\left(\frac{b}{m},\sqrt{\frac{b}{p}}\right)\right.$$

$$\left.\left.+k\text{ Elliptic } P_i\left(b,\sqrt{\frac{b}{p}}\right)-\text{Elliptic } K\left(\sqrt{\frac{b}{p}}\right)\right]\right\}. \tag{1.209}$$

Consider the boundary of the physical area $BM_1M_2HA_1A_2PD_2$ that the interval of a real variable $b \leq t \leq p$ corresponds to. This range contains the singular points M and A, where the integrals a_2 and a_4 have logarithmic discontinuities for $t = m$ as well as the integrals b_2 and b_4 for $t = 1$. Special types of integrals are summarized in Table 1.8.

TABLE 1.8

Calculation Formulas for the Boundary of the Flow Range $BM_1M_2HA_1A_2P$

Interval, Points	Formulas
1	**2**

$B \le t \le p$

$$a_1 = -i \arccos \frac{p+b-2t}{p-b}$$

M, H

$$a_{1M} = -i \arccos \frac{p+b-2m}{p-b}, \quad a_{1H} = -i \arccos \frac{p+b-2h}{p-b}$$

A, P

$$a_{1A} = -i \arccos \frac{p+b-2}{p-b}, \quad a_{1P} = -i\pi.$$

$B \le t < m$

$$a_{2BM_2} = \frac{i}{\sqrt{p-m}\sqrt{m-b}} \ln\left|\frac{1+\beta(t)}{1-\beta(t)}\right|, \quad \beta(t) = \sqrt{\frac{t-b}{p-t}}\sqrt{\frac{p-m}{m-b}};$$

M

$$a_{2M} = \lim_{t \to m} a_{2BM_2} = i \cdot \infty l;$$

$m < t \le p$

$$a_{2M_1P} = \Delta a_{2m} + \frac{i}{\sqrt{p-m}\sqrt{m-b}} \ln\left|\frac{1+\beta(t)}{1-\beta(t)}\right|, \quad a_{2M_1P} = \Delta a_{2m} + \frac{i}{\sqrt{p-m}\sqrt{m-b}} \ln\left|\frac{1+\beta(t)}{1-\beta(t)}\right|,$$

H

$$a_{2H} = \Delta a_{2m} + \frac{i}{\sqrt{p-m}\sqrt{m-b}} \ln\left|\frac{1+\beta(h)}{1-\beta(h)}\right|,$$

A, P

$$a_{2A} = \Delta a_{2m} + \frac{i}{\sqrt{p-m}\sqrt{m-b}} \ln\left|\frac{1+\beta(1)}{1-\beta(1)}\right|, \quad a_{2P} = \Delta a_{2m}.$$

$B \le t \le p$

$$a_3 = \frac{2i}{\sqrt{p}}\left[\text{Elliptic } F(z_3, k_3) - \text{Elliptic } K(k_3)\right], z_3 = \sqrt{\frac{p-t}{p-b}}; \ k_3 = \sqrt{1 - \frac{b}{p}}, \ \text{Re}(a_3) = 0:$$

$\text{Im}(a_3) = -V$, V is the real value that increases with the increase in t

M

$$a_{3M} = \lim_{t \to m}(a_3), \quad \left(z_3 = \sqrt{\frac{p-m}{p-b}}\right),$$

H

$$a_{3H} = \lim_{t \to h}(a_3), \quad \left(z_3 = \sqrt{\frac{p-h}{p-b}}\right),$$

A

$$a_{3A} = \lim_{t \to 1}(a_3), \quad \left(z_3 = \sqrt{\frac{p-1}{p-b}}\right),$$

P

$$a_{3P} = -\frac{2i}{\sqrt{p}} \text{Elliptic } K(k_3), \quad (z_3 = 0).$$

$B \le t \le p$

$$a_4 = \frac{2i}{\sqrt{p}(p-m)}\left[\text{Elliptic } P_i(z_3, v_4, k_3) - \text{Elliptic } P_i(1, v_4, k_3)\right] v_4 = \frac{p-b}{p-m}; \ \text{Re}(a_4) = 0 \text{ for}$$

$b \le t < m$ and $\text{Re}(a_4) = \Delta a_{4m}$ for $p \ge t > m$; $\text{Im}(a_4)$ increases from zero (for $t = b$) up to $+I$
(for $t = m$) and then decreases from $+I$ (for $t = m$) down to $\text{Im}(a_{4p})$ (for $t = p$); $I \gg 1$;

M_1

$$a_{4M_1} = \lim_{t \to m-\varepsilon}(a_4), \quad m \gg \varepsilon > 0;$$

M_2

$$a_{4M_2} = \lim_{t \to m+\varepsilon}(a_4), \text{ or } a_{4M_2} = a_{4M_1} + \Delta a_{4m}; \Delta a_{4m} = -\frac{\pi}{\sqrt{m}\sqrt{m-b}\sqrt{p-m}},$$

H, A

$$a_{4H} = \lim_{t \to h}(a_4), \quad a_{4A} = \lim_{t \to 1}(a_4),$$

P

$$a_{4P} = \Delta a_{4m} - \frac{2i}{\sqrt{p}(p-m)} \text{Elliptic } P_i(1, v_4, k_3).$$

(Continued)

TABLE 1.8 (*Continued*)

Calculation Formulas for the Boundary of the Flow Range $BM_1M_2HA_1A_2P$

Interval, Points	Formulas
1	**2**

$1 > t \geq b$ $b_2 = \dfrac{i}{\sqrt{p-1}\sqrt{1-b}} \ln\left|\dfrac{1+\alpha(t)}{1-\alpha(t)}\right|, \quad \alpha(t) = \sqrt{\dfrac{t-b}{p-t}}\sqrt{\dfrac{p-1}{1-b}},$

M, H $b_{2M} = \lim\limits_{t \to m}(b_2), \quad b_{2H} = \lim\limits_{t \to h}(b_2),$

A_1 $b_{2A_1} = \lim\limits_{t \to 1-\varepsilon_a}(b_2) = i \cdot \infty, \quad 1 \gg \varepsilon_a > 0,$

A_2 $b_{2A_2} = b_{2A_1} + \Delta b_{2A}, \quad \Delta b_{2A} = -\dfrac{\pi}{\sqrt{p-1}\sqrt{1-b}},$

$P \geq t > 1$ $b_{2A_2P} = b_{2A_2} + \lim\limits_{t \to 1+\varepsilon_a}(b_2) - b_2\big|_{t>1},$

P $b_{2P} = \lim\limits_{t \to p}(b_{2A_2P}) = 0 \cdot i + \Delta b_{2A}.$

$1 > t \geq b$ $b_4 = \dfrac{2i}{(p-1)\sqrt{p}}\left[\text{Elliptic } P_i(z_3, \text{v}_3, k_3) - \text{Elliptic } P_i(1, \text{v}_3, k_3)\right], \quad \text{v}_3 = \dfrac{p-b}{p-1}; \text{Re}(b_4) = 0 \text{ for}$

    $b \leq t < 1$ and $\text{Re}(b_1) = \Delta b_{4A}$ for $p \geq t > 1$, $\text{Im}(b_4)$ increases from zero (for $t = b$) up to $+\infty$

    (for $t = 1-\varepsilon_a$) and then decreases from $+\infty$ (for $t = 1 + \varepsilon_a$) down to zero (for $t = p$);

M, H $b_{4M} = \lim\limits_{t \to m}(b_4), \quad b_{4H} = \lim\limits_{t \to h}(b_4),$

A_1 $b_{4A_1} = \lim\limits_{t \to 1-\varepsilon_a}(b_4),$

A_2 $b_{4A_2} = \lim\limits_{t \to 1+\varepsilon_\infty}(b_4), \text{ or } b_{4A_2} = b_{4A_1} + \Delta b_{4A}, \quad \Delta b_{4A} = -\dfrac{\pi}{\sqrt{1-b}\sqrt{p-1}} = \Delta b_{2A},$

$P \geq t > 1$ $b_{4A_2P} = b_{4A_2} + \lim\limits_{t \to 1+\varepsilon_a}(b_4) - b_4\big|_{t>1},$

P $b_{4P} = \lim\limits_{t \to p}(b_{4A_2P}) = 0 \cdot i + \Delta b_{4A}.$

$+\infty > t \geq p$ $a_1 = 2\ln\dfrac{\sqrt{t-b}+\sqrt{t-p}}{\sqrt{p-b}},$

D_2 $a_{1D_2} = \lim\limits_{t \to \infty}(a_1) = +\infty,$

    $a_2 = \dfrac{2}{\sqrt{p-m}\sqrt{m-b}}\,\text{arctg}\left(\sqrt{\dfrac{t-p}{t-b}}\sqrt{\dfrac{m-b}{p-m}}\right),$

    $a_{2D_2} = \lim\limits_{t \to \infty}(a_2) = \dfrac{2}{\sqrt{p-m}\sqrt{m-b}}\,\text{arctg}\left(\sqrt{\dfrac{m-b}{p-m}}\right),$

    $a_3 = \dfrac{2}{\sqrt{p}}\,\text{Elliptic } F\left(\sqrt{\dfrac{t-p}{t-b}}, \sqrt{\dfrac{b}{p}}\right),$

$+\infty > t \geq p$ $a_{3D_2} = \lim\limits_{t \to \infty}(a_3) = \dfrac{2}{\sqrt{p}}\,\text{Elliptic } K\left(\sqrt{\dfrac{b}{p}}\right),$

D_2 $a_4 = \dfrac{2}{\sqrt{p(m-b)}}\left(\dfrac{p-b}{p-m}\,\text{Elliptic } P_i\left(\sqrt{\dfrac{t-p}{t-b}}, -\dfrac{m-b}{p-m}, \sqrt{\dfrac{b}{p}}\right) - \text{Elliptic } F\left(\sqrt{\dfrac{t-p}{t-b}}, \sqrt{\dfrac{b}{p}}\right)\right),$

    $a_{4D_2} = \lim\limits_{t \to \infty}(a_4) = \dfrac{2}{\sqrt{p}(m-b)}\left(\dfrac{p-b}{p-m}\,\text{Elliptic } P_i\left(-\dfrac{m-b}{p-m}, \sqrt{\dfrac{b}{p}}\right) - \text{Elliptic } K\left(\sqrt{\dfrac{b}{p}}\right)\right)$

(Continued)

TABLE 1.8 (Continued)

Calculation Formulas for the Boundary of the Flow Range $BM_1M_2HA_1A_2P$

Interval, Points	Formulas
1	2

$$b_2 = \frac{2}{\sqrt{p-1}\sqrt{1-b}} \operatorname{arctg}\left(\sqrt{\frac{t-p}{t-b}}, \sqrt{\frac{1-b}{p-1}}\right),$$

$$b_{2D_2} = \lim_{t\to\infty}(b_2) = \frac{2}{\sqrt{p-1}\sqrt{1-b}} \operatorname{arctg}\left(\sqrt{\frac{1-b}{p-1}}\right),$$

$$b_4 = \frac{2}{\sqrt{p}(1-b)}\left(\frac{p-b}{p-1} \text{ Elliptic } P_i\left(\sqrt{\frac{t-p}{t-b}}, -\frac{1-b}{p-1}, \sqrt{\frac{b}{p}}\right) - \text{Elliptic } F\left(\sqrt{\frac{t-p}{t-b}}, \sqrt{\frac{b}{p}}\right)\right),$$

$$b_{4D_2} = \lim_{t\to\infty}(b_4) = \frac{2}{\sqrt{p}(1-b)}\left(\frac{p-b}{p-1} \text{ Elliptic } P_i\left(-\frac{1-b}{p-1}, \sqrt{\frac{b}{p}}\right) - \text{Elliptic } K\left(\sqrt{\frac{b}{p}}\right)\right).$$

Using the data from Table 1.8 and on the basis of Equation 1.194, we can write the equation of correspondence of the physical boundary points $BMHAPD_2$. Thus, for the point H, we have $z_H = z_B + \frac{\delta_\infty}{\pi} \cdot F_H$ where for simplicity we can assume

$$F_H = a_{1H} + \left(\sqrt{p} + \sqrt{b}\right)a_{3H} + (1-k)\left[\left(m + \sqrt{b}\sqrt{p}\right)a_{2H} + m\left(\sqrt{p} + \sqrt{b}\right)a_{4H}\right]$$

$$+ k\left[\left(1 + \sqrt{b}\sqrt{p}\right)b_{2H} + \left(\sqrt{p} + \sqrt{b}\right)b_{4H}\right].$$

Given that $z_H = M + iR$ and $z_B = S + i$, we obtain $-M + iR = S + i + \frac{\delta_\infty}{\pi}\operatorname{Re}(F_H) + i\frac{\delta_\infty}{\pi}\operatorname{Im}(F_H)$. Here, we have the system of equations

$$\left.\begin{aligned} M &= -S - \frac{\delta_\infty}{\pi}\operatorname{Re}(F_H) \\ R &= 1 + \frac{\delta_\infty}{\pi}\operatorname{Im}(F_H) \end{aligned}\right\}. \tag{1.210}$$

Since the values a_{1H}, a_{3H}, b_{2H}, b_{4H} are purely imaginary and

$$\operatorname{Re}(a_{2H}) = -\frac{\pi}{\sqrt{p-m}\sqrt{m-b}}; \quad \operatorname{Re}(a_{4H}) = -\frac{\pi}{\sqrt{m}\sqrt{p-m}\sqrt{m-b}}$$

it is easy to verify that the real part of the function F_H at the point H is

$$\operatorname{Re}(F_H) = -(1-k)\pi\frac{\sqrt{p} + \sqrt{m}}{\sqrt{p-m}}\frac{\sqrt{m} + \sqrt{b}}{\sqrt{m-b}}$$

and that is why the first equation of Equation 1.210 becomes a simple relation

$$M + S = \delta_\infty (1-k) \frac{\sqrt{p}+\sqrt{m}}{\sqrt{p-m}} \frac{\sqrt{m}+\sqrt{b}}{\sqrt{m-b}},$$

(1.211)

and, as will be shown later, is identically equal to the equation for the flow rate of airflow intercepted from the main flow with a vertical board HM_2 and the wall BM_1.

As for the imaginary part, after some simple transformations are made, we obtain

$$Im(F_H) \equiv F_4 = -\arccos\left(\frac{p+b-2h}{p-b}\right) + c_1 \ln\left|\frac{1+\beta(h)}{1-\beta(h)}\right| + c_2 \ln\left|\frac{1+\alpha(h)}{1-\alpha(h)}\right|$$

$$+ 2\left(1+\sqrt{\frac{b}{p}}\right)\left(\text{Elliptic } F\left(\sqrt{\frac{p-h}{p-b}}, \sqrt{1-\frac{b}{p}}\right)\right) + 2\frac{k}{p-1}\left(1+\sqrt{\frac{b}{p}}\right)$$

$$\cdot \text{Re}\left(\text{Elliptic } P_i\left(\sqrt{\frac{p-h}{p-b}}, \frac{p-b}{p-1}, \sqrt{1-\frac{b}{p}}\right) - \text{Elliptic } P_i\left(\frac{p-b}{p-1}, \sqrt{1-\frac{b}{p}}\right)\right)$$

$$+ 2\frac{1-k}{p-m}m\left(1+\sqrt{\frac{b}{p}}\right)\text{Re}\left(\text{Elliptic } P_i\left(\sqrt{\frac{p-h}{p-b}}, \frac{p-b}{p-m}, \sqrt{1-\frac{b}{p}}\right)\right)$$

$$- \text{Elliptic } P_i\left(\frac{p-b}{p-m}, \sqrt{1-\frac{b}{p}}\right),$$

where

$$c_1 = (1-k)\frac{m+\sqrt{bp}}{\sqrt{p-m}\sqrt{m-b}}$$

$$c_2 = k\frac{1+\sqrt{bp}}{\sqrt{p-1}\sqrt{1-b}}$$

Thus, on the basis of the second equation of Equation 1.210, we can write the important relation

$$R = 1 + \frac{\delta_\infty}{\pi} F_4$$

(1.212)

that relates the hole size R and the parameter of the problem h (if the values of the other parameters are known—k, b, m, p).

Performing similar operations using Equation 1.194 and the data of Table 1.8, we obtain the values for the point A_1 (if the point $t = 1$ is approached from the left) due to the fact that $a_{1A} = ie_1$; $a_{2A} = \Delta a_{2m} + ie_2$; $a_3 = ie_3$; $a_{4A} = \Delta a_{4m} + ie_4$; $b_{2A_1} = i \cdot \infty$; $b_{4A_1} = ie_6$, where $e_1, \ldots, e_6$ are real finite values and, therefore,

$$Re(z_{A_1}) = S + \frac{\delta_\infty}{\pi}(1-k)\left[\left(m+\sqrt{bp}\right)\Delta a_{2m} + m\left(\sqrt{p}+\sqrt{b}\right)\Delta a_{4m}\right] = S + \frac{\delta_\infty}{\pi}Re(F_H) = -M;$$

$$Im(z_{A_1}) = +\infty,$$

which corresponds to the point A_1 of the physical area.

For the point A_2 (if approaching the point $t = 1$ from the right), the value of the integrals a_{1A}, ..., a_{4A} does not change. There is only a jump in the value of the integrals: $b_{2A_2} = \Delta b_{2A} + i\infty$; $b_{4A_2} = \Delta b_{2A} + ie_6$. Therefore,

$$\text{Re}\left(z_{A_2}\right) = S + \frac{\delta_\infty}{\pi}\left\{(1-k)\left[\left(m+\sqrt{bp}\right)\Delta a_{2m} + m\left(\sqrt{p}+\sqrt{b}\right)\Delta a_{4m}\right] + k\left[\left(1+\sqrt{bp}\right)+\sqrt{p}+\sqrt{b}\right]\Delta b_{2A}\right\}$$

and as can be readily shown

$$\text{Re}\left(z_{A_2}\right) = \text{Re}\left(P\right) = -G,$$

and

$$\text{Im}\left(z_{A_2}\right) = +\infty.$$

The integrals at the point $P(t = p > 1)$ in accordance with Table 1.8 can be written as

$$a_{1P} = -i\pi, \quad a_{2P} = \Delta a_{2m}, \quad a_{3P} = -i\frac{2}{\sqrt{p}}\text{Elliptic } K\left(\sqrt{1-\frac{b}{p}}\right),$$

$$a_{4P} = \Delta a_{4m} - i\frac{2}{\sqrt{p(p-m)}}\text{Elliptic } P_i\left(\frac{p-b}{p-m}, \sqrt{1-\frac{b}{p}}\right), \quad b_{2P} = b_{4P} = \Delta b_{2A}.$$

And then, by Equation 1.194 we can find

$$\text{Re}\left(z_P\right) = S - \delta_\infty\left[(1-k)\frac{\sqrt{m}+\sqrt{b}}{\sqrt{m}-\sqrt{b}}\frac{\sqrt{p}+\sqrt{m}}{\sqrt{p}-\sqrt{m}} + k\frac{1+\sqrt{b}}{\sqrt{1-b}}\frac{\sqrt{p}+1}{\sqrt{p}-1}\right]$$

whence we have the following relation taking into account the position of the point P in the physical area ($x_P = -G$; $y_P = 0$):

$$G+S = \delta_\infty(1-k)\frac{\sqrt{m}+\sqrt{b}}{\sqrt{m}-\sqrt{b}}\frac{\sqrt{p}+\sqrt{m}}{\sqrt{p}-\sqrt{m}} + \delta_\infty k\frac{1+\sqrt{b}}{\sqrt{1-b}}\frac{\sqrt{p}+1}{\sqrt{p}-1}, \quad (1.213)$$

as will be shown, which is identical to the equation for the airflow rate.

Write the latter initially as obvious equalities:

$$(1-k)q = u_M(S+M); \quad kq = u_A(G-M), \quad q = \delta_\infty^0 u_0 \quad (1.214)$$

or

$$\frac{u_M}{u_0}(S+M) + \frac{u_A}{u_0}(G-M) = \delta_\infty.$$

Find the air velocities in order to obtain the design ratios to determine the airflow rate. As owing to Equation 1.187

$$\frac{u}{u_0} = \text{Re}\left(\frac{\sqrt{t-b}}{\sqrt{t}+\sqrt{b}}\frac{\sqrt{t-p}}{\sqrt{t}+\sqrt{p}}\right),$$

we have for the velocities at the point $M(t = m)$ and the point $A(t = 1)$,

$$\frac{u_M}{u_0} = \frac{\sqrt{m-b}}{\sqrt{m}+\sqrt{b}}\frac{\sqrt{p-m}}{\sqrt{m}+\sqrt{b}}, \quad \frac{u_A}{u_0} = \frac{\sqrt{1-b}}{\sqrt{1}+\sqrt{b}}\frac{\sqrt{p-1}}{\sqrt{1}+\sqrt{p}}. \tag{1.215}$$

It is easy to verify that Equation 1.214 in expanded form is identical to the relations shown in Equations 1.211 and 1.213.

Thus, the direct problem that consists in the definition of the auxiliary parameters b, m, h, and p using the known dimensions of the physical area S, M, R, and G is technically solvable by simultaneously solving the system of Equations 1.209, 1.211, 1.212, and 1.213 taking into account (Equations 1.189, 1.191, 1.203, and 1.204). However, finding the general solution of this system of four transcendental equations can be very difficult. Especially with a practically relevant configuration of the flow boundaries when the impermeable wall is located at "infinity" ($G \geq 10$) and the screen with a central opening $R = 1$ ($0.2 < M < 1$) is placed in close vicinity, when the jet thickness δ_∞ is significantly changed. We managed to get the solution of the direct problem by simultaneously setting one auxiliary parameter k ($0 < k < 1$) and three main dimensions of the physical area (G, M and S).

This made it possible to express the desired parameter h in terms of the parameter m (in accordance with Equation 1.191), it is necessary to find the function $B_1\left(b, m, p, h(m)\right)$ defined by Equation 1.204 and use the formula of Equation 1.203 to determine the jet thickness δ_∞ as a function of the other required parameters b, m, p. A system of three equations is created based on Equations 1.209, 1.211, and 1.213 to determine such required parameters. Having solved the system to find the parameters b, m, p and using the parameter h practically preset, we define the fourth dimension of the flow range R (after calculating δ_∞ and the value of the function F_4) using the formula shown as in Equation 1.213. The problem is solved naturally in case of successful setting of the parameter of the flow k. The program developed in *Maple* enabled to find the parameters of the problem for some flow range dimensions of interest to us (Table 1.9).

In the solution, in some cases, there was "ambiguity" of the parameters found (Figure 1.61); for the same value k, there were several triples of roots (m, b and p) that were obtained by dividing the total interval $0 < b < 1$ into several narrow intervals (the total intervals $0 < m < 1$ and $1 < p < 1.5$ remaining unchanged).

However, the analysis of changes in the values of these parameters for different values of all the geometric dimensions, including R, showed that unambiguous parameters had been found. Four unique geometrical dimensions G, M, S and R correspond to every four parameters b, m, p and h (this can be seen in Table 1.9).

1.7.2 FINDINGS OF THE STUDY

An orthogonal grid of equipotentials and flow lines can be easily built using the obtained parameters (Figure 1.62). They should be carefully built in the area of singular points P and B, avoiding assigning an exact flow line $\psi = 0$ and $\psi = q$ near these points (in other words, it is necessary to avoid the points $t = b$ and $t = 1$ of the auxiliary upper semiplane t).

As can be seen from the map of flows shown in Figure 1.62, a board with a central hole separates the suction flare into two parts, and, although the flow rate in the cutoff section nearest to the hole is low, its velocity is higher than in the other part, which increases the inertial jet contraction to the axis of the channel and, as a consequence, reduces the jet thickness δ_∞.

The value δ_∞ is affected by both the location of the board M and the size of the hole R (Figures 1.63 and 1.64). So when the board ($M > 0.5$) is at long range, with an increase in R, the jet thickness increases steadily from $\delta_\infty(0)$ to $\delta_\infty(\infty)$. In this case, $\delta_\infty(0) \equiv \delta_\infty|_{R\to 0}$ and $\delta_\infty(\infty) \equiv \delta_\infty|_{R\to\infty}$ are ultimate jet thicknesses without a board. In the first case, this corresponds to the situation where a board is combined with an impermeable wall (and thus $M = G$), and in the second case a board "vanishes" (disappears into infinity), and the ultimate values of $\delta_\infty(0)$ and $\delta_\infty(\infty)$ match the special case of the projecting hole ($S \neq 0$) and one impermeable screen (discussed in Chapter 6).

TABLE 1.9

Parameters of the Problem at a Great Distance of the Impermeable Wall ($G = 10$) and Close Location of the Board with a Central Hole

k	R	δ_∞	b	m	p
At $S = 0.5$; $M = -0.2$					
0.99999	1.003375	0.543887	0.191052	0.191052	1.029461
0.9999	1.010677	0.543861	0.191175	0.191175	1.029463
0.999	1.033949	0.543603	0.192409	0.192416	1.029485
0.995	1.077803	0.542474	0.197917	0.198083	1.029584
0.990	1.113580	0.541106	0.204851	0.205549	1.029711
0.985	1.143772	0.539783	0.211845	0.213500	1.029842
0.980	1.171826	0.538507	0.218903	0.222009	1.029978
0.975	1.199162	0.537277	0.226033	0.231166	1.030119
0.970	1.226622	0.536092	0.233245	0.241080	1.030268
0.965	1.254845	0.534953	0.240551	0.251889	1.030424
0.960	1.284425	0.533860	0.247970	0.263772	1.030591
0.950	1.350400	0.531818	0.263257	0.291787	1.030956
0.940	1.432547	0.529977	0.279474	0.328437	1.031430
0.930	1.550895	0.528369	0.297623	0.382216	1.032076
0.920	1.875415	0.527082	0.325877	0.519060	1.033647
0.9199	2.017378	0.527071	0.332505	0.570251	1.034232
0.920	2.038155	0.527079	0.333299	0.574255	1.034313
0.930	2.765311	0.527648	0.347143	0.755168	1.036464
0.940	3.453450	0.527980	0.350909	0.843866	1.037688
0.950	4.420387	0.528175	0.352784	0.908925	1.038731
0.960	6.138465	0.528276	0.353781	0.958508	1.039687
0.965	7.841262	0.528300	0.354082	0.978513	1.040145
0.970	12.625839	0.528310	0.354291	0.995623	1.040595
At $S = 0.5$; $M = 0.0$					
0.987	1.025855	0.578074	0.020242	0.020266	1.017493
0.985	1.030151	0.575863	0.023491	0.023529	1.017684
0.98	1.041214	0.571006	0.031729	0.031824	1.018117
0.960	1.090096	0.556900	0.066045	0.069937	1.019530
0.940	1.147491	0.547109	0.102218	0.105711	1.020736
0.930	1.180297	0.543147	0.120978	0.126968	1.021316
0.920	1.216704	0.539645	0.140234	0.149929	1.021901
0.910	1.257773	0.536533	0.160072	0.175148	1.022504
0.900	1.305201	0.533764	0.180650	0.203479	1.023142
0.890	1.361935	0.531306	0.202262	0.236366	1.023844
0.880	1.433895	0.529144	0.225516	0.276611	1.024660
0.870	1.536691	0.527285	0.252016	0.331313	1.025714
0.860	1.783563	0.525791	0.293100	0.448800	1.027870
0.860	1.968219	0.525783	0.310996	0.523515	1.029220
0.870	2.420617	0.526533	0.333324	0.663305	1.031821
0.880	2.768680	0.527062	0.341098	0.737947	1.033312
0.890	3.139240	0.527458	0.345672	0.795680	1.034554
0.900	3.569050	0.527755	0.348671	0.843519	1.035671
0.910	4.099331	0.527973	0.350724	0.884190	1.036712
0.920	4.798200	0.528128	0.352152	0.919069	1.037703

(Continued)

TABLE 1.9 (*Continued*)
Parameters of the Problem at a Great Distance of the Impermeable Wall (*G* = 10)
and Close Location of the Board with a Central Hole

k	R	δ_∞	b	m	p
0.930	5.808107	0.528229	0.353143	0.948948	1.038659
0.940	7.535794	0.528287	0.353817	0.974327	1.039591
0.950	12.764299	0.528309	0.354259	0.995538	1.040506
At $S = 0.5$; $M = 0.2$					
0.7995	1.002603	0.522059	0.020215	0.022932	1.009610
0.796	1.005720	0.522422	0.021017	0.023850	1.009691
0.797	1.012259	0.523134	0.022766	0.025853	1.009862
0.798	1.019284	0.523825	0.024744	0.028124	1.010045
0.799	1.026925	0.524496	0.027010	0.030732	1.010244
0.800	1.035373	0.525145	0.029651	0.033781	1.010464
0.801	1.044935	0.525769	0.032807	0.037436	1.010712
0.802	1.056166	0.526368	0.036726	0.041997	1.011002
0.803	1.070267	0.526936	0.041945	0.048108	1.011362
0.804	1.091223	0.527468	0.050237	0.057912	1.011888
0.8045	1.113401	0.527714	0.059567	0.069096	1.012431
0.8045	1.137682	0.527716	0.070279	0.082157	1.013007
0.804	1.163890	0.527511	0.082070	0.096836	1.013599
0.803	1.193025	0.527135	0.095341	0.113791	1.014230
0.802	1.215732	0.526786	0.105621	0.127255	1.014700
0.801	1.236046	0.526457	0.114692	0.139424	1.015105
0.800	1.255228	0.526145	0.123104	0.150965	1.015474
0.798	1.292477	0.525563	0.138901	0.173382	1.016160
0.796	1.330490	0.525039	0.154170	0.196105	1.016820
0.794	1.371587	0.524539	0.169619	0.220337	1.017494
0.792	1.418989	0.524090	0.186039	0.247717	1.018225
0.790	1.479970	0.523682	0.205009	0.281888	1.019105
0.788	1.603800	0.523325	0.236783	0.347256	1.020721
0.788	1.674964	0.523324	0.251532	0.382298	1.021565
0.790	1.816383	0.523582	0.274839	0.446516	1.023093
0.792	1.894772	0.523811	0.285001	0.479100	1.023868
0.794	1.9599222	0.524022	0.292179	0.504349	1.024470
0.796	2.016735	0.524220	0.297820	0.525750	1.024984
0.798	2.070061	0.524406	0.302488	0.544674	1.025441
0.800	2.120624	0.524582	0.306474	0.561830	1.025859
0.820	2.578863	0.525965	0.329429	0.686991	1.029035
0.840	3.061939	0.526896	0.340273	0.774414	1.031488
0.860	3.653732	0.527526	0.346465	0.843140	1.033657
0.880	4.461279	0.527936	0.350205	0.898887	1.035676
0.900	5.730816	0.528179	0.352474	0.944208	1.037606
0.920	8.440285	0.528291	0.353806	0.980460	1.039486
0.930	12.873892	0.528309	0.354226	0.995450	1.040414
At $S = 0.5$; $M = 0.25$					
0.7405	0.975382	0.508463	0.020049	0.023735	1.007965
0.741	0.977398	0.508790	0.020476	0.024244	1.008010
0.7415	0.979435	0.509115	0.020914	0.024766	1.008057

(Continued)

TABLE 1.9 (*Continued*)
Parameters of the Problem at a Great Distance of the Impermeable Wall (*G* = 10)
and Close Location of the Board with a Central Hole

k	R	δ_∞	b	m	p
0.742	0.981495	0.509439	0.021364	0.025302	1.008103
0.744	0.989974	0.510711	0.023284	0.027595	1.008297
0.746	0.998885	0.511951	0.025425	0.030155	1.008502
0.748	1.008307	0.513155	0.027824	0.033031	1.008719
0.750	1.018340	0.514323	0.030530	0.036285	1.008952
0.755	1.047126	0.517076	0.039125	0.046690	1.009621
0.760	1.085088	0.519605	0.052135	0.062670	1.010498
0.765	1.149689	0.521709	0.077338	0.094572	1.011937
0.770	1.770723	0.522764	0.263344	0.425084	1.021417
0.775	1.916621	0.523345	0.283414	0.485988	1.022942
0.780	2.036259	0.523846	0.295955	0.530747	1.024073
0.785	2.045304	0.524291	0.305065	0.567766	1.025023
0.790	2.249370	0.524690	0.312179	0.599987	1.025865
0.80	2.452939	0.525381	0.322631	0.655150	1.027348
0.81	2.659995	0.525957	0.330087	0.702037	1.028666
0.82	2.878313	0.526441	0.335654	0.743164	1.029879
0.84	3.375876	0.527187	0.343275	0.812994	1.032107
0.86	4.011915	0.527699	0.348021	0.870334	1.034172
0.88	4.908717	0.528032	0.351026	0.917947	1.036141
0.90	6.386274	0.528223	0.352903	0.957090	1.038050
0.92	10.066514	0.528303	0.354024	0.988653	1.039925
At S = 0.5; M = 0.5					
0.42	0.753111	0.445663	0.020590	0.030124	1.002075
0.44	0.789571	0.453414	0.024466	0.035743	1.002421
0.46	0.828001	0.461047	0.029231	0.042654	1.002820
0.48	0.868860	0.468509	0.035124	0.051213	1.003281
0.50	0.912781	0.475746	0.042461	0.061903	1.003817
0.52	0.960662	0.482696	0.051662	0.075386	1.004445
0.54	1.013829	0.489295	0.063294	0.092589	1.005188
0.56	1.074298	0.495473	0.078123	0.114846	1.006078
0.58	1.145232	0.501158	0.097163	0.144099	1.007162
0.60	1.231670	0.506275	0.121653	0.183143	1.008506
0.62	1.341296	0.510753	0.152658	0.235567	1.010196
0.64	1.483496	0.514539	0.189628	0.304129	1.012296
0.66	1.662983	0.517631	0.228242	0.386205	1.014745
0.68	1.873497	0.520097	0.262088	0.472176	1.017322
0.70	2.105967	0.522053	0.288125	0.553173	1.019830
0.75	2.786215	0.525359	0.326127	0.719669	1.025524
0.80	3.738144	0.527175	0.343215	0.842518	1.030646
0.85	5.444320	0.528049	0.350960	0.932489	1.035502
0.90	12.999868	0.528309	0.354173	0.995311	1.040268
At S = 0.5; M = 0.75					
0.005	0.082248	0.388388	0.020722	0.036522	1.0000002
0.01	0.116691	0.389602	0.021055	0.037094	1.000001
0.020	0.166097	0.392042	0.021745	0.038276	1.000004
0.025	0.186310	0.393267	0.022101	0.038887	1.000006

(Continued)

TABLE 1.9 (*Continued*)
Parameters of the Problem at a Great Distance of the Impermeable Wall (*G* = 10)
and Close Location of the Board with a Central Hole

k	R	δ_∞	b	m	p
0.03	0.204762	0.394495	0.022466	0.039511	1.000009
0.06	0.295446	0.401935	0.024839	0.043566	1.000037
0.08	0.345879	0.406956	0.026614	0.046595	1.000069
0.10	0.392193	0.412019	0.028567	0.049919	1.000112
0.15	0.498375	0.424819	0.034371	0.059770	1.000276
0.20	0.598729	0.437723	0.041845	0.072412	1.000543
0.25	0.698881	0.450575	0.051578	0.088843	1.000942
0.30	0.802969	0.463180	0.064383	0.110488	1.001518
0.35	0.915354	0.475296	0.081389	0.139415	1.002329
0.40	1.041946	0.486648	0.104094	0.178651	1.003465
0.42	1.098465	0.490905	0.115170	0.198154	1.004037
0.44	1.159491	0.494971	0.127560	0.220326	1.004692
0.46	1.226029	0.498825	0.141357	0.245539	1.005442
0.48	1.299298	0.502447	0.156605	0.274169	1.006301
0.50	1.380739	0.505820	0.173266	0.306551	1.007281
0.52	1.471988	0.508929	0.191166	0.342897	1.008397
0.54	1.574765	0.511763	0.209952	0.383186	1.009656
0.56	1.690716	0.514317	0.229078	0.427048	1.011059
0.58	1.821223	0.516593	0.247851	0.473706	1.012597
0.60	1.967353	0.518599	0.265562	0.522040	1.014252
0.62	2.130024	0.520349	0.281635	0.570793	1.015997
0.64	2.310382	0.521863	0.295735	0.618800	1.017805
0.66	2.510268	0.523161	0.307767	0.665153	1.019654
0.68	2.732650	0.524265	0.317825	0.709236	1.021527
0.70	2.982087	0.525194	0.326104	0.750691	1.023412
0.75	3.776696	0.526875	0.340556	0.841959	1.028149
0.80	5.037044	0.527829	0.348702	0.915621	1.032907
0.85	7.821910	0.528249	0.352952	0.972615	1.037710
0.88	19.147255	0.528310	0.354305	0.999245	1.040629
At *S* = 0.5; *M* = 1.0					
0.0096	0.147392	0.433897	0.059574	0.117170	1.000001
0.01	0.150471	0.433976	0.059635	0.117283	1.000002
0.02	0.214195	0.435968	0.061170	0.120166	1.000006
0.04	0.306991	0.439951	0.064410	0.126244	1.000026
0.06	0.381189	0.443928	0.067892	0.132768	1.000060
0.08	0.446437	0.447895	0.071635	0.139777	1.000110
0.10	0.506479	0.451845	0.075663	0.147317	1.000177
0.15	0.644829	0.461604	0.087141	0.168810	1.000435
0.20	0.776975	0.471108	0.100983	0.194820	1.000847
0.25	0.910844	0.480231	0.117681	0.226466	1.001456
0.28	0.994578	0.485465	0.129284	0.248720	1.001939
0.282	1.000295	0.485807	0.130103	0.250302	1.001975
0.30	1.052654	0.488835	0.137744	0.265122	1.002318
0.32	1.113041	0.492099	0.146813	0.282898	1.002750
0.34	1.176225	0.495248	0.156506	0.302155	1.003238
0.36	1.242742	0.498272	0.166825	0.322996	1.003790
0.38	1.313192	0.501116	0.177758	0.345514	1.004412

(Continued)

TABLE 1.9 (*Continued*)
Parameters of the Problem at a Great Distance of the Impermeable Wall ($G = 10$)
and Close Location of the Board with a Central Hole

k	R	δ_∞	b	m	p
0.40	1.388246	0.503910	0.189265	0.369783	1.005109
0.42	1.468646	0.506508	0.201281	0.395845	1.005889
0.44	1.555203	0.508951	0.213704	0.423699	1.006757
0.46	1.648795	0.511233	0.226394	0.453281	1.007719
0.48	1.750352	0.513351	0.239177	0.484456	1.008777
0.50	1.860848	0.515302	0.251846	0.517003	1.009932
0.52	1.981296	0.517087	0.264184	0.550626	1.011184
0.55	2.183026	0.519460	0.281607	0.602268	1.013233
0.60	2.586479	0.522645	0.306424	0.688194	1.017042
0.65	3.101580	0.524968	0.325058	0.769113	1.021225
0.70	3.790135	0.526568	0.337939	0.841370	1.025665
0.75	4.797689	0.527579	0.346271	0.903317	1.030292
0.80	6.575789	0.528123	0.351307	0.954511	1.035073
0.83	13.134026	0.528308	0.354078	0.995060	1.040005
At $S = 0.5$; $M = 1.25$					
0.007	0.153661	0.459507	0.101926	0.216768	1.000001
0.008	0.164382	0.459668	0.102137	0.217191	1.000001
0.01	0.184034	0.459991	0.102560	0.218039	1.000002
0.02	0.262049	0.461600	0.104713	0.222352	1.000009
0.04	0.375819	0.464798	0.109204	0.231354	1.000036
0.06	0.466984	0.467965	0.113954	0.270882	1.000083
0.08	0.547346	0.471096	0.118976	0.250972	1.000152
0.10	0.621494	0.474187	0.124855	0.261660	1.000245
0.12	0.691821	0.477233	0.129896	0.272986	1.000365
0.14	0.759800	0.480228	0.135820	0.284989	1.000512
0.16	0.826446	0.483167	0.142072	0.297713	1.000690
0.18	0.892522	0.486045	0.148661	0.311201	1.000902
0.20	0.958652	0.488858	0.155598	0.325496	1.001151
0.22	1.025372	0.491598	0.162888	0.340642	1.001439
0.24	1.093176	0.494261	0.170533	0.356683	1.001771
0.26	1.162535	0.496841	0.178530	0.373658	1.002150
0.28	1.233917	0.499334	0.186871	0.391603	1.002580
0.30	1.307799	0.501734	0.195537	0.410547	1.003065
0.32	1.384680	0.504036	0.204501	0.430508	1.003610
0.34	1.465085	0.506237	0.213727	0.451495	1.004218
0.36	1.549573	0.508331	0.223164	0.473495	1.004895
0.38	1.638743	0.510317	0.232751	0.496479	1.005643
0.40	1.733238	0.512190	0.242412	0.520391	1.006467
0.42	1.833749	0.513948	0.252061	0.545150	1.007370
0.45	1.997448	0.516370	0.266304	0.583629	1.008876
0.50	2.312002	0.519829	0.288607	0.650067	1.011793
0.55	2.693592	0.522590	0.307962	0.716911	1.015194
0.60	3.167806	0.524708	0.323576	0.781256	1.019020
0.65	3.780386	0.526254	0.335362	0.840759	1.023200
0.70	4.625815	0.527308	0.343742	0.893904	1.027670
0.75	5.944118	0.527951	0.349360	0.939903	1.032385
0.80	8.709402	0.528258	0.352874	0.978454	1.037318
0.82	11.637014	0.528302	0.353823	0.991779	1.039349

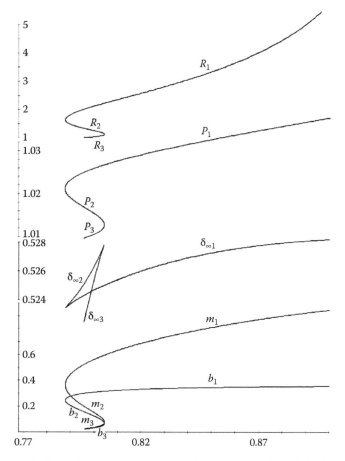

FIGURE 1.61 Change in the problem parameters (b, m, p) in the size of the hole in the screen R and the jet thickness δ_∞ (for $G = 10$; $S = 0.5$; $M = 0.2$) depending on the value K.

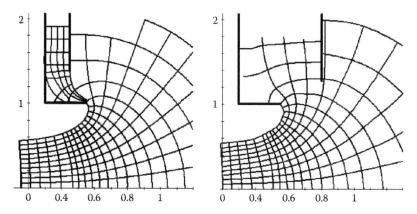

FIGURE 1.62 Orthogonal grid at different positions of a vertical board.

Similarly, for a distance $-S \leq M \leq 0$ and $R = 1$, the value $\delta_\infty(1) \equiv \delta_\infty|_{R \to 1}$ reaches its maximum value $\delta_\infty = 0.6098$ (for $G = 10$), which corresponds to the case $S = 0$, and the minimum in the range $M \approx 0.75$ (Figure 1.64). In the area $0.25 \geq M \geq 0$ (close location of a screening board to the suction hole), a slight increase δ_∞ is seen due to the screening of the suction hole, which reduces the effect of the projection S on the jet thickness δ_∞. This screening ceases its effect at $R \geq 1.5$ and the value δ_∞ tends to $\delta_\infty(\infty)$ with a further increase in R.

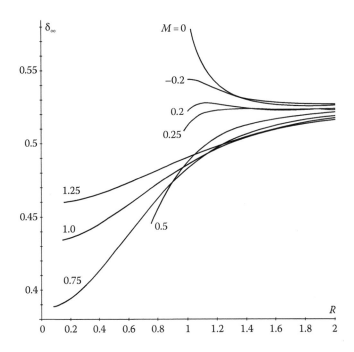

FIGURE 1.63   Change in the jet thickness δ_∞ depending on the size of the central hole in the board R and the distance of the board from the suction slot M (for $G = 10$; $S = 0.5$).

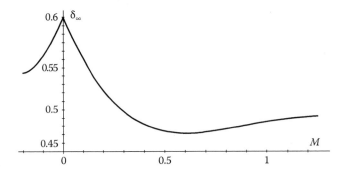

FIGURE 1.64   Change in the jet thickness δ_∞ depending on the distance of the board with a central hole $R = 1$ from the suction slot M (for $G = 10$; $S = 0.5$).

As we know, the value δ_∞ has a significant effect on the coefficient of resistance to air inlet into a suction hole. This is due to the fact that the resistance can be suppressed by the Borda–Carnot effect, which consists in the fact that the pressure loss is determined by inelastic "impact" of a fast flow against a slow one when a flow is contracted and then expands. As early as in 1944, Idelchik in [152] proposed on the basis of this effect the following dependence of the local drag coefficient on the thickness of the contracted jet*:

$$\zeta_{Bo} = \left(\frac{1}{\delta_\infty} - 1\right)^2,$$

---

* In the previous chapters, this coefficient is named the coefficient of inertia; thereby, we emphasize the braking effect of the inertial flow which is separated from the walls at the channel entrance.

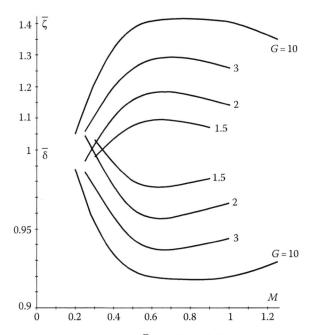

FIGURE 1.65 Change in the relative jet thickness $\bar{\delta} = \delta_\infty(1)/\delta_\infty(\infty)$ and the relative air inlet drag coefficient $\bar{\zeta} = \zeta_{Bo}(1)/\zeta_{Bo}(\infty)$ depending on the distance of the board M (with a central hole $R = 1$) and the impermeable wall G for $S = 0.5$ from the suction slot.

where the following formula is used for the loss of total pressure:

$$\Delta P = A\zeta_{Bo}\frac{u^2}{2}\rho.$$

Here, the index "Bo" emphasizes the connection with the Borda–Carnot effect; A is the correction factor offered by Idelchik (most often, $A \approx 1$); u is the flow-averaged velocity in the channel, m/s; ρ is the air density, kg/m³.

As shown by our study (Figure 1.65), a change in the relative thickness $\bar{\delta} = \delta_\infty(1)/\delta_\infty(\infty)$ and in the relative drag coefficient $\bar{\zeta} = \zeta_{Bo}(1)/\zeta_{Bo}(\infty)$ (here $\zeta_{Bo}(1) = \zeta_{Bo}|_{R=1}$; $\zeta_{Bo}(\infty) = \zeta_{Bo}|_{R\to\infty}$) have distinct extrema in case of a distant board $M \approx 0.75$ not only at the big location of an impermeable wall but also when this wall is located significantly close to the suction hole. In this case, the value $\bar{\delta}$ has a minimum and $\bar{\zeta}$ has a maximum.

Despite a small change $\bar{\delta}$ ($0.9 < \bar{\delta} < 1$) the value $\bar{\zeta}$ increases markedly and reaches about 1.5 in the area $M \approx 0.75$, which indicates a significant increase in the air inlet drag when a holed board $R = 1$ is placed in front of the channel (this hole designed for process purposes, e.g., for the passage of transport).

Thus, screening of the projecting channel suction hole with a closely located ($M = 0.25 - 0.75$) board with a central hole causes a reduction in the jet thickness in the channel and increase in the air inlet drag coefficient. The increase in the inertia of the separated flow contributes to this, which increases the impact of the Borda–Carnot effect on the pressure losses at the channel inlet.

With a process hole in the board ($R = 1$, $M = 0.75$, $S = 0.5$, $G \geq 10$), the jet thickness is decreased by 21% and the local drag coefficient increases by 2.75 times and the airflow rate in the channel is reduced almost by 40% compared with an unscreened channel ($R \to \infty$, $G \geq 10$) flush with the wall ($S = 0$). In a differentiated assessment, the extension of the channel entry section up to $S = 0.5$ increases the amount of pressure loss by 54% (ζ_{Bo} increases from 0.41 up to 0.7972), and the installation of a board with a hole $R = 1$ for $M = 0.75$, $S = 0.5$ increases it by 46% (ζ_{Bo} increases from 0.7972 to 1.1275).

2 Calculation of Flat Potential Flows by the Boundary Integral Equation Method

2.1 FLOW SUPERPOSITION METHOD

2.1.1 LINE AND POINT SINKS

Consider the flow at a line sink. A line sink (source) is a flow occurring when a perfect fluid runs down (flows out) into a straight line (of the straight line). It is used for the simulation of flows at slot-type suction units, that is, for a rectangular LVE, one side of which is larger than the other by 10 times or more. In this case, the suction unit is in infinite space. The line sink can be obtained as follows: Gas is sucked out of an infinitely long pipe. Inlets are equispaced on the surface of the pipe. If the number of inlets tends to infinity and the pipe radius tends to zero, we obtain a flow that is named line sink. The flow is identical in any plane perpendicular to the line sink. Therefore, the velocity field of this flow must be determined in the plane. That is, we have a flow at a point in the plane, which perfect fluid drains to. Assume that the sink is located at the point $\xi(\xi_1, \xi_2)$ and determine the air velocity at the point $x(x_1, x_2)$ (Figure 2.1).

Obviously, the value of the velocity is the same at any point of a circle with a radius r centered at $\xi(\xi_1, \xi_2)$. Denote the volumetric flow rate of air flowing to the point $\xi(\xi_1, \xi_2)$ as $Q(\xi)$. The strength of the source at the point $\xi(\xi_1, \xi_2)$ will mean the value $q(\xi) = +Q(\xi)$, and the strength of sink will mean the value $q(\xi) = -Q(\xi)$. The volume of air flowing through any circle at a unit time is a constant, which follows from the law of conservation of flow rate. Thus, we have the following equation:

$$Q(\xi) = 2\pi r \cdot v(x) \cdot 1 \text{ m}, \tag{2.1}$$

where $v(x)$ is the velocity modulus at the point $x(x_1, x_2)$.

Flow is taken per unit length in this equality. Cofactor 1 is omitted thereafter. The projections on the coordinate axes are as follows:

$$v_1(x) = \frac{q(\xi) \cdot \cos\alpha}{2\pi r} = \frac{q(\xi) \cdot (x_1 - \xi_1)}{2\pi\left((x_1 - \xi_1)^2 + (x_1 - \xi_1)^2\right)},$$

$$v_1(x) = \frac{q(\xi) \cdot \sin\alpha}{2\pi r} = \frac{q(\xi) \cdot (x_2 - \xi_2)}{2\pi\left((x_1 - \xi_1)^2 + (x_1 - \xi_1)^2\right)}.$$

Define the velocity along a single direction $\vec{n} = \{n_1, n_2\}$:

$$v_n(x) = \vec{v} \cdot \vec{n} = v_1 n_1 + v_2 n_2 = \frac{(x_1 - \xi_1)n_1 + (x_2 - \xi_2)n_2}{2\pi\left((x_1 - \xi_1)^2 + (x_2 - \xi_2)^2\right)} q(\xi).$$

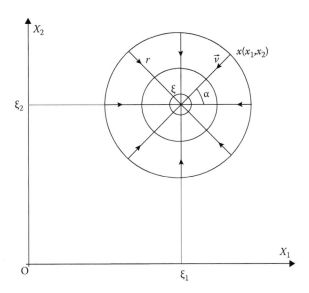

FIGURE 2.1 Line sink.

Denote

$$F_2(x, \xi) = \frac{1}{2\pi} \frac{(x_1 - \xi_1)n_1 + (x_2 - \xi_2)n_2}{(x_1 - \xi_1)^2 + (x_2 - \xi_2)^2}. \qquad (2.2)$$

The function $F_2(x, \xi)$ expresses the influence of a single source located at the point $\xi(\xi_1, \xi_2)$ along the unit vector $\vec{n} = \{n_1, n_2\}$ on the point $x(x_1, x_2)$.

Then, we can define the velocity at the point x along the direction $\vec{n} = \{n_1, n_2\}$, which is induced by the sink (source) of a strength $q(\xi)$, located at the point ξ, by the formula

$$v_n(x) = F_2(x, \xi)q(\xi). \qquad (2.3)$$

As proposed in [90–93], we name the point x an observation point and the point ξ the load center by analogy with the theory of elasticity.

Consider the flow at a point sink. The point sink can be obtained as follows: Assume there is a semi-infinite pipe with a hollow sphere secured on its end. Identical inlets are equispaced on the surface of the sphere. If the number of inlets tends to infinity and the radii of the pipe and the sphere tend to zero, we obtain a point sink. Thus, the point sink is a point perfect fluid flows down to. Assume a point sink of a strength $q(\xi)$ is at $\xi(\xi_1, \xi_2, \xi_3)$. Define the air velocity at the point $x(x_1, x_2, x_3)$ (Figure 2.2).

In this case, the law of conservation of flow rate can be written as follows:

$$Q(\xi) = 4\pi r^2 \cdot v(x).$$

The air velocity at the point $x(x_1, x_2, x_3)$ in the direction $\vec{n} = \{n_1, n_2, n_3\}$ can be determined from the formula

$$v_n(x) = F_3(x, \xi)q(\xi), \qquad (2.4)$$

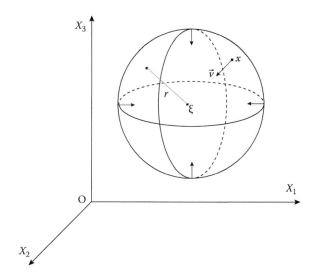

FIGURE 2.2   Point sink.

where

$$F_3(x, \xi) = \frac{1}{4\pi} \frac{(x_1 - \xi_1)n_1 + (x_2 - \xi_2)n_2 + (x_3 - \xi_3)n_3}{\left((x_1 - \xi_1)^2 + (x_2 - \xi_2)^2 + (x_3 - \xi_3)^2\right)^{3/2}}. \tag{2.5}$$

The method of point sinks is used to calculate the LVE with a compact suction inlet in infinite space.

Note that the values of the velocity and LVE found by the method of line or point sinks are in significant errors in comparison with experimental data especially near the suction inlets of LVE and those constrained with production equipment, as the characteristics of air flows are significantly affected by the flow boundaries.

2.1.2   SIMPLEST EXAMPLES OF THE APPLICATION OF FLOW SUPERPOSITION METHOD

The flow superposition method implies that the air velocity for a *complex* flow at a given point is defined as the sum of velocities caused by *simple* flows:

$$\vec{v} = \sum_{k=1}^{n} \vec{v}_k, \tag{2.6}$$

where

$\vec{v}_k$ is the velocities initiated at this point by flows components
n is their number

Define, for example, the velocity at the point x, caused by suction above an unlimited plane. Replace the suction with a line sink if it is slot type and with a point sink if it is of a compact shape. To account for the impenetrable plane, the respective sinks of the same strength are placed symmetrically about the axis OX_1 for the line sink (Figure 2.3a) and point sink (Figure 2.3b).

The point ξ' has coordinates $(\xi_1, -\xi_2)$ for a line sink and $(\xi_1, \xi_2, -\xi_3)$ for a point sink.

Then the velocity at the point x can be determined from the formula

$$v_n(x) = F(x, \xi)q(\xi) + F(x, \xi')q(\xi'),$$

where the functions $F(x, \xi)$, $F(x, \xi')$ are found from Equations 2.2 and 2.5, respectively.

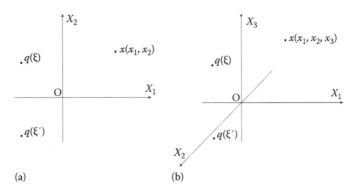

FIGURE 2.3 Model of a sink above an impermeable plane: (a) for a line sink and (b) for a point sink.

We use the flow superposition method to obtain a formula to calculate the velocity at the point $x(x_1, x_2)$ induced by a dipole located at a point $\xi(\xi_1, \xi_2)$ along the unit vector $\vec{n} = \{n_1, n_2\}$. Assume that the sink is at the point ξ'' and the source at the point ξ' (Figure 2.4).

Their intensities are equal in the absolute value; the distance between them is $2h$. Letting h to zero and their strength to infinity, we obtain

$$v_n(x) = \lim_{\substack{h \to 0, \\ q \to \infty}} \left[F(x, \xi'')q(\xi'') + F(x, \xi')q(\xi') \right] = m(\xi)D(x, \xi), \tag{2.7}$$

where

$$D(x, \xi) = \frac{\left[r_1^2 - r_2^2 \right]\left(n_1 \cos\alpha - n_2 \sin\alpha \right) + 2r_1 r_2 (n_1 \sin\alpha + n_2 \cos\alpha)}{2\pi \left[r_1^2 + r_2^2 \right]^2} \tag{2.8}$$

$m(\xi) = \lim_{\substack{h \to 0, \\ q \to \infty}} (2hq)$ is the dipole moment

$\xi_1 = (\xi_1' + \xi_1'')/2, \quad \xi_2 = (\xi_2' + \xi_2'')/2 \; r_1 = x_1 - \xi_1, \quad r_2 = x_2 - \xi_2$

α is the angle between the dipole axis and the positive direction of the axis OX

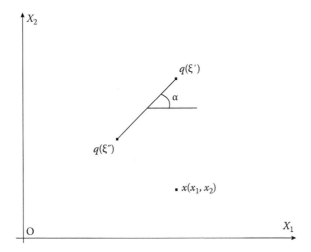

FIGURE 2.4 To the development of the formula of dipole influence.

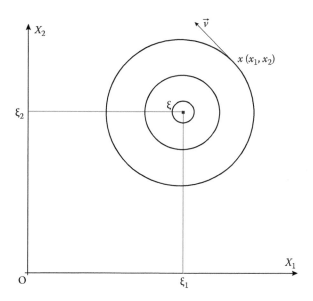

FIGURE 2.5 2D Vortex.

Superposing a dipole on a 2D parallel flow we can obtain a flow around a cylinder [95].

A 2D vortex can be used to account for the rotation effect of the cylindrical parts on the air flow sucked by the LVE.

Assume that the vortex is located at the point $\xi(\xi_1, \xi_2)$ (Figure 2.5).

The following formula is used to define the velocity caused by a 2D vortex at the point $x(x_1, x_2)$ along the predetermined unit direction $\vec{n} = \{n_1, n_2\}$:

$$v_n(x) = G(x, \xi) \cdot \Gamma(\xi), \tag{2.9}$$

$$G(x, \xi) = \frac{(x_1 - \xi_1)n_2 - (x_2 - \xi_2)n_1}{2\pi\left((x_1 - \xi_1)^2 + (x_2 - \xi_2)^2\right)}, \tag{2.10}$$

$\Gamma(\xi)$ is the vortex strength (circulation).

For example, a model of flow at a suction unit built in an infinite flat wall, whose area of effect contains a rotating cylinder, can be obtained by superposing flows caused by a sink located at the origin of coordinates (Figure 2.6) and two vortices with circulations $\Gamma(\xi') = -\Gamma(\xi'') = \Gamma$ symmetrized about the axis OX_1.

The velocity at the point $x(x_1, x_2)$ along a unit direction $\vec{n} = \{n_1, n_2\}$ can be determined from the formula

$$v_n(x) = G(x, \xi') \cdot \Gamma(\xi') + G(x, \xi'') \cdot \Gamma(\xi'') + F_2(x, O) \cdot q. \tag{2.11}$$

This flow model is used in [106] to calculate the LVE from lathes. However, this model does not account for the boundaries of a cylindrical part (the cylinder is permeable) and other components of process equipment.

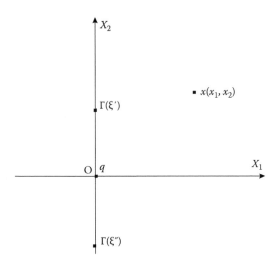

FIGURE 2.6   Model of the flow from a rotating cylinder at an LVE.

Shepelev, Altynova, and Logachev used the flow superposition method and integrated the actions of the sinks to obtain formulas to calculate the axial velocity near suction units of different geometrical shapes (slot-type, circular, elliptical, ring-shaped, square, rectangular, and regular polygon type) built in a flat infinite wall.

Note that the flow superposition method does not apply to areas with complex boundaries where it is difficult to distribute sources and sinks in such a way that the boundary conditions are satisfied. The development of this method is the method of boundary integral equations, which enables to measure the strength of the sources (sinks) distributed over the area boundary so that the boundary conditions are satisfied.

In addition, the following interpretation of the flow superposition method is used in the practice of industrial ventilation. The square (cube) of the velocity is equal to the sum of the squares (cubes) of velocities caused by the flow components [12]. This method was used to calculate ventilation supply air jets. However, this method has no physical or mathematical justification.

2.1.3 Calculation of the Axial Air Velocity at Suction Inlets Built in a Flat Wall

Changing the geometry of the exhaust inlet results in a change in velocity in the suction flare, the increase of which leads to a decrease in the performance of the aspiration system and, in turn, to a decrease in energy consumption. We use the flow superposition method to define the shape of the inlet that has the greatest range.

We assume that the air velocity at all the points with exhaust vents of equal sizes is constant and equal to V_0.

We define the axial velocity V_z at a regular n-gon–shaped suction inlet with an area S (Figure 2.7).

Calculate the velocity V_{zOAB} at some point M on the axis OZ caused by the action of the triangle suction inlet OAB. Since the angle $\angle BOA = \pi/n$, the area of the polygon is $S = 1/2 \cdot |OB|^2 \cdot n \cdot \sin 2\pi/n$ and the length is

$$|OA| = |OB| \cdot \cos \frac{\pi}{n} = \sqrt{\frac{S}{n} \cdot ctg \frac{\pi}{n}}.$$

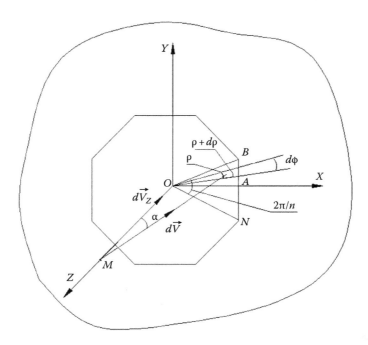

FIGURE 2.7 Regular polygon built in a flat infinite wall.

Use the polar coordinate system (the polar axis coincides with OX) and separate an elementary area $\rho d\rho d\phi$ in the plane OAB, assuming that point sink is located there. Then the elementary rate of flow is

$$dL = V_0\rho d\rho d\phi, \text{ the velocity is } dV_z = dV\cos\alpha = \frac{dL}{2\pi\left(\rho^2 + Z^2\right)}\frac{z}{\sqrt{\rho^2 + z^2}} \text{ and}$$

$$V_{z\,OAB} = \frac{V_0 Z}{2\pi}\int_0^{\frac{\pi}{n}} d\phi \int_0^{\frac{|OA|}{\cos\phi}} \frac{\rho \cdot d\rho}{\left(\rho^2 + Z^2\right)^{\frac{3}{2}}}. \tag{2.12}$$

Integrate Equation 2.12 and multiply the result by the number of such triangles $2n$ to obtain the dependence for the axial velocity near a regular polygon:

$$V_z = \frac{nV_0}{\pi}\left[\frac{\pi}{n} - \arcsin\frac{Z\cdot\sin\frac{\pi}{n}}{\sqrt{\frac{S}{n}\text{ctg}\frac{\pi}{n} + Z^2}}\right]. \tag{2.13}$$

Letting $n \to \infty$, we obtain the well-known formula to calculate the axial velocity at a circular suction inlet [29]:

$$V_z = V_0\left[1 - \frac{Z}{\sqrt{R^2 + Z^2}}\right], \tag{2.14}$$

where R is the radius of the circle.

For $n = 4$ we have a formula to calculate the velocity at the square inlet, which was obtained by Shepelev:

$$V_z = \frac{2V_0}{\pi} \arctan \frac{A^2/4}{\sqrt{A^2/2 + Z^2}}, \tag{2.15}$$

where A is the length of the side square.

Here are also formulas [12,29] to calculate the axial velocities of air:

At a rectangular $2A \times 2B$ inlet

$$V_z = \frac{2V_0}{\pi} \arctan \frac{B \cdot A}{Z\sqrt{B^2 + A^2 + Z^2}}, \tag{2.16}$$

at a circular inlet with an inner radius R_1 and the outer radius R_2

$$V_z = ZV_0 \left(\frac{1}{\sqrt{R_1^2 + Z^2}} - \frac{1}{\sqrt{R_2^2 + Z^2}} \right) \tag{2.17}$$

and at an elliptical inlet

$$V_z = \frac{2V_0}{\pi} \int_0^{\pi/2} \left[1 - \frac{Z}{\sqrt{\dfrac{a^2 \cdot b^2}{b^2 \cos^2 \varphi + a^2 \sin^2 \varphi} + Z^2}} \right] d\varphi, \tag{2.18}$$

where

a is the major semiaxis of ellipse

b is the minor one

Analysis of changes in the axial velocity (for $V_0 = 1$) at the regular polygons, circle and the ring (Table 2.1) suggest that the circular inlet is the longest ranged.

Converting a circular inlet into an elliptical one (Figure 2.8) and a rectangular one into a square one (Figure 2.9) produces no effect; the range decreases.

TABLE 2.1
Axial Air Velocity at Regular Figures

			n			
Z	3	4	6	10	Circle	Ring
0.25	0.8343	0.8414	0.8433	0.8437	0.8437	0.1065
0.5	0.6825	0.6944	0.6977	0.6982	0.6982	0.1823
1	0.4478	0.4609	0.4646	0.4652	0.4652	0.2256
5	0.4592	0.4637	0.0464	0.0464	0.0464	0.0409
10	0.01221	0.01225	0.01225	0.01225	0.01225	0.0112

Note: Z is a distance from the suction in absolute length units; n is the number of the polygon sides.

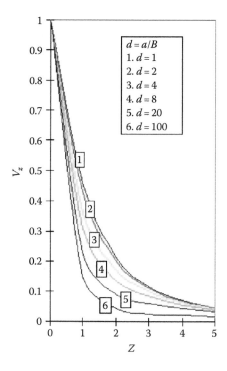

FIGURE 2.8   Axial air velocity at elliptical inlets.

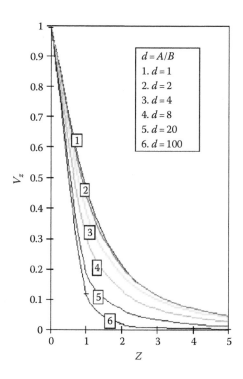

FIGURE 2.9   Axial air velocity at rectangular inlets.

Some engineering publications erroneously conclude that a slot-type exhaust unit has a greater range than a circular one [97] or that a square exhaust outlet has a greater range than a circular one [6]. In the first case, a slot-type suction unit and a circular one both built in a flat infinite wall having the same absorption rate but different flow rates were compared, and the difference in the flow rates led to the erroneous conclusion. In the second case, the comparison is incorrect, since the distance from the suction unit was measured in arbitrary units—fractions of the hydraulic diameters. The hydraulic radius of a circle is greater than that of a square, so the velocities were compared at different points.

It is of interest to study the influence of a circular suction inlet area at the same flow rate of exhausted air exerted on the axial air velocity.

The velocity V_z in Equation 2.14 is a function of Z and R. Fix Z. As $V_0 \pi R^2 = L$, where L is the airflow, then Equation 2.14 is transformed as follows:

$$V_z = \frac{L}{\pi R^2} \left(1 - \frac{Z}{\left(R^2 + Z^2 \right)^{1/2}} \right). \tag{2.19}$$

Differentiating Equation 2.19 with respect to R and equating to 0, we find R that could have an extremum:

$$\frac{dV_z}{dR} = \frac{L}{\pi R} \left[\frac{-2 \left(R^2 + Z^2 \right)^{3/2} + 2Z \left(R^2 + Z^2 \right) + Z \cdot R^2}{R^2 \left(R^2 + Z^2 \right)^{3/2}} \right] = 0. \tag{2.20}$$

Expanding the numerator by the factors, we obtain

$$(Z - t)(- Z^2 - Zt + 2t^2) = 0, \tag{2.21}$$

where $t = \sqrt{R^2 + Z^2}$.

Expression of Equation 2.19 has one real root, satisfying the physical description of the problem, $Z = t => R = 0$. There are no extrema. The velocity on the suction axis decreases with increase in the circle radius and tends to zero for $R \rightarrow \infty$ (Figure 2.10). Thus, a circular suction inlet of a smaller radius and higher absorption rate is always longer ranged than a circle of a larger radius with a slower velocity (the flow rates in both cases are identical), which is confirmed by the calculations.

But in reality, the radius of the inlet can be reduced to a certain limit. In addition, the influence of the viscous forces, which are not accounted for in the calculations, greatly increases for small inlets.

It should be remembered that in actual practice, the suction range most often *serves* some environmentally unfriendly area rather than a point of maximum emission of harmful impurities.

In this case, the average velocity is a more convenient measure to assess the range of a flare

$$V_{avg} = \frac{Q_x}{S_0}, \tag{2.22}$$

where Q_x is the airflow through the area located at the distance X and having the same area as the suction inlet S_0.

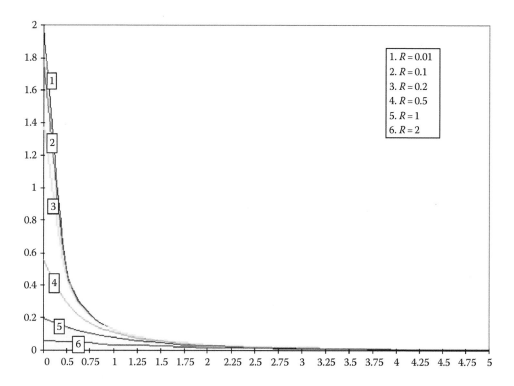

FIGURE 2.10   Change in the velocity on the axis of a circular inlet with change in its radius R and fixed flow rate $L = 1$.

2.1.4   SCREENING OF A LOCAL SUCTION BY SUPPLY AIR JETS

First, consider a 2D problem. Place the sink with a capacity Q_0 at the coordinate origin, and the jets at the points $(0, -a)$ and $(0, a)$ (Figure 2.11).

For the sink, the flow function is

$$\psi = -\frac{Q_0}{\pi}\, \text{arctg}\, \frac{y}{x}, \tag{2.23}$$

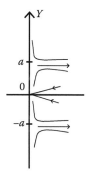

FIGURE 2.11   Outflow pattern of two submerged jets at $(0, a)$ and $(0, -a)$ and the sink at the point $(0, 0)$.

and for flat laminar jets [95]

$$\psi = A\sqrt[3]{x}\, \text{th}\left(B\frac{y\mp a}{\sqrt[3]{x^2}} \right),$$ (2.24)

where

$$A = 1.651\sqrt[3]{\frac{\nu J_0}{\rho}}$$

$$B = 0.2752\sqrt[3]{\frac{J_0}{\rho\nu^2}}$$

ρ is the air density, kg/m^3

ν is the coefficient of air kinematic viscosity, m^2/s

J_0 is the initial momentum of the jet, which is determined by the expression

$$J_0 = \frac{Q_c^3\rho}{\nu\cdot 3.3019^3\cdot x},$$ (2.25)

where Q_c is the volume flow per second in the section of the jet at a distance $x = 1$ m from the origin of coordinates. In our calculations, $Q_c = 0.001$ m^2/s. The low value of Q_c is related to computational difficulties as it increases.

We introduce n vertical lines on which the sinks and jets are located (Figure 2.12).

This location of linear sinks and submerged jets will allow getting an impenetrable line in case of unlimited increase in n; such line being the model of the *table* for $x = H/2$ and the model of the *ceiling* for $x = 0$. In this case, the flow function is described by the equation

$$\psi = \sum_{n=-\infty}^{\infty}\left(-\frac{Q_0}{\pi}\arctg\frac{y}{x-Hn} + A\sqrt[3]{x-Hn}\cdot\text{th}\left(B\frac{y-a}{\sqrt[3]{(x-Hn)^2}} \right) \right.$$

$$\left. + A\sqrt[3]{x-Hn}\cdot\text{th}\left(B\frac{y+a}{\sqrt[3]{(x-Hn)^2}} \right) \right),$$ (2.26)

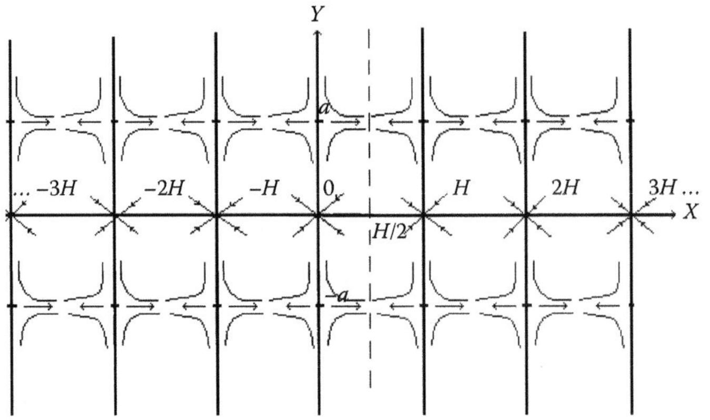

FIGURE 2.12   Chain of sinks and submerged jets.

and the horizontal component of the air velocity is described by the following equation:

$$v_x = \frac{\partial \psi}{\partial y} = \sum_{n=-\infty}^{\infty} \left(\frac{AB}{(x-Hn)^{1/3} \, ch^2 \left(\dfrac{B(y-a)}{(x-Hn)^{2/3}} \right)} \right.$$

$$\left. + \frac{AB}{(x-Hn)^{1/3} \, ch^2 \left(\dfrac{B(y+a)}{(x-Hn)^{2/3}} \right)} - \frac{Q_0(x-Hn)}{\pi \left((x-Hn)^2 + y^2 \right)} \right). \qquad (2.27)$$

In particular, for $n = 0$ the axial air velocity (on the axis OX) is defined by an obvious correlation

$$v_{ox} = \frac{2AB}{x^{1/3} \cdot ch^2 \left(\dfrac{Ba}{x^{2/3}} \right)} - \frac{Q_0}{\pi x}, \qquad (2.28)$$

which implies that when the sink is confined with two submerged jets in infinite space the range of the sink decreases (Figure 2.13). It can be seen more clearly when we consider two submerged jets without a line sink (Figure 2.14); in this case, the axial velocity is positive. No reverse flow occurs between the jets. The range of the sink between them cannot be increased.

When we consider n submerged jets located on n-vertical lines, we can see that when a jet flows over the impenetrable line $x = H/2$ (table) there is a reverse flow of air (Figure 2.15), and, on any line parallel to OY, there are three velocity extrema v_x, one maximum (along the jet axis) and two minima. Finding these extrema is of some interest.

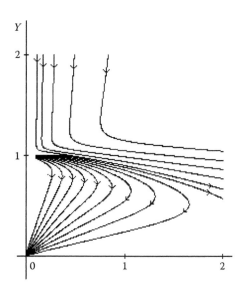

FIGURE 2.13   Flow lines from the sink and two submerged jets at (0, 1) and (0, −1).

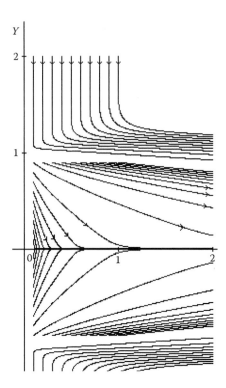

FIGURE 2.14   Flow lines from two symmetric flat jets from the points (0, 1) and (0, −1).

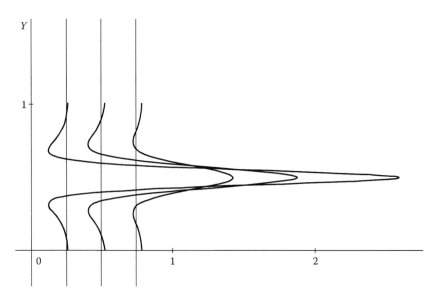

FIGURE 2.15   Hodograph of the velocity v_x in different sections ($x = 0.25$; $x = 0.5$; $x = 0.75$) caused by jets with a strength $Q_c = 0.001$ m²/s at the points ($\pm H \cdot n$; 0.5), where $H = 2$ and $n = 1, 2, ..., 15,000$.

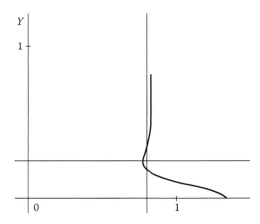

FIGURE 2.16   Velocity hodograph in a straight line $x = 0.8$ m when the jets outflow from $(x\text{-}nH, 0)$, $n = 1$, 2, ..., 30,000.

Consider n jets on an unlimited number of vertical lines flowing out of the intersections of these lines with the horizontal axis OX. Obviously, we have the following equation for a family of extreme points:

$$\frac{\partial v_x}{\partial y} = \sum_{n=-\infty}^{\infty} \frac{-2AB^2 \, \text{sh}\left(\dfrac{By}{(x-Hn)^{2/3}}\right)}{(x-Hn)\cdot \text{ch}^3\left(\dfrac{By}{(x-Hn)^{2/3}}\right)} = 0. \tag{2.29}$$

Solving Equation 2.29, we can find extreme points, for example, determine the ordinate a for which we have the highest reverse flow velocity v_x for a fixed value of x (Figure 2.16).

Note that the larger jet flow Q_c is, the closer the points of maximum reverse flow are to the jet axis (for a fixed abscissa x and the distance H), that is, the value a is inversely proportional to the flow rate Q_c. With increase in the distance to the *ceiling H/2* (for a constant flow rate Q_c and the fixed abscissa x), these points become more distant from the axis of the jet. Finally, for the constant Q_c and H, the minimum points approach OX when x decreases from $H/2$ down to 0.

Thus, the value of the ordinate of the highest reverse flow velocity depends on three values:

$$a = f\left(\frac{Hx}{Q_c}\right). \tag{2.30}$$

Note that when solving real problems related to the containment of dust release, the suction unit is located as close as possible (so allowed by the production process) to the table (H is fixed) and the representative point with the maximum dust emission is defined, in which the required airflow velocity must be provided (e.g., for welding it is a point with the highest developmental rate of convective flow).

Assume, for example, that the table is located at a distance of 1 m from the local suction unit and the representative point is 0.2 m away from the table. The flow rate of the jets symmetric about OX is $Q_c = 0.001$ m²/s. When the jets are moved at a distance $a = 0.24$ m (defined in the numerical solution of Equation 2.29) from the axis of symmetry, we get a vortex that contributes to increase in the range of the sink located at the point 0 (Figure 2.17).

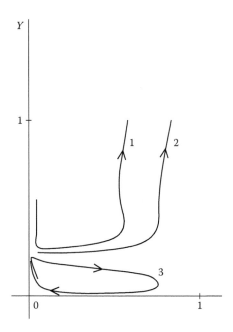

FIGURE 2.17   Flow lines of jets symmetric about the axis OX and located on 60,000 vertical lines spaced $H = 2$ m apart: $1 - \psi = 0.000202$, $2 - \psi = 0.0003212$, $3 - \psi = -0.00000765$.

At the point $x = 0.8$ m, we have the velocity $v_{ox} = -0.00016$ m/s with only a sink $Q_0 = 0.001$ m²/s and $v_{ox} = -0.00032$ m/s when the sink is screened with jets. That is, the range in the test point is increased by 2 times.

Now, consider a more complicated case of an axially symmetric flow. We use formula [29] to determine the air velocity at any arbitrary point of the compact isothermal jet:

$$u = \frac{u_0 \sqrt{F_0}}{\sqrt{\pi} cx} e^{-\frac{1}{2}\left(\frac{r}{cx}\right)^2},$$ (2.31)

where
  u_0 is the flow velocity of the compact jet, which is uniform across the intake area of F_0
  r is the distance from the jet axis to an arbitrary space point
  x is the distance from the jet efflux plane
  $c = 0.082$ is the experimental constant

We apply the approach [29], which is often used in engineering practice, to determine the expression to calculate the axial velocity of an annular jet in infinite space:

$$u_x^2 = \frac{u_0^2}{x^2 \pi c^2} \int\limits_{0}^{2\pi} \int\limits_{R_1}^{R_2} e^{-\left(\frac{r}{cx}\right)^2} r\,dr\,d\beta = u_0^2 \left[e^{-\left(\frac{R_1}{cx}\right)^2} - e^{-\left(\frac{R_2}{cx}\right)^2} \right],$$ (2.32)

where
  R_1 is the inner radius of the ring
  R_2 is the outer one

Assume that the jet located at the coordinate origin is directed along the axis OX. All jets located to the left are directed in the same direction; those located to the right are directed in the opposite direction. The distance between the jet efflux planes is H. When the number of such planes increases unrestrictedly, we have a model of airflow in the annular jet that flows from the *ceiling* and over the *floor* where the distance between them being H. Then,

$$u_x = u_0 \sum_{i=1}^{\infty} \left\{ \sqrt{e^{-\left(\frac{R_1}{c(x+iH)}\right)^2} - e^{-\left(\frac{R_2}{c(x+iH)}\right)^2}} - \sqrt{e^{-\left(\frac{R_1}{c(x-iH)}\right)^2} - e^{-\left(\frac{R_2}{c(x-iH)}\right)^2}} \right\} + u_0 \sqrt{e^{-\left(\frac{R_1}{cx}\right)^2} - e^{-\left(\frac{R_2}{cx}\right)^2}}. \tag{2.33}$$

Determine the values R_1 and R_2, for which a reverse flow of air with the greatest velocity occurs on the axis of the annular jet flowing over an impermeable surface.

To this end, we consider the interaction of an unlimited number of compact jets located similarly to the above annular ones. In this case, the velocity is

$$u = \frac{u_0 \sqrt{F_0}}{\sqrt{\pi}c} \sum_{i=-\infty}^{\infty} \frac{1}{x-iH} e^{-\frac{1}{2}\left(\frac{r}{c(x-iH)}\right)^2}. \tag{2.34}$$

We set the values $x = 0.8$ m, $H = 2$ m, and $u_0 = 1$ m/s to build the velocity hodograph (Figure 2.18), which shows that the reverse flow is observed in the area $a < y < b$. It can be naturally supposed that when we consider an annular jet with radii $R_1 = a$, $R_2 = b$, we get the highest velocity of the reverse flow. Calculations have shown that $R_1 = 0.0945$ m, $R_2 = 0.243$ m, and $u_x = -0.1125$ m/s. After a series of numerical experiments we concluded that the outer radius R_2 needs a little adjustment. For example, for $R_2 = 0.2$ m, we have the highest velocity $u_x = -0.1199$ m/s. Here, the airflow rate is $L_1 = u_0 \cdot \pi \left(R_2^2 - R_1^2 \right) = 0.0976$ m³/s.

Determine the effect the reverse flow has where a circular suction inlet with $R = 0.05$ m is screened with an annular jet.

The axial air velocity at a circular inlet built in a flat infinite wall is defined by Equation 2.14.

In the confined space between the *ceiling* and the *floor* it is

$$u_x = u_0 \sum_{i=1}^{\infty} \left(\frac{(x-iH)}{\sqrt{(x-iH)^2 + R^2}} - \frac{(x+iH)}{\sqrt{(x+iH)^2 + R^2}} \right) - u_0 \left(1 - \frac{x}{\sqrt{x^2 + R^2}} \right). \tag{2.35}$$

In this case, for the above parameters $u_x = -0.0011$ m/s, $L_2 = u_0 \pi r^2 = 0.007854$ m³/s.

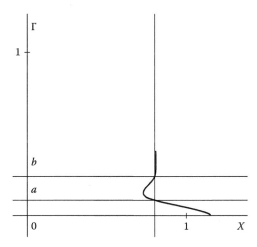

FIGURE 2.18 Velocity hodograph for the counterflowing of 20,000 compact jets spaced 2 m apart in the section $x = 0.8$ m.

When we consider an exhaust outlet with a flow rate $L_1 + L_2$, we obtain the velocity on the suction axis = 0.0152 m/s, that is, with screening by an annular jet above the table, the range of the suction flare at $x = 0.8$ m increases by $0.1199/0.0152 \approx 8$ times.

The calculation of screened local suction by the flow superposition method has a major drawback: the formation of vortex zones is not taken into account. When two parallel jets flowing into the unlimited space interact, then, as shown by experimental studies, a zone of reverse flows of gas occurs, which does not occur in the potential-flow model (see Figure 2.14). The resulting vortex zone also facilities improving the efficiency of the exhaust pipe for a particular location of supply inlets. The determination of the velocity field in the specified vortex zone is set out in the viscous incompressible gas model in Chapter 3 and is based on the discrete vortex method.

2.2 BOUNDARY INTEGRAL EQUATIONS METHOD

2.2.1 FLAT FLOWS IN MULTIPLY CONNECTED REGIONS WITHOUT SINGULARITIES

2.2.1.1 Derivation of the Basic Relations and the Construction of Stages of Solution

The generalized method of flow superposition or *singularities* [37,53], later named the boundary integral equations method (BIEM) or indirect boundary element method [54–59,90], is based on the theorem derived by H. Lamb [99] "... any generally potential flow can be obtained from a specific system of sources and sinks that are distributed along the area boundary." The solution of the Laplace equation written as integral, for given boundary conditions, enables to determine such a system.

Let there be given a certain airflow region A bounded by the contour S (Figure 2.19). There are no sources or sinks inside the area. The normal component of velocity $\vec{v}_n(\xi)$ is set on the boundary S, where ξ is an arbitrary point of S (the normal component of the velocity is equal to 0 on hard sections of the flow area boundary, that is, the condition of impermeability is established).

It is necessary to determine the air velocity in the inner points of the area along any given direction $\vec{v}_n(x)$, $x \in A$.

The Laplace equation is true for the potential ϕ; we write this equation in the polar coordinates:

$$\frac{\partial}{\partial r}\left(r\frac{\partial \varphi}{\partial r}\right) + \frac{1}{r}\frac{\partial^2 \varphi}{\partial \psi^2} = 0, \tag{2.36}$$

where r and ψ are the polar radius and angle.

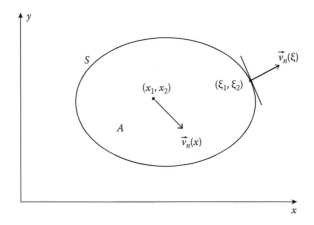

FIGURE 2.19   Airflow diagram in the area A.

Where the potential is not dependent upon the angle ψ (circular symmetry), Equation 2.36 is simplified:

$$\frac{d}{dr}\left(r\frac{d\varphi}{dr}\right) = 0. \tag{2.37}$$

We integrate Equation 2.37 twice to have

$$\varphi = C_1 \ln r + C_2. \tag{2.38}$$

In the course of mathematical physics [98], expression Equation 2.38 is called a fundamental solution of the Laplace equation in the plane for $C_1 = -1$ and $C_2 = 0$.

If the pole of the polar coordinate system is placed at the point (x_1, x_2) and the source (sink) with a strength $q(\xi)$ at the point (ξ_1, ξ_2) and $C_1 = q(\xi)/2\pi$, $C_2 = 0$ are taken as the constants in Equation 2.38, then Equation 2.38 is as follows:

$$\varphi(x) = \frac{q(\xi)}{2\pi}\ln r, \tag{2.39}$$

where the distance between the points x and ξ is $r = \sqrt{(x_1 - \xi_1)^2 + (x_2 - \xi_2)^2}$.

Thus, the fundamental solution of the Laplace equation (accurate to a constant) is the value of the potential at an arbitrary point x, which is caused by a source (sink) of a strength $q(\xi)$ at some point ξ.

The value of the air velocity v_n at the point (x_1, x_2) along any given direction $\vec{n}\{n_1, n_2\}$ caused by the source (sink) at the point (ξ_1, ξ_2) is defined by the following equation:

$$v_n = \vec{v}\cdot\vec{n} = v_1 n_1 + v_2 n_2 = \frac{\partial\varphi}{\partial x_1}n_1 + \frac{\partial\varphi}{\partial x_2}n_2 = \frac{q(\xi)}{2\pi r^2}\left[n_1(x_1 - \xi_1) + n_2(x_2 - \xi_2)\right]. \tag{2.40}$$

Assume that the sources (sinks) of an unknown strength $q(\xi)$, $\xi \in S$ are distributed on the boundary S of the area A under study. On integrating their effect, we obtain the value of the air velocity at the inner point (x_1, x_2) of the area A along a predetermined direction $\vec{n}$:

$$v_n(x) = \int_S q(\xi)F(x,\xi)dS(\xi), \tag{2.41}$$

where $F(x,\xi) = \frac{1}{2\pi r^2}\left[n_1(x_1 - \xi_1) + n_2(x_2 - \xi_2)\right]$.

In order to determine the unknown value of the strength $q(\xi)$, we let the point $x(x_1, x_2)$ to the point x_0 of the boundary S in the direction of the unit vector of the outward normal $\vec{n}$. Then Equation 2.41 is as follows:

$$v_n(x_0) = \int_S q(\xi)F(x_0,\xi)dS(\xi), \tag{2.42}$$

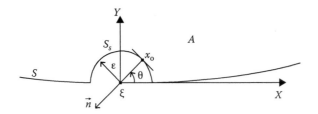

FIGURE 2.20   Boundary of the area without the point ξ.

where the integral is the principal-value integral as the function $F(x_0, \xi)$ increases without limit as x_0 tends to ξ. To get rid of this singularity, we encircle the point ξ (Figure 2.20) with a semicircle of a radius $\varepsilon \to 0$. Then, the integral in Equation 2.42 can be expressed as the sum of two integrals:

$$\int_S = \int_{S-S_\varepsilon} + \int_{S_\varepsilon} ,$$

where the first integral is taken along the contour S without this semicircle and the second one is an integral along the arc S_ε of the semicircle.

Integrating along the arc of a semicircle S_ε and assuming that the value of $q(\xi)$ on this arc is constant, we obtain

$$\int_{S_\varepsilon} F(x_0, \xi)dS = \frac{1}{2\pi} \int_{S_\varepsilon} \frac{(x_{01} - \xi_1)n_1 + (x_{02} - \xi_2)n_2}{r^2} dS$$

$$= \left\{ \begin{array}{l} x_{01} - \xi_1 = \varepsilon\cos\theta, \ x_{02} - \xi_2 = \varepsilon\sin\theta, \ r = \varepsilon \\ n_1 = -\cos\theta, \ n_2 = -\sin\theta, \ dS = \varepsilon d\theta. \end{array} \right\} = -\frac{1}{2},$$

and, therefore, Equation 2.42 can be written as

$$v_n(x_0) = -\frac{1}{2}q(x_0) + \int_S q(\xi)F(x_0, \xi)dS(\xi). \tag{2.43}$$

The first summand Equation 2.43 corresponds to the case where x_0 coincides with ξ and the second summand is an integral in the usual sense, whose integration contour does not contain the point $\xi = x_0$.

If the integral in Equation 2.43 were calculated analytically, then by solving this equation for $q(\xi)$ and substituting this value in Equation 2.41 we would obtain a solution to the problem. However, such cases practically do not exist in real problems and we need a numerical solution of Equation 2.43.

We divide the boundary S on N right-lined intervals. We assume that for each of the resulting intervals, for example, k-m, the strength of the sources (sinks) is constant and equal to q^k. Summarizing the effects of all these sources on the center of the pth interval, we obtain a discrete analogue of Equation 2.43:

$$v_n(x_0^p) = -\frac{1}{2}q^p + \sum_{\substack{k=1, \\ k \neq p}}^N q^k \int_{S^k} F(x_0^p, \xi^k)dS(\xi^k), \tag{2.44}$$

where ξ^k is an arbitrary point of the kth interval.

The integrals in Equation 2.44 are taken by the kth intervals (S^k). The velocity $v_n(x_0^p)$ is known from the boundary conditions of the set problem. By varying p from 1 to N, we obtain a system of N linear equations with N unknowns. Solving the resulting system of Equation 2.44, we find q^k, $k = 1$, 2, ..., N. The velocity at the inner point x along a given direction is determined by

$$v_n(x) = \sum_{k=1}^{N} q^k \int_{S^k} F(x, \xi^k) dS(\xi^k). \tag{2.45}$$

Thus, the algorithm for solving a flat problem consists of three steps:

1. Dividing the boundaries (discretization) of the airflow area under study into straight intervals (boundary elements).
2. Solving the system of linear equations (Equation 2.44) and the definition of the strength of sources q^k, $k = 1, 2, ..., N$.
3. Determining the air velocity at the desired point by formula (Equation 2.45).

2.2.1.2   Discretization of the Area Boundary

The discretization of the area boundary S into the boundary elements is made by means of a set of right-lined intervals (Figure 2.21) defined in the GCCS. Each boundary interval has its own number and coordinates beginning at a and ending at b. a and b are chosen so that the unit normal vector to the given interval is directed outward the area.

2.2.1.3   Calculation of the Strength of Sources (Sinks)

In order to determine the strength of the sources (sinks), we must first create a matrix $F = \left(F^{pk} \right) = \left[\int_{S^k} F(x_0^p, \xi^k) dS(\xi^k) \right]$ composed of the coefficients for the unknowns q^k in the system of linear algebraic equations (Equation 2.44), where $p = 1, 2, ..., N$ and $q = 1, 2, ..., N$.

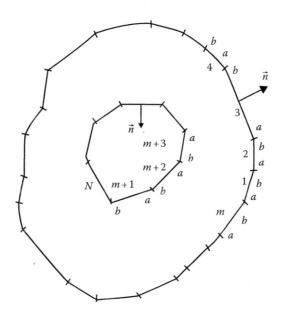

FIGURE 2.21   Discretization of the boundary into N right-lined intervals.

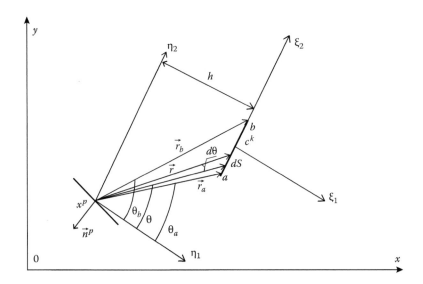

FIGURE 2.22   To the construction of local coordinate systems.

The integrals F^{pk} are computed analytically. Assume that the center of pth interval, the point x^p, is set in the global Cartesian coordinate system (GCCS) XOY and the center of the kth interval $[a, b]$, the point c^k, is also set in the GCCS. We construct a local rectangular Cartesian coordinate system centered at c^k, direct the vertical axis ξ_2 along the vector $\overrightarrow{ab}$, and locate the horizontal axis ξ_1 so that the resulting system of coordinates is a right-handed one (Figure 2.22). The local Cartesian coordinate system (LCCS) $\eta_1\eta_2$ is obtained from $\xi_1\xi_2$ by parallel shift to the point x^p.

In the LCCS $\eta_1\eta_2$, the equalities $\eta_1 = h, \eta_2 = h\,\mathrm{tg}\,\theta$ are true for the coordinates of an arbitrary point ξ^k of the kth interval and for the point x^p : $\eta_1 = 0, \eta_2 = 0$. If we use $\{n_1, n_2\}$ to denote the coordinates of the unit outward normal vector to the pth interval in the coordinate system $\eta_1\eta_2$, we obtain for $F(x_0^p, \xi^k)$ the following expression:

$$F(x_0^p, \xi^k) = \frac{(0-h)n_1 + (0-h\,\mathrm{tg}\,\theta)n_2}{2\pi\left((0-h)^2 + (0-h\,\mathrm{tg}\,\theta)^2\right)} = -\frac{n_1 + n_2\,\mathrm{tg}\,\theta}{2\pi h}\cos^2\theta.$$

Therefore, the integral is

$$F^{pk} = \int_{S^k} F(x_0^p, \xi^k)dS(\xi^k) = \left\{ dS = d\eta_2 = \frac{h}{\cos^2\theta}d\theta \right\} = -\frac{1}{2\pi}\int_{\theta_a}^{\theta_b} (n_1 + n_2 \cdot \mathrm{tg}\,\theta)d\theta.$$

On integrating, we obtain

$$F^{pk} = -\frac{1}{2\pi}\left[n_1(\theta_b - \theta_a) + n_2 \cdot \ln\frac{r_b}{r_a} \right], \tag{2.46}$$

where r_b, r_a are the lengths of the corresponding radius vectors of the ends a and b of the kth interval. Consider in detail the calculation of the parameters of the formula given by Equation 2.46.

Let the coordinates in the GCCS of the interval $[a, b]$: $a(a_1, a_2)$; $b(b_1, b_2)$ be given. We construct an LCCS on this interval, as shown in Figure 2.4 and calculate the coordinates of the unitary vectors of the resulting system of coordinates. As

$$\vec{e}_2 = \frac{\vec{b} - \vec{a}}{\left|\vec{b} - \vec{a}\right|},$$

where $\vec{a}, \vec{b}$ are the radius vectors of the points a and b in the GCCS, then

$$\begin{cases} e_{2x} = \dfrac{b_1 - a_1}{\sqrt{(b_1 - a_1)^2 + (b_2 - a_2)^2}}, \\[4mm] e_{2y} = \dfrac{b_2 - a_2}{\sqrt{(b_1 - a_1)^2 + (b_2 - a_2)^2}}. \end{cases} \tag{2.47}$$

The coordinates of the unitary vector of the horizontal axis can be found from the conditions of vector perpendicularity $\vec{e}_1$, $\vec{e}_2$ and the equality of the vector $\vec{e}_1 \times \vec{e}_2$ to the unitary vector $\vec{k}$ of the GCCS, as the LCCS is right handed:

$$\vec{e}_1 \cdot \vec{e}_2 = e_{1x} \cdot e_{2x} + e_{1y} \cdot e_{2y} = 0,$$

$$(\vec{e}_1 \times \vec{e}_2) = e_{1x} \cdot e_{2y} - e_{1y} \cdot e_{2x} = 1.$$

Thus, we obtain

$$e_{1x} = e_{2y}, \quad e_{1y} = -e_{2x}. \tag{2.48}$$

Assume that the coordinates of a certain vector $\vec{m} = \{m_1, m_2\}$ in the GCCS are known. We define its position in the LCCS. In the GCCS

$$\begin{cases} \vec{e}_1 = e_{1x}\vec{i} + e_{1y}\vec{j}, \\ \vec{e}_2 = e_{2x}\vec{i} + e_{2y}\vec{j}. \end{cases}$$

Whence

$$\begin{cases} \vec{i} = e_{1x}\vec{e}_1 + e_{2x}\vec{e}_2, \\ \vec{j} = e_{1y}\vec{e}_1 + e_{2y}\vec{e}_2. \end{cases}$$

The vector

$$\vec{m} = m_1\vec{i} + m_2\vec{j} = m_1(e_{1x}\vec{e}_1 + e_{2x}\vec{e}_2) + m_2(e_{1y}\vec{e}_1 + e_{2y}\vec{e}_2)$$

$$= (m_1 e_{1x} + m_2 e_{1y})\vec{e}_1 + (m_1 e_{2x} + m_2 e_{2y})\vec{e}_2.$$

Thus, the coordinates of the vector $\vec{m}$ in the LCCS are

$$\begin{cases} m_1' = m_1 e_{1x} + m_2 e_{1y}, \\ m_2' = m_1 e_{2x} + m_2 e_{2y}. \end{cases} \tag{2.49}$$

Given the point (a_1, a_2) in the GCCS, we define its coordinates in the LCCS $\eta_1 \eta_2$. When the global coordinate system is rotated in such a way that the vertical axis becomes codirectional with $\vec{e}_2$, the coordinates of the point become

$$\begin{cases} a_1' = a_1 e_{1x} + a_2 e_{1y}, \\ a_2' = a_1 e_{2x} + a_2 e_{2y}. \end{cases} \tag{2.50}$$

After a parallel shift of the coordinate origin to the point $x^p(x_1^p, x_2^p)$, the coordinates of the point a will be, accordingly,

$$\begin{cases} a_1'' = a_1' - x_1^p, \\ a_2'' = a_2' - x_2^p. \end{cases} \tag{2.51}$$

Then, using Equations 2.12 through 2.16 the algorithm to calculate F^{pk} consists of the following steps:

1. The initial number of the pth interval is set: $p = 1$.
2. We calculate the global coordinates of the unitary vectors of the LCCS built on the pth interval:

$$e_{2x}^p = \frac{b_1^p - a_1^p}{\sqrt{\left(b_1^p - a_1^p\right)^2 + \left(b_2^p - a_2^p\right)^2}}, \quad e_{2y}^p = \frac{b_2^p - a_2^p}{\sqrt{\left(b_1^p - a_1^p\right)^2 + \left(b_2^p - a_2^p\right)^2}}, \quad e_{1x}^p = e_{2y}^p, \quad e_{1y}^p = -e_{2x}^p.$$

   Then, the unit outward normal vector to the pth interval is $\vec{n}^p = \vec{e}_1^p$, if $\vec{e}_1^p$ is directed outward the flow area, that is, the flow area is to the left when passing from a to b. If a and b are chosen in such a manner that the area is to the right when passing the boundary from a to b, then $\vec{n}^p = -\vec{e}_1^p$.

3. The initial number of the kth interval is set: $k = 1$.
4. If $p = k$, then F^{pk} and go to Step 10.
5. For the normal vector $\vec{n}^p$, the coordinates are calculated in the LCCS of the kth interval. To do this, the unitary vectors are first defined,

$$e_{2x}^k = \frac{b_1^k - a_1^k}{\sqrt{\left(b_1^k - a_1^k\right)^2 + \left(b_2^k - a_2^k\right)^2}}, \quad e_{2y}^k = \frac{b_2^k - a_2^k}{\sqrt{\left(b_1^k - a_1^k\right)^2 + \left(b_2^k - a_2^k\right)^2}}, \quad e_{1x}^k = e_{2y}^k, \quad e_{1y}^k = -e_{2x}^k,$$

$$\tag{2.52}$$

   and then the desired coordinates are calculated:

$$n_1^{pk} = n_1^p e_{1x}^k + n_2^p e_{1y}^k, \quad n_2^{pk} = n_1^p e_{2x}^k + n_2^p e_{2y}^k.$$

6. We calculate the coordinates of the radius vector of the center of the pth interval in the GCCS:

$$x_1^p = \frac{a_1^p + b_1^p}{2}, \quad x_2^p = \frac{a_2^p + b_2^p}{2},$$

and its coordinates are determined in the LCCS of the kth interval:

$$x_1^{pk} = x_1^p e_{1x}^k + x_2^p e_{1y}^k, \quad x_2^{pk} = x_1^p e_{2x}^k + x_2^p e_{2y}^k.$$

7. We calculate the coordinates of the vertexes a^k, b^k of the kth interval in the coordinate system of the pth interval ($\eta_1 \eta_2$ in Figure 2.22):

$$\begin{cases} a_1^{kp} = a_1^k e_{1x}^k + a_2^k e_{1y}^k - x_1^{pk}, \\ a_2^{kp} = a_1^k e_{2x}^k + a_2^k e_{2y}^k - x_2^{pk}, \\ b_1^{kp} = b_1^k e_{1x}^k + b_2^k e_{1y}^k - x_1^{pk}, \\ b_2^{kp} = b_1^k e_{2x}^k + b_2^k e_{2y}^k - x_2^{pk}. \end{cases} \tag{2.53}$$

8. We calculate the lengths of the radius vectors of the points a^k, b^k in the coordinate system of the pth interval,

$$r_a = \sqrt{\left(a_1^{kp}\right)^2 + \left(a_2^{kp}\right)^2}, \quad r_b = \sqrt{\left(b_1^{kp}\right)^2 + \left(b_2^{kp}\right)^2}, \tag{2.54}$$

and the angles,

$$\theta_a = \begin{cases} \operatorname{arctg} \dfrac{a_2^{kp}}{a_1^{kp}}, & \text{if } a_1^{kp} \neq 0; \\ \dfrac{\pi}{2}, & \text{if } a_1^{kp} = 0, \end{cases} \qquad \theta_b = \begin{cases} \operatorname{arctg} \dfrac{b_2^{kp}}{b_1^{kp}}, & \text{if } b_1^{kp} \neq 0; \\ \dfrac{\pi}{2}, & \text{if } b_1^{kp} = 0. \end{cases} \tag{2.55}$$

9. We calculate the matrix element

$$F^{pk} = -\frac{1}{2\pi} \left[n_1^{pk}(\theta_b - \theta_a) + n_2^{pk} \ln \frac{r_b}{r_a} \right].$$

10. Step for k: $k = k + 1$ and go to Step 4 until $k \leq N$.
11. Step for p: $p = p + 1$ and go to Step 2 until $p \leq N$.

When the matrix (F^{pk}) elements have been formed, we solve the system of equations:

$$\begin{cases} F^{11}q^1 + F^{12}q^2 + \cdots + F^{1N}q^N = v^1, \\ F^{21}q^1 + F^{22}q^2 + \cdots + F^{2N}q^N = v^2, \\ \vdots \\ F^{N1}q^1 + F^{N2}q^2 + \cdots + F^{NN}q^N = v^N, \end{cases}$$

whence we determine the unknown strengths of the sources (sinks) $q^1, q^2, \ldots, q^N$ distributed along N boundary intervals. The boundary values of the normal velocity component $v^1, v^2, \ldots, v^N$ are given in the statement of the problem.

2.2.1.4   Calculation of the Air Velocity inside the Flow Area

In order to determine the air velocity at an arbitrary point $x(x_1, x_2)$ of the flow area A along any given direction $\vec{n} = \{n_1, n_2\}$, the following steps must be performed:

1. Define $k = 1$.
2. Calculate the coordinates of the vectors $\vec{e}_1^k$, $\vec{e}_2^k$ by Equation 2.52, then the coordinates of the vector $\vec{n}$ in the LCCS of the kth interval:

$$n_1^k = n_1 e_{1x}^k + n_2 e_{1y}^k, \quad n_2^k = n_1 e_{2x}^k + n_2 e_{2y}^k.$$

3. Calculate the coordinates of the point x in the LCCS of the kth interval:

$$x_1^k = x_1 e_{1x}^k + x_2 e_{1y}^k, \quad x_2^k = x_1 e_{2x}^k + x_2 e_{2y}^k.$$

4. Define the coordinates of the points a^k, b^k in the LCCS centered at x with the unitary vectors $\vec{e}_1^k$, $\vec{e}_2^k$ by the formula shown as Equation 2.53.
5. The values r_a, r_b must be calculated by Equation 2.54 and the angles θ_a, θ_b by the formula shown as Equation 2.55.
6. Calculate the value

$$f^k = -\frac{1}{2\pi}\left[n_1^k(\theta_b - \theta_a) + n_2^k \ln \frac{r_b}{r_a} \right].$$

7. Step for k: $k = k + 1$ and go to Step 2 until $k \le N$.
8. Calculation of the velocity

$$v_n(x) = f^1 q^1 + f^2 q^2 + \cdots + f^N q^N.$$

The above algorithm can be used to calculate the flow lines. To do so, it is necessary to define a point at which we begin the calculation of the flow line; to calculate the horizontal v_x and vertical v_y components of the flow velocity in the selected GCCS; to determine the direction of the flow velocity $\vec{v}$; and to take a step in this direction, and perform again the described calculation procedure. The calculation stops when the predetermined section is reached (e.g., air suction line). This algorithm for calculation of flow lines corresponds to Euler's method for solution of an ordinary differential equation (ODE)

$$\frac{dx}{v_x} = \frac{dy}{v_y}$$

for given initial conditions. Note, however, that the calculation of the flow lines can be made by more accurate methods for solving ODEs (e.g., the Runge–Kutta method).

2.2.1.5   Test Problem: Calculation of the Axial Air Velocity at Slot-Type Suction Units

To determine the accuracy of the calculations by BIEM, as well as to find the most economical method for dividing the area boundary into boundary intervals, we calculated the axial air velocity at freely spaced slot-type suction units built in a flat infinite wall as discussed in Section 2.1. The flow area in this case is shown in Figure 2.23.

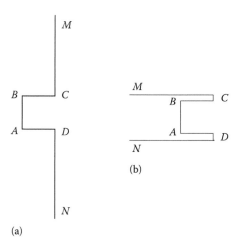

FIGURE 2.23 The flow area is (a) at a slot-type suction built in a flat infinite wall and (b) at freely spaced slot-type suction.

One of the boundary conditions is the condition of impenetrability, that is, the normal velocity component is equal to zero at the boundary (polygonal lines BCM and ADN). The second condition is the velocity in the slot on the interval AB (the velocity at *infinity*): $v_\infty = 1$. In this case, an infinite section is as close as one gauge to the slot entry ($AB = AD$). The area was discretized into the boundary intervals in the following manner. The intervals AB, BC, AD are divided equally. The division is denser at the edges than near the center. The law is as follows: $0.001i$, where $i = 1, 2, 4, 8, \ldots$ until the end of the boundary element crosses the middle of the interval. A part of the boundary element is cut off; the midpoint of the interval to be divided is taken as its terminal point. The interval on the other end is divided likewise. We use the same law to divide the rays CM and DN: $0.001i$, where $i = 1, 2, 4, 8, \ldots$ until the points M and N are located at a distance greater than 10 gauges from D. We obtained 88 boundary elements with the described discretization of the area.

Comparison of calculations made by the BIE method and the conformal mapping method (CMM) showed that the relative error is

$$\varepsilon = 100\% \cdot \frac{(v_{BIE} - v_{CMM})}{v_{CMM}},$$

where v_{BIE}, v_{CMM} are the velocities calculated using the BIE and CMM, respectively, that do not exceed 5% (Table 2.2).

2.2.1.6 Dedusting the Casting Roller Processing Procedure

In machining cast iron casting rollers, the greatest emissions of dust occur during grinding of the casting skin. This is due to the high hardness of the surface layer, the presence of cavities, flux inclusions, and metal penetrations. The main sources of dust emissions are the cutting zone and the chip transport train. The strength of dust emissions is significantly affected by a high content of graphite in cast iron rolls and the temperature factor. The dust dispersion from the cutting zone is determined by the operation features of roll-turning machines, the main of which is the small rotational speed of billets. In this regard, the main role in the process is played by convective currents produced by the switching of the mechanical energy of deformation and friction during cutting into heat. Dust contents of the air in the workplaces of roll-turning machine operators during roughing rollers is 7–30 mg/m^3 with the MAC equal to 2 mg/m^3. The highest dust content of 1700–3200 mg/m^3 is in the cutting zone. Outside the cutting zone, the dust content of air drops sharply down to 60 mg/m^3, reaching at a certain distance the value of the ambient concentration in the workshop.

TABLE 2.2

Comparison of the Calculated Values of Axial Air Velocity by the CMM and Boundary Integral Equation Method

At a Slot-Type Suction Unit in Infinite Space

x/B	v_{BIE}	v_{CMM}	ε, %
0.5	0.545	0.557	−2.15
1.0	0.378	0.378	0
1.5	0.272	0.269	1.12
2.0	0.207	0.203	1.97
2.5	0.165	0.161	2.48
3.0	0.136	0.133	2.26
3.5	0.116	0.112	3.57
4.0	0.100	0.097	3.09
4.5	0.089	0.085	4.71
5.0	0.079	0.076	3.95

At a Slot-Type Suction Unit Built in a Flat Infinite Wall

0.5	0.631	0.645	−2.17
1.0	0.470	0.471	−0.21
1.5	0.362	0.365	−0.08
2.0	0.290	0.290	0
2.5	0.240	0.240	0
3.0	0.205	0.202	1.49
3.5	0.176	0.176	1.14
4.0	0.158	0.155	1.94
4.5	0.141	0.139	1.44
5.0	0.128	0.125	2.40

Note: x is the distance from the suction entry, B is the half width of the suction slot.

To contain dust release, we suggest that the machine slide should be equipped with a local suction shelter equipped with a flap on a machine slide to prevent convective dust discharge (Figure 2.24).

Assume that air is sucked at a velocity $u_0 = 1$ m/s. We define the air velocity in the representative areas with an intense emission of dust: dust under the roller results from convection air currents from chips accumulated on the frame and that over the roller results from convection air currents from the cutting zone and falling chips.

The area boundary is discretized into a set of 280 right-lined intervals. Segmentation is shorter near jogs of the boundary. The roller is broken more densely near the local suction and flap.

Consider the case where the radius of the roller is $R = 525$ mm. The machine slide moves along the axis l. We assume that the length of the slide removal is $l = 0$, where the length of the interval AB (Figure 2.24) is equal to 60 mm, which corresponds to the start of roughing a roller with a radius of 525 mm with a standard cutter position (the distance from the roller to the entrance of the suction slot is 110 mm). The shortest distance between the flap and the roller is 20 mm. The flap length is 400 mm. The unit length of the slide removal is taken as 100 mm, that is, if $l = 1$, then $|AB| = 160$ mm; if $l = 2$, then $|AB| = 260$ mm, etc.

The flow pattern is seen from the flow lines flowing around the roller as built using a computer (Figures 2.25 and 2.26). When the slide is moving away, the rear critical point (the branching point of flow lines) moves downward counterclockwise.

The calculation of the profile of horizontal velocity component (Figure 2.27) demonstrates the qualitative and quantitative flow pattern. The velocities are significantly larger under the roller than

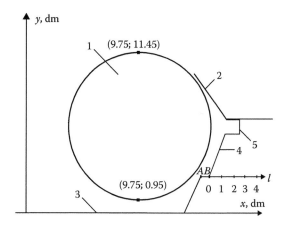

FIGURE 2.24 Scheme of a roll-turning machine with a local suction shelter: 1, roller; 2, flap; 3, frame; 4, slide; 5, local suction.

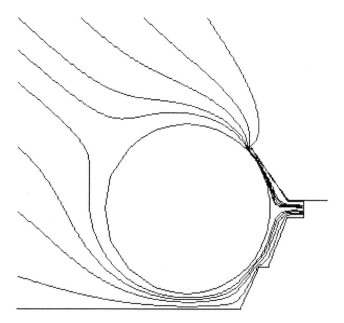

FIGURE 2.25 Flow lines in the flow around a roller with a radius of 525 mm and a length of the slide removal $l = 0$.

those above it as, in the first case, the airflow is confined by the frame and the roller while in the second case, the flow only has a lower limit; the upper one is at infinity.

For a fixed removal distance of the slide l, the value of the dimensionless horizontal velocity component v_x/v_o under the roller is almost constant and increases only slightly as it approaches the surface of the roller (Figure 2.28). The maximum velocity is achieved with the standard position of the cutter, that is, for $l = 0$. When the slide is moving away, the velocity at the point under the roller decreases sharply (Figure 2.29). Indeed, the suction flow is divided into two streams: an overflow and an underflow. The overflow passes between the flap and the roller and the underflow goes between the frame and the roller.

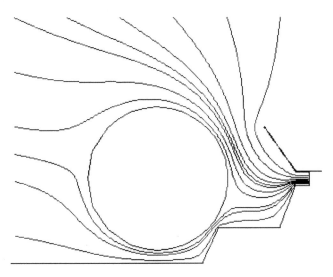

FIGURE 2.26 Flow lines in the flow around a roller with a radius of 525 mm and a length of the slide removal $l = 4$.

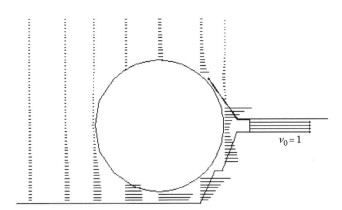

FIGURE 2.27 Velocity profile of flow lines in the flow around the roller with a radius of 525 mm and the length of the slide removal $l = 0$.

The smaller the cross section of the overflow is, the greater the flow velocity is in the underflow, as it follows from the equation of continuity.

And conversely, the greater the slot between the roller and the flap is, the slower the velocity is under the roller. The horizontal velocity component above the roller varies slightly depending on the stroke of the slide (Figures 2.30 and 2.31), but it decreases sharply at a greater distance from the roller.

The calculation results for a local exhaust unit without a flap (Figures 2.32 through 2.35) showed that when processing the same roller and with different distances of the slide removal, velocities under and above the roller change identically to the previous case, but the values of velocities are significantly lower.

The calculation results can be explained as follows. The value of the velocity under the roller is affected by two factors: the distance from the point to the local suction and the velocity balance of the underflow and the overflow.

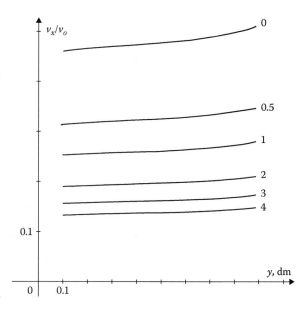

FIGURE 2.28 Change in the horizontal velocity component in a section under the roller for different distances of the slide removal (0; 0.5; 1; ...; 4).

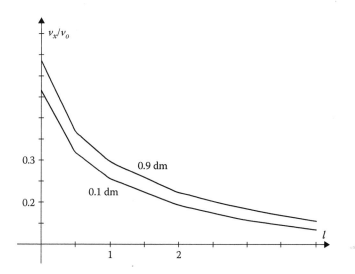

FIGURE 2.29 Change in the horizontal velocity component under the roller depending on the stroke of the slide.

For small radii of a roller processed, the first factor has a greater effect: the closer the point is to the suction slot, the higher the velocity at it is. When the radius of the roll increases, the distance between the point under study and the suction slot increases and the velocity decreases, but the second factor comes into operation: the thickness of the overflow decreases much faster than that of the underflow; therefore, the flow rate of the overflow decreases and that of the underflow increases (the velocity curve goes up). The value of the velocity above the roller is affected by the same factors, but in this case, they act in the same direction, so the velocity curve goes down.

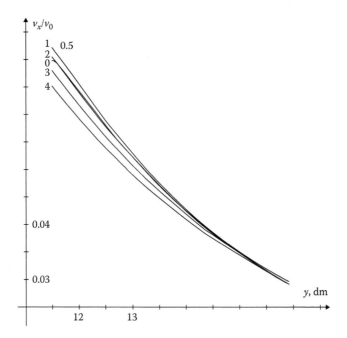

FIGURE 2.30 Change in the horizontal velocity component above the roller for various distances of the slide removal (0; 0.5; 1; …; 4).

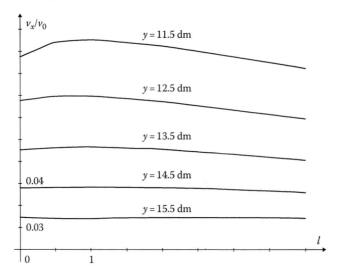

FIGURE 2.31 Change in the horizontal velocity component above the roller depending on the slide removal.

The obtained calculated values of the velocity under the roller have good experimental evidence (Figure 2.36) for different distances of the slide removal. The relative error

$$d = \frac{v_{BIE} - v_{exp}}{v_{exp}} \cdot 100\%$$

of the velocities v_{BIE} calculated by the BIE method compared to the experimental values v_{exp} ranges within 10%–30%, which is acceptable in engineering calculations.

The calculated speed above the roller is significantly different from the experimental measurements for small distances of the slide removal, which is due to the influence of the viscous forces

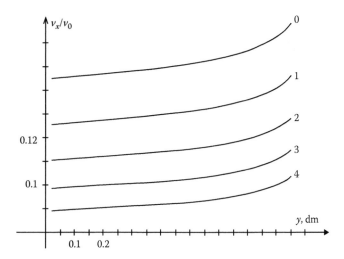

FIGURE 2.32   Change in the horizontal velocity component under the roller (nonflapped slot).

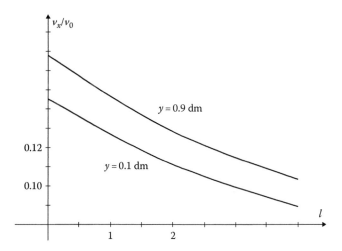

FIGURE 2.33   Change in the horizontal velocity component under the roller depending on the slide removal (nonflapped slot).

between the flap and the roller that are not accounted for in the model of potential flows of gas. With long distances of the slide removal, the effect of the viscosity forces decreases noticeably; therefore, the relative error is greatly reduced and is within the accuracy of the aerodynamic experiment. The linear correlation coefficient r for velocities above the roller has the smallest value of 0.86 for $l = 0$ and the greatest value of 0.992 for $l = 2, 3, 4$. For velocities under the roller $r = 0.998$. Thus, the behavior of the airflow rate calculated by the BIE method and measured experimentally is similar. Using linear regression equations, we can calculate the actual velocities based on the calculated values, when we have a large discrepancy with the experimental measurements.

Experimental studies suggest that air movement of 0.5 m/s must be created under and above the roller in order to contain dust release in roughing of casting rollers.

The cross-sectional area of the suction slot under study is 0.046 m^2. The velocity calculated for a suction unit without a flap by the BIE method at a distance of 5 cm (Figures 2.14 through 2.17) is $v_{p1} \approx 0.17v_0$ under the roller and $v_{p1} \approx 0.06v_0$ above it. Therefore, the necessary air suction velocities are $v_{01} = 0.5/0.17 \approx 2.94$ m/s, $v_{02} = 0.5/0.06 \approx 8.5$ m/s. We choose the highest rate and define the aspiration volume $L = v_{02} \cdot S \approx 1400$ m^3/h.

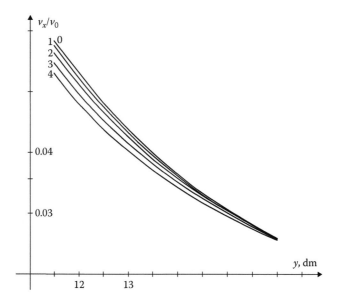

FIGURE 2.34   Change in the horizontal velocity component above the roller (nonflapped slot).

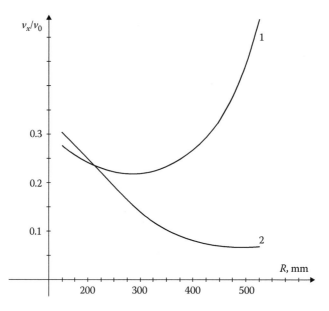

FIGURE 2.35   Change in the horizontal velocity component depending on the radius of the roller: 1, under the roller, 2, above the roller.

The calculation of the velocity near the local suction-shelter should only be made for a point under the roller as the flap prevents convective dust discharge. According to the graph shown in Figure 2.28, the calculated velocity is $v_p \approx 0.5 v_0$, so the necessary suction velocity is 1 m/s and the aspiration volume is $L \approx 170$ m³/h.

2.2.2   FLAT FLOWS IN MULTIPLY CONNECTED REGIONS WITH CUTS

To solve a number of problems related to industrial ventilation requires the calculation of the air velocity field near the suction inlets of local suctions with thin hoods in their effect area. Such hoods (mechanical screens) have a small thickness (a few millimeters) and help enhance

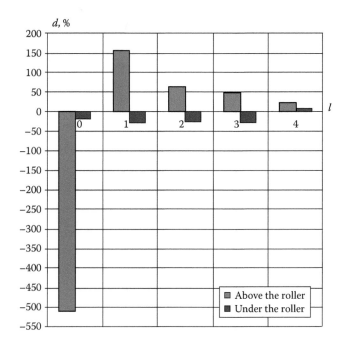

FIGURE 2.36 Relative error chart of the calculated values of dimensionless velocity compared with the experimental values for the different distances of slide removal.

the effectiveness of the local suction. The CMM as a classical method for calculating potential flows allows accounting for the effect of thin hoods only in simply connected domains [16]. BIEM is used to solve a number of problems related to potential flows (Sections 2.2.1.5 and 2.2.16) bounded by thin hoods, where both sides of the hoods and the interval bonding them are divided into boundary elements. In this case, sources (sinks) were distributed at each element, the strengths of which were assumed to be constant. Let us assume that the hoods are infinitely thin, which is quite acceptable as their thickness is significantly less than the dimensions of other parts. Thus, the problem is to determine the velocity of a potential flow within a multiply connected region with cuts for the given values of boundary normal velocity component. We will place dipoles at each of the boundary elements of the cut and conventionally sources (sinks) on the rest of the boundaries and will prove this possibility.

Theorem: The effect of the cut on the velocity within the area of the potential flow is determined by the location of dipoles on this cut.

Actually, consider the flow area whose boundary has a projection of a finite thickness (Figure 2.37).

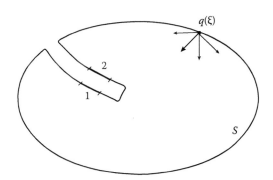

FIGURE 2.37 Area with a projection.

We divide the area boundary S into boundary intervals and place sources (sinks) with strength $q(\xi)$, $\xi \in S$ on them. For simplicity, assume that they are constant on each of the boundary elements. We separate two opposite intervals on the projection. The effect of all area intervals on them is defined by the following equations:

$$v_n(x^1) = -\frac{1}{2}q(x^1) + \int_S q(\xi)F(x^1, \xi)dS(\xi),$$

$$v_n(x^2) = -\frac{1}{2}q(x^2) + \int_S q(\xi)F(x^2, \xi)dS(\xi),$$

(2.56)

where $v_n(x^1)$, $v_n(x^2)$ are the velocities along the direction of the outward normal at the points x^1, x^2, located in the center of Intervals 1 and 2; $q(x^1)$, $q(x^2)$ are the strengths of the sources (sinks) on these intervals; $\vec{n} = \{n_1, n_2\}$ is the unit outward normal vector;

$$F(x, \xi) = \frac{n_1(x_1 - \xi_1) + n_2(x_2 - \xi_2)}{2\pi\left[(x_1 - \xi_1)^2 + (x_2 - \xi_2)^2\right]},$$

(x_1, x_2), (ξ_1, ξ_2) are the coordinates of x and ξ.

The condition of impermeability implies that $v_n(x^1) = v_n(x^2) = 0$. As the outward normals of Intervals 1 and 2 are in opposition to each other, then, letting the thickness of the projection to zero, we obtain $F(x^1, \xi) = -F(x^2, \xi)$. Then Equation 2.56 is as follows:

$$0 = -\frac{1}{2}q(x^1) + \int_S q(\xi)F(x^1, \xi)dS(\xi),$$

$$0 = -\frac{1}{2}q(x^2) - \int_S q(\xi)F(x^1, \xi)dS(\xi).$$

Combining these equations, we obtain

$$q(x^2) = -q(x^1) = q.$$

(2.57)

Thus, there will be a sink on one interval and a source on the other one and the strengths thereof are equal.

Now consider the point x_0 of the boundary which is not on the intervals under review. We define the effect exerting on this point in the direction $\vec{n} = \{n_1, n_2\}$ by the source and the sink that are located at the points ξ^1, ξ^2 of Intervals 1 and 2 located on a line perpendicular to them (Figure 2.38). If we introduce an LCCS centered at x_0 with the vertical axis parallel to the intervals under review, letting the distance between the intervals $2\varepsilon \to 0$ and ignoring values of the order ε^2, we obtain

$$v_n(x_0) = q\left[F(x_0, \xi^2) - F(x_0, \xi^1)\right] = \frac{2q\varepsilon}{2\pi}\left[\frac{n_1 \cdot (h^2 - \xi^2) - n_2 \cdot 2h\xi}{(h^2 + \xi^2)^2}\right].$$

In this case, the strength $q \to \infty$, as otherwise the point x_0 would not be affected by the point (h, ξ) resulted from the merger of ξ^1, ξ^2, and, therefore, the entire cut as both the points ξ^1, ξ^2 and Intervals 1, 2 are chosen arbitrarily. The value $m = 2\varepsilon q$ is called a dipole moment and, thus, the theorem is proved.

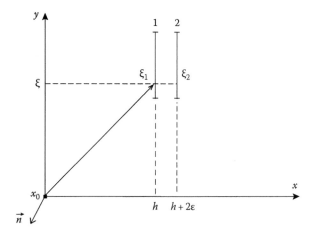

FIGURE 2.38   To the theorem about the effect of the cut.

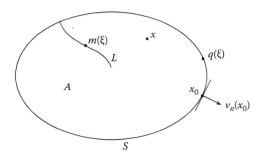

FIGURE 2.39   Area with a cut.

Construct an algorithm to calculate the velocity of the potential flow in areas with cuts.

Let there be given a flow region A bounded by the contour S and having a cut L (Figure 2.39). We place the sources (sinks) with a strength $q(\xi)$ on the contour S and dipoles with a moment $m(\xi)$ on the cut L, where ξ is an arbitrary point of the boundary ($\xi \in S + L$). The values $q(\xi)$ and $m(\xi)$ are unknown in advance. Then the velocity at the inner point $x(x_1, x_2)$ along the direction $\vec{n}$ is defined by the following equation:

$$v_n(x) = \int_S q(\xi)F(x, \xi)dS(\xi) + \int_L m(\xi)D(x, \xi)dL(\xi), \qquad (2.58)$$

where

$$D(x, \xi) = \frac{\left[r_1^2 - r_2^2 \right]\left(n_1 \cos\alpha - n_2 \sin\alpha \right) + 2 r_1 r_2 (n_1 \sin\alpha + n_2 \cos\alpha)}{2\pi \left[r_1^2 + r_2^2 \right]^2}$$

$$r_1 = x_1 - \xi_1$$
$$r_2 = x_2 - \xi_2$$

α is the angle between the dipole axis and the positive direction of the axis OX

When the inner point x tends to the boundary point ($x \to x_0 \in S + L$), we will obtain the following boundary integral equation:

$$v_n(x_0) = -\frac{1}{2}q(x_0) + c \cdot m(x_0) + \int_S q(\xi)F(x_0, \xi)dS(\xi) + \int_L m(\xi)D(x_0, \xi)dL(\xi) \qquad (2.59)$$

where the integrals do not contain the point $x_0 = \xi$, which is taken into account in the first two summands. We will determine the value c later. We divide the boundary $S + L$ into N parts (S into $M < N$ parts). The discrete analog of Equation 2.59 is as follows:

$$v^p = \sum_{k=1}^{M} q^k F^{pk} + \sum_{k=M+1}^{N} m^k D^{pk}, \qquad (2.60)$$

where

$F^{pk} = -1/2$

$D^{pk} = c$

$p = 1, 2, ..., N$ are the numbers of boundary intervals resulted from the segmentation of the area boundary

$v^p = v_n(x_0^p)$ is the normal velocity component in the center of the pth interval (known in the setting the problem)

$q^k = q(\xi^k)$ is the strength of the source (sink) at an arbitrary point of the kth interval

$m^k = m(\xi^k)$ is the dipole moment at the point of the kth interval

$F^{pk} = \int_{\Delta S^k} F(x_0^p, \xi^k)dS(\xi^k), \quad D^{pk} = \int_{\Delta L^k} D(x_0^p, \xi^k)dL(\xi^k)$ are the integrals by the lengths of the kth intervals—$\Delta S^k, \Delta L^k$

Here, we assumed that q^k, m^k are constant on each of the intervals.

By varying p from 1 to N in Equation 2.60, we obtain a system of N linear algebraic equations with N unknown $q^1, q^2, ..., q^M, m^{M+1}, m^{M+2}, ..., m^N$. After determining the unknowns, the air velocity at the inner point x along a given direction $\vec{n}$ is determined by

$$v_n(x) = \sum_{k=1}^{M} q^k F^k + \sum_{k=M+1}^{N} m^k D^k, \qquad (2.61)$$

where

$F^k = \int_{\Delta S^k} F(x, \xi^k)dS(\xi^k),$

$D^k = \int_{\Delta L^k} D(x, \xi^k)dL(\xi^k).$

Consider in detail the calculation of the integrals F^{pk} and D^{pk}. They are analytically calculated, which can be conveniently done in an LCCS of the kth interval (Figure 2.40) centered in the middle of the pth interval—the point x_0^p; we direct the vertical axis along the kth interval (with vertexes $a(a_1, a_2)$, $b(b_1, b_2)$) and the horizontal axis so that the system of coordinates is right-handed.

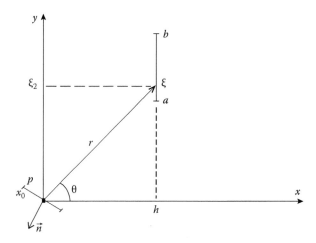

FIGURE 2.40 To the integration of F^{pk}, D^{pk}.

As can be seen from Figure 2.23, $\xi_1^k = h, \xi_2^k = h\,\mathrm{tg}\,\theta, x_1 = x_2 = 0$.
Therefore,

$$F^{pk} = -\frac{1}{2\pi}\left[n_1(\theta_b - \theta_a) - n_2 \ln\frac{r_a}{r_b}\right], \tag{2.62}$$

if $p \neq k$. For $p = k$, $F^{pk} = -1/2$.

$$D^{pk} = \frac{\cos\alpha}{4\pi}\left[2n_1\left(\frac{b_2}{r_b^2} - \frac{a_2}{r_a^2}\right) - \frac{n_2}{h}\left(\frac{2h^2 - r_b^2}{r_b^2} - \frac{2h^2 - r_a^2}{r_a^2}\right)\right] \quad \text{for } h \neq 0, \tag{2.63}$$

$$D^{pk} = \frac{n_1}{2\pi}\cos\alpha\left[\frac{1}{b_2} - \frac{1}{a_2}\right] \quad \text{for } h = 0. \tag{2.64}$$

The case $p = k$ is accounted for in formula (Equation 2.64); α is the angle of the dipole axis to the positive direction of the horizontal axis. Since we do not know in advance the dipole moment, and, accordingly, $\alpha = 0$ or $\alpha = \pi$, then we take a horizontal component of the normal vector to the kth interval as $\cos\alpha$, calculated from its vertexes in the GCCS (in a coordinate system where all vertexes of boundary intervals are preset) and then converted into the LCCS of the kth interval. Moreover, for all the intervals of the cut, the vertexes a and b should be identically oriented (the vertex b of the previous interval must coincide with the vertex a of the succeedent one). This technique is offset by the fact that the dipole moment obtained by solving the system of linear algebraic equations given as Equation 2.60 can be either positive or negative depending on how the direction of the dipole axis has been chosen.

A programmed algorithm for calculating potential flows in areas with cuts was used to solve a test problem related to determination of axial velocity at a slot-type suction in infinite space (Figure 2.41).

As the calculation of the axial velocity (Table 2.3) has shown, the location of the dipoles on infinitely thin hoods provides a significant increase in accuracy in addition to a significantly reduced consumption of computer resources spent to solve a problem (134 boundary elements compared to 200 in case the sources and sinks are located on both sides of the hood). The average relative error of calculation where sources (sinks) are used is 27.5% relative to the calculation using the CMM and 7.7% where dipoles are used.

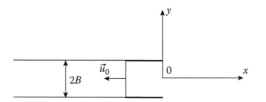

FIGURE 2.41   Slot-type suction unit in infinite space.

TABLE 2.3

Axial Velocities Calculated by the BIE and CMMs

Hoods of a Thickness of 0.0000001 with Sources (Sinks)

x/B	0	0.5	1	1.5	2	2.5	3	4	5	6
v_x/u_0	0.84	0.497	0.277	0.178	0.127	0.0998	0.0795	0.0623	0.045	0.037

Hoods Cuts with Dipoles

v_x/u_0	0.66	0.336	0.191	0.131	0.0986	0.0793	0.0662	0.0498	0.0399	0.0332

Calculations by the CMM

v_x/u_0	0.78	0.378	0.203	0.133	0.097	0.076	0.062	0.045	0.036	0.029

The developed calculation algorithm was used to simulate the airflows near the aspirating inlet of a closed aspirated discharge chamber in a powder material unpacking unit of the Semiluksky refractory plant (Figure 2.42). A container is entered into the chamber through its multileafed doors using a telpher, then the doors are closed. Powder pouring out of the container is monitored and adjusted through the bottom opening (between the door and the grid of the hopper). A local suction is connected to the rear wall of the discharge chamber.

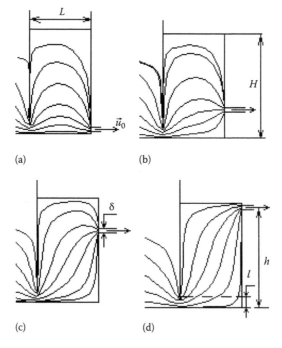

FIGURE 2.42   Flow lines in the shelter at different heights of the suction axis from the floor: (a) $h = 0.1$ m, (b) $h = 0.65$ m, (c) $h = 1.65$ m, and (d) $h = 2.3$ m.

To prevent the removal of dust from the aspiration chamber, a *defensive* velocity, that is, the air counter flow velocity, must be maintained in the bottom opening at the level of at least $v_a = 0.5$ m/s (for dust openings not exposed to significant convective airflows).

Flow lines (Figure 2.42) were built, and the patterns of velocity variations at the entrance to the chamber were identified. The chamber width is $L = 1.46$ m, the height is $H = 2.4$ m, the height of the suction slot is $\delta = 0.1$ m, and the height of entry opening is $l = 0.2$ m.

The velocity of the cross flow throughout the height of the opening is increased: its minimum value at the bottom for $y = 0.025$ m (y is the distance from the floor) and the maximum value at the top for $y = 0.175$ m. The cross flow was at maximum velocity when the local suction is installed at $h = 0.4$ m.

The obtained results of numerical simulation make it possible to define the main parameter of the effective dust containment—the required performance of the local suction unit

$$Q_a = v_0 \cdot F_0, \text{ m}^3/\text{s},$$

where

v_0 is the required air velocity in the local suction, m/s
$F_0 = \delta \cdot L$ is the cross-sectional area of the local suction, m^2

For the velocity $v_0 = 1$ m/s, the minimum velocity in the inlet aperture of the chamber for accepted structural dimensions amounts to

$$v_x\big|_{y=0.025} \equiv v_x^{\min} = 0.335 \text{ m/s}.$$

The required air velocity in the local suction is determined from

$$v_0 = v_0 \frac{v_a}{v_x^{\min}} = 1.5 \text{ m/s}.$$

Therefore, the required performance is

$$Q_a = 0.225 \text{ m}^3/\text{s}.$$

A closed aspiration discharge chamber designed in the Belgorod State Technological University was installed in the powder unpacking department of the periclase-carbonaceous shop floor of the Semiluksky refractory plant.

Note that the offered algorithm for the calculation of flat potential flows near the suction inlets can be applied for spatial problems with a certain adjustment. Obviously, here, the increase in accuracy is even more pronounced as in this case, the integrals F^{pk}, D^{pk} are numerically evaluated, and where only sources (sinks) are used, significant errors occur in the calculation of the mutual influence of the opposite boundary elements.

2.2.3 FLAT FLOWS IN MULTIPLY CONNECTED REGIONS WITH ROTATING CYLINDERS

When calculating the performance of the local suction from machined cylindrical parts required containing dust releases, either the results of special experimental studies [106–109] or analytical data [110] are used for the air velocity field near a cylinder rotating in a viscous fluid. Moreover, [111] shows that an airflow initiated by this rotating cylinder is potential. We superposed this flow on the airflow sucked in by local suction to obtain the desired velocity field and studied the movement of dust particles in it. However, the influence of both the cylinder (in this approach it is permeable) and other pieces of process equipment was not taken into account.

Within the framework of the flat potential fluid flow model, this book describes a method for determining the air velocity field near local suction units for an arbitrary geometry of the air movement area boundaries whose area of effect may comprise rotating cylindrical parts.

Assume that the fluid flow range is bounded by the contour S, on which a normal component of the air velocity $v_n(x)$ is preset. An impermeable cylinder of a radius r rotating at a linear velocity v is inside the range. Generally speaking, there can be any number of such rotating cylinders, but we assume that we have only one cylinder to simplify the essence of the method being described. We place a linear vortex with a circulation $\Gamma = 2\pi r \cdot v$ at the center of the cylinder. To determine the air velocity in the area under review, we place sources (sinks) of a previously unknown strength $q(\xi)$, $\xi \in S$ on the area boundary. The strength $q(\xi)$ is defined assuming that the given limit values of the normal velocity component (impermeability on the walls of the boundary, the known velocity at the entrance of the suction units) are met and a linear vortex is present.

The effect on the inner point of the flow x along the direction $\vec{n}$ exerted by the sources (sinks) and the vortex continuously distributed along the boundary S is expressed by the following integral equation:

$$v_n(x) = \int_S q(\xi)F(x, \xi)dS(\xi) + \Gamma(a) \cdot G(x, a), \tag{2.65}$$

where

$$F(x, \xi) = \frac{n_1(x_1 - \xi_1) + n_2(x_2 - \xi_2)}{2\pi\left[(x_1 - \xi_1)^2 + (x_2 - \xi_2)^2\right]}$$

$$G(x, a) = \frac{n_2(x_1 - a_1) - n_1(x_2 - a_2)}{2\pi\left[(x_1 - a_1)^2 + (x_2 - a_2)^2\right]}$$

$\{n_1, n_2\}$ are the coordinates of the unit vector $\vec{n}$

(x_1, x_2), (ξ_1, ξ_2), (a_1, a_2) are the coordinates of inner point x, boundary point ξ, and the location point of the vortex a, respectively

We divide the contour S into N right-lined intervals. Assuming that the strength of the sources and sinks on each of the intervals is constant, we obtain the discrete analogue of Equation 2.65:

$$v_n(x) = \sum_{k=1}^{N} q(\xi^k) \int_{\Delta S^k} F(x, \xi^k)dS + \Gamma(a) \cdot G(x, a), \tag{2.66}$$

where the integrals are taken by intervals ΔS^k and ξ^k is the arbitrary point of the kth interval.

Letting the inner point x to the boundary one x_0, we obtain the boundary integral equation:

$$v_n(x_0) = -\frac{1}{2}q(x_0) + \sum_{k=1}^{N} q(\xi^k) \int_{\Delta S^k} F(x_0, \xi^k)dS + \Gamma(a) \cdot G(x_0, a), \tag{2.67}$$

where the first summand results from the coincidence of the points x_0 and ξ^k. Under the summation sign in Equation 2.67, there is no summand that corresponds to the case in which x_0 is on the

kth interval. If the center x_0^p of the pth interval is used as x_0, then, if we sort out all the N boundary intervals, we obtain a system of N linear algebraic equations with N unknowns:

$$-\frac{1}{2}q(x_0^p) + \sum_{\substack{k=1, \\ k \neq p}}^{N} q(\xi^k) \int_{\Delta S^k} F(x_0^p, \xi^k)dS = v_n(x_0^p) - \Gamma(a) \cdot G(x_0^p, a), \qquad (2.68)$$

where $p = 1, 2, ..., N$.

When we have solved the system of Equations 2.68 using one of the numerical methods, for example, the Gauss method with pivoting, and determined the unknowns $q(\xi^1), q(\xi^2), ..., q(\xi^N)$, we can find the desired air velocity at x along this direction $\vec{n}$ by Equation 2.66.

Consider the example of an airflow near a 0.2 m wide suction slot, the area of effect of which comprises a cylinder with a radius of 1 m rotating counterclockwise at a velocity of 1 m/s (Figure 2.43). Accordingly, the vortex located in the center of the cylinder has a circulation $\Gamma = 2\pi$. The distance from the cylinder axis to the suction is 1.5 m. The parameters of this flow are similar to the aspiration scheme of a roll-turning machine where air is sucked from a chip-catching channel.

The flow boundary was divided into 110 right-lined intervals. The circumference is evenly sampled into 40 intervals. The suction inlet and the impenetrable part of the frame are divided nonuniformly. The division is denser at the edges of the suction than near the center. The law is as follows: $0.001i$, where $i = 1, 2, 4, 8, ...$ until the end of the boundary element crosses the middle of the interval. A part of this boundary element is cut off; the midpoint of the interval to be divided is taken as its terminal point. The interval on the other end is divided likewise. The law for dividing the rays defining the frame is the same: $0.001i$, where $i = 1, 2, 4, 8, ...$ until the terminal points are located from the suction center at a distance greater than 50 gauges (a gauge is the suction width).

If the cylinder is immovable, the flow pattern is symmetrical (Figure 2.43a). If there is no suction, a circulatory airflow is around the cylinder (Figure 2.43b). If the suction is connected and the velocity in it increases, the area of the circulatory flow decreases (Figure 2.43c and d) and even

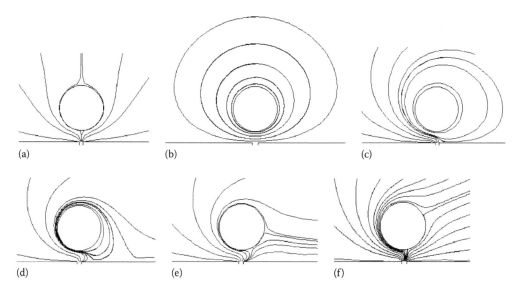

(a)   (b)   (c)

(d)   (e)   (f)

FIGURE 2.43 Flow lines in the flow around a cylinder located in the area of effect of a slot-type suction: (a) $\Gamma = 0$; (b) $\Gamma = 2\pi$, $v_0 = 0$; (c) $\Gamma = 2\pi$, $v_0 = 4$ m/s; (d) $\Gamma = 2\pi$, $v_0 = 8$ m/s; (e) $\Gamma = 2\pi$, $v_0 = 10$ m/s; and (f) $\Gamma = 2\pi$, $v_0 = 15$ m/s.

disappears at a suction velocity close to 10 m/s. Here, there is only one critical point. With further increase in the velocity v_0, two critical points are found and the distance between them increases (Figure 2.43e and f).

When the velocity v_0 increases infinitely, then, obviously, the flow pattern will be as shown in Figure 2.43a as the effect of the cylinder rotation becomes vanishingly small.

In the velocity field found, we investigated trajectories of dust particles in the Stokes regime of flowing by an air current, as described by the equation:

$$\frac{\pi d_e^3}{6}\rho_1 \frac{d\vec{v}_1}{dt} = \frac{\pi d_e^3}{6}\rho_1\vec{g} - 3\pi\chi\mu d_e(\vec{v}_1 - \vec{v}), \tag{2.69}$$

where

d_e is the equivalent diameter of the particle, m
$\rho_1 = 2600$ kg/m³ is the particle density
v_1 is the particle velocity, m/s
$\mu = 1.78 \cdot 10^{-5}$ $Pa \cdot s$ is the dynamic viscosity of the air
$\chi = 1$ is the particle dynamic shape coefficient (particles were assumed to be spherical)
$g = 9.81$ m/s² is the acceleration of gravity
v is the air velocity, m/s

We built the flight path of the dust particles by reducing Equation 2.69 to a system of ODEs and solving it numerically by the Runge–Kutta method, determining at each step the air velocity according to the above algorithm.

The built trajectories of dust particles (Figure 2.44) formed at the contact of the cutter and the work piece being processed show that it is more difficult to trap finest dust particles. This agrees with experimental data [109]. Large fractions of dust particles reach the dust collecting bag due to inertia. But fine fractions are transported by airflow sucked away. Their trajectories are close to the flow lines (Figure 2.44b), and they are trapped only after a considerable period of time or settle on the process equipment components. Therefore, the velocity in the suction unit and, therefore, the necessary volume of aspiration will be chosen under the condition that there will be no circulatory area of airflow around the cylinder (Figure 2.43e).

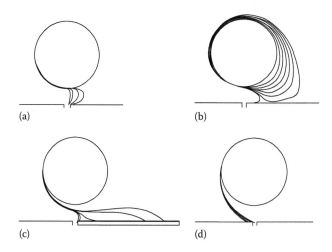

(a)   (b)   (c)   (d)

FIGURE 2.44 Trajectories of dust particles: (a) $d = 10$ μm, v_0 [m/s] = 25, 15, 10, 9 (from left to right); (b) $d = 10$ μm, v_0 [m/s] = 8; (c) $d = 50$ μm, v_0 [m/s] = 15, 8, 4, 1, 0 (from left to right); and (d) $d = 100$ μm, v_0 [m/s] = 0, 1, 4, 8 (from left to right).

When calculating the trajectories of dust particles, their initial velocity of escape from the cutting zone was considered to be zero, which is quite acceptable for the given range of particle sizes and the cutting speed. As shown by the numerical experiment, the initial velocity of dust particles attenuates rapidly and has no significant effect on their flight. For coarse particles, the influence of their initial velocity of escape can be significant.

The travel direction of the chips and dust from the cutter should be taken into account in choosing the necessary volumes of aspiration for different cutting processes. As stated in [106], this direction depends on the physical and chemical properties of the material being processed, the type of processing, cutting mode, geometrical parameters of the cutting tool. With data on the direction and velocity of the dust particles and chips, their size, density, coefficient of drag [106,108,110]; adjusting the equation of motion given in Equation 2.69; and setting the appropriate initial conditions for dust and chip travels, this method can be used to determine the necessary volume of aspiration from various lathes, drilling, grinding, milling, woodworking, and other machines with rotating cylindrical parts.

2.2.4 THREE-DIMENSIONAL FLOWS IN MULTIPLY CONNECTED REGIONS WITHOUT SINGULARITIES

2.2.4.1 Derivation of the Basic Relations and the Construction of Main Stages of Solution

Assume that a homogeneous region A is bounded by the surface S, where the velocity $v_n(\xi)$ of airflow toward the outward normal $\vec{n}(\xi)$ at an arbitrary point ξ of the boundary is predetermined. It is necessary to determine the air velocity $v_n(x)$ at an arbitrary inner point x along the given direction $\vec{n}(x)$.

The procedure for the solution of a spatial problem is similar to the solution of a flat problem. At first, we define the fundamental solution of the Laplace equation:

$$\frac{\partial^2 \varphi}{\partial x^2} + \frac{\partial^2 \varphi}{\partial y^2} + \frac{\partial^2 \varphi}{\partial z^2} = 0, \tag{2.70}$$

which is as follows in a spherical coordinate system:

$$\frac{1}{r^2}\frac{\partial}{\partial r}\left(r^2\frac{\partial \varphi}{\partial r}\right) + \frac{1}{r^2 \sin\theta}\frac{\partial}{\partial \theta}\left(\sin\theta\frac{\partial \varphi}{\partial \theta}\right) + \frac{1}{r^2 \sin^2\theta}\frac{\partial^2 \varphi}{\partial \psi^2} = 0, \tag{2.71}$$

where r, θ, ψ are the spherical coordinates.

In the case of spherical symmetry, that is, for $\theta = const$ and $\psi = const$, the potential φ will only depend on r. Here, the solution of Equation 2.71 will be determined from an ODE

$$\frac{d}{dr}\left(r^2\frac{d\varphi}{dr}\right) = 0, \tag{2.72}$$

which having been twice integrated becomes

$$\varphi = -\frac{1}{r}C_1 + C_2.$$

Assuming that $C_1 = -1$, $C_2 = 0$, we obtain the fundamental solution of the Laplace equation in space:

$$\varphi = \frac{1}{r}.$$

Accurate to a constant, the fundamental solution coincides with the flow potential at the point x resulting from the action of the source (sink) with a strength $q(\xi)$ at the point ξ:

$$\varphi(x) = \frac{q(\xi)}{4\pi r}, \tag{2.73}$$

where $r = \sqrt{(x_1 - \xi_1)^2 + (x_2 - \xi_2)^2 + (x_3 - \xi_3)^2}$ is the distance between the points $x(x_1, x_2, x_3)$ and $\xi(\xi_1, \xi_2, \xi_3)$ set in a rectangular Cartesian coordinate system.

In order to determine the air velocity along the direction defined by unit vector $\vec{n}\{n_1, n_2, n_3\}$, it is necessary to differentiate φ along this direction:

$$v_n(x) = \frac{\partial \varphi}{\partial n} = \frac{\partial \varphi}{\partial x_1}n_1 + \frac{\partial \varphi}{\partial x_2}n_2 + \frac{\partial \varphi}{\partial x_3}n_3 = -\frac{(x_1 - \xi_1)n_1 + (x_2 - \xi_2)n_2 + (x_3 - \xi_3)n_3}{4\pi r^3}q(\xi)$$

or, introducing the denotation

$$F(x, \xi) = -\frac{(x_1 - \xi_1)n_1 + (x_2 - \xi_2)n_2 + (x_3 - \xi_3)n_3}{4\pi r^3}, \tag{2.74}$$

we obtain a formula to calculate the air velocity at the point x along a predetermined direction $\vec{n}$ caused by the action of a source (or sink) with a strength $q(\xi)$ at the point ξ as follows:

$$v_n(x) = F(x, \xi)q(\xi). \tag{2.75}$$

Assume that sources (sinks) of a strength $q(\xi)$ are continuously distributed on the boundary S of the flow area A, where ξ is an arbitrary point S. At each point of the boundary the strength $q(\xi)$ can have different values, and this distribution must satisfy the limit values for the velocity.

After integrating the effects of all the sources (sinks) with strength $q(\xi)$ unknown in advance, we obtain the required air velocity

$$v_n(x) = \iint_S F(x, \xi)q(\xi)dS(\xi). \tag{2.76}$$

Note that ξ is the variable of integration in this integral.

Letting the inner point x to the boundary one x_0, we obtain the boundary integral equation

$$v_n(x_0) = \iint_S F(x_0, \xi)q(\xi)dS(\xi), \tag{2.77}$$

where $F(x_0, \xi) \to \infty$ for $x_0 \to \xi$. To get rid of this singularity, we encircle the point ξ with a semisphere of a radius ε. Then, the integral in Equation 2.77 can be expressed as the sum of two integrals:

$$\iint_S = \iint_{S - S_\varepsilon} + \iint_{S_\varepsilon},$$

where the first integral is taken along the surface S without this semisphere and the second one along the surface S_ε of the semisphere.

Letting $\varepsilon \to 0$ and considering that the strength $q(\xi)$ of the sources and sinks is constant on the semisphere, we have

$$\iint_{S_\varepsilon} F(x, \xi)q(\xi)dS = -\iint_{S_\varepsilon} \frac{|\vec{r}| \cdot |\vec{n}| \cdot \cos\pi}{4\pi r^3} q(\xi)dS = \begin{Bmatrix} |\vec{r}| = \varepsilon \\ |\vec{n}| = 1 \end{Bmatrix} = \frac{q(\xi)}{4\pi\varepsilon^2} \iint_{S_\varepsilon} dS = \frac{q(\xi)}{4\pi\varepsilon^2} 2\pi\varepsilon^2 = \frac{1}{2}q(\xi).$$

Thus, Equation 2.77 can be written as

$$v_n(x_0) = \frac{1}{2}q(x_0) + \iint_S F(x_0, \xi)q(\xi)dS(\xi), \tag{2.78}$$

where the integrand $F(x_0, \xi)$ has no singularity for $x_0 = \xi$, since this case is eliminated and taken into account in the first summand.

Divide the boundary surface S into N plane triangles; in each of them the strength of the sources and sinks is assumed to be constant. Assume the point x_0^p is the center of gravity of the pth triangle and ξ^k is an arbitrary point of the kth triangle. Then the discrete analog of Equation 2.78 is as follows:

$$v_n(x_0^p) = \frac{1}{2}q^p + \sum_{\substack{k=1, \\ k\neq p}}^{N} q^k \iint_{\Delta S^k} F(x_0^p, \xi^k)dS(\xi^k), \tag{2.79}$$

where q^p, q^k is the strength of the sources (sinks) of the pth and kth triangles, respectively, and surface integrals are taken over the triangles ΔS^k. By varying p from 1 to N, we obtain a system of linear algebraic equations with unknown $q^1, q^2, ..., q^N$. If we solve it we can calculate the air velocity at the inner point x of the area A along a given direction by the formula

$$v_n(x) = \sum_{k=1}^{N} q^k \iint_{\Delta S^k} F(x, \xi^k)dS(\xi^k). \tag{2.80}$$

Thus, the main steps to calculate the air velocity include

1. Discretization of the boundary surface of the flow area into boundary elements
2. Solution of the system of linear equations given in Equation 2.79 and the definition of the unknown strengths of sources (sinks) $q^1, q^2, ..., q^N$
3. Determination of the air velocity at a given point inside the flow area by Equation 2.80

2.2.4.2 Discretization of the Boundary Surface Local Coordinates

The boundary is divided into plane triangles. The coordinates of the triangle vertices $r(r_1, r_2, r_3)$; $s(s_1, s_2, s_3)$; $t(t_1, t_2, t_3)$ are defined in the selected coordinate system, that is, the GCCS. The vertexes are chosen in such a way that the shortest rotation of the vector $\vec{rs}$ to the vector $\vec{rt}$ is clockwise when viewed from the end of the vector of the outward normal to this boundary triangle. This is to facilitate the calculation of the coordinates of the outer normal on these vectors, as will be explained in more detail in the following. There can be other options, but the vertexes must be preset for all the triangles according to the same rule, and, therefore, the calculation algorithm must be adjusted.

All the triangular elements must be numbered. For example, $r^k(r_1^k, r_2^k, r_3^k)$; $s^k(s_1^k, s_2^k, s_3^k)$; $t^k(t_1^k, t_2^k, t_3^k)$ are the coordinates of the vertexes of the kth triangle.

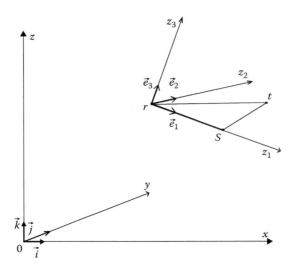

FIGURE 2.45 Global and local systems of coordinates.

We construct a local rectangular Cartesian coordinate system on a triangle with the vertexes r, s, and t as follows: the center of the coordinate system coincides with the vertex r; the horizontal axis z_1 is directed along the vector $\vec{rs}$; the vertical axis z_2 is in the plane of the triangle perpendicular to z_1 and is directed in such a way that if viewed from the applicate axis z_3, perpendicular to the plane of the triangle rst, the shortest rotation of the axis z_1 to z_2 is counterclockwise (Figure 2.45). Then the unitary vector of the horizontal axis given,

$$\left|\vec{rs}\right| = \sqrt{(s_1 - r_1)^2 + (s_2 - r_2)^2 + (s_3 - r_3)^2},$$

is calculated by the formula

$$\vec{e}_1 = \{e_{11}, e_{12}, e_{13}\} = \frac{\vec{rs}}{\left|\vec{rs}\right|} = \left\{\frac{s_1 - r_1}{\left|\vec{rs}\right|}, \frac{s_2 - r_2}{\left|\vec{rs}\right|}, \frac{s_3 - r_3}{\left|\vec{rs}\right|}\right\}. \tag{2.81}$$

The unitary vector of the applicate axis $\vec{e}_3$ is determined from the vectorial vector of $\vec{rs}$ and $\vec{rt}$:

$$\vec{e}_3 = \frac{\vec{rs} \times \vec{rt}}{\left|\vec{rs} \times \vec{rt}\right|}.$$

As $\vec{rs} \times \vec{rt} = \begin{vmatrix} \vec{i} & \vec{j} & \vec{k} \\ s_1 - r_1 & s_2 - r_2 & s_3 - r_3 \\ t_1 - r_1 & t_2 - r_2 & t_3 - r_3 \end{vmatrix}$, then the coordinates of the vector $\vec{e}_3\{e_{31}, e_{32}, e_{33}\}$ are calculated by the formulas

$$e_{31} = \frac{(s_2 - r_2)(t_3 - r_3) - (s_3 - r_3)(t_2 - r_2)}{\left| \vec{rs} \times \vec{rt} \right|},$$

$$e_{32} = \frac{(s_3 - r_3)(t_1 - r_1) - (s_1 - r_1)(t_3 - r_3)}{\left| \vec{rs} \times \vec{rt} \right|},$$ (2.82)

$$e_{33} = \frac{(s_1 - r_1)(t_2 - r_2) - (s_2 - r_2)(t_1 - r_1)}{\left| \vec{rs} \times \vec{rt} \right|},$$

where

$$\left| \vec{rs} \times \vec{rt} \right| = \sqrt{\left(E_{31} \right)^2 + \left(E_{32} \right)^2 + \left(E_{33} \right)^2}, \; E_{31} = \left[(s_2 - r_2)(t_3 - r_3) - (s_3 - r_3)(t_2 - r_2) \right]$$

$$E_{32} = \left[(t_1 - r_1)(s_3 - r_3) - (s_1 - r_1)(t_3 - r_3) \right], \; E_{33} = \left[(s_1 - r_1)(t_2 - r_2) - (s_2 - r_2)(t_1 - r_1) \right]$$

The unit vector of the vertical axis $\vec{e}_2$ is defined by the vectorial product

$$\vec{e}_2 = \vec{e}_1 \times \vec{e}_3 = \begin{vmatrix} \vec{i} & \vec{j} & \vec{k} \\ e_{11} & e_{12} & e_{13} \\ e_{31} & e_{32} & e_{33} \end{vmatrix} = \left\{ e_{21}, e_{22}, e_{23} \right\},$$ (2.83)

where

$$e_{21} = e_{12}e_{33} - e_{13}e_{32}, \quad e_{22} = e_{31}e_{13} - e_{11}e_{33}, \quad e_{23} = e_{11}e_{32} - e_{12}e_{31}.$$

Thus, the vectors $\vec{e}_1, \vec{e}_2, \vec{e}_3$ can be decomposed on the unitary vectors of the GCCS $\vec{i}, \vec{j}, \vec{k}$ as follows:

$$\vec{e}_1 = e_{11}\vec{i} + e_{12}\vec{j} + e_{13}\vec{k},$$

$$\vec{e}_2 = e_{21}\vec{i} + e_{22}\vec{j} + e_{23}\vec{k},$$ (2.84)

$$\vec{e}_3 = e_{31}\vec{i} + e_{32}\vec{j} + e_{33}\vec{k},$$

or in matrix form

$$\begin{pmatrix} \vec{e}_1 \\ \vec{e}_2 \\ \vec{e}_3 \end{pmatrix} = \begin{pmatrix} e_{11} & e_{12} & e_{13} \\ e_{21} & e_{22} & e_{23} \\ e_{31} & e_{32} & e_{33} \end{pmatrix} \begin{pmatrix} \vec{i} \\ \vec{j} \\ \vec{k} \end{pmatrix}.$$ (2.85)

Since we have a rectangular Cartesian coordinate system, the inverse transform is as follows:

$$\begin{pmatrix} \vec{i} \\ \vec{j} \\ \vec{k} \end{pmatrix} = \begin{pmatrix} e_{11} & e_{21} & e_{31} \\ ej_{12} & e_{22} & e_{32} \\ e_{13} & e_{23} & e_{33} \end{pmatrix} \begin{pmatrix} \vec{e}_1 \\ \vec{e}_2 \\ \vec{e}_3 \end{pmatrix}.$$ (2.86)

Assume that the point $x(x_1, x_2, x_3)$ is set in the GCCS. We define its coordinates in the LCCS. Consider the vector starting at r and ending at x, having the coordinates $\vec{rx} = \{x_1 - r_1, x_2 - r_2, x_3 - r_3\}$, that is,

$$\vec{rx} = (x_1 - r_1)\vec{i} + (x_2 - r_2)\vec{j} + (x_3 - r_3)\vec{k} = (x_1 - r_1)(e_{11}\vec{e}_1 + e_{21}\vec{e}_2 + e_{31}\vec{e}_3)$$

$$+ (x_2 - r_2)(e_{12}\vec{e}_1 + e_{22}\vec{e}_2 + e_{32}\vec{e}_3) + (x_3 - r_3)(e_{13}\vec{e}_1 + e_{23}\vec{e}_2 + e_{33}\vec{e}_3)$$

$$= \vec{e}_1\left[(x_1 - r_1)e_{11} + (x_2 - r_2)e_{12} + (x_3 - r_3)e_{13}\right]$$

$$+ \vec{e}_2\left[(x_1 - r_1)e_{21} + (x_2 - r_2)e_{22} + (x_3 - r_3)e_{23}\right]$$

$$+ \vec{e}_3\left[(x_1 - r_1)e_{31} + (x_2 - r_2)e_{32} + (x_3 - r_3)e_{33}\right] = z_1\vec{e}_1 + z_2\vec{e}_2 + z_3\vec{e}_3.$$

Thus, the coordinates of the vector $\vec{rx}$ in the LCCS are

$$z_1 = (x_1 - r_1)e_{11} + (x_2 - r_2)e_{12} + (x_3 - r_3)e_{13},$$

$$z_2 = (x_1 - r_1)e_{21} + (x_2 - r_2)e_{22} + (x_3 - r_3)e_{32}, \qquad (2.87)$$

$$z_3 = (x_1 - r_1)e_{31} + (x_2 - r_2)e_{32} + (x_3 - r_3)e_{33},$$

or in matrix form

$$\begin{pmatrix} z_1 \\ z_2 \\ z_3 \end{pmatrix} = \begin{pmatrix} e_{11} & e_{12} & e_{13} \\ e_{21} & e_{22} & e_{23} \\ e_{31} & e_{32} & e_{33} \end{pmatrix} \begin{pmatrix} x_1 - r_1 \\ x_2 - r_2 \\ x_3 - r_3 \end{pmatrix}. \qquad (2.88)$$

As $\vec{rx}$ is the radius vector of the point x in the LCCS, then we use Equation 2.87 to determine the coordinates of the point x in this coordinate system.

2.2.4.3   Determination of the Boundary Strengths of Sources and Sinks

To determine the strength of the sources (sinks) we need, at first, to form a matrix

$$F = \left(F^{pk}\right) = \left(\iint\limits_{\Delta S^k} F(x_0^p, \xi^k) dS(\xi^k)\right),$$

consisting of the coefficients for the unknowns q^k in the system of linear algebraic equations given in Equation 2.79.

The integrals F^{pk} can be found only numerically. Assume the coordinates of the vertexes of the pth triangle $r^p(r_1^p, r_2^p, r_3^p)$; $s^p(s_1^p, s_2^p, s_3^p)$; $t^p(t_1^p, t_2^p, t_3^p)$ are predetermined in the GCCS. Their coordinates in the LCCS of the pth triangle can be calculated using Equation 2.87. We denote them as follows: $r^{pp}\left(r_1^{pp}, r_2^{pp}, r_3^{pp}\right)$; $s^{pp}\left(s_1^{pp}, s_2^{pp}, s_3^{pp}\right)$; $t^{pp}\left(t_1^{pp}, t_2^{pp}, t_3^{pp}\right)$. The coordinates of the center of gravity of the pth triangle in the LCCS of this triangle are

$$c^{pp} = \left(\frac{1}{3}\left(s_1^{pp} + t_1^{pp}\right), \frac{1}{3}t_2^{pp}, 0\right) = \left(c_1^{pp}, c_2^{pp}, c_3^{pp}\right). \tag{2.89}$$

The coordinates of the center of gravity of pth triangle in the GCCS, as follows from Equations 2.88 and 2.89, are calculated by

$$\begin{pmatrix} c_1^p \\ c_2^p \\ c_3^p \end{pmatrix} = \begin{pmatrix} r_1^p \\ r_2^p \\ r_3^p \end{pmatrix} + \begin{pmatrix} e_{11} & e_{21} & e_{31} \\ e_{12} & e_{22} & e_{32} \\ e_{13} & e_{23} & e_{33} \end{pmatrix} \begin{pmatrix} \frac{1}{3}\left(s_1^{pp} + t_1^{pp}\right) \\ \frac{1}{3}t_2^{pp} \\ 0 \end{pmatrix}. \tag{2.90}$$

The coordinates of the center of gravity of the pth triangle in the LCCS of the kth triangle are calculated using Equation 2.88 from the expression

$$\begin{pmatrix} c_1^{pk} \\ c_2^{pk} \\ c_3^{pk} \end{pmatrix} = \begin{pmatrix} e_{11} & e_{12} & e_{13} \\ e_{21} & e_{22} & e_{23} \\ e_{31} & e_{32} & e_{33} \end{pmatrix} \begin{pmatrix} c_1^p - r_1^k \\ c_2^p - r_2^k \\ c_3^p - r_3^k \end{pmatrix}, \tag{2.91}$$

where $\left(r_1^k, r_2^k, r_3^k\right)$ are the coordinates of the vertex r of the kth triangle predetermined in the GCCS.

The coordinates of the unit vector of the outward normal to the pth triangle are calculated from Equation 2.82 given that $\vec{n}^p = -\vec{e}_3^p$. In the LCCS of the kth triangle, the coordinates $\vec{n}^p$ are transformed to

$$\begin{pmatrix} n_1^{pk} \\ n_2^{pk} \\ n_3^{pk} \end{pmatrix} = \begin{pmatrix} e_{11} & e_{12} & e_{13} \\ e_{21} & e_{22} & e_{23} \\ e_{31} & e_{32} & e_{33} \end{pmatrix} \begin{pmatrix} n_1^p \\ n_2^p \\ n_3^p \end{pmatrix}. \tag{2.92}$$

The integral $F^{pk} = \iint_{\Delta S^k} F\left(x_0^p, \xi^k\right) dS(\xi^k)$ is calculated from the kth triangle. Therefore, given the above, we have

$$F^{pk} = -\iint_{\Delta S^k} \frac{\left(c_1^{pk} - \xi_1^k\right)n_1^{pk} + \left(c_2^{pk} - \xi_2^k\right)n_2^{pk} + \left(c_3^{pk} - \xi_3^k\right)n_3^{pk}}{4\pi\left[\left(c_1^{pk} - \xi_1^k\right)^2 + \left(c_2^{pk} - \xi_2^k\right)^2 + \left(c_3^{pk} - \xi_3^k\right)^2\right]^{3/2}} dS(\xi^k), \tag{2.93}$$

where $\xi_1^k = z_1^k$, $\xi_2^k = z_2^k$, $\xi_3^k = 0$ (Figure 2.46).

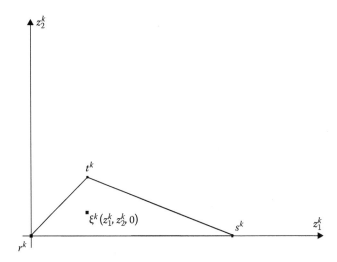

FIGURE 2.46   Integration domain of the *k*th triangle.

Reducing the integral in Equation 2.93 to an iterated one, we obtain

$$F^{pk} = \int_0^{t_2^k} dz_2^k \int_{\frac{t_1^k}{t_2^k}z_2^k}^{\frac{s_1^k-t_1^k}{s_2^k-t_2^k}\left(z_2^k-t_2^k\right)+t_1^k} F(c^{pk}, \xi^k)dz_1^k. \tag{2.94}$$

The integral in Equation 2.94 can be calculated by the Gauss quadrature formulas

$$\int_{-1}^1 dx_1 \int_{-1}^1 f(x_1, x_2)dx_2 = \sum_{i=1}^N \sum_{j=1}^N A_i A_j f(x_i, x_j), \tag{2.95}$$

where
A_i are the weighting factors
x_i are the nodes of the quadrature formula determined from tables (e.g., [101,102])
N is the number of nodes

For $N = 5$, the values x_i and Ai are shown in Table 2.4.

TABLE 2.4
Gaussian Weight Factors and Nodes

I	x_i	A_i
1	0.090617984593866399	0.23692688505618908751
2	0.538493101568309103	0.47862867049936645690
3	0.000000000000000000	0.56888888888888888889
4	−0.090617984593866399	0.23692688505618908751
5	−0.538493101568309103	0.47862867049936645690

The integral in Equation 2.94 is the integral of the form $\int_a^b dy \int_{\varphi(y)}^{\psi(y)} f(x, y)dx$, which can be reduced to the form of the integral in Equation 2.95 as follows:

$$\int_a^b dy \int_{\varphi(y)}^{\psi(y)} f(x,y)dx = \left\{ \begin{array}{l} x = \dfrac{\psi(y)-\varphi(y)}{2}x_1 + \dfrac{\psi(y)+\varphi(y)}{2} \\[3mm] dx = \dfrac{\psi(y)-\varphi(y)}{2}dx_1 \end{array} \right\} =$$

$$\int_a^b dy \int_{-1}^1 f\left(\frac{\psi(y)-\varphi(y)}{2}x_1 + \frac{\psi(y)+\varphi(y)}{2}, y \right) \frac{\psi(y)-\varphi(y)}{2}dx_1 = \left\{ \begin{array}{l} y = \dfrac{b-a}{2}y_1 + \dfrac{b+a}{2} \\[3mm] dy = \dfrac{b-a}{2}dy_1 \end{array} \right\}$$

$$= \int_{-1}^1 dy_1 \int_{-1}^1 f\left(\frac{\psi\left(\dfrac{b-a}{2}y_1+\dfrac{b+a}{2}\right)-\varphi\left(\dfrac{b-a}{2}y_1+\dfrac{b+a}{2}\right)}{2}x_1 \right.$$

$$+ \frac{\psi\left(\dfrac{b-a}{2}y_1+\dfrac{b+a}{2}\right)+\varphi\left(\dfrac{b-a}{2}y_1+\dfrac{b+a}{2}\right)}{2}, \left. \frac{b-a}{2}y_1+\frac{b+a}{2} \right)$$

$$\times \frac{\psi\left(\dfrac{b-a}{2}y_1+\dfrac{b+a}{2}\right)-\varphi\left(\dfrac{b-a}{2}y_1+\dfrac{b+a}{2}\right)}{2}dx_1. \tag{2.96}$$

Thus, the algorithm of the formation of the matrix elements (F^{pk}) will be as follows:

1. The initial number of the pth interval is set: $p = 1$.
2. The coordinates of the vertexes of the pth triangle are predetermined in its LCCS. To this end, first calculate the coordinates of the unitary vectors of the pth triangle LCCS (formulas in Equations 2.81 through 2.83), then find the coordinates of the vertexes s and t of the pth triangle in its LCCS Equation 2.87.
3. The coordinates of the center of gravity of the pth triangle are predetermined in its LCCS Equation 2.89.
4. The coordinates of the center of gravity of the pth triangle are predetermined in the GCCS Equation 2.90.
5. The initial number of the kth triangle is predetermined.
6. If $p = k$, then $F^{pk} = 0.5$ and go to Step 12.
7. If $p \ne k$, then calculate the coordinates of the unitary vectors of the kth triangle LCCS and, afterward, the coordinates of the center of gravity of the pth triangle in the coordinate system of the kth triangle (using Equation 2.91).
8. The coordinates of the vertexes s and t of the kth triangle are predetermined in its LCCS Equation 2.87.
9. The coordinates of the unit vector of the outward normal to the pth triangle are predetermined in the GCCS: $\vec{n}^p = -\vec{e}_3^p$.

10. The coordinates of the unit vector of the outward normal to the pth triangle are determined in the LCCS of the kth triangle using Equation 2.92.
11. Calculate the matrix element F^{pk} (using Equations 2.94 through 2.96).
12. Step for k: $k = k + 1$ and go to Step 7. Calculate until $k \leq N$.
13. Step for p: $p = p + 1$ and go to Step 2. Calculate until $p \leq N$.

After the matrix elements (F^{pk}) have been formed, we solve the system of linear algebraic equations:

$$\begin{cases} F^{11}q^1 + F^{12}q^2 + \cdots + F^{1N}q^N = v^1, \\ F^{21}q^1 + F^{22}q^2 + \cdots + F^{2N}q^N = v^2, \\ \vdots \\ F^{N1}q^1 + F^{N2}q^2 + \cdots + F^{NN}q^N = v^N, \end{cases}$$

whence we determine the unknown strengths of the sources (sinks) q^1, q^2, ..., q^N distributed on N-boundary triangles. The boundary values of the normal velocity component v^1, v^2, ..., v^N are given in the statement of the problem.

The following algorithm is used to determine the velocity in the inner flow region along this direction:

1. Preset the coordinates of the point x and the direction $\vec{n}$ in which the projection of the velocity is determined.
2. The initial number $k = 1$ of the boundary triangle is predetermined.
3. The coordinates of the vertexes s and t of the kth triangle and the point x are predetermined in the LCCS of the kth triangle Equation 2.87.
4. The coordinates of the direction vector $\vec{n}$ are determined in the LCCS of the kth triangle Equation 2.92.
5. Calculate the integral F^{pk} (using Equations 2.94 through 2.96).
6. Step for k: $k = k + 1$ and go to Step 3. Calculate until $k \leq N$.
7. Calculate the required value of velocity $v_n(x) = \sum_{k=1}^{n} F^{pk}q^k$.

To determine the value of the velocity at a given point, we must calculate the air velocity components along the coordinate axes, and then compute $v = \sqrt{v_x^2 + v_y^2 + v_z^2}$.

2.2.4.4   Test Problem: Calculation of the Axial Air Velocity at Rectangular Suction Inlet Built in a Flat Infinite Wall

Define the air velocity on the axis of a flow sinking down to a rectangular suction inlet of a size $2A \times 2B$ built in a flat infinite wall. The suction velocity is assumed to be the same and equal to $u_0 = 1$ across the suction rectangle with sides $A = 1$ m, $B = 0.5$ m.

We used the flow superposition method (see Equation 2.16) and the BIE method in the calculation. We discretize the suction rectangle into a set of N plane triangles (Figure 2.47). There is no point of dividing the remaining boundary of the flow area into boundary elements, since $F(x, \xi) = 0$ (see Equation 2.74), as the boundary triangles are in one plane; the normal velocity is zero and as it follows from Equation 2.79, the strength of the sources is $q(x_o^p) = 0$ everywhere on the impermeable wall.

The calculation of the axial velocity (Table 2.5) shows that even for $N = 16$ there is a fairly good agreement of the results obtained.

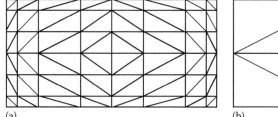
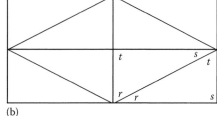

(a)   (b)

FIGURE 2.47   Discretization of the suction rectangle into N plane triangles: (a) $N = 96$ and (b) $N = 8$.

TABLE 2.5
Comparison of Calculations by the Flow Superposition Method and the BIE Method

Distance from the Suction z, m

	0.2	1.0	1.8	2.6	3.4	4.2	5
Velocity Calculated by the Flow Superposition Method, m/s							
	0.85941	0.43591	0.23511	0.14048	0.09139	0.06350	0.04641
N	**Velocity Calculated by the BIE Method, m/s**						
2	0.86156	0.20489	0.08298	0.04319	0.02614	0.01743	0.01242
8	0.87719	0.43590	0.23511	0.14048	0.09139	0.06350	0.04641
16	0.85743	0.43591	0.23511	0.14048	0.09139	0.06350	0.04641
32	0.85918	0.43591	0.23511	0.14048	0.09139	0.06350	0.04641
48	0.85852	0.43591	0.23511	0.14048	0.09139	0.06350	0.04641
96	0.85932	0.43591	0.23511	0.14048	0.09139	0.06350	0.04641

2.2.4.5   Calculation of the Air Flow Sucked by a Rectangular Suction That Flows around a Cylinder (Suction from a Roll-Turning Machine)

Assume the air moves to a rectangular inlet of size 400×300 mm (at an average velocity in the inlet $u_0 = 1$ m/s) built in a flat infinite wall. The absorption spectrum of the rectangle contains a cylinder of radius R, the axis of which is projected onto the axis of symmetry of the rectangle (along its longer side). The distance from the cylinder axis to the suction plane is 670 mm (Figure 2.48).

The described flow area corresponds to the model of a 1A-825–type roll-turning machine as to the assumption that the frame is flat and infinite. Air is sucked from the chip-catching channel (a rectangle in Figure 2.48). No other parts of the machine, other than the roller (cylinder), are taken into account. Now, determine the velocity field of the airflow around the roller and sucked from the chip catcher.

The flow area boundary was discretized by a set of 296 plane triangles. The suction area and its adjacent area are sampled the most densely (Figure 2.49).

We calculated the values of the air velocity along the cylinder surface (at a distance of 5 mm, 55 mm, and 105 mm) for different sections ($x = 0$ mm, $x = 200$ mm, and $x = 400$ mm) and the radii of the roller ($R = 300$ mm, $R = 375$ mm, $R = 450$ mm, and $R = 525$ mm). The calculation was made from the angle $\varphi = 0$ that corresponds to the lowest point of the cylinder until $\varphi = \pi$ (the highest point) with an increment $\pi/100$.

The graphical representation of the calculation results (Figures 2.51 through 2.53) shows that, regardless of the section x, the velocities for $\varphi < \pi/2$ are higher when the radius of the roller increases. We see an inverse dependence in the cutting area ($\varphi = \pi/2$), although the difference between the velocities is small. In the upper part of the roller ($\varphi > \pi/2$), the velocities are practically independent of the radius of the roller.

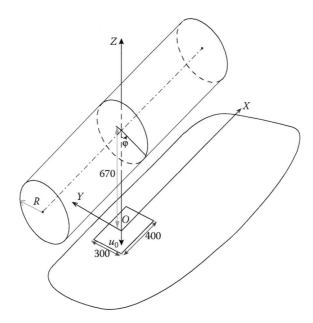

FIGURE 2.48 Scheme of the flow area boundary.

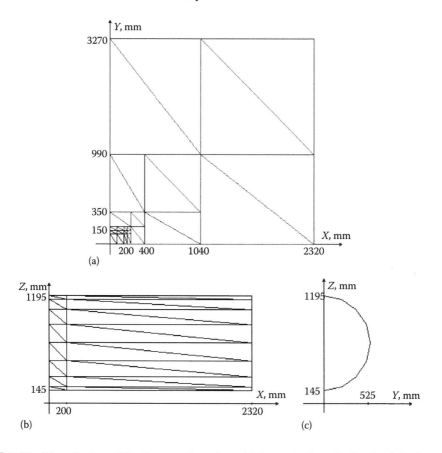

FIGURE 2.49 Discretization of the flow area boundary: (a) the projection of a fourth of the frame onto a horizontal plane, (b) the projection of the surface of a fourth of the cylinder with a radius of 525 mm onto the plane *XOZ*, and (c) the projection of the surface of a half-cylinder onto the plane *YOZ*.

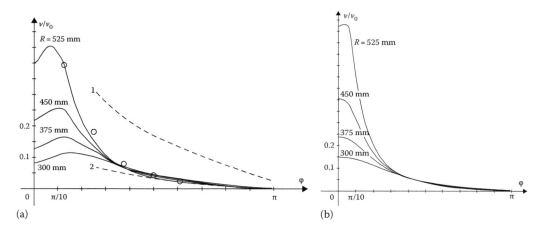

FIGURE 2.50 Change in the value of the dimensionless air velocity on the circumference, each point of which is located at a distance δ from the surface of rollers of different radii R in the cross section $x = 0$ mm: $\circ$ are the experimental measurements of the velocity for $R = 525$ mm; Curve 1 is the calculation using the CMM; Curve 2 is the calculation using the CMM with a conventional width of the slot; (a) $\delta = 55$ mm; (b) $\delta = 105$ mm.

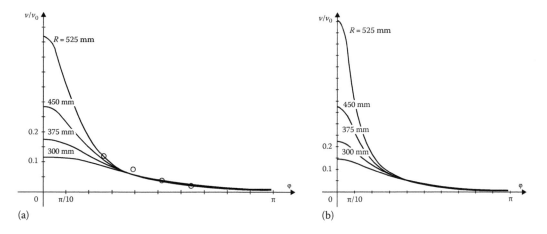

FIGURE 2.51 Change in the value of dimensionless air velocity on the circumference, each point of which is located at a distance δ from the surface of rollers of different radii R in the cross section $x = 200$ mm: $\circ$ are the experimental measurements of the velocity for $R = 525$ mm; (a) $\delta = 55$ mm and (b) $\delta = 105$ mm.

As can be seen from Figures 2.50 through 2.52, the BIE method has a good experimental support. The error of the calculated velocities in respect to the experimental findings does not exceed 25%. Replacing the spatial flow for a flat one (Figure 2.50, a) provides overestimated results (Curve 1). The introduction of the conventional width of the slot (Curve 2) helps to achieve a satisfactory agreement with the experimental data only in a small area.

The cross-sectional area of the suction from the chip-catching channel is $S = 0.12$ m². In this scheme of aspiration, the point under the roller is in close vicinity to the suction; therefore, the suction performance will be based on the point situated where the cutter touches the roller. Air movement required for capturing dust at this representative point must have a value equal to the velocity of a convective flow emerging above the cutter and be of the opposite direction. This velocity is $v = 0.51$ m/s.

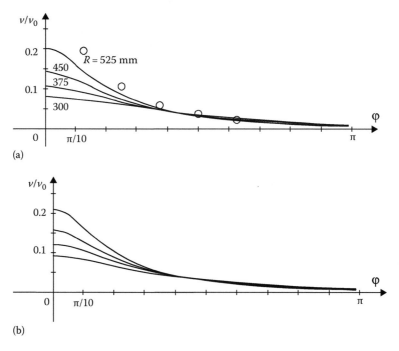

(a)

(b)

FIGURE 2.52 Change in the value of the dimensionless air velocity on the circumference, each point of which is located at a distance δ from the surface of rollers of different radii R in the cross section $x = 400$ mm: ○ are the experimental measurements of the velocity for $R = 525$ mm; (a) δ = 55 mm and (b) δ = 105 mm.

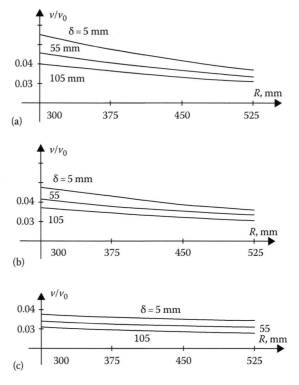

(a)

(b)

(c)

FIGURE 2.53 Change in the dimensionless value of the velocity in the cutting area (φ = π/2) for different distances δ from the cylinder surface depending on its radius R in the cross section: (a) $x = 0$ mm, (b) $-x = 200$ mm, and (c) $-x = 400$ mm.

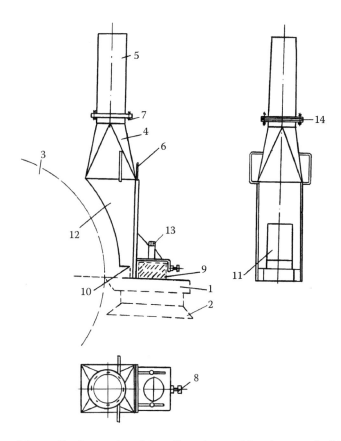

FIGURE 2.54 Local dust-collecting suction of the roll-turning machine: 1, cutter; 2, slide; 3, roller; 4, local suction; 5, connection pipe; 6, sliding gate; 7, bolt; 8, securing clamp; 9, clamping strip; 10, horizontal suction inlet; 11, curvilinear suction inlet; 12, sidewall; 13, locking bolts; 14, gasket.

For the radius of the roller $R = 525$ mm, $\varphi = \pi/2$, and $x = 400$ mm, the calculated velocity at the representative point (Figure 2.53c) is $v_p \approx 0.04$ m/s. Therefore, the necessary air suction velocity is $v_0 = v_n/v_p \approx 12.75$ m/s. The suction performance is $L = v_0 \cdot S \approx 5500$ m³/h. Here, the radius $R = 525$ mm and the section $x = 400$ mm are chosen as the most unfavorable parameters for the containment of dust release. With other parameters, the calculated velocity will be higher, and hence, the performance of the local suction can be reduced.

For currently used roll-turning machines, it is recommended using a local dust-collecting suction (Figure 2.54) that has undergone industrial tests on a roll-turning machine by the Lutuginskiy turning roller production association. The local dust-collecting suction is located directly above the cutting area and is rigidly secured with a clamping bolt and locking bolts to the front clamping strip. It has two rectangular articulated suction inlets, one of which is flat and horizontal and provides dust extraction directly from where the cutter contacts the roller.

The second one is made on the cylindrical surface and designed to contain a convective dust and air flare generated during the cutting process. To prevent clogging of the horizontal inlet with cuttings, the sidewalls of the dust-collecting device have rectangular cut outs, and the cylindrical inlet is distant from the sidewalls in order to reduce harmful air inflowing. The dust-collecting device is mounted on the axis of the cutter. When the cutter is installed at the end of the slide, the position of the dust-collecting device must be changed. The support bracket has an opening to insert the locking nut of the slide. The connection pipe is connected to the hinged duct system and can be moved in a vertical plane in order to disconnect and remove the system from the machine. Abroad, a suspended flexible duct is used instead of the hinged system.

For existing machines, a slot-type local suction is also recommended, which is rigidly secured to the clamping strip of the slide and has a rectangular suction inlet. The length of the suction inlet equals the length of the slide clamping strip. Replacing the cutter does not require the removal of the suction. A lattice is installed at the entrance to the slot-type suction to prevent cuttings from the ingress into the ductwork. The connection pipe of the local suction is connected by a telescopic duct system to a suction device, which allows moving it freely in two mutually perpendicular directions. Unlike the local dust-collecting suction, it allows a better overview of the cutting process and a less laden slide.

It is recommended that designed machines should be equipped with local suction shelter built in the machine slide (under the clamping strip). The local suction shelter is connected to a flexible underfloor duct. The lid of the shelter is put down on the clamping strips and the side walls are pressure tightly pressed against the slide housing. The front end wall has an open aperture enabling the cutter to move forward and perform dust suction from the cutting area and the rear one is a flap wall that enables to install the cutter inside the shelter. A connection pipe with a telescopic duct is attached to one of the lateral suction hood; the last portion of lateral suction hood is connected to an underfloor flexible duct. The telescopic system allows lateral movement of the slide, while the flexible duct allows its longitudinal movement. A flap transparent shield is mounted on top of the front end wall to visually monitor the processing of the roller and increase the efficiency of the suction of the contaminated air convective flow. The advantages of the shelter are the lack of interferences in tightening the screws of the clamping plates when the cutter is being installed, no upper overhead distribution of ducts that facilitates the installation and removal of rollers with an overhead crane, and the local suction need no transposition when the position of the cutter on slide is changed. Among the disadvantages of the suction unit include the fact that the cutting area is cluttered with shields and, inconvenience in monitoring the processing of a roller.

A suction unit from the chip-catching channel in the machine frame is recommended for new machines. There are a number of chip catchers mounted parallel to the generant of the roller under it. When the slide travels along the machine, these chip catchers automatically open up. In this case, the open aperture of the catcher serves not only to remove chips but also to suck dusty air. When the slide moves away at a certain preset distance from the catcher, the latter will close and the next catcher will open up, which is closer to the cutter at that time. This preset distance for a 1A-825–type machine is 400 mm (distance between the centers of two adjacent chip catchers). In order to reduce the volumes of exhaust air, the chip catching chutes are equipped with sealing flaps that close the cavities of idle chutes. Inlets are provided in the walls of the frame to connect the chute cavities with the aspiration channel. The undoubted advantage of this suction is that nothing disturbs the roll turner at work. The combined aspiration system that includes a local suction shelter and a suction unit from the chip-catching channel is shown in Figure 2.55.

2.2.4.6 Determination of the Optimal Geometric Parameters of Local Suction Units of Automatic Presses

Dust while pressing at automatic presses is released through loose joints along the bottom edge of the loading cassette sliding along the surface of the bed: through the gap around the perimeter of the moving die when it enters into the lower die and due to the spills of charge (Figure 2.56).

As a result of dust releases, dust concentration in the working area of the pressing shop area exceeds the maximum permissible concentration (MPC). For example, in the service area of K-8130 press, concentration reaches 15 mg/m³ during pressing the pieces of copper-based powders while

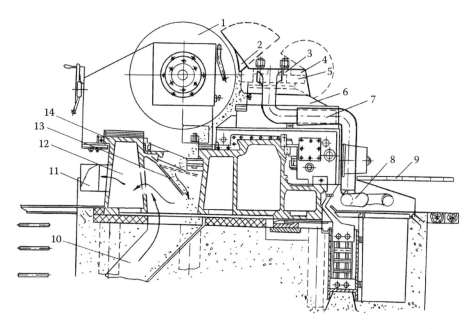

FIGURE 2.55 Combined aspiration system for programmable machines: 1, roller; 2, flap; 3, clamping strip; 4, shelter suction; 5, cutter; 6, slide; 7, telescopic connection pipe; 8, flexible duct; 9, wood flooring; 10, chip removal channel; 11, local suction; 12, aspiration channel; 13, upper pressure-sealing shield; 14, frame.

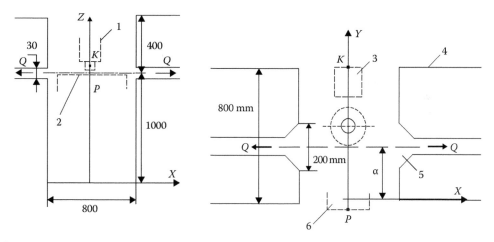

FIGURE 2.56 Model of the effect area of suction flares of local suctions of K-8130 automatic presses: 1, moving die; 2, bed; 3, cassette; 4, frame; 5, suction inlet; 6, parts catcher.

the MAC is 1 mg/m³. Usually, a local suction is placed at the area where pressed products are unloaded. However, as there are no reasonable recommendations related to the choice of schemes and the definition of local suction performance, it is difficult to achieve the performance of the dust release containment.

The location of the local suction unit must be chosen in such a way as to minimize the volume of exhaust air, that is, the distance to zones of intense dust emissions must be minimal. The required volumes of aspiration for open-type local suction units should, in turn, provide the necessary air movement that prevents dust from dispersing from these zones outside the bed

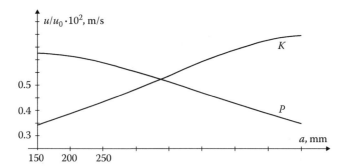

FIGURE 2.57   Change in the value of the relative velocity depending on the ordinate a of the local suction.

of press. Given that dust in this case is transferred by air currents resulting from the travel of the cassette, the desired air movement will be

$$u_T \geq 1.2 v_K,$$

where v_K is the travel velocity of the cassette at a frequency of 25 min and the stroke of 150 mm for a K-8130 type press, $v_K = 0.125$ m/s.

Thus, the problem consists in describing the absorption spectrum of local suction units built in the frame at the level of the bed. We must determine how the air velocity varies in the area of the cassette [point K with the coordinates (0, 800, 1060)] and in the compacted products unloading zone [point P(0; −86; 852)] when the local suction travels in the horizontal plane.

The flow area boundary was discretized into a set of 280 plane triangles. The area boundary is most densely divided in and near the inlet section of the suction inlet, as well as in the breaks of the boundary surface.

As can be seen from the results obtained (Figure 2.57), the air movement decreases in the area of the parts catcher (Curve P) and increases in the area of the cassette (Curve K) as the inlets move away from the front wall of the press. When the center of the suction inlets is located at a distance of 335 mm from the front wall, the velocities in these areas will be identical and amount to $u = 0.00525 \cdot u_0$.

For the required air movement $u_T = 1.2 \cdot 0.125 = 0.15$ m/s, the entry velocity in the inlets must be maintained in the level $u_0 = 0.15/0.00525 = 28.6$ m/s and the exhaust airflow must be maintained at the level $2 \cdot 28.6 \cdot 0.03 \cdot 0.2 = 0.42$ m³/s.

As the distance between the frames decreases from 670 down to 450 mm, air movement in the areas of the points K and P will be slightly higher: $u = 0.0074 \cdot u_0$, and desired exhaust airflow decreases down to 0.243 m³/s.

In conclusion, it should be noted that the results obtained have some margin, as in the calculation of the velocity field we did not take into account such factor as cluttering the space between the frames, parts of the bed, moving die, lower die, cassette, etc.

2.2.5   THREE-DIMENSIONAL FLOWS IN MULTIPLY CONNECTED REGIONS WITH ROTATING CYLINDERS

The calculation of the air velocity field near the local suction units from the rotating cylindrical parts is required to determine properly the volumes of aspiration in different types of turning, drilling, grinding, milling, woodworking and other machines. Section 2.2.2 examines algorithms for numerical calculation of dusty and airflows at a narrow class of the slot-type suctions from rotating cylinders, and only when the problem is reduced to a plane (i.e., the axis of the cylinder is parallel to axis of suction). This subsection generalizes earlier results for 3D space within the model of potential flows.

Assume that the flow area is bounded by the surface S. For simplicity, we assume that within the area there is only one cylinder with a radius R rotating at a velocity v_{rot}. The normal velocity component amounts to $v_n = 0$ on the impermeable boundary surface S including the cylinder surface as well. The velocity along the direction of the outward normal is known in the suction openings and air openings. Determine the velocity $\vec{v} = \{v_1, v_2, v_3\}$ at an arbitrary point $x(x_1, x_2, x_3)$ along the predetermined direction $\vec{n} = \{n_1, n_2, n_3\}$, where $|\vec{n}| = 1$.

Place sources (sinks) of a previously unknown strength $q(\xi)$ on the boundary surface S, where $\xi(\xi_1, \xi_2, \xi_3)$ is the arbitrary point of S. Place an infinitely long concentrated vortex with circulation $\Gamma = v_{rot} \cdot 2\pi R$ along the axis of the cylinder. Define the unknown variables $q(\xi)$ based on the given boundary conditions for the normal velocity component. To this end, we apply the previously described solution procedure by the boundary integral equation method.

The air velocity at the point $x(x_1, x_2, x_3)$ along the direction $\vec{n} = \{n_1, n_2, n_3\}$ caused by the action of unknown sources (sinks) of $q(\xi)$ and the concentrated vortex is defined by the relation:

$$v_n(x) = \iint_S F(x, \xi) q(\xi) dS(\xi) + 2\pi R v_1 G_n(x, r), \qquad (2.97)$$

where the function

$$F(x, \xi) = \frac{\sum_{i=1}^{3} (\xi_i - x_i) n_i}{4\pi \left[\sum_{i=1}^{3} (\xi_i - x_i) n_i \right]^{3/2}}$$

expresses the influence of a single point source located at the point $\xi(\xi_1, \xi_2, \xi_3)$ on the point x along the direction $\vec{n}$.

We obtain the expression for the function $G_n(x, r)$ expressing the influence on the point x along the direction $\vec{n}$ of the concentrated vortex with a single circulation, which is located on the line

$$\frac{x - x_0}{m} = \frac{y - y_0}{l} = \frac{z - z_0}{n},$$

where (x_0, y_0, z_0) is the given point on the line; $\vec{a} = \{m, l, n\}$ is the directing vector of the line. Rotation around the concentrated vortex is counterclockwise when viewed from the end of the vector $\vec{a}$.

Initially, we determine the distance from the point $x(x_1, x_2, x_3)$ to the concentrated vortex:

$$r = \sqrt{\left(x_1 - mt - x_0\right)^2 + \left(x_2 - lt - y_0\right)^2 + \left(x_3 - nt - z_0\right)^2}, \qquad (2.98)$$

where

$$t = \frac{(x_1 - x_0)m + (x_2 - y_0)l + (x_3 - z_0)n}{m^2 + n^2 + l^2}.$$

Calculate the unit vector $\vec{\tau}$ tangental to the circle of a radius r centered at $(mt + x_0, lt + y_0, nt + z_0)$:

$$\vec{\tau} = \left[\vec{a} \times \vec{r} \right] / |\vec{a} \times \vec{r}|,$$

where

$$\vec{r} = \{r_1, r_2, r_3\} = \{x_1 - mt - x_0, x_2 - lt - y_0, x_3 - nt - z_0\},$$

Thus, the coordinates of the vector $\vec{\tau}$ are

$$\vec{\tau} = \left\{ \frac{lr_3 - nr_2}{|\vec{a} \times \vec{r}|}, \quad \frac{nr_1 - mr_3}{|\vec{a} \times \vec{r}|}, \quad \frac{mr_2 - lr_1}{|\vec{a} \times \vec{r}|} \right\}, \tag{2.99}$$

$$|\vec{a} \times \vec{r}| = \sqrt{\left(lr_3 - nr_2\right)^2 + \left(nr_1 - mr_3\right)^2 + \left(mr_2 - lr_1\right)^2}.$$

Then the value of the velocity caused by the cylinder with a radius R rotating at a speed v_1 at the point x located at a distance r from its axis, along the direction $\vec{n}$ is expressed by the formula

$$v_n(x) = \frac{R}{r} v_1 \left(\tau_1 n_1 + \tau_2 n_2 + \tau_3 n_3 \right) \tag{2.100}$$

and, therefore

$$G_n(x,r) = \frac{1}{2\pi r} \left(\tau_1 n_1 + \tau_2 n_2 + \tau_3 n_3 \right).$$

In order to determine the unknown variables $q(\xi)$ in Equation 2.97, we let the inner point x to the boundary one x_0. Then we obtain the boundary integral equation:

$$v_n(x_0) = \frac{1}{2} q(x_0) + \iint_S F(x_0, \xi) q(\xi) dS(\xi) + 2\pi R v_1 G_n(x_0, r), \tag{2.101}$$

where the first summand corresponds to the case $x_0 = \xi$ and the integral does not contain this point. The surface integral of the first kind in Equation 2.101 is integrated over the variable ξ.

Since integral Equation 2.101 cannot be solved analytically, we use the numerical method. Divide the boundary surface S into N plane triangles, along each of them the strength $q(\xi)$ is assumed to be constant. Then the discrete analog of Equation 2.101 is as follows:

$$v_n(x_0^p) = \frac{1}{2} q(x_0^p) + \sum_{\substack{k=1, \\ k \neq p}}^{N} q(\xi^k) \iint_{\Delta S} F(x_0^p, \xi^k) dS(\xi^k) + 2\pi R v_1 G_n(x_0^p, r), \tag{2.102}$$

where

x_0^p is the center of gravity of the pth triangle; $p = 1, 2, \ldots, n$

ξ^k is the point on the kth triangle

Denote: $v_n(x_0^p) = v^p$; $q(x_0^p) = q^p$; $\quad q(\xi^k) = q^k$; $2\pi R v_1 G_n(x_0^p, r) = G^p$; $\displaystyle\iint_{\Delta S} F(x_0^p, \xi^k) dS(\xi^k) = F^{pk}$.

Then, solving a system of N linear algebraic equations with N unknown q^p:

$$\frac{1}{2}q^p + \sum_{\substack{k=1,\\k\neq p}}^{N} q^k F^{pk} = v^p - G^p,$$ (2.103)

where $p = 1, 2, \ldots, n$, we calculate the velocity at the inner point x along a given direction $\vec{n}$ by

$$v_n(x) = \sum_{k=1}^{N} q^k F^k + 2\pi R v_1 G_n(x,r),$$ (2.104)

where $F^k = \iint_{\Delta S} F(x, \xi^k) dS(\xi^k)$.

The algorithms for discretization of the boundary surface and for forming the matrix $F = \left(F^{pk} \right) = \left(\iint_{\Delta S^k} F(x_0^p, \xi^k) dS(\xi^k) \right)$ are discussed in detail in Section 2.2.4.5.

As an example of the application of the proposed method, consider airflow near a rectangular inlet whose spectrum of action has a rotating cylinder (Figure 2.58). The case of a motionless cylinder is considered in Section 2.2.4.5.

The boundary surface was discretized into 296 plane triangles (Figure 2.49).

The calculation results presented in Figure 2.59 show that a circulatory flow of air occurs in case of a fixed suction velocity and increase in the velocity of the cylinder around the latter; neglecting this

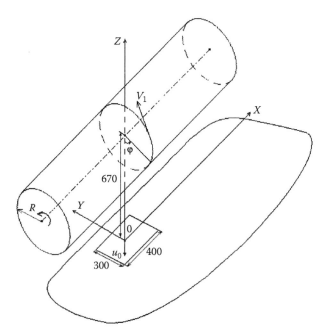

FIGURE 2.58 Rotating cylinder in the spectrum of action of the rectangular suction.

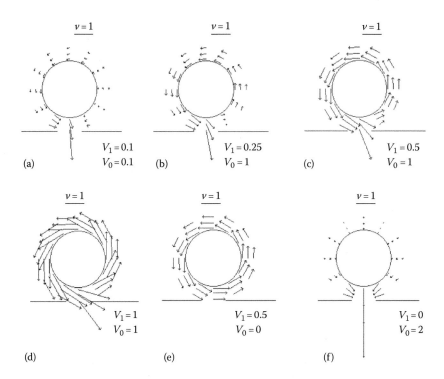

FIGURE 2.59 Velocity field in the plane $x = 0$ (the interval named $V = 1$, indicates the scale of the velocity value): (a) $V_1 = 0.1$, $V_0 = 0.1$; (b) $V_1 = 0.25$, $V_0 = 1$; (c) $V_1 = 0.5$, $V_0 = 1$; (d) $V_1 = 1$, $V_0 = 1$; (e) $V_1 = 0.5$, $V_0 = 0$ and (f) $V_1 = 0$, $V_0 = 2$.

flow can lead to significant errors in the calculation of local suction units (Figure 2.59f). Note that the flow pattern is slightly asymmetric relative to the cylinder axis where there is no suction (Figure 2.59e).

The velocity is higher under the cylinder than above it, indicating the effect of an impermeable plane.

The proposed algorithm can be applied with minor modifications to flows not only containing several rotating cylinders but also having a finite length. In this case, a concentrated vortex of finite length must be used to simulate them.

3 Calculation of Vortex Flows

3.1 FRICTIONAL FLOWS

3.1.1 BASIC EQUATIONS: GENERAL ALGORITHM OF NUMERICAL CALCULATION

Let us assume that liquid is viscous, incompressible, and its flow is steady state. Using the equation of continuity

$$\frac{\partial u}{\partial x} + \frac{\partial v}{\partial y} = 0 \tag{3.1}$$

and the vorticity definition

$$\vec{\omega} = \vec{\nabla} \times \vec{v} = \begin{vmatrix} \vec{i} & \vec{j} & \vec{k} \\ \dfrac{\partial}{\partial x} & \dfrac{\partial}{\partial y} & \dfrac{\partial}{\partial z} \\ u & v & 0 \end{vmatrix} = \vec{k}\left(\frac{\partial v}{\partial x} - \frac{\partial u}{\partial y}\right) \tag{3.2}$$

that conform to the boundary conditions, integral approximation [59] of the same can be obtained:

$$\vec{v}(\xi) = -\frac{1}{2\pi}\left[\iint_{R} \frac{\vec{\omega}(x) \times (\vec{r}(x) - \vec{r}(\xi))}{\vec{r}(x) - \vec{r}(\xi)^2}\, dR(x) - \int_{B} \frac{\left\{\vec{v}(x) \cdot \vec{n}(x) - \left(\vec{v}(x) \times \vec{n}(x)\right) \times\right\}\left(\vec{r}(x) - \vec{r}(\xi)\right)}{\left|\vec{r}(x) - \vec{r}(\xi)\right|^2}\, dB(x) \right], \tag{3.3}$$

where
    R is an area occupied by liquid
    B is the flow area boundary
    x point is a variable of integration
    ξ is the point at which $\vec{v}(\xi) = \{u(\xi), v(\xi)\}$ velocities are determined
    $\vec{n}(x)$ is the outward normal
    $\vec{\omega}(x)$ *is x point vorticity vector*
    $\vec{r}(x), \vec{r}(\xi)$ are radius vectors of points x and ξ

The vorticity can be determined from Navier–Stokes equation expressed in terms of vorticity:

$$\vec{\nabla} \times \vec{\omega} = \frac{1}{\nu}(\vec{v} \times \vec{\omega} - \vec{\nabla} h_0), \tag{3.4}$$

where

$h_0 = \dfrac{p}{\rho_{med}} + \dfrac{|\vec{v}|^2}{2}$ is the total head

p is the relative static pressure

ρ_{med} is the medium density

$|\vec{v}|$ is the velocity modulus

ν is the kinematic viscosity coefficient

An integral countertype of this equation is given by

$$\vec{\omega}(\xi) = -\frac{1}{2\pi}\left[\frac{1}{\nu} \iint\limits_R \frac{\left[\vec{v}(x) \times \vec{\omega}(x)\right] \times \left(\vec{r}(x) - \vec{r}(\xi)\right)}{|\vec{r}(x) - \vec{r}(\xi)|^2} \, dR \right.$$

$$\left. - \int\limits_B \frac{\left\{ h_0(x)/\nu \cdot \vec{n}(x) \times + \vec{\omega}(x) \cdot \vec{n}(x) - [\vec{\omega}(x) \times \vec{n}(x)] \times \right\}\left(\vec{r}(x) - \vec{r}(\xi)\right)}{|\vec{r}(x) - \vec{r}(\xi)|^2} \, dB \right]. \tag{3.5}$$

Given the correct boundary values of vertical $v(x)$ and horizontal $u(x)$ velocity components, the field of velocities within the flow area can be found using the following iterative procedure:

1. By the given initial approximation for the vorticity within the flow area, determine the corresponding boundary values of vorticity from Equation 3.3 written at the range boundary.
2. From Equation 3.3, determine the vertical and horizontal velocity components at reference points within the flow area.
3. From Equation 3.5 written at the range boundary, determine the boundary value for total head h_0.
4. By the calculated values of velocities u and v, vorticity ω, and total head h_0, determine the new distribution of vorticity within the flow area using the formula given as Equation 3.5.
5. Pass to the first step of this iterative procedure and continue calculations until differences between old and new vorticity values become insignificant.

3.1.2 FLOW AREA DISCRETIZATION DISCRETE COUNTERTYPES OF INTEGRAL EQUATIONS

Let us write Equations 3.3 and 3.5 as coordinate axes components and write

$$\vec{\rho}(x) = \vec{r}(x) - \vec{r}(\xi) = \left\{ r_1(x) - r_1(\xi), r_2(x) - r_2(\xi), 0 \right\} = \left\{ \rho_1, \rho_2, 0 \right\}.$$

Then the vector product in the first integral (Equation 3.3) is

$$\vec{\omega}(x) \times \vec{\rho}(x) = \begin{vmatrix} \vec{i} & \vec{j} & \vec{k} \\ 0 & 0 & \omega \\ \rho_1 & \rho_2 & 0 \end{vmatrix} = -\rho_2 \cdot \omega \cdot \vec{i} + \rho_1 \cdot \omega \cdot \vec{j},$$

where ω is the OZ-component of vorticity.

As far as

$$\vec{v}(x) \cdot \vec{n}(x) = u \cdot n_1 + v \cdot n_2, \quad \vec{v}(x) \times \vec{n}(x) = \begin{vmatrix} \vec{i} & \vec{j} & \vec{k} \\ u & v & 0 \\ n_1 & n_2 & 0 \end{vmatrix} = \vec{k}(un_2 - vn_1),$$

$$[\vec{v}(x) \times \vec{n}(x)] \times \vec{\rho} = \begin{vmatrix} \vec{i} & \vec{j} & \vec{k} \\ 0 & 0 & un_2 - vn_1 \\ \rho_1 & \rho_2 & 0 \end{vmatrix} = -\rho_2(un_2 - vn_1)\vec{i} + \rho_1(un_2 - vn_1)\vec{j},$$

the subintegral function numerator in the second integral in view of

$$\vec{\rho}(x) = \vec{r}(x) - \vec{r}(\xi) = \{r_1(x) - r_1(\xi), r_2(x) - r_2(\xi), 0\} = \rho_1 \vec{i} + \rho_2 \vec{j}$$

will be

$$\{\vec{v}(x) \cdot \vec{n}(x) - (\vec{v}(x) \times \vec{n}(x)) \times\} (\vec{r}(x) - \vec{r}(\xi)) = (un_1 + vn_2)\vec{\rho} - \left[-\rho_2(un_2 - vn_1)\vec{i} + \rho_1(un_2 - vn_1)\vec{j} \right]$$

$$= \left[u(n_1\rho_1 + n_2\rho_2) + v(n_2\rho_1 - n_1\rho_2) \right]\vec{i} + \left[v(n_1\rho_1 + n_2\rho_2) + u(n_1\rho_2 - \rho_1 n_2) \right]\vec{j}.$$

Based on the foregoing, we obtain

$$\vec{v}(\xi) = -\frac{1}{2\pi} \left[\iint_R \frac{-\rho_2\omega\vec{i} + \rho_1\omega\vec{j}}{\rho^2} dR - \int_B \left\{ \frac{[u(n_1\rho_1 + n_2\rho_2) + v(n_2\rho_1 - n_1\rho_2)]\vec{i}}{\rho^2} \right. \right.$$

$$\left. \left. + \frac{\vec{j}[v(n_1\rho_1 + n_2\rho_2) + u(n_1\rho_2 - \rho_1 n_2)]}{\rho^2} \right\} dB \right].$$

Thus, coordinate axes component of velocity can be written as

$$\begin{cases} u(\xi) = \dfrac{1}{2\pi} \iint_R \dfrac{\rho_2}{\rho^2} \omega dR + \dfrac{1}{2\pi} \int_B \dfrac{n_1\rho_1 + n_2\rho_2}{\rho^2} u dB - \dfrac{1}{2\pi} \int_B \dfrac{n_1\rho_2 - n_2\rho_1}{\rho^2} v dB, \\[4mm] v(\xi) = -\dfrac{1}{2\pi} \iint_R \dfrac{\rho_1}{\rho^2} \omega dR + \dfrac{1}{2\pi} \int_B \dfrac{n_1\rho_1 + n_2\rho_2}{\rho^2} v dB + \dfrac{1}{2\pi} \int_B \dfrac{n_1\rho_2 - n_2\rho_1}{\rho^2} u dB. \end{cases} \tag{3.6}$$

Likewise, Equation 3.5 will be transformed. By calculating vector and scalar products of vectors in the integral (Equation 3.5) we obtain

$$\vec{v} \times \vec{\omega} = \begin{vmatrix} \vec{i} & \vec{j} & \vec{k} \\ u & v & 0 \\ 0 & 0 & \omega \end{vmatrix} = \vec{i}v\omega - \vec{j}u\omega,$$

$$[\vec{v} \times \vec{\omega}] \times \vec{\rho} = \begin{vmatrix} \vec{i} & \vec{j} & \vec{k} \\ v\omega & -u\omega & - \\ \rho_1 & \rho_2 & 0 \end{vmatrix} = \vec{k}[v\omega\rho_2 + u\omega\rho_1],$$

$$\vec{n} \times \vec{\rho} = \begin{vmatrix} \vec{i} & \vec{j} & \vec{k} \\ n_1 & n_2 & 0 \\ \rho_1 & \rho_2 & 0 \end{vmatrix} = \vec{k}\left[n_1\rho_2 - n_2\rho_1 \right], \quad \vec{\omega} \cdot \vec{n} = 0 \cdot n_1 + 0 \cdot n_2 + \omega \cdot 0 = 0,$$

$$\vec{\omega} \times \vec{n} = \begin{vmatrix} \vec{i} & \vec{j} & \vec{k} \\ 0 & 0 & \omega \\ n_1 & n_2 & 0 \end{vmatrix} = -n_2\omega\vec{i} + n_1\omega\vec{j},$$

$$[\vec{\omega} \times \vec{n}] \times \vec{\rho} = \begin{vmatrix} \vec{i} & \vec{j} & \vec{k} \\ -n_2\omega & n_1\omega & 0 \\ \rho_1 & \rho_2 & 0 \end{vmatrix} = \vec{k}[-n_2\omega\rho_2 - n_1\omega\rho_1].$$

By inserting the resulting values in Equation 3.5, writing out the integrals of the sum by the sums of integrals, and projecting the equation on OZ-axis we obtain

$$\omega(\xi) = -\frac{1}{2v\pi}\iint\limits_R \frac{\rho_2}{\rho^2} v\omega dR - \frac{1}{2\pi v}\iint\limits_R \frac{\rho_1}{\rho^2} u\omega dR + \frac{1}{2\pi v}\int\limits_B \frac{n_1\rho_2 - n_2\rho_1}{\rho^2} h_0 dB + \frac{1}{2\pi}\int\limits_B \frac{n_2\rho_2 + n_1\rho_1}{\rho^2} \omega dB,$$

$$(3.7)$$

where

n_1, n_2 are $\vec{n}(x)$ outward normal coordinates

ρ_1, ρ_2 are $\vec{\rho}(x) = \vec{r}(x) - \vec{r}(\xi)$ vector coordinates

$|\vec{\rho}|$ vector modulus is denoted through ρ

A computational solution of Equations 3.6 and 3.7 requires the flow area to be discretized into inner cells and boundary elements. Let N be a number of inner nodes within the range; N_b is a number of boundary nodes where vorticity is other than 0; V_n is the total number of inner cells (e.g., flat triangles); G_e is a number of boundary intervals.

It is convenient to assume some i-node where i may vary from 1 to $N + N_b$ as point ξ at which unknown velocity values are to be determined. Assume that the vertical and horizontal velocity components and vorticity vary linearly by the corresponding boundary elements and inner cells; therefore, we will use linear interpolating functions. Write integrals in the expressions as sums of integrals over triangular cells (integrals over R) and over linear elements nodes (integrals over B):

$$\frac{1}{2\pi}\iint_R \frac{\rho_2}{\rho^2}\omega dR = \sum_{j=1}^{N+N_b} f_{ij}\omega_j, \quad \frac{1}{2\pi}\iint_R \frac{\rho_1}{\rho^2}\omega dR = \sum_{j=1}^{N+N_b} g_{ij}\omega_j,$$

$$\frac{1}{2\pi}\int_B \frac{n_1\rho_1 + n_2\rho_2}{\rho^2}u dB = \sum_{j=N+1}^{N+N_b} h_{ij}u_j, \quad \frac{1}{2\pi}\int_B \frac{n_1\rho_1 + n_2\rho_2}{\rho^2}v dB = \sum_{j=N+1}^{N+N_b} h_{ij}v_j,$$

$$\frac{1}{2\pi}\int_B \frac{n_1\rho_2 - n_2\rho_1}{\rho^2}u dB = \sum_{j=N+1}^{N+N_b} l_{ij}u_j, \quad \frac{1}{2\pi}\int_B \frac{n_1\rho_2 - n_2\rho_1}{\rho^2}v dB = \sum_{j=N+1}^{N+N_b} l_{ij}v_j,$$

where

f_{ij}, g_{ij} are sums of integrals over triangular cells

h_{ij}, l_{ij} are sums of integrals over boundary intervals with a vertex at j-node

$$f_{ij} = \frac{1}{2\pi}\sum\iint_{R_j} \frac{\rho_2}{\rho^2}\varphi dR, \quad g_{ij} = \frac{1}{2\pi}\sum\iint_{R_j} \frac{\rho_1}{\rho^2}\varphi dR;$$

$$h_{ij} = \frac{1}{2\pi}\sum\int_{B_j} \frac{n_1\rho + n_2\rho_2}{\rho^2}\psi dB, \quad l_{ij} = \frac{1}{2\pi}\sum\int_{B_j} \frac{n_1\rho_2 - n_2\rho_1}{\rho^2}\psi dB,$$

where

$\{\rho_1, \rho_2\} = \{r_1(j)-r_1(i), r_2(j)-r_2(i)\}$

$\{r_1(i), r_2(i)\}$ are radius vector coordinates for i-node

$\{r_1(j), r_2(j)\}$ are radius vector coordinates for a random point at a boundary interval or in a triangular cell incorporating j-node

φ, ψ are linear interpolating functions in triangular cells and boundary intervals, respectively, calculated based on which vertex of these elements lies in j-node

$\vec{n} = \{n_1, n_2\}$ is the external normal to the boundary interval incorporating the boundary j-node

Thus, the discrete countertype of Equations 3.6 and 3.7 is as follows:

$$\begin{cases} u_i = \displaystyle\sum_{j=1}^{N+N_b} f_{ij}\omega_j + \sum_{j=N+1}^{N+N_b} h_{ij}u_j - \sum_{j=N+1}^{N+N_b} l_{ij}v_j, \\[2em] v_i = -\displaystyle\sum_{j=1}^{N+N_b} g_{ij}\omega_j + \sum_{j=N+1}^{N+N_b} h_{ij}v_j + \sum_{j=N+1}^{N+N_b} l_{ij}u_j, \end{cases} \tag{3.8}$$

$$\omega_i = -\frac{1}{\nu}\sum_{j=1}^{N+N_b} f_{ij}(v\omega)_j - \frac{1}{\nu}\sum_{j=1}^{N+N_b} g_{ij}(u\omega)_j + \frac{1}{\nu}\sum_{j=N+1}^{N+N_b} l_{ij}(h_0)_j + \sum_{j=N+1}^{N+N_b} h_{ij}\omega_j. \tag{3.9}$$

Using the formulas given in Equations 3.8 and 3.9, the computation algorithm for the field of velocities in inner nodes within the flow area will be as follows:

1. The initial vorticity ω_i, $i = 1, 2, 3, ..., N$ in inner nodes within the flow area is given.
2. Boundary vorticity values are calculated from the first or the second equation of the set given in Equation 3.8 at $i = N + 1, N + 2, ..., N + N_b$, expressed as

$$\sum_{j=N+1}^{N+N_b} f_{ij}\omega_j = u_i - \sum_{j=1}^{N} f_{ij}\omega_j - \sum_{j=N+1}^{N+N_b} \left[h_{ij}u_j - l_{ij}v_j \right], \quad (3.10)$$

$$\sum_{j=N+1}^{N+N_b} g_{ij}\omega_j = -v_i - \sum_{j=1}^{N} g_{ij}\omega_j + \sum_{j=N+1}^{N+N_b} \left[h_{ij}v_j + l_{ij}u_j \right]. \quad (3.11)$$

Once this iteration of the new (starting from the second) value of boundary vorticity ω_{ni} has been defined, it is redefined by

$$\omega_i = \frac{\omega_{-i} + k \cdot \omega_{ni}}{1 + k}, \quad (3.12)$$

where
ω_{-i} is the previous iteration boundary vorticity value
k is a coefficient from 0 to 1 determined by means of a numerical experiment and used to enable the iteration process convergence

3. Calculating the vertical and horizontal velocity components in the inner nodes of the range suing formulas of Equation 3.8 where i varies from 1 to N.
4. Determining the boundary values of total head $\left(h_0 \right)_j$ from Equation 3.9 written at the range boundary

$$\sum_{j=N+1}^{N+N_b} l_{ij} \left(h_0 \right)_j = v\omega_i + \sum_{j=1}^{N} \left[f_{ij} \left(v\omega \right)_j + g_{ij} \left(u\omega \right)_j \right] - v \sum_{j=N+1}^{N+N_b} h_{ij}\omega_j, \quad (3.13)$$

where $i = N + 1, N + 2, ..., N + N_b - 2$.
As the principal determinant of the set given in Equation 3.13 turns out to be equal to 0 at $i = N + 1, N + 2, ..., N + N_b$ and for symmetric regions even at $i = N + 1, N + 2, ..., N + N_b - 1$, the total head is taken to be equal to 0 in the boundary node $N + N_b$ and is calculated as the arithmetic average between the adjacent nodes in $N + N_b - 1$.
5. Calculating a new set of vorticity values within the flow area using the formula given in Equation 3.9 for $i = 1, 2, ..., N$ and passing to step 2 whereas vorticity values need to be redefined by the formula given in Equation 3.12, where $i = 1, 2, ..., N$.

The iteration procedure will be executed until the maximum modulus of difference between the new set of vorticity values and the old one becomes less than the required accuracy.

3.1.3 Calculating F and G Matrix Elements

Build a local oblique system of coordinates (LOCS) on a triangle with vertexes r, s, and t given in the global rectangular coordinate system (GRCS). The LOCS center is placed at r vertex, X-axis η_1

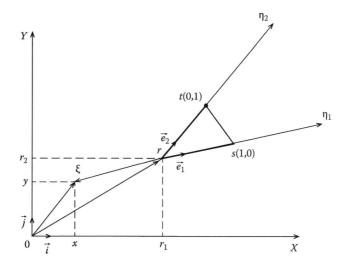

FIGURE 3.1 Rectangular coordinate system XOY and local oblique coordinate system $\eta_1 r \eta_2$.

is directed along $\vec{rs}$, and Y-axis η_2 is directed along $\vec{rt}$ (Figure 3.1). The unit of length will be the length of interval $|\vec{rs}| = l_1$ along η_1 axis and the length of interval $|\vec{rt}| = l_2$ along η_2 axis.

Correlate coordinates of point ξ in the LOCS (η_1, η_2) and GRCS (x, y).

The radius vector of point ξ

$$\vec{r} = x\vec{i} + y\vec{j} = r_1\vec{i} + r_2\vec{j} + \eta_1 l_1 \vec{e_1} + \eta_2 l_2 \vec{e_2}.$$

As

$$\vec{e_1} = \frac{s_1 - r_1}{l_1}\vec{i} + \frac{s_2 - r_2}{l_1}\vec{j}, \quad \vec{e_2} = \frac{t_1 - r_1}{l_2}\vec{i} + \frac{t_2 - r_2}{l_2}\vec{j},$$

equaling the respective coordinates we obtain

$$\begin{cases} x = r_1 + \eta_1(s_1 - r_1) + \eta_2(t_1 - r_1), \\ y = r_2 + \eta_1(s_2 - r_2) + \eta_2(t_2 - r_2). \end{cases} \tag{3.14}$$

The Jacobian of the coordinate system mapping will take the following form:

$$J = \begin{vmatrix} \dfrac{\partial x}{\partial \eta_1} & \dfrac{\partial x}{\partial \eta_2} \\ \dfrac{\partial y}{\partial \eta_1} & \dfrac{\partial y}{\partial \eta_2} \end{vmatrix} = \begin{vmatrix} s_1 - r_1 & t_1 - r_1 \\ s_2 - r_2 & t_2 - r_2 \end{vmatrix} = (s_1 - r_1)(t_2 - r_2) - (t_1 - r_1)(s_2 - r_2). \tag{3.15}$$

The interpolating functions for vertexes s, t and r will be, respectively,

$$\varphi_s = \eta_1, \quad \varphi_t = \eta_2, \quad \varphi_r = 1 - \eta_1 - \eta_2. \tag{3.16}$$

Then, double integrals over inner cells will be calculated from the following formulas:

$$\iint\limits_{R_j} \frac{\rho_1}{\rho^2} \varphi dR = \int\limits_0^1 d\eta_1 \int\limits_0^{1-\eta_1} \frac{(x-\xi_1)|J|}{(x-\xi_1)^2 + (y-\xi_2)^2} \varphi d\eta_2,$$

$$\iint\limits_{R_j} \frac{\rho_2}{\rho^2} \varphi dR = \int\limits_0^1 d\eta_1 \int\limits_0^{1-\eta_1} \frac{(y-\xi_2)|J|}{(x-\xi_1)^2 + (y-\xi_2)^2} \varphi d\eta_2,$$

where
  x, y are coordinates of the point on R_j-triangle calculated by the formulas of Equation 3.14
  $|J|$ is the modulus of the Jacobian of mapping calculated by Equation 3.15
  ϕ is the interpolating function
  (ξ_1, ξ_2) are i-node coordinates

Using the resulting relations we will obtain an algorithm for the calculation of $F = \{f_{ij}\}$ and $G = \{g_{ij}\}$ matrix elements comprising the following steps:

1. A number of i-node is given. At the first stage $i = 1$.
2. A number of j-node is given. Initially $j = 1$.
3. Matrix elements are assigned to zero: $f_{ij} = 0$, $g_{ij} = 0$.
4. An initial number for the inner cell is given: $k = 1$.
5. Vertex r for the inner cell is given.
6. The triangle cell vertex coordinate is compared with j-node coordinates and if they coincide with each other the following is calculated:

$$g_{ij} = g_{ij} + \frac{1}{2\pi} \int\limits_0^1 d\eta_1 \int\limits_0^{1-\eta_1} \frac{(x-\xi_1)|J|}{(x-\xi_1)^2 + (y-\xi_2)^2} \varphi d\eta_2,$$

$$f_{ij} = f_{ij} + \frac{1}{2\pi} \int\limits_0^1 d\eta_1 \int\limits_0^{1-\eta_1} \frac{(y-\xi_2)|J|}{(x-\xi_1)^2 + (y-\xi_2)^2} \varphi d\eta_2,$$

  where (ξ_1, ξ_2) are i-node coordinates; x and y are determined from Equation 3.14, the interpolating function ϕ corresponds to the vertex and is determined using Equation 3.16.
7. There are actions described in step 6 executed for vertexes s and t.
8. The following inner cell is chosen: $k = k + 1$, then the calculation is repeated from step 5 until $k \leq V_n$.
9. j-node number is increased by unit and the calculation is repeated from step 3 until inequation $j \leq N + N_b$ remains true.
10. The value of i is increased by 1, and the calculation is repeated from step 2 until condition $i \leq N + N_b$ remains true.

Integrals in formation of $\{f_{ij}\}$ and $\{g_{ij}\}$ matrix elements must be calculated numerically using Gauss quadrature rules. To this effect, they should be traced to the form of the integral in Equation 2.95. The integral in question is $\int_0^1 dx \int_0^{1-x} f(x,y)dy$. Substitute the variables and convert this integral as follows:

$$\int_0^1 dx \int_0^{1-x} f(x,y)dy = \left\{ \begin{matrix} y = \dfrac{1-x}{2}(x_2+1) \\[3mm] dy = \dfrac{1-x}{2}dx_2 \end{matrix} \right\} = \int_0^1 dx \int_{-1}^1 f\left(x, \dfrac{1-x}{2}(x_2+1)\right)\dfrac{1-x}{2}dx_2$$

$$= \left\{ \begin{matrix} x = \dfrac{1}{2}(x_1+1) \\[3mm] dx = \dfrac{1}{2}dx_1 \end{matrix} \right\} = \dfrac{1}{8}\int_{-1}^1 dx_1 \int_{-1}^1 f\left(\dfrac{1}{2}(x_1+1), \dfrac{(1-x_1)(1+x_2)}{4}\right)(1-x_1)dx_2.$$

Using Gauss quadrature rule (Equation 2.95), we have

$$\int_0^1 dx \int_0^{1-x} f(x,y)dy = \dfrac{1}{8}\sum_{i=1}^N \sum_{j=1}^N A_i A_j f\left(\dfrac{1}{2}(x_i+1), \dfrac{(1-x_i)(1+x_j)}{4}\right)(1-x_i),$$

where
$\quad x_i$ are nodes
$\quad A_i$ are the Gauss quadrature rule weight coefficients

3.1.4 CALCULATING L AND H MATRIX ELEMENTS

Build a local rectangular coordinate system (LRCS) on the interval with vertexes $a(a_1, b_1)$ and $b(b_1, b_2)$. The LRCS center is placed at the interval center point, X-axis η_1 is directed along $\vec{ab}$ vector, and Y-axis η_2 is directed opposite to the outward normal $\vec{n}$ (Figure 3.2).

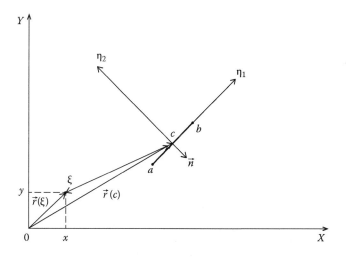

FIGURE 3.2 Building the local coordinate system on the interval.

Correlate coordinates of point ξ in LRCS (ξ_1, ξ_2) and GRCS (x, y).

The LRCS center coordinates are calculated by the following formulas:

$$c_1 = \frac{a_1 + b_1}{2}, \quad c_2 = \frac{a_2 + b_2}{2}. \tag{3.17}$$

Denote the interval length as $l = \sqrt{(b_1 - a_1)^2 + (b_2 - a_2)^2}$. X-axis unit vector can be presented as follows:

$$\vec{e}_1 = \frac{b_1 - a_1}{l}\vec{i} + \frac{b_2 - a_2}{l}\vec{j} = \{e_{11}, e_{12}\}. \tag{3.18}$$

Determine $\vec{e}_2$ axis unit vector based on its perpendicularity condition $\vec{e}_1$: $\vec{e}_1 \cdot \vec{e}_2 = e_{11}e_{21} + e_{12}e_{22} = 0$ and due to the fact that the coordinate system is right-handed: $\vec{e}_1 \times \vec{e}_2 = \vec{k}$, that is, $e_{11}e_{22} - e_{12}e_{21} = 1$. By inserting the vector components $\vec{e}_1 = \{e_{11}, e_{12}\}$ in these equations and resolving the resulting set of equations

$$\begin{cases} \dfrac{b_1 - a_1}{l}e_{21} + \dfrac{b_2 - a_2}{l}e_{22} = 0, \\[3mm] \dfrac{b_1 - a_1}{l}e_{22} - \dfrac{b_2 - a_2}{l}e_{21} = 1, \end{cases}$$

we determine the required components:

$$e_{21} = -\frac{b_2 - a_2}{l}, \quad e_{22} = \frac{b_1 - a_1}{l}. \tag{3.19}$$

Write the radius vector of point ξ in the GRCS as a sum of the radius vector of point c in the GRCS and the radius vector of point ξ in the LRCS:

$$\vec{r}(\xi) = x\vec{i} + y\vec{j} = c_1\vec{i} + c_2\vec{j} + \xi_1\vec{e}_1 + \xi_2\vec{e}_2.$$

By inserting the LRCS unit vector expressions (Equations 3.18 and 3.19) in this formula and equating the corresponding coordinates we will obtain the formula of correlation of the point coordinates in the GRCS and LRCS:

$$\begin{cases} x = c_1 + \xi_1\dfrac{b_1 - a_1}{l} - \xi_2\dfrac{b_2 - a_2}{l}, \\[3mm] y = c_2 + \xi_1\dfrac{b_2 - a_2}{l} + \xi_2\dfrac{b_1 - a_1}{l}. \end{cases} \tag{3.20}$$

The inverse transformation has the following form:

$$\begin{cases} \xi_1 = \dfrac{1}{l}\left[(x - c_1)(b_1 - a_1) + (b_2 - a_2)(y - c_2)\right], \\[3mm] \xi_2 = \dfrac{1}{l}\left[(y - c_2)(b_1 - a_1) - (b_2 - a_2)(x - c_1)\right]. \end{cases} \tag{3.21}$$

The interpolating functions for vertexes a and b, respectively, are calculated from the formulas

$$\psi_a = \frac{1}{l}\left(\frac{l}{2} - \eta_1\right), \quad \psi_b = \frac{1}{l}\left(\frac{l}{2} + \eta_1\right). \tag{3.22}$$

Integrals $\dfrac{1}{2\pi}\displaystyle\int_B \dfrac{n_1\rho_1 + n_2\rho_2}{\rho^2}\psi dB$ and $\dfrac{1}{2\pi}\displaystyle\int_B \dfrac{n_1\rho_2 - n_2\rho_1}{\rho^2}\psi dB$ are calculated analytically. Let point ξ have coordinates (x, y) in the GRCS and $(\xi_1\xi_2)$ in the LRCS. Let the coordinates of the random point η on ab interval be denoted as (η_1, η_2). In the LRCS $\eta_2 = 0$ and the unit outward normal coordinates are $\vec{n} = \{0, -1\}$. Vector $\vec{\rho} = \vec{r}(\eta) - \vec{r}(\xi) = \{\eta_1 - \xi_1, -\xi_2\}$, $n_1\rho_1 + n_2\rho_2 = \xi_2$, $\rho = \sqrt{(\eta_1 - \xi_1)^2 + \xi_2^2}$. By inserting the resulting formulas in subintegral functions we obtain the values of integrals for various interpolating functions.

For vertex a

$$\frac{1}{2\pi}\int_B \frac{n_1\rho_1 + n_2\rho_2}{\rho^2}\frac{1}{l}\left(\frac{l}{2} - \eta_1\right)dB = \frac{1}{2\pi}\int_{-l/2}^{l/2} \frac{\xi_2}{(\eta_1 - \xi_1)^2 + \xi_2^2}\frac{1}{l}\left(\frac{l}{2} - \eta_1\right)d\eta_1$$

$$= \frac{1}{2\pi}\frac{\xi_2}{l}\left[\frac{l - 2\xi_1}{2\xi_2}\operatorname{arctg}\frac{l - 2\xi_1}{2\xi_2} - \frac{1}{2}\ln\left|\left(1 - \frac{2}{l}\xi_1\right)^2 + \left(\frac{2\xi_2}{l}\right)^2\right|\right]$$

$$-\frac{1}{2\pi}\frac{\xi_2}{l}\left[\frac{l - 2\xi_1}{2\xi_2}\operatorname{arctg}\frac{-l - 2\xi_1}{2\xi_2} - \frac{1}{2}\ln\left|\left(-1 - \frac{2}{l}\xi_1\right)^2 + \left(\frac{2\xi_2}{l}\right)^2\right|\right]. \tag{3.23}$$

$$\frac{1}{2\pi}\int_B \frac{n_1\rho_2 - n_2\rho_1}{\rho^2}\frac{1}{l}\left(\frac{l}{2} - \eta_1\right)dB = \frac{1}{2\pi}\int_{-l/2}^{l/2} \frac{\eta_1 - \xi_1}{(\eta_1 - \xi_1)^2 + \xi_2^2}\frac{1}{l}\left(\frac{l}{2} - \eta_1\right)d\eta_1$$

$$= \frac{1}{4\pi}\left[\frac{2\xi_2}{l}\operatorname{arctg}\frac{l - 2\xi_1}{2\xi_2} - 1 + \frac{l - 2\xi_1}{2l}\ln\left|\left(1 - \frac{2}{l}\xi_1\right)^2 + \left(\frac{2\xi_2}{l}\right)^2\right|\right]$$

$$-\frac{1}{4\pi}\left[\frac{2\xi_2}{l}\operatorname{arctg}\frac{-l - 2\xi_1}{2\xi_2} + 1 + \frac{l - 2\xi_1}{2l}\ln\left|\left(-1 - \frac{2}{l}\xi_1\right)^2 + \left(\frac{2\xi_2}{l}\right)^2\right|\right]. \tag{3.24}$$

For vertex b

$$\frac{1}{2\pi}\int_B \frac{n_1\rho_1 + n_2\rho_2}{\rho^2}\frac{1}{l}\left(\frac{l}{2} + \eta_1\right)dB = \frac{1}{2\pi}\int_{-l/2}^{l/2} \frac{\xi_2}{(\eta_1 - \xi_1)^2 + \xi_2^2}\frac{1}{l}\left(\frac{l}{2} + \eta_1\right)d\eta_1$$

$$= \frac{1}{2\pi}\frac{\xi_2}{l}\left[\frac{l + 2\xi_1}{2\xi_2}\operatorname{arctg}\frac{l - 2\xi_1}{2\xi_2} + \frac{1}{2}\ln\left|\left(1 - \frac{2}{l}\xi_1\right)^2 + \left(\frac{2\xi_2}{l}\right)^2\right|\right]$$

$$-\frac{1}{2\pi}\frac{\xi_2}{l}\left[\frac{l + 2\xi_1}{2\xi_2}\operatorname{arctg}\frac{-l - 2\xi_1}{2\xi_2} + \frac{1}{2}\ln\left|\left(-1 - \frac{2}{l}\xi_1\right)^2 + \left(\frac{2\xi_2}{l}\right)^2\right|\right]. \tag{3.25}$$

$$\frac{1}{2\pi}\int_B \frac{n_1\rho_2 - n_2\rho_1}{\rho^2}\frac{1}{l}\left(\frac{l}{2}+\eta_1\right)dB = \frac{1}{2\pi}\int_{-l/2}^{l/2}\frac{\eta_1-\xi_1}{(\eta_1-\xi_1)^2+\xi_2^2}\frac{1}{l}\left(\frac{l}{2}+\eta_1\right)d\eta_1$$

$$=\frac{1}{4\pi}\left[-\frac{2\xi_2}{l}\operatorname{arctg}\frac{l-2\xi_1}{2\xi_2}+1+\frac{l+2\xi_1}{2l}\ln\left|\left(1-\frac{2}{l}\xi_1\right)^2+\left(\frac{2\xi_2}{l}\right)^2\right|\right]$$

$$-\frac{1}{4\pi}\left[-\frac{2\xi_2}{l}\operatorname{arctg}\frac{-l-2\xi_1}{2\xi_2}-1+\frac{l+2\xi_1}{2l}\ln\left|\left(-1-\frac{2}{l}\xi_1\right)^2+\left(\frac{2\xi_2}{l}\right)^2\right|\right]. \qquad (3.26)$$

In order to calculate $H = \{h_{ij}\}$ and $L = \{l_{ij}\}$ matrix elements we will take i-node as point ξ. To determine elements h_{ij}, l_{ij}, we have to find integrals $\dfrac{1}{2\pi}\displaystyle\int_B \dfrac{n_1\rho_1 + n_2\rho_2}{\rho^2}\psi dB$ and $\dfrac{1}{2\pi}\displaystyle\int_B \dfrac{n_1\rho_2 - n_2\rho_1}{\rho^2}\psi dB$ taken over the boundary intervals with a vertex in j-node, and then summarize the respective integrals.

A separate consideration is needed in the case when $i = j$. Envelop $i = j$ node with an arc of a circle with radius $\varepsilon \to 0$ (Figure 3.3).

We assume that $\psi = 1$ and $|\vec{\rho}| = \varepsilon$ on the circular arc. Then the integrals will be calculated as follows:

$$\frac{1}{2\pi}\int_B \frac{\rho_1 n_1 + \rho_2 n_2}{\rho^2}dB = \frac{1}{2\pi}\int_{-\pi}^{\alpha}\frac{\varepsilon\cdot|\vec{n}|\cos 0}{\varepsilon^2}\varepsilon d\theta = \frac{1}{2\pi}\int_{-\pi}^{\alpha}d\theta = \frac{\pi+\alpha}{2\pi},$$

$$\frac{1}{2\pi}\int_B \frac{n_1\rho_2 - n_2\rho_1}{\rho^2}d\theta = \frac{1}{2\pi}\int_{-\pi}^{\alpha}\frac{|\vec{n}|\cos\theta\cdot\varepsilon\sin\theta - |\vec{n}|\sin\theta\cdot\varepsilon\cos\theta}{\varepsilon^2}\varepsilon d\theta = 0.$$

Thus, the matrix elements in the case of i- and j-nodes convergence are calculated from the following formulas:

$$h_{ij} = \frac{1}{2\pi}(\alpha + \pi), \quad l_{ij} = 0, \qquad (3.27)$$

where α is an angle between the previous and the next boundary intervals (Figure 3.3).

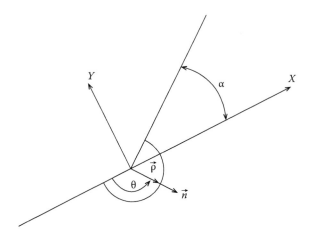

FIGURE 3.3   Regarding the calculation of elements of the main diagonal of H and L matrixes.

Generalizing the foregoing we obtain an algorithm for the calculation of $H = \{h_{ij}\}$ and $L = \{l_{ij}\}$ matrix elements, which includes the following steps:

1. An initial number of i-node is given (initially $i = 1$).
2. An initial number of j-node is given. At the first stage $j = N + 1$.
3. Matrix elements are assigned to zero:

$$h_{ij} = 0, \quad l_{ij} = 0.$$

4. An initial number for the boundary interval is given: $k = 1$.
5. There are k-interval a and b vertex coordinates compared with j-node coordinates. If j-node coincides with the interval vertexes at $i \neq j$, the elements are reassigned for a vertex as follows:

$$h_{ij} = h_{ij} + \frac{1}{2\pi} \int_{-l/2}^{l/2} \frac{\xi_2}{(\eta_1 - \xi_1)^2 + \xi_2^2} \frac{1}{l}\left(\frac{l}{2} - \eta_1\right) d\eta_1,$$

$$l_{ij} = l_{ij} + \frac{1}{2\pi} \int_{-l/2}^{l/2} \frac{\eta_1 - \xi_1}{(\eta_1 - \xi_1)^2 + \xi_2^2} \frac{1}{l}\left(\frac{l}{2} - \eta_1\right) d\eta_1,$$

For vertex b

$$h_{ij} = h_{ij} + \frac{1}{2\pi} \int_{-l/2}^{l/2} \frac{\xi_2}{(\eta_1 - \xi_1)^2 + \xi_2^2} \frac{1}{l}\left(\frac{l}{2} + \eta_1\right) d\eta_1,$$

$$l_{ij} = l_{ij} + \frac{1}{2\pi} \int_{-l/2}^{l/2} \frac{\eta_1 - \xi_1}{(\eta_1 - \xi_1)^2 + \xi_2^2} \frac{1}{l}\left(\frac{l}{2} + \eta_1\right) d\eta_1,$$

where integrals are taken from formulas given in Equations 3.23 through 3.26; (ξ_1, ξ_2) are i-node coordinates in the LRCS for k-interval as determined from Equation 3.21 wherein (x, y) are j-node coordinates in the GRCS; (a_1, a_2) and (b_1, b_2) are coordinates of k-interval vertexes in the GRCS; (c_1, c_2) are k-interval center coordinates determined from Equation 3.17; l is the k-interval length.

When $i = j$, the matrix elements are determined from Equation 3.27.

6. k increment: $k = k + 1$ and calculations are repeated from step 5 as long as $k \leq Ge$.
7. j increment: $j = j + 1$ and calculations are repeated from step 3 as long as $j \leq N + N_b$.
8. i is increased by unit: $i = i + 1$ and calculations are repeated from step 2 as long as $i \leq N + N_b$.

3.1.5 Test Case: Backward Step Flow

For the purpose of validating the algorithm of numerical calculation there was a problem of experimentally studied backward step flow resolved [96]. The boundary conditions upstream and downstream of the channel are shown in Figure 3.4c. There is a slipping condition set over the channel upper boundary and the velocity varies linearly from the inlet section velocity to the outlet section velocity. The velocity is equal to zero on the remaining boundary. Kinematic viscosity coefficient is $\nu = 0.01$ m²/s. There were 744 inner cells (rectangular isosceles triangles and 104 boundary intervals) used for discretizing the flow area (Figure 3.4a).

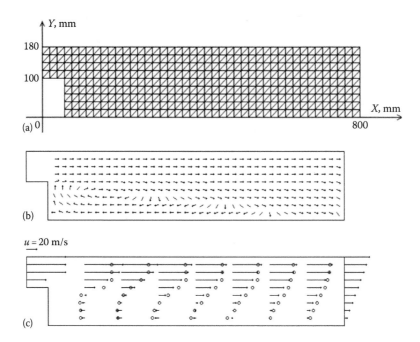

FIGURE 3.4 Gas flow behind the backward step: (a) the flow area discretization; (b) the quality flow pattern, air directional field; and (c) design and experimental velocity profiles.

The average inlet section velocity is $u_{av} = 50.7$ m/s. The arithmetic mean average departure of the estimated velocity values from the experimental values (Figure 3.4c)

$$\frac{\sum_{i=1}^{N} \dfrac{u_i - (u_э)_i}{(u_э)_i}}{N} = 0.194$$

($N = 56$, which is the number of reference points; u_i, $(u_э)_i$ are estimated and experimental velocity values), which is acceptable for engineering design.

3.1.6 Calculation of the Mutual Effect of Two Straight Supply Air Jets

We assume that two slot-type outlets, each 0.04 m wide, in a plane boundless wall discharge gas with the same velocity $u_{av} = 20$ m/s. It is assumed that the gas is viscous, incompressible, and no bending of the axes of supply air jets takes place. It is required to determine the gas velocity on the symmetry axis of the obtained system of two flat jets and find the geometric parameters at which the reverse flow axial velocity is maximal.

The field of velocities was calculated at various distance d of inlet slots from the symmetry axis: from 0.02 to 0.088 m (Figures 3.5 and 3.6). The vortex region length and the reverse flow velocity are proportional to d while the maximum velocity $u_{max} = 9.64$ m/s (Figures 3.6 and 3.7) is observed at $d = 0.08$ m at the symmetry axis point located at the distance of 0.13 m from the wall where jets are flowing from. In the last case, the flow area was discretized by a set of 1600 inner cells and 120 boundary intervals (Figure 3.6a).

If we calculate 0.1 m wide slot-type exhaust outlet in a plane boundless wall (see Section 1.3) using the conformal mapping method provided that the exhaust gas flow is equal to the aggregate flow of both supply air jets, it will emerge that the axial velocity at the point located at the distance of 0.15 m away from the exhaust inlet is 3.25 m/s, which is 2.9 times as little as for supply air jets (Figure 3.8).

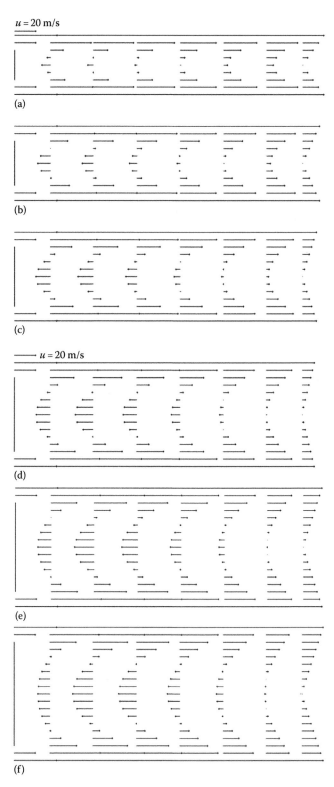

FIGURE 3.5 Velocity profiles in the interaction of two parallel supply air jets at distance d from the symmetry axis. (a) $d = 0.02$ m, (b) $d = 0.03$ m, (c) $d = 0.04$ m, (d) $d = 0.05$ m, (e) $d = 0.06$ m, and (f) $d = 0.07$ m.

(Continued)

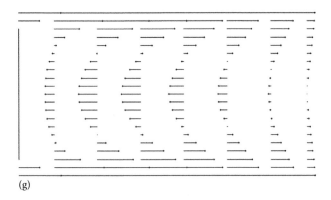

(g)

FIGURE 3.5 (*Continued*)   Velocity profiles in the interaction of two parallel supply air jets at distance d from the symmetry axis. (g) $d = 0.088$ m.

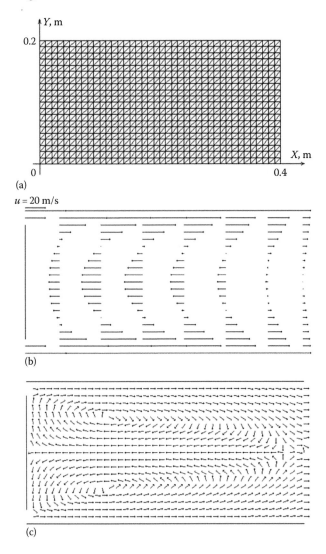

FIGURE 3.6   Interaction of two parallel supply air jets at the distance of 0.08 m from the symmetry axis: (a) the range discretization, (b) horizontal velocity component profiles, and (c) the air directional field.

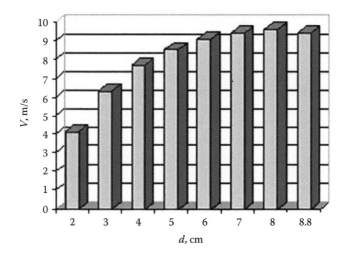

FIGURE 3.7   Relation between the axial air velocity and the supply air jet distance (d) to the symmetry axis.

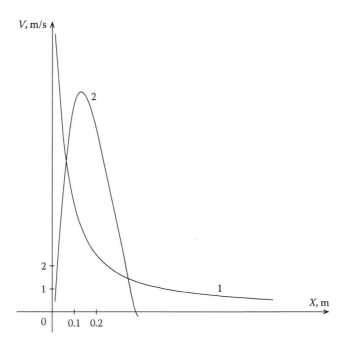

FIGURE 3.8   Relation between the axial air inflow velocity at slot-type suction (curve 1) and two parallel supply air jets at the distance of 0.08 m from the symmetry axis (curve 2); the distance to the suction inlet.

Thus, gas backflow region occurs by the interaction of two parallel jets flowing from slot-type outlets into a boundless space. The resulting vortex region contributes to an acceleration of the axial velocity of gas at the slot-type suction inlet screened off with supply air jets across the length of gas backflows. At certain points, the axial velocity may be exceeded 2 or 3 times.

In order to enhance the efficiency of a screened local suction unit one ought to use the elements of process equipment within the exhaust plume action range. Where the ventilation system is arranged above a plane table, the inflow velocity at the local suction inlet can be increased several dozens of times due to a screening effect of supply air jets and the table effect (see Section 3.5) [80].

3.2   FLOW SIMULATION USING A VORTEX LAYER

When simulating flows close to inlet ports of local exhausts using the boundary integral approach, we used sources and sinks distributed along the flow area boundary, that is, there was a so-called simple layer used. It is of interest to distribute vortices over the boundary surface or, in other words, to apply a vortex layer. Such vortex layer as described in the study by Professor N.Ya. Fabrikant [53] is kinematically equivalent to the boundary layer. This is explained by the fact that there is a quick variation in velocity along the normal direction and, accordingly, rotation of particles taking place in close proximity to the impermeable surface.

The critical importance of the notion of vortex (vorticity) is referred to in the article by Professor O.G. Goman [79] who points out that there are seemingly no reasons for particularly distinguishing a gradient of the field of velocities $\vec{\omega} = \text{rot } \vec{v}$ among others. However, it is appropriate to take such gradient for the initial notion for representing a field of liquid velocities. But the nonrotational flow condition $\text{rot } \vec{v} = 0$ over the range actually results in liquid ceasing to be as such, since it loses the infinite number of degrees of freedom that predetermine the endless variety of liquid motion patterns.

Resolving problems of aerodynamics using vortices are called the vortex method while their numerical implementation is commonly called the discrete vortex method [79,112,113]. The discrete vortex method is applied in aerodynamics to study flows at bearing surfaces. A continuous vortex layer simulating such surface as well as a trace it leaves behind is substituted with a system of discrete vortices. There are reference points chosen on a bearing surface so that the condition of impermeability is met. Unknown circulations are determined by resolving a set of linear algebraic equations. The discrete vortex method had been formulated for the first time in 1955 by Professor S.M. Belotserkovsky. For practical experience, perspectives, and mathematical validation of the method, refer to [79,112,113].

There is a suction inlet airflow model developed using a continuously distributed vortex layer. Instead of cumulative vortices, the range discretization utilizes vortex intervals, that is, intervals at which such vortices are continuously distributed.

Let S contour line limit A flow area. The boundary normal air velocity $\vec{v}_n(x)$ is given. Let us distribute linear vortices with circulation $\Gamma = \Gamma(\xi)$ along the flow area boundary. It is necessary to determine the air velocity at inner points of the range along any given direction. Air velocity $\vec{v}_n(x)$ at point x (x_1, x_2) along direction $\vec{n} = \{n_1, n_2\}$ induced by the linear vortex effect at point $\xi(\xi_1, \xi_2)$ is determined using the following formula:

$$\vec{v}_n(x) = \Gamma(\xi)G(x,\xi), \tag{3.28}$$

where

$$G(x, \xi) = \frac{n_2(x_1 - \xi_1) - n_1(x_2 - \xi_2)}{2\pi\left[(x_1 - \xi_1)^2 + (x_2 - \xi_2)^2\right]}$$

$\{n_1, n_2\}$ are $\vec{n}$ unit vector coordinates
$\{x_1, x_2\}$ are x inner point coordinates
(ξ_1, ξ_2) are the coordinates of point ξ which is at the range boundary

Having summarized the effect of all such vortices continuously distributed along the boundary contour on point x, we obtain the following integral equation:

$$v_n(x) = \int_S \Gamma(\xi)G(x, \xi)dS(\xi). \tag{3.29}$$

We will divide boundary S into N straight intervals. Let us assume that linear vortex circulation $\Gamma(\xi)$ is constant across the length of each interval, then a discrete countertype of Equation 3.29 will be as follows:

$$v_n(x) = \sum_{k=1}^{N} \Gamma^k \int_{\Delta S^k} G(x, \xi^k) dS(\xi^k), \tag{3.30}$$

where

integrals are taken over intervals ΔS^k

ξ^k means that a random point of k-interval is a variable of integration

Γ^k is the circulation at any point of k-interval

By directing inner point x to boundary point x_0 we obtain the boundary integral equation

$$v_n(x_0) = \sum_{k=1}^{N} \Gamma^k \int_{\Delta S^k} G(x_0, \xi^k) dS(\xi^k). \tag{3.31}$$

Note that in the case of convergence of x_0 and ξ^k points the integral becomes zero because the vortex has no effect on itself [53]. Using point x_0 as the center of p-interval x_0^p and taking all N of boundary intervals, we obtain a system of N linear algebraic equations with N unknown variables:

$$\sum_{k=1}^{N} \Gamma^k \int_{\Delta S^k} G(x_0^p, \xi^k) dS(\xi^k) = v_n(x_0^p), \tag{3.32}$$

where $p = 1, 2, 3, \ldots, N$.

We add the condition of impermeable surface irrotational flow to this set:

$$\sum_{k=1}^{N} \Gamma^k l^k = 0, \tag{3.33}$$

where l^k is k-interval length.

We obtain an overdetermined system of $N + 1$ equations with N unknown variables, which is typically inconsistent. Therefore, following a study [113] by Professor I.K. Lifanov we will introduce a new additional variable (regulating factor) γ.

Denoting

$$\int_{\Delta S^k} G(x_0^p, \xi^k) dS(\xi^k) = G^{pk}, \quad v_n(x_0^p) = v^p$$

and using this, we will obtain a system of $N + 1$ linear algebraic equations with $N + 1$ unknown variables:

$$\begin{cases} \Gamma^1 G^{11} + \Gamma^2 G^{12} + \Gamma^3 G^{13} + \cdots + \Gamma^N G^{1N} + \gamma = v^1, \\ \Gamma^1 G^{21} + \Gamma^2 G^{22} + \Gamma^3 G^{23} + \cdots + \Gamma^N G^{2N} + \gamma = v^2, \\ \Gamma^1 G^{31} + \Gamma^2 G^{32} + \Gamma^3 G^{33} + \cdots + \Gamma^N G^{3N} + \gamma = v^3, \\ \vdots \\ \Gamma^1 G^{N1} + \Gamma^2 G^{N2} + \Gamma^3 G^{N3} + \cdots + \Gamma^N G^{NN} + \gamma = v^N, \\ \Gamma^1 l^1 + \Gamma^2 l^2 + \Gamma^3 l^3 + \cdots + \Gamma^N l^N = 0, \end{cases}$$

solving which using, for example, the Gauss method with the basic element selection, we will determine the unknown variables in circulation $\Gamma^1, \Gamma^2, \Gamma^3, \ldots, \Gamma^N$.

G^{pk} integral is calculated analytically in the local rectangular Cartesian coordinate system (LRCCS). We assume that the center of p-interval, that is, point x_0^p, is given in XOY GRCCS and the center of k-interval $[a, b]$, that is, point c^k is also given in the GRCCS. We will build the local rectangular Cartesian coordinate system with the center at point c^k, direct X-axis ξ_2 along $\overrightarrow{ab}$ vector, and position Y-axis ξ_1 so that the resulting coordinate system is right handed (Figure 2.4). LRCCS $\eta_1\eta_2$ is obtained from $\xi_1\xi_2$ by means of a parallel shift to point x_0^p. For coordinates of k-interval random point ξ in LRCCS $\eta_1\eta_2$, it is true that $\eta_1 = h$, $\eta_2 = htg\theta$, while for point x^p: $\eta_1 = 0$, $\eta_2 = 0$. Denoting the coordinates of the unit vector of the outward normal to p-interval in $\eta_1\eta_2$ coordinate system as $\{n_1, n_2\}$, we will obtain the following expression for $G(x_0^p, \xi^k)$ function:

$$G(x_0^p, \xi^k) = (n_2(0-h) - n_1(0 - htg\theta))/(2\pi[(0-h)^2 + (0 - htg\theta)^2])$$

$$= [(n_1 tg\theta - n_2)/2\pi h]\cos^2\theta.$$

It follows that

$$G^{pk} = \int_{\Delta S^k} G(x_0^p, \xi^k)dS(\xi^k) = \{dS = d\xi_2 = (h/\cos^2\theta)d\theta\} = -1/2\pi(n_1 \ln|\cos\theta| + n_2)\Big|_{\theta_a}^{\theta_b}.$$

Finally, we obtain

$$G^{pk} = 1/2\pi[n_1 \ln(r_b/r_a) - n_2(\theta_b - \theta_a)], \tag{3.34}$$

where r_a and r_b are lengths of the respective radius vectors of a and b ends of k-interval.

The algorithm for calculation of G^{pk} includes the following stages:

1. An initial number of p-interval is given: $p = 1$.
2. Global coordinates of unitary vectors in the LRCCS built on p-interval are calculated:

$$e_{2x}^p = \frac{b_1^p - a_1^p}{\sqrt{\left(b_1^p - a_1^p\right)^2 + \left(b_2^p - a_2^p\right)^2}}, \quad e_{2y}^p = \frac{b_2^p - a_2^p}{\sqrt{\left(b_1^p - a_1^p\right)^2 + \left(b_2^p - a_2^p\right)^2}},$$

$$e_{1x}^p = e_{2y}^p, \quad e_{1y}^p = -e_{2x}^p.$$

Then the outward normal to p-interval is $\vec{n}^p = \vec{e}_1^p$ if $\vec{e}_1^p$ is directed beyond the flow area, that is, in case when passing from $\dot{a}$ to b the flow area is leftward. If a and b are chosen in such a way that when passing along the boundary from a to b the flow area is rightward, then $\vec{n}^p = -\vec{e}_1^p$.

3. An initial number of k-interval is given: $k = 1$.
4. k-interval LRCCS coordinates are calculated for $\vec{n}^p$. To this effect, the unitary vectors are found first:

$$e_{2x}^k = \frac{b_1^k - a_1^k}{\sqrt{\left(b_1^k - a_1^k\right)^2 + \left(b_2^k - a_2^k\right)^2}}, \, e_{2y}^k = \frac{b_2^k - a_2^k}{\sqrt{\left(b_1^k - a_1^k\right)^2 + \left(b_2^k - a_2^k\right)^2}}, \, e_{1x}^k = e_{2y}^k, \, e_{1y}^k = -e_{2x}^k, \tag{3.35}$$

then the required coordinates are calculated:

$$n_1^{pk} = n_1^p e_{1x}^k + n_2^p e_{1y}^k, \quad n_2^{pk} = n_1^p e_{2x}^k + n_2^p e_{2y}^k.$$

5. GRCCS coordinates are calculated for p-interval center radius vector:

$$x_1^p = \frac{a_1^p + b_1^p}{2}, \quad x_2^p = \frac{a_2^p + b_2^p}{2}$$

and k-interval LRCCS coordinates are determined for the same:

$$x_1^{pk} = x_1^p e_{1x}^k + x_2^p e_{1y}^k, \quad x_2^{pk} = x_1^p e_{2x}^k + x_2^p e_{2y}^k.$$

6. Coordinates for k-interval vertexes a^k, b^k are calculated in p-interval coordinate system ($\eta_1 \eta_2$ in Figure 2.4):

$$\begin{cases} a_1^{kp} = a_1^k e_{1x}^k + a_2^k e_{1y}^k - x_1^{pk}, \\ a_2^{kp} = a_1^k e_{2x}^k + a_2^k e_{2y}^k - x_2^{pk}, \\ b_1^{kp} = b_1^k e_{1x}^k + b_2^k e_{1y}^k - x_1^{pk}, \\ b_2^{kp} = b_1^k e_{2x}^k + b_2^k e_{2y}^k - x_2^{pk}. \end{cases} \tag{3.36}$$

7. Then, there is a calculation performed to determine the lengths of the radius vectors of points a^k, b^k in p-interval coordinate system:

$$r_a = \sqrt{\left(a_1^{kp}\right)^2 + \left(a_2^{kp}\right)^2}, \quad r_b = \sqrt{\left(b_1^{kp}\right)^2 + \left(b_2^{kp}\right)^2} \tag{3.37}$$

and angles

$$\theta_a = \begin{cases} \text{arctg}\, \dfrac{a_2^{kp}}{a_1^{kp}} \text{ if } a_1^{kp} \neq 0; \\ \dfrac{\pi}{2} \text{ if } a_1^{kp} = 0, \end{cases} \qquad \theta_b = \begin{cases} \text{arctg}\, \dfrac{b_2^{kp}}{b_1^{kp}} \text{ if } b_1^{kp} \neq 0; \\ \dfrac{\pi}{2} \text{ if } b_1^{kp} = 0. \end{cases} \tag{3.38}$$

8. The matrix element is calculated:

$$G^{pk} = -1/2\pi \left\{ n_2^{pk}(\theta_b - \theta_a) - n_1^{pk} \ln(r_b/r_a) \right\}.$$

9. k increment: $k = k + 1$ and the calculation is repeated from step 4 as long as $k \leq N$.
10. p increment: $p = p + 1$ and the calculation is repeated from step 2 as long as $p \leq N$.

The following steps should be performed in order to determine the air velocity at random point $x(x_1, x_2)$ of flow area A along any preset direction $\vec{n} = \{n_1, n_2\}$.

1. Define $k = 1$.
2. Find coordinates of vectors $\vec{e}_1^k$, $\vec{e}_2^k$ using the formula given as Equation 3.35 and then the coordinates of vector $\vec{n}$ in k-interval LRCCS:

$$n_1^k = n_1 e_{1x}^k + n_2 e_{1y}^k, \quad n_2^k = n_1 e_{2x}^k + n_2 e_{2y}^k.$$

3. Find coordinates of point x in k-interval LRCCS:

$$x_1^k = x_1 e_{1x}^k + x_2 e_{1y}^k, \quad x_2^k = x_1 e_{2x}^k + x_2 e_{2y}^k.$$

4. Determine the coordinates of points a^k, b^k in the LRCCS with the center at point x and unitary vectors $\vec{e}_1^{\,k}$, $\vec{e}_2^{\,k}$ from the formula given in Equation 3.36.
5. Find r_a, r_b from the formulas given in Equation 3.37 and angles θ_a, θ_b from the formulas given in Equation 3.38.
6. Calculate

$$g^k = 1/2\pi \cdot \left\{ n_1^k \ln(r_b/r_a) - n_2^k(\theta_b - \theta_a) \right\}.$$

7. Increase k: $k = k + 1$ and repeat the calculation from step 2 as long as $k \le N$.
8. Determine the required velocity value:

$$v_n(x) = g^1 \Gamma^1 + g^2 \Gamma^2 + \cdots + g^N \Gamma^N.$$

As an example we consider airflow next to 0.1 m wide slot-type suction unit in which operating range there is a fixed cylinder with the radius of 0.25 m. The distance from the cylinder axis to the suction unit is 0.5 m (Figure 3.9). The origin of coordinates is in the cylinder center while X-axis is parallel to the impermeable surface. The flow lines were plotted starting from points with y-coordinate −0.49 and x-coordinates −0.05; −0.03; −0.015; 0; −0.001 or 0.001; 0.015; 0.03; 0.05. Their plotting algorithm is similar to the Euler method of resolution of an ordinary differential equation. There were horizontal and vertical velocity components calculated with a further increment in a direction opposite to the velocity vector. Then the velocity direction was calculated at the resulting point once again followed by another increment and so until the count interruption condition was met. The flow line increment was chosen to be min 0.0002. The number of points on the line was max 35,000.

The flow was simulated using a simple and a vortex layer. When sources and sinks located on the flow boundary are used (Figure 3.9), a smooth flow past cylinder occurs. It should be noted that here the range was divided into intervals nonuniformly: the discretization was more frequent where the problem boundary conditions were changed.

The similar discretization of the range using a vortex layer (Figure 3.10a) disregarding the condition Equation 3.33 shows jogging of the flow lines around the cylinder. Besides, the flow lines extending from points (−0.05; −0.49) and (0.05; −0.49) form closed curves that cross the impermeable wall. Thus, in modeling using a vortex layer it is necessary to use either a linear variation of

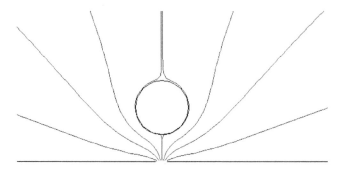

FIGURE 3.9 The lines of flow past cylinder in the range of a slot-type suction built in a plane wall plotted using a simple layer (110 boundary intervals).

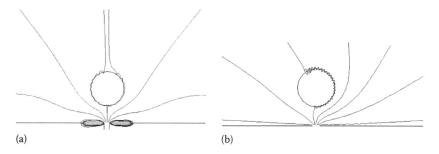

(a)   (b)

FIGURE 3.10 The lines of flow past cylinder in the range of a slot-type suction built in a plane wall plotted using a vortex layer without any noncircular flow condition observed (Equation 3.33): (a) 138 boundary intervals and (b) 500 boundary intervals.

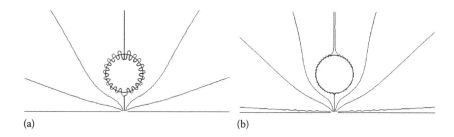

(a)   (b)

FIGURE 3.11 The lines of the flow past cylinder in the range of a slot-type suction built in a plane wall plotted using a vortex layer in the noncircular flow conditions (Equation 3.33): (a) 250 boundary intervals and (b) 500 boundary intervals.

circulation along the boundary interval or just a uniform discretization of the range. If the condition of Equation 3.33 is disregarded in the case of a uniform partition of the range, a flow past cylinder with a small circulation occurs (Figure 3.10b). If there is noncircular flow condition (Equation 3.33) used when a regulating variable is introduced, the flow will be symmetrical. There is a sharp change in the flow direction observed next to the cylinder (Figure 3.11). In case of a considerable increase in the number of boundary intervals (Figure 3.11b) these changes are observed at the cylinder surface itself and can be used as a model of the boundary layer around an aerodynamic body.

Note that the vortex method allows for solving not only 2- and 3D stationary problems with bound vortex formation but also nonstationary problems where the configuration is changed over time, and, accordingly, it is possible to obtain nonstationary vortex patterns near sharp edges, which will be discussed in the following paragraphs.

3.3 VORTEX FLOW AROUND A SLOT-TYPE SUCTION ABOVE THE RIGHT DIHEDRAL ANGLE

3.3.1 NUMERICAL COMPUTATION ALGORITHM

The problem of a flow around a slot-type suction unit above the right dihedral angle was solved analytically in [114] based on the Joukowski method and experimentally [115]. It is of interest to compare the research results obtained by different means. The set problem can be deemed a model (or a test) problem that is an element of much more complicated problems of industrial ventilation.

We assume that there is velocity v_0 given in *BC* intake opening (Figure 3.12) and the no-fluid-loss condition given for the rest part of the boundary. It is required to determine the air velocity at any point of the flow area, and plot the flow lines and the flow vortex pattern.

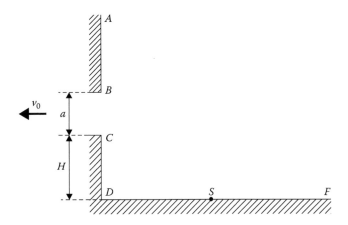

FIGURE 3.12   Flow area boundaries.

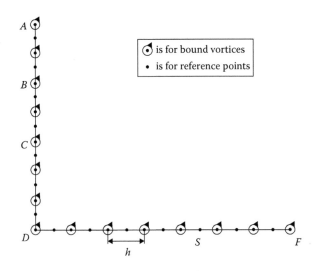

FIGURE 3.13   Flow area boundary discretization.

Let vortices be placed on the flow area boundary (Figure 3.13) and reference points be put right in between these bound vortices where the boundary conditions for normal air velocity component be fulfilled (it is $v_n = 0$ everywhere except BC interval where $v_n = v_0$). The distance between any two adjacent reference points or two adjacent bound vortices is the same and is equal to h.

We consider the initial instant $t = 0$. There, only bound vortices exist with $t = 0$. There are no free vortices detached from point S (flow separation point). The velocity at reference point x_i along the direction of the external normal to the boundary will be determined by summing up the action on it by all vortices:

$$v_n(x_i) = \sum_{j=1}^{n} \Gamma(\xi_j)G(x_i, \xi_j), \tag{3.39}$$

where

$\Gamma(\xi_j)$ is a circulation of the vortex at point ξ_j

n is the number of bound vortices

$$G(x_i, \xi_j) = \frac{n_2(x_1 - \xi_1) - n_1(x_2 - \xi_2)}{2\pi \cdot [(x_1 - \xi_1)^2 + (x_2 - \xi_2)^2]}$$

$\{n_1, n_2\}$ are the coordinates of outward normal $\vec{n}$ at point $x_i(x_1, x_2)$

(ξ_1, ξ_2) are the coordinates of vortex location point ξ_j

Function $G(x_i, \xi_j)$ expresses the value of velocity at point x_i induced by the singular circulation vortex action at point ξ_j.

Note that in the case of discretization of the flow area shown in Figure 3.13 the number of reference point is 1 less than the number of bound vortices. Therefore, by changing i in the formula given in Equation 3.39 from 1 to $n - 1$ we will obtain $n - 1$ system of linear algebraic equations with n unknown variables. Let us add the condition of vortex circulation stability to Equation 3.39:

$$\sum_{j=1}^{n} \Gamma(\xi_j) = 0. \tag{3.40}$$

By solving the resulting set of equations we will find unknown variables $\Gamma_1, \Gamma_2, ..., \Gamma_n$.

We consider the following instant $t = \Delta t$.

The first free vortex of γ_1 intensity equal to the circulation of the bound vortex located at point S is separating from point S along the direction of the normal to the boundary into the interior of the range. Note that determining the point of vortex layer separation from the smooth surface is a problem that is quite difficult to solve—in view of viscosity and using boundary layer equations. We put forward the following kinematic condition for the flow separation as a hypothesis: the vortex layer is separated between the reference points of different tangential velocity signs. In the strict sense, such separation must take place at a tangent to the surface. However, due to the model discreteness it cannot be achieved since the separated vortex may fly out of the flow area boundary. Therefore, the first free vortex will be placed above the separation point S at the distance equal to the discreteness increment $h/2$. Then it is moving along a liquid particle path. Eventually, point S would probably change its position, and thus, its position must be redefined at each instant of time. It would appear reasonable that in case of a significant increase at t point S will be *floating* no longer.

Thus, the first free vortex has separated from point S. Then the systems in Equations 3.39 and 3.40 will look as follows:

$$\begin{cases} \sum_{j=1}^{n} \Gamma(\xi_j) G(x_i, \xi_j) = v_n(x_i) - \gamma_1 G(x_i, y_1), \\ \\ \sum_{j=1}^{n} \Gamma(\xi_j) + \gamma_1 = 0, \end{cases} \tag{3.41}$$

where y_1 is the first free vortex location point (at distance $h/2$ from point S along the normal into the interior of the range). This free vortex intensity γ_1 remains unchanged, that is, the vortex will be moving with this intensity until it reaches BC suction point.

Then, the system of Equation 3.41 is resolved and circulations $\Gamma_1, \Gamma_2, ..., \Gamma_n$ are found again.

The next instant of time is $t = 2 \cdot \Delta t$. The vortex of γ_2 intensity is separating from point S (it will also be at distance $h/2$ from point S along the normal into the interior of the range). It is required to determine γ_1 vortex position at the given instant of time. To this effect, it is necessary to find velocity $\{v_x, v_y\}$ at the point of its previous position from the formula of Equation 3.39, where $x_i = \{x_1, x_2\}$ is the point of the previous position of γ_1 vortex. The direction for calculating v_x and v_y is $\vec{n} = \{1,0\}$ and $\vec{n} = \{0,1\}$, respectively. New vortex position $\{x_1', x_2'\}$ will be determined from the formulas

$$x_1' = x_1 + v_x \Delta t, \quad x_2' = x_2 + v_y \Delta t. \tag{3.42}$$

Thus, at the instant of time $t = 2 \cdot \Delta t$ there are already two free vortices γ_1 and γ_2. Herewith, the system given in Equation 3.41 will be converted to the following form:

$$\begin{cases} \sum_{j=1}^{n} \Gamma(\xi_j)G(x_i, \xi_j) = v_n(x_i) - \gamma_1 G(x_i, y_1) - \gamma_2 G(x_i, y_2), \\ \sum_{j=1}^{n} \Gamma(\xi_j) + \gamma_1 + \gamma_2 = 0, \end{cases} \tag{3.43}$$

where y_2 is γ_2 vortex position point. Having solved the system given in Equation 3.43 and determined intensities $\Gamma_1, \Gamma_2, ..., \Gamma_n$ we will redefine positions of vortices γ_1 and γ_2 at another instant of time $t = 3 \cdot \Delta t$. Velocity components for each vortex will be determined from the formula

$$v_n(x_i) = \sum_{j=1}^{n} \Gamma(\xi_j)G(x_i, \xi_j) + \gamma_r \cdot G(x_i, \xi_r), \tag{3.44}$$

where the direction for calculating v_x and v_y is again $\vec{n} = \{1,0\}$ and $\vec{n} = \{0,1\}$, respectively; $r = 1$ for γ_2 and $r = 2$ for γ_1 (none of the vortices has any effect on itself). New coordinates for the vortices will be calculated using the formulas given in Equation 3.42. At the same instant of time $t = 3 \cdot \Delta t$, new free vortex γ_3 is separating from point S.

Let us consider a random instant of time $t = \mathrm{m} \cdot \Delta t$ where we have m free vortices. Herewith, the system from Equation 3.43 will look as follows:

$$\begin{cases} \sum_{j=1}^{n} \Gamma(\xi_j)G(x_i, \xi_j) = v_n(x_i) - \sum_{k=1}^{m} \gamma_k G(x_i, y_k), \tag{3.45} \\ \\ \sum_{j=1}^{n} \Gamma(\xi_j) + \sum_{k=1}^{m} \gamma_k = 0. \tag{3.46} \end{cases}$$

Having solved this system and determined $\Gamma_1, \Gamma_2, ..., \Gamma_n$ we will define positions of vortices γ_1, $\gamma_2, ..., \gamma_m$ at the instant of time $t = (m + 1)\Delta t$. Velocity components for each vortex will be determined from the formula

$$v_n(x_i) = \sum_{j=1}^{n} \Gamma(\xi_j)G(x_i, \xi_j) + \sum_{\substack{r=1, \\ x_i \neq \xi_r}}^{m} \gamma_r \cdot G(x_i, \xi_r). \tag{3.47}$$

New coordinates for the vortices will be calculated using the formulas given in Equation 3.42. At the next instant of time vortex γ_{m+1} is separating from point S.

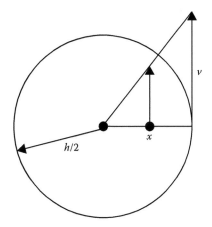

FIGURE 3.14 Regarding the determination of mutual effect of vortices.

If at a certain instant of time the vortex is reaching a distance to a solid wall of less than $h/2$, the vortex will move aside the wall along the normal so that the distance to the flow boundary becomes equal to $h/2$. If the same happens to the vortex and the opening, the vortex disappears.

If two vortices are reaching a distance of less than $h/2$ (Figure 3.14) the mutual effect of these vortices will be determined by means of the linear approximation

$$\frac{v(x)}{v} = \frac{x}{h/2}. \tag{3.48}$$

Longitudinal and transverse velocity pulsations will be determined from the formulas

$$v'_x = v_x - \langle v_x \rangle = v_x - \frac{1}{M} \sum_{k=m}^{M+m} v_{kx}, \quad v'_y = v_y - \langle v_y \rangle = v_y - \frac{1}{M} \sum_{k=m}^{M+m} v_{ky}, \tag{3.49}$$

where

$\langle v_x \rangle$, $\langle v_y \rangle$ are the velocity components averaged for M instants of time

v_{kx}, v_{ky} are the velocity components at k instant of time

v_x, v_y are the velocity components at the given instant of time

m, $M + m$ is the initial and final instants of time for calculation of pulsations

Root-mean-square longitudinal and transverse velocity pulsations will be determined from the formulas

$$\left(\langle v'^2_x \rangle \right)^{1/2} = \left[\frac{1}{M} \sum_{k=m}^{M+m} \left(v'_{kx} \right)^2 \right]^{1/2}, \quad \left(\langle v'^2_y \rangle \right)^{1/2} = \left[\frac{1}{M} \sum_{k=m}^{M+m} \left(v'_{ky} \right)^2 \right]^{1/2}, \tag{3.50}$$

where v'_{kx}, v'_{ky} are the longitudinal and transverse velocity pulsations at k instant of time.

3.3.2 CALCULATION DATA

There was a software program developed based on the computation algorithms built to enable plotting a vortex flow pattern, flow lines, determining velocity fields, calculating velocity pulsations and root-mean-square pulsations, and varying geometric and kinematic flow parameters.

The methodical research conducted using the program allowed for using $h = 0.01$ as the discreteness increment and obtaining the following formula for the time increment:

$$\Delta t = \frac{h}{4v_0}\left(\frac{H}{a}\right)^2. \tag{3.51}$$

The number of bound vortices is 70 for AB interval, 10 for BC interval, and 300 for DF interval. There are 20, 30, and 40 vortices on CD interval that corresponds to the cases $H/a = 2, 3, 4$. In calculations, $v_0 = 1$. As is clear from Figure 3.15 ($H/a = 2$), the separation point is being shifted rightward in time but its motion velocity is decelerated and at a certain instant of time it stops.

The vortex flow area obtained using the program (Figures 3.16a and 3.17) has boundaries similar to the experimental data (Figure 3.18) [114,115].

The visualized vortex area (Figure 3.18) was obtained in [115] in the following manner: The 2D flow was experimentally simulated by means of two horizontal planes. The required exhaust geometry was achieved using vertical plates. Air was sucked off from the space formed that way using a vacuum cleaner. At the beginning of the test, the lower plane was covered with fine onion shells. The vacuum cleaner made all shells except those in the vortex area whirl away to the exhaust. The dotted line in Figure 3.18 denotes the vortex range boundary calculated by the formulas obtained using the Joukowski method [114].

The study [115] suggests the formula to determine the separation point x-coordinate

$$S_x/a = 1.316 \cdot H/a + 0.47. \tag{3.52}$$

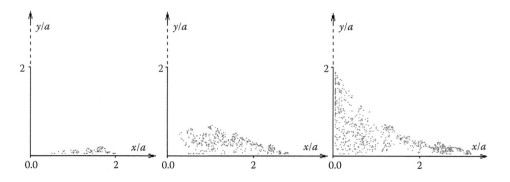

FIGURE 3.15   Vortex pattern development in time ($t = 1, 4, 8$ from left to right).

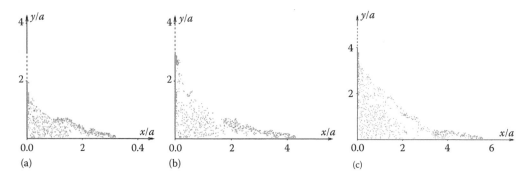

FIGURE 3.16   Vortex flow pattern at (a) $H/a = 2$; $t = 10.04$; $S_x = 0.32$; (b) $H/a = 3$; $t = 21.02$; $S_x = 0.44$; and (c) $H/a = 4$; $t = 33.88$; $S_x = 0.56$.

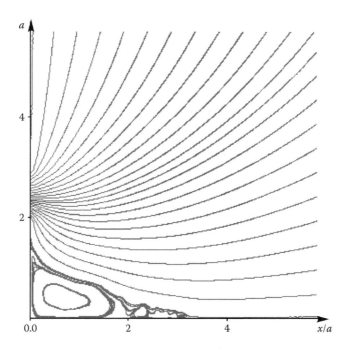

FIGURE 3.17   Flow lines at $H/a = 2$, $t = 10.04$.

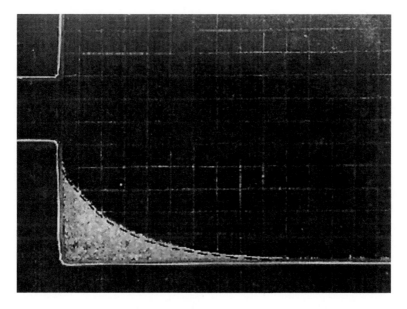

FIGURE 3.18   Photograph of the vortex range boundaries around a lateral suction unit at $H/a = 2$.

The separation point x-coordinate values obtained using the program satisfactorily agree with the experimental values and formula (Equation 3.52) (Table 3.1).

It should be noted, however, that the comparison with the experiment is quite conventional because onion shells are not air particles after all.

Even though the vortex range boundaries are not significant they are constantly varying with time. Vortices occur both in the corner and near the impermeable wall to the left from the separation point (Figure 3.17). Turbulent pulsations are observed both inside (Figure 3.19) and outside

TABLE 3.1

X-Coordinates of Separation Point X_S/a

	$H/a = 1$	$H/a = 2$	$H/a = 3$	$H/a = 4$
Program calculation	2	3.2	4.3	5.6
Experiment	2	3.65	4.7	6.0
Calculation by formula (Equation 3.52)	1.79	3.1	4.2	5.7

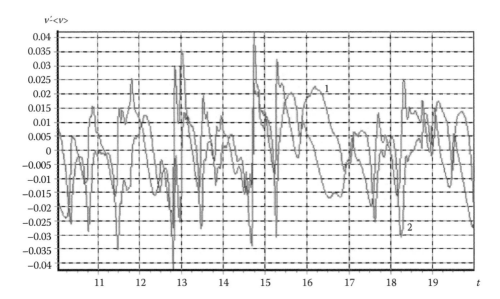

FIGURE 3.19 Longitudinal (1) and transverse (2) velocity pulsations at point $(0, 1; 0, 1)$ $\langle v_x \rangle = -0{,}187$; $\langle v_y \rangle = -0{.}0075$; $\left(\langle v_x'^2 \rangle \right)^{1/2} = \left(\langle v_y'^2 \rangle \right)^{1/2} = 0{.}013.$

the vortex region. But in the latter case, pulsations are of significantly less intensity. The degree of turbulence characterized by $\left(\langle v_x'^2 \rangle \right)^{1/2}$, $\left(\langle v_y'^2 \rangle \right)^{1/2}$ is by several orders less in the vortex-free region than inside it.

The developed program can be used for practical purposes of industrial ventilation for the design of open lateral suction units. In addition, the proposed approach to the determination of a separation point can be used in the design of vortex flows that occur when air is leaking onto obstacles. Such flows considered in the practice of industrial ventilation can be observed in flows past buildings, screens, as well as in operation of local exhausts in cabin-type suction units.

3.4 NONSTATIONARY FLOWS AROUND SLOT-TYPE AND ROUND SUCTION UNITS

3.4.1 Flow around a Slot-Type Suction Unit in an Unlimited Space

A flow of ideal incompressible liquid upstream of a slot-type suction unit was analyzed using the methods of conformal mappings and boundary integral equations [22] in Chapter 1 (intact model) using the Joukowski method [16,89] (separated flow) and the discrete vortex method (DVM) [117]. The most promising in our opinion is the DVM that enables defining boundaries of flow vortex regions as well as distributing velocities inside them, particularly turbulent flow characteristics. The study [117] analyzed the flow based on the DVM superposition and conformal mappings with the precise fulfillment of the boundary conditions. However, such strict approach is applicable to a narrow class of problems where it is possible to find a function that could map the physical flow area over the geometrical one. Such ranges do not include 2D multiply connected or 3D flow areas.

We resolve the problem of the flow of the slot-type suction unit where the intake cross section is located at the end distance from the inlet opening so the boundary conditions are set approximately. It is of interest to compare the solutions obtained by different methods and to determine the geometrical parameters at which they correlate well.

As is known, air velocity at point $x(x_1, x_2)$ along $\vec{n} = \{n_1, n_2\}$ induced by a sink of $q(\xi)$ intensity at point $\xi(\xi_1, \xi_2)$ will be determined from the following formula:

$$v_n(x) = -\frac{(x_1 - \xi_1)n_1 + (x_2 - \xi_2)n_2}{(x_1 - \xi_1)^2 + (x_2 - \xi_2)^2} \frac{q(\xi)}{2\pi}.$$

We will obtain air velocity at the same point induced by the interval (Figure 3.20) with continuously distributed sinks having the intensity of

$$q(\xi) = \frac{u_0 \cdot 2B}{2B}.$$

By integrating over the interval length given that $\xi_1 = L$ and $\xi_2 = \xi$ we will obtain the following expression:

$$v_n(x) = -\frac{u_0}{2\pi} \int_{-B}^{B} \frac{(x_1 - L)n_1 + (x_2 - \xi)n_2}{(x_1 - L)^2 + (x_2 - \xi)^2} d\xi$$

$$= -\frac{u_0}{2\pi} \left[n_1 \cdot \left\{ \operatorname{arctg} \frac{x_2 + B}{x_1 - L} - \operatorname{arctg} \frac{x_2 - B}{x_1 - L} \right\} - \frac{n_2}{2} \ln \frac{(x_1 - L)^2 + (x_2 - B)^2}{(x_1 - L)^2 + (x_2 + B)^2} \right].$$

We consider the flow around a slot-type suction unit (Figure 3.21).

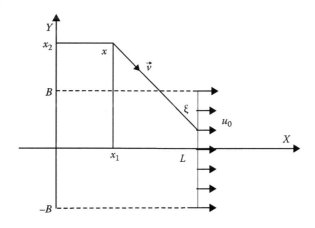

FIGURE 3.20 Suction interval.

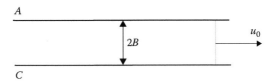

FIGURE 3.21   Slot-type suction unit located freely in space.

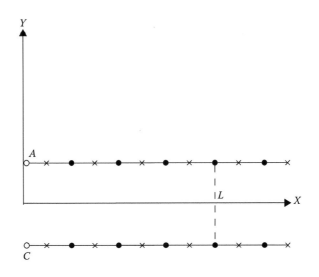

FIGURE 3.22   Flow area discretization: × is for reference points, ● is for bound vortices, ○ is for free vortices.

The flow separation (vortex wake bound) from points A and C occurs. We will observe the vortex pattern development over time. We assume the walls to be infinitely thin. We will place reference points and bound vortex position points along the range boundary as shown in Figure 3.22.

The discretization interval is h (the distance between the adjacent reference point and vortex). The number of reference points is $2N$, the number of bound vortices is $2N - 2$, and the number of free vortices is 2. Such a pattern of vortices is observed at the initial instant of time: $t = 0$. Unknown vortex circulations will be determined subject to the no-fluid-loss condition at the boundary:

$$\sum_{k=1}^{2N} G^{pk}\Gamma^k + F^p = 0, \tag{3.53}$$

where

$$G^{pk} = \frac{1}{2\pi} \frac{(x_1 - \xi_1)n_2 - (x_2 - \xi_2)n_1}{(x_1 - \xi_1)^2 + (x_2 - \xi_2)^2} \tag{3.54}$$

$$F^p = -\frac{u_0}{\pi} \left[n_1 \cdot \left\{ \arctan\frac{x_2 + B}{x_1 - L} - \arctan\frac{x_2 - B}{x_1 - L} \right\} - \frac{n_2}{2} \ln \frac{(x_1 - L)^2 + (x_2 - B)^2}{(x_1 - L)^2 + (x_2 + B)^2} \right]$$

$x(x_1, x_2)$ is the reference point
$\xi(\xi_1, \xi_2)$ is the vortex position point
L is the distance from the suction interval to the origin of coordinates

At the next instant of time $t = \Delta t = h/u_0$ first free vortices will be displaced into the interior of the flow area. Their new position will be determined from the formulas

$$x' = x + v_x\Delta t, \quad y' = y + v_y\Delta t, \tag{3.55}$$

where
(x, y) is the old free vortex position
(x', y') is the new free vortex position
(v_x, v_y) are the velocity vector components at point (x, y)

At this instant of time system of Equation 3.53, considering two new free vortices, will look as follows:

$$\sum_{k=1}^{2N} G^{pk}\Gamma^k + F^p + G_1^{p1}\gamma_1^1 + G_2^{p1}\gamma_2^1 = 0, \tag{3.56}$$

where γ_1^1, γ_2^1 are circulations of two detached free vortices at the first instant of time equal to the circulations of free vortices at points A and C at the previous instant of time. These circulations will remain unchanged at subsequent instants of time. G_1^{p1}, G_2^{p1} will be determined by the formula given in Equation 3.55 where (x_1, x_2) are coordinates in p-reference point; (ξ_1, ξ_2) are coordinates of the first or the second free vortex, respectively. Having resolved the system in Equation 3.56, we will obtain a new distribution of circulations for the boundary vortices. Another two vortices with circulations γ_1^2, γ_2^2 are separating from the sharp edges. Vortices γ_1^1, γ_2^1 acquire a new position.

At the instant of time $t = m \cdot \Delta t$, system of Equation 3.53 will be transformed into the following form:

$$\sum_{k=1}^{2N} G^{pk}\Gamma^k + F^p + \sum_{k=1}^{m}\left(G_1^{pk}\gamma_1^k + G_2^{pk}\gamma_2^k\right) = 0. \tag{3.57}$$

The inferior indices of G_1^{pk}, G_2^{pk} mean that vortices are separating from the first (upper) and the second (lower) edges, respectively. Index k means that the vortex had separated at the instant of time k.

Having resolved this system of equations we will find the distribution of boundary vortices at the instant of time m. Vortices are separating from the sharp edges again. The rest of the vortices will acquire new positions. As soon as the vortices reach the suction interval they disappear.

Based on the elaborated algorithms there had been a software program developed by means of which it was found that the symmetry of free vortices is distorted over time. Such effect is stipulated by a calculation error. For instance, circulations of symmetric bound vortices differ for a value of about 10^{-14}–10^{-16} after the first calculation. This results in the first free vortices also having different circulations and coordinates that differ by almost the same order. Eventually, after 1 s of modeling, differences in the location of vortices in the upper and lower halves become visually observable. Therefore, it was decided to maintain the symmetry of vortices synthetically by means of algorithm modification. In this case, there were only the coordinates of upper vortices calculated while lower vortices were positioned symmetrically about the OX-axis.

There was a case studied when the distance between the adjacent reference points was $h = 0.0075$ m, the total number of reference points was 320 (i.e., 160 points per interval), the suction unit half-width was 0.15 m, the suction interval x-axis was 1 m, the suction interval air velocity was 1 m/s, and the time increment $\Delta t = 0.005$ s.

The pattern of free vortices and flow lines after 3 s of modeling are shown in Figure 3.23.

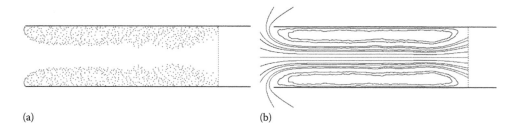

(a)   (b)

FIGURE 3.23   (a) Vortex flow pattern and (b) flow lines.

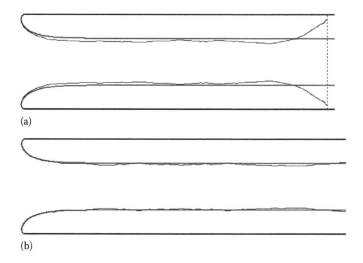

(a)

(b)

FIGURE 3.24   Vortex region contours obtained using the Joukowski method and using the program: (a) $t = 3$ s; suction interval X-coordinate $L = 1$ m and (b) $t = 3.5$ s; $L = 1.5$ m.

Plotting contours for the vortex region obtained using various methods (Figure 3.24) demonstrates that they actually coincide at the distance of $L/2B = 10$ from the entrance opening. The difference in the axial velocities determined using the Joukowski method and DVM does not exceed 5% (Table 3.2).

There were time-averaged velocity profiles calculated for various channel cross sections. The time-averaged velocity in this case meant an average longitudinal (horizontal) velocity component

TABLE 3.2

Comparing Axial Velocities Calculated Using the Program and Using the Joukowski Method

X-Coordinate, m	Axial Velocity Calculated Using the Program, m/s	Axial Velocity Calculated Using the Joukowski Method, m/s
0	1.342	1.285
−0.1	0.650	0.646
−0.2	0.350	0.352
−0.3	0.224	0.226
−0.4	0.160	0.163
−0.5	0.124	0.126
−0.6	0.100	0.102

in reference points (the transverse component is small and has no significant effect on the velocity) calculated in the following manner:

$$\langle u_x \rangle = \frac{1}{T} \int_{\tau}^{\tau+T} u_x(t)dt,$$

where
  subintegral function $u_x(t)$ is the longitudinal component of instantaneous velocity at the instant of time t
  τ is the initial averaging instant of time
  $\tau + T$ is the final averaging instant of time

In a discrete form, the time-averaged velocity (its longitudinal component) is calculated in the following manner:

$$\langle u_x \rangle = \frac{1}{M} \sum_{i=m}^{m+M} u_x(t_i),$$

where
  m is the initial averaging instant of time
  $m + M$ is the final averaging instant of time

Figure 3.25 shows the comparison of average velocity profiles plotted using the program with the profiles obtained in [117]. The calculation was made in four cross sections at $x/h = 0.3 \div 2.5$. Figure 3.25a shows no profiles for cross sections $x/h = 0.6$ and $x/h = 2.5$ because they actually coincide with the profile at $x/h = 1.3$. Experimental data for a round sharp-edged channel at $x/h = 0.7$, 1.5, and 2.7 can be found in [118].

As in [117], the calculations made using the developed program show a reverse flow around the walls.

Root-mean-square longitudinal velocity pulsations will be determined using the formula

$$\left(\langle u_x'^2 \rangle\right)^{1/2} = \left[\frac{1}{M} \sum_{k=m}^{M+m} \left(u_{kx}'\right)^2 \right]^{1/2},$$

where $u_{kx}' = u_x(k) - \langle u_x \rangle$ is the longitudinal pulsation at the instant of time k.

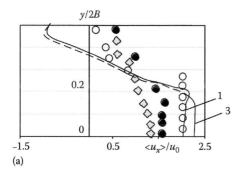 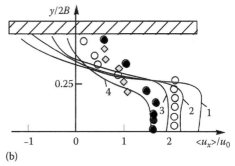

FIGURE 3.25 Average velocity profiles in channel cross sections: (a) program-based calculation and (b) data presented in [116]. Calculation: 1, $x/2B = 0.3$; 2, $x/2B = 0.6$; 3, $x/2B = 1.3$; 4, $x/2B = 2.5$. The experiment for a round channel: $\bigcirc$ is $x/2B = 0.7$; $\bullet$ is $x/2B = 1.5$; $\diamond$ is $x/2B = 2.7$.

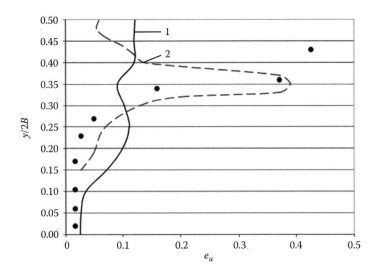

FIGURE 3.26   Profile of longitudinal velocity pulsations in the cross section $x/2B = 0.6$: 1 is for a program-based calculation, 2 is for profile presented in [116], and ● is for experimental data.

The comparison of the profiles of root-mean-square longitudinal velocity pulsations $e_u = (<u'^2>)^{1/2}/u_0$ in the cross section $x/2B = 0.6$ obtained using the developed program based on the design data [116] and experimental data [118] is shown in Figure 3.26.

The circulation of a vortex separating from a sharp edge changes significantly only at the initial interval of time (Figure 3.27), then it varies about a certain value. A similar observation is made in [116].

The proposed approach to resolution of the problem in question with an approximate fulfillment of the boundary conditions allows for solving a number of new problems of airflow around suction inlets in multiply connected 2D and 3D spatial domains.

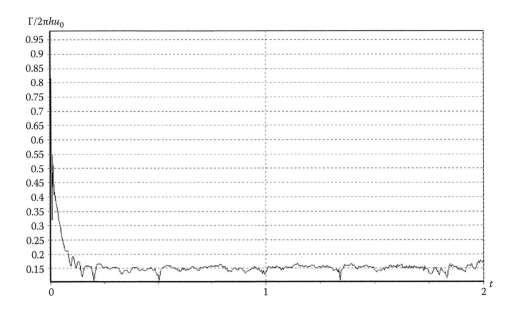

FIGURE 3.27   Graph of the behavior of circulations of vortices separating from the suction unit edge.

3.4.2 Flows around Slot-Type Bell-Shaped Suction Units

An extensive use of local bell-shaped exhausts (suction units) in industrial ventilation stipulated a considerable interest in the analysis of flows around them. The best consistency with the experimental data is demonstrated by the results obtained with consideration for the separation of flows from sharp edges of suction units. Based on the complex variable theory methods (the Joukowski method), the study [124] and paragraph 4.4 define the contours of the first vortex region that occurs upstream of the bell-shaped suction unit. However, such method does not allow for calculating the field of velocities within the vortex region and track a temporal development of the vortex pattern.

This study objective is to calculate the field of velocities around bell-shaped suction units with consideration for separation of flows from all sharp edges of suction units and to define the contours of large-scale vortex patterns upstream of the bell-shaped suction unit.

We will resolve the problem within the framework of the model of an ideal incompressible liquid based on the discrete vortex method. We will place the TS suction opening (Figure 3.28) at the end distance from the opening inlet, that is, the boundary condition for velocity at infinity v_0 will be fulfilled approximately, which will allow for further switching to an axially symmetric problem. The flow parameters that are of interest to us are shown in Figure 3.29.

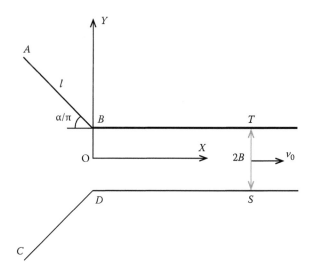

FIGURE 3.28 Bell-shaped suction unit.

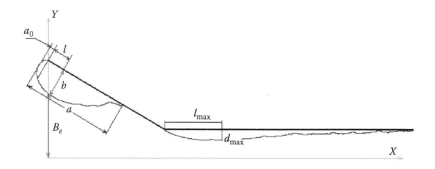

FIGURE 3.29 Required flow parameters.

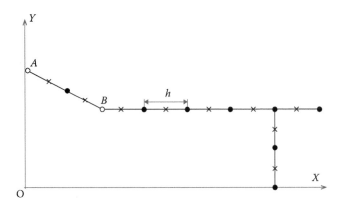

FIGURE 3.30 Region boundary discretization: × is for reference points, ● is for bound vortices, ○ is for free vortices.

We will discrete the region boundary with reference $N - 1$ points and bound vortices as shown in Figure 3.30. The reference points are located between the bound vortices. The discreteness increment is h (the distance between two adjacent bound vortices). There are normal boundary velocity values given in the reference points. The normal boundary velocity is equal to zero everywhere but at the TS suction opening where $v_n = v_0$.

At the initial instant of time liquid exists at all points in space. At the next instant of time, the TS opening exhaust is activated and vortices are separating from all sharp edges A, B, C, and D in the flow direction.

The velocity along outward normal $\vec{n} = \{n_1, n_2\}$ at p-reference point at the instant of time m maybe expressed by composing the action on it by all bound and free vortices:

$$v_n^p = \sum_{k=1}^{N} \Gamma^k G^{pk} + \sum_{\tau=1}^{m} \sum_{l=1}^{4} G_l^{p\tau} \gamma_l^\tau, \tag{3.58}$$

where
Γ^k is bound vortex k-bound vortex circulation

$$G^{pk} = \frac{n_2(x_1 - \xi_1) - n_1(x_2 - \xi_2)}{2\pi \cdot [(x_1 - \xi_1)^2 + (x_2 - \xi_2)^2]} \tag{3.59}$$

(x_1, x_2) are the p-reference point coordinates
(ξ_1, ξ_2) are the k-bound vortex coordinates
m is the instant of time
$G_l^{p\tau}$ is the function expressing the impact of a free vortex of γ_l^τ intensity separated from l-point at the instant of time $t = \tau \Delta t$ on the same p-point
$G_l^{p\tau}$ is calculated from the formula given in Equation 3.59 where (ξ_1, ξ_2) are the coordinates of the vortex of γ_l^τ intensity.

There should be the circulation stability condition added to Equation 3.58:

$$\sum_{k=1}^{N} \Gamma^k + \sum_{\tau=1}^{m} \sum_{l=1}^{4} \gamma_l^\tau = 0. \tag{3.60}$$

At the instant of time $m + 1$, the number of free vortices will be increased by four due to the separation from four points of the flow. All vortices that were in the flow before that instant of time will acquire new positions:

$$x' = x + v_x \Delta t, \quad y' = y + v_y \Delta t,$$

where v_x, v_y are components of the velocity (calculated by Equation 3.58) induced by the action on the previous vortex position point (x, y) by the entire vortex system at the instant of time m except for the vortex itself.

Then the systems of Equations 3.58 and 3.60 are solved again and there are unknown circulation variables determined. In those systems, new vortices appear, old ones change their positions, etc.

If at a certain instant of time the vortex is reaching a distance of less than $h/2$ to a solid wall, the vortex will move aside the wall along the normal so that the distance to the flow boundary becomes equal to $h/2$. If the same happens to the vortex and the opening, the vortex disappears.

If two vortices are reaching a distance $x < h/2$, the mutual effect of these vortices will be determined by means of linear approximation:

$$\frac{v(x)}{v} = \frac{x}{h/2},$$

where $v(x)$, v are velocities induced by the vortex at a point at x and $h/2$ distance, respectively.

The discretization interval in the calculations is $h = 0.01$; the distance to the TS suction interval is 2; the suction unit half-width (gauge) is 0.15; velocity $v_0 = 1$. In further calculations, we will change to dimensionless units. All linear dimensions will be referred to the suction unit half-width while the velocity will be referred to v_0. The first and the second vortex regions will mean the regions that occur in the result of the flow separation from point A and point B, respectively. The vortex regions that occur in the result of the flow separation from point C and point D are symmetrical to those mentioned above around the OX-axis.

The design parameters were determined using the developed program to plot the flow line with the origin in sharp edge A as soon as free vortices have completely filled the design region: free vortices separated from edges A and C start penetrating the second vortex region; free vortices separated from edges B and D reach the TS suction interval; dimensions of the vortex regions are no longer changing in one direction but start varying in time. Then, five instants of time were randomly selected and the resulting values were averaged. The computation duration is quite long. At the time increment of $\Delta t = h/v_0 = 0.01$, the calculation was performed until $4 < t < 17$ based on α angle and bell length d.

The resulting flow vortex pattern (Figure 3.31) contours are similar to the experimental data (Figure 3.32) and computations by the Joukowski method [124].

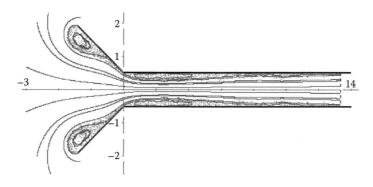

FIGURE 3.31 Flow lines and flow vortex pattern upstream of a slot-type bell-shaped suction unit with opening angle 45° and length four gauges.

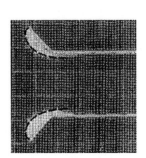
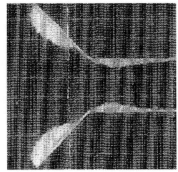

FIGURE 3.32 Photographs of the vortex regions in the experiments with onion shells. (From Katkov, M.V., The study of slotted drain flows: Thesis for the degree of Cand. Sc. (Engineering), Kazan. KSTU n.a. A.N. Tupolev, Tatarstan, Russia, 2000, 153p.)

The following equation is proposed to define contours of the first vortex region:

$$\begin{cases} \dfrac{(x+c)^2}{m^2} + \dfrac{(y-r)^2}{n^2} = 1, & -b \le x \le 0, \quad r \le y \le a, \\ x = -\dfrac{b}{2}\left(1 - \cos\dfrac{\pi y}{r}\right), & 0 \le y \le r, \end{cases} \tag{3.61}$$

where

$r = -l - a_0 + a$

$n = l + a_0$

$m = b - c$

$c = \dfrac{b}{1 + \dfrac{n}{\sqrt{n^2 - l^2}}}$

This formula is applicable to so-called long bells, that is, to those cases when there are clearly defined first and second vortex regions. Given Equation 3.61 and using Figure 3.33 it is possible to plot the contours of the first vortex regions in a system of coordinates with the origin in the flow line and bell convergence point. Herewith, the *Y*-axis is directed along the bell (Figure 3.34).

Contouring bell-shaped suction units along the resulting lines will allow for reducing the suction unit power consumption out of decreasing the aerodynamic drag at the suction inlet. Note that in the case with small lengths of the bell or in case with small α slope angles, the first and the second vortex regions merge (Figure 3.35).

An effective suction dimension of the bell-shaped suction unit (Figure 3.36) is proportional to the bell length and opening angle. The experimental data from Figure 3.32 mapped on Figure 3.36 are well consistent with the test data.

The analysis of the relation between the axial velocity and the bell slope angle shows that the velocity around the suction unit is proportional to α angle and at long distances is minimal at ≈45°–60° (Figure 3.37a). The highest axial velocity is observed in all cases at α = 90°. Note that the last fact was observed in the intact flow too. A suction unit with a bell is more effective than one

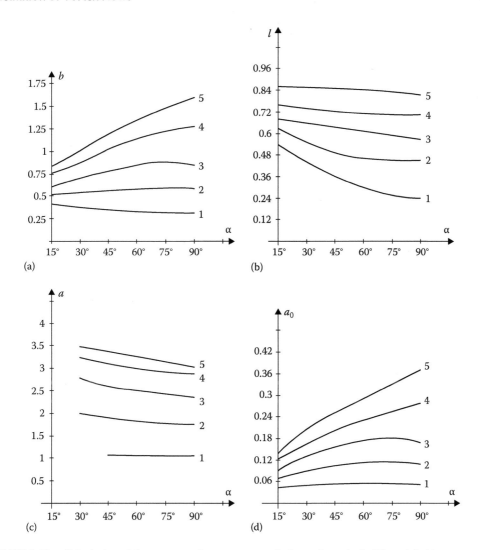

FIGURE 3.33 Calculation of the vortex region parameters (1, 2, ..., 5 are the bell lengths): (a) parameter b, (b) parameter l, (c) parameter a, and (d) parameter a_0.

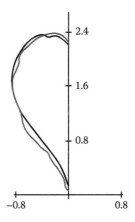

FIGURE 3.34 Comparison of vortex region contours obtained using the program and using formula (4) at $\alpha = 90°$ and bell length four gauges.

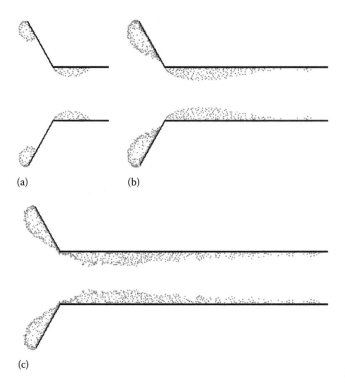

FIGURE 3.35   Temporal development of the vortex pattern for a slot-type bell-shaped suction unit at $\alpha = 60°$ and $d = 2$: (a) $t = 0.5$, (b) $t = 1.7$, (c) $t = 3$.

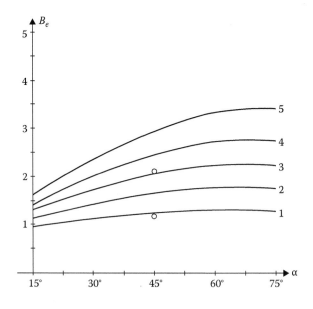

FIGURE 3.36   Change in the bell effective dimension based on angle α and length d (1, 2, …, 5).

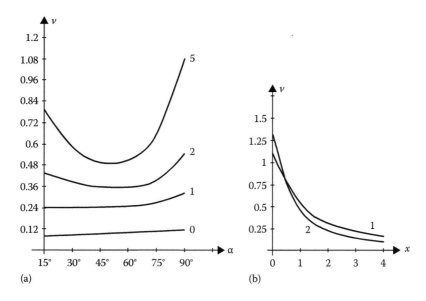

FIGURE 3.37 Velocity versus (a) α slope angle and length $d = 3$ at various distances from the suction inlet (0; 1; 2; 5) and (b) suction opening distance: 1, $d = 3$, $\alpha = 90°$; 2, $d = 0$.

without a bell at distances exceeding a half of the gauge (Figure 3.37b). With an increase in the bell length and constant α, the suction inlet velocity is decreased, but at a distance exceeding two gauges the relation is inverse (Figure 3.38).

3.4.3 ROUND BELL-SHAPED SUCTION UNITS

In this case, we have an axially symmetric problem. The computation algorithm is developed in the same way as for the 2D problem. We will discrete the boundary with a system of infinitely thin vortex rings and reference points as shown in Figure 3.39.

At each instant of time, two infinitely thin ring vortices are separating from sharp edges. It should be noted that for modeling vortex wakes we will use a system of *truncated* infinitely thin noninductive vortex rings, the correct use of which was described in [126]. This means that when some point of the vortex ring approaches a distance shorter than the discretization interval, the velocity at this point will be equal to zero.

The action on the given M point located at a longer distance along the direction $\vec{n} = \{n_1, n_2\}$ of the vortex ring (Figure 3.40) is defined by the expression

$$v_n = v_1 n_1 + v_2 n_2 = \Gamma \cdot \frac{\xi_2}{4\pi} \int_0^{2\pi} \frac{(\xi_2 - x_2 \cos\theta)n_1 + (x_1 - \xi_1)n_2 \cos\theta}{\left[(x_1 - \xi_1)^2 + \xi_2^2 + x_2^2 - 2x_2\xi_2 \cos\theta\right]^{3/2}} d\theta.$$

After simple transformations, the expression for velocity at point (x_1, x_2) induced by the infinitely thin vortex ring at point (ξ_1, ξ_2) with circulation $\Gamma(\xi)$ will take the following form:

$$v_n = \frac{\gamma \xi_2^2 n_1}{4\pi} \int_0^{2\pi} \frac{d\theta}{\left[a - b\cos\theta\right]^{3/2}} + \frac{\gamma \xi_2}{4\pi} \left[(x_1 - \xi_1)n_2 - x_2 n_1\right] \int_0^{2\pi} \frac{\cos\theta d\theta}{\left(a - b\cos\theta\right)^{3/2}}$$

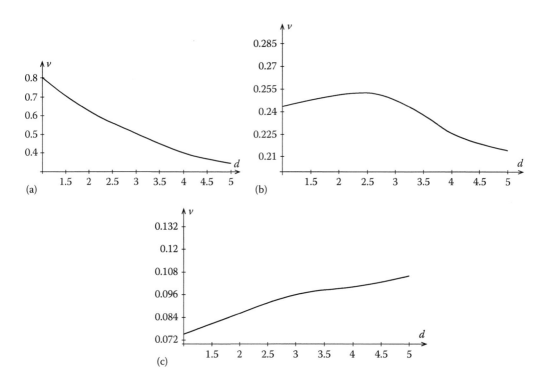

FIGURE 3.38   Change in the axial velocity with an increase in the bell length at $\alpha = 60°$: (a) $x = 0$, (b) $x = 2$, and (c) $x = 5$.

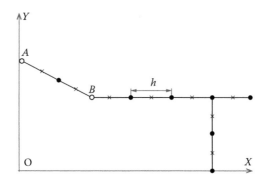

FIGURE 3.39   Region boundary discretization in the meridional semiplane: × is for reference points, ● is for bound vortices, and ○ is for free vortices.

or alternatively

$$v_n(x) = \Gamma(\xi) \cdot G(x,\xi),$$
(3.62)

where

$$
\begin{cases}
G(x,\xi) = \dfrac{(A_1 b + A_2 a)}{b} \cdot \dfrac{4}{(a-b)\sqrt{a+b}} E(t) - \dfrac{A_2}{b} \cdot \dfrac{4}{\sqrt{a+b}} K(t) & \text{at } b \neq 0, \\[2ex]
G(x,\xi) = \dfrac{\xi_2^2 \cdot n_1}{2a\sqrt{a}} & \text{at } b = 0
\end{cases}
$$
(3.63)

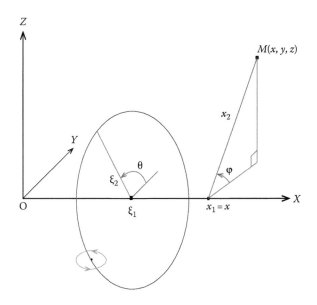

FIGURE 3.40 For the definition of velocity at a point induced by the vortex ring effect.

$$2x_2\xi_2 = b > 0, \quad a = (x_1 - \xi_1)^2 + \xi_2^2 + x_2^2 > 0, \quad A_1 = \frac{\xi_2^2 n_1}{4\pi}, \quad A_2 = \frac{\xi_2}{4\pi}\left[(x_1 - \xi_1)n_2 - x_2 n_1\right]$$

$$K(t) = \int_0^{\pi/2} \frac{d\theta}{\sqrt{1 - t^2 \sin^2 \theta}}, \quad E(t) = \int_0^{\pi/2} \sqrt{1 - t^2 \sin^2 \theta}\, d\theta$$

are complete elliptic integrals of 1 and 2 kind;

$$K(t) = \sum_{i=0}^{4} c_i (1-t)^i + \sum_{i=0}^{4} d_i (1-t)^i \ln\frac{1}{1-t}, \quad E(t) = 1 + \sum_{i=1}^{4} c_i (1-t)^i + \sum_{i=1}^{4} d_i (1-t)^i \ln\frac{1}{1-t}, \quad t = \frac{2b}{a+b},$$

c_i, d_i were taken from the table [123].

Qualitatively, the flow pattern in an axially symmetric problem is similar to the one in a 2D problem. However, velocities induced by a round bell-shaped suction unit (3D) are significantly lower than those induced by a slot-type suction unit (2D). As a consequence, parameters for the flow vortex region are lower (Figures 3.41 and 3.42). By the parameters calculated for the first vortex region, it is possible to define its contours using the same formula (Equation 3.61) and Figure 3.41. Note the slightly conservative axial velocity values (Figure 3.43) calculated using the developed software program as compared with the experimental data presented in [16,127]. With an increase in the bell length and constant bell opening angle, the axial velocity is decreased but at a distance of up to five gauges and is increased at a longer distance (Figure 3.44).

Contouring local ventilation exhaust openings along the resulting lines will allow for reducing the local exhaust ventilation power consumption.

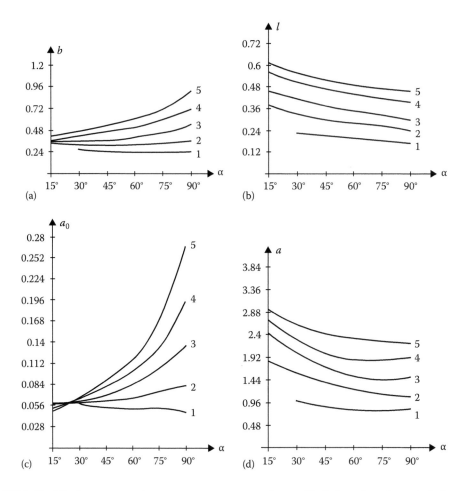

FIGURE 3.41   The first vortex region parameters upstream of a round bell-shaped suction unit (1, 2, …, 5 are the bell lengths): (a) parameter b, (b) parameter l, (c) parameter a, and (d) parameter a_0.

3.4.4   Experimental Determination of Local Drag Factors of Profiled Local Exhausts

It was assumed in the previous subsection that the contouring entrance lengths of exhausts along the boundaries of vortex regions would significantly improve aerodynamic characteristics of exhausts and, consequently, reduce the required intensity of the same. Let us verify this thesis experimentally.

There were detachable profiles fabricated for the bell-shaped suction unit based on the boundaries defined by the formula of Equation 3.61 and the local drag factor (LDF) variation analyzed against a suction unit without such profiles.

The general diagram of the experimental arrangement is shown in Figure 3.45, and the slot-type bell-shaped suction unit structures in question are shown in Figure 3.46.

The bell-shaped suction unit is clamped between two planes. The lower plane is 565 × 550 mm wood board. The upper plane is 480 × 495 mm glass. The distance between the glass and the plane (the suction unit wall height) is 60 mm. The suction unit walls that are 324 mm long are made of 2.5 mm thick smooth metal. Air tightness at joints between the suction unit walls and glass is achieved by means of thin rubber strips attached to the upper wall faces. There was no noticeable air leakage observed at joints. The detachable profiles are made of cardboard and smooth paper with quite accurate reproduction of the theoretical shape of such profiles.

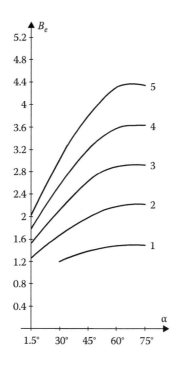

FIGURE 3.42 Relation between the effective dimension of a round bell-shaped suction unit with $\alpha = 60°$ and various bell lengths (1, 2, …, 5).

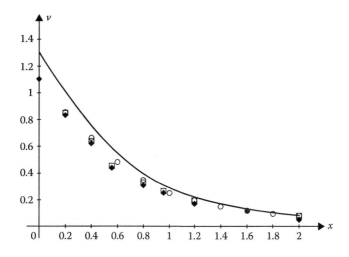

FIGURE 3.43 Change in the axial air velocity at a round suction unit without a bell.

The part of the suction unit extending the glass and wood board is sealed with a pressboard stuck with scotch. The cardboard face of the suction unit (located at the distance of 58 mm away from the glass and installed perpendicularly to the glass) has Ø 30 mm hole for the tube connected to a fan (vacuum cleaner Vityaz-2M).

Measurements in CD cross section (Figure 3.45) were carried out using micromanometer MMN-2400(5)-1.0 TU 25-01-816-79 and a pneumatic tube tightly connected to the micromanometer with flexible hoses.

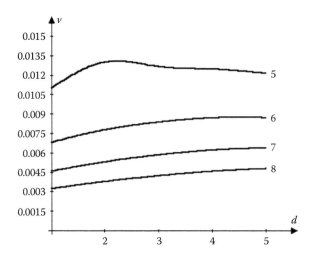

FIGURE 3.44 Change in the axial velocity around a bell-shaped suction unit of round cross section with α = 60° based on the bell length and various distances (5, 6, 7, and 8 gauges) from the suction inlet.

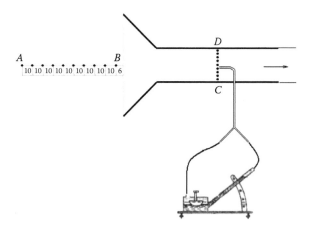

FIGURE 3.45 Experimental arrangement diagram.

The reading of the micromanometer with an inclined tube ΔL proportional to the manometric pressure being measured is

$$\Delta P = \Delta L \cdot K, \quad K = \rho_{sp} g \sin \alpha \left(1 + \frac{f}{F \sin \alpha} \right),$$

where
$\rho_{sp} = 0.8095$ g/cm^3 is the density of the micromanometer spirit solution
α is the micromanometer tube inclination angle
$F \approx 4771$ mm^2 is the micromanometer reservoir cross-section area
$f \approx 28$ mm^2 is the micromanometer tube cross-section area
g is the free fall acceleration

The full pressure variation in the low LDF suction units is also very low. In order to measure the full pressure variation with a micromanometer, it is necessary to incline the spirit tube so that its angle α is as small as possible. The micromanometer has the following geometric characteristics:

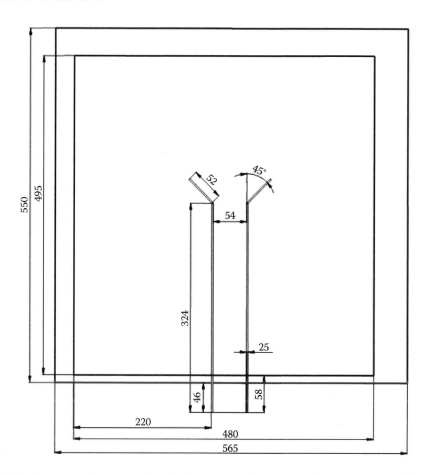

FIGURE 3.46 Structure of two gauges long bell-shaped suction unit installed at the angle of 45°.

L_1 = 307 mm is the micromanometer tube length; in the lowermost tube position h_1 = 59.5 mm, which is the distance from the table surface to the lower tube end; h_2 = 73.5 mm, which is the distance from the table surface to the upper tube end; L_2 = 260 mm is the micromanometer base length; H_1 = 44 mm is the height of the lower (left) edge of the micromanometer base; and H_2 = 48 mm is the height of the upper (right) edge of the micromanometer base.

Since $H_1 \neq H_2$, but the micromanometer base is set in a horizontal position, there is an angle between the planes of the micromanometer base and the table planes:

$$\alpha_1 = \arcsin\left(\frac{H_2 - H_1}{L_2}\right) = \arcsin\left(\frac{48 - 44}{260}\right) = 0.88°, \quad \sin\alpha = \sin\left(\arcsin\left(\frac{h_2 - h_1}{L_1}\right) - \alpha_1\right).$$

When measuring the full pressure, the micromanometer tube is set in the lowermost position and

$$\sin\alpha_f = \sin\left(\arcsin\left(\frac{73.5 - 59.5}{260}\right) - 0.88°\right) \approx 0.039,$$

$$K_f = \rho_{sp}\sin\alpha_f g\left(1 + \frac{f}{F\sin\alpha_{\text{попн}}}\right) = 0.8095 \cdot 0.039 \cdot 9.81 \cdot \left(1 + \frac{28}{4771 \cdot 0.039}\right) = 0.353.$$

When measuring the dynamic pressure the tube is in the fixed position and $K = 0.2$, that is, $K_{din} = 0.2$.

Let us lay down the test procedure according to [129].

In general terms, LDF is determined from the formula

$$\zeta = \frac{\int_S (-p_f)vdS}{\frac{\rho v_{av}^3}{2} S}, \quad (3.64)$$

where the average velocity in section S (wherein the full pressure is measured) is calculated from the relation

$$v_{av} = \int_S \frac{vdS}{S}. \quad (3.65)$$

In the case with Figure 3.45 and nonuniform partition of CD interval the discrete countertype of Equation 3.64 looks as follows:

$$\zeta = \frac{\sum_{i=1}^n (-p_{fi})v_i \Delta b_i a}{\frac{\rho v_{av}^3}{2} ab}, \quad (3.66)$$

where

p_{fi} is the full pressure at i-point, Pa
v_i is the velocity at i-point, m/s
Δb_i is the length of i-part of CD interval, m
a is the suction width, m
ρ is the air density, kg/m^3
v_{av} is the average suction velocity, m/s

$$v_{av} = \sum_{i=1}^n \frac{v_i \Delta b_i a}{ab}. \quad (3.67)$$

CD interval is uniformly divided into $n = 9$ parts. CD section measuring points are positioned as follows: one point is on the suction axis; the rest of the points are at the distance of 6, 12, 18, and 24 (mm) from the suction axis (rightward and leftward).

The left suction wall has Ø 6 mm hole drilled for the pneumatic tube 120 mm away from the glass edge at the height of 235 mm from the board.

Measuring the full pressure at the walls shows wide fluctuations because of which measurements were carried out in the following manner: there was a vacuum cleaner powered on and 1 min noted during which the manometer readings were not taken (so that fluctuations become more or less steady); then the manometer readings were taken every 3 s during the second and the third minute; thus, three tests were conducted and the arithmetic mean was calculated. In other words, the manometer readings were taken 180 times when the full pressure was being measured.

The dynamic pressure was measured three times, averaged, and the LDF was determined from the formula

$$\zeta = \frac{\frac{1}{n}\sum_{i=1}^n (-\Delta P_{fi} \cdot v_i)}{\frac{\rho_{air} \cdot v_{av}^3}{2}} = \frac{\frac{2}{n}\sum_{i=1}^n (-\Delta P_{fi} \cdot v_i)}{\rho_{air} \cdot \left(\frac{1}{n}\sum_{i=1}^n v_i\right)^3}.$$

Air velocity in i-point: $v_i = \sqrt{2P_{fi}/\rho_{air}}$, where air density $\rho_{air} = 1.293$ kg/m^3.

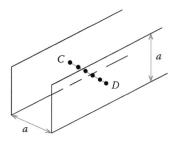

FIGURE 3.47   Slot-type suction unit w/o a bell in an unlimited space.

In order to check the obtained results for adequacy, a slot-type suction unit without a bell or profiles, (Figure 3.47) for which the LDF is given in [129], was studied initially. There was a hole drilled in the lateral wall so that the pneumatic tube end was at the distance of about 100 mm from the suction unit edge. The suction unit width and height was $a = 33$ mm. CD interval was divided into five parts, the width of each part was $\Delta B_i = 6.6$ mm.

The average suction velocity and LDF:

$$v_{av} = \frac{1}{n} \sum_{i=1}^{n} v_i = \frac{1}{5}(2 \cdot 14.53 + 2 \cdot 24.13 + 27.51) = 20.97 \text{ m/s.}$$

$$\zeta = \frac{\sum_{i=1}^{n}\left(-\Delta P_{fi} \cdot v_i \cdot \Delta B_i \cdot a\right)}{\dfrac{\rho_{air} \cdot v_{av}^3}{2} \cdot a^2} = \frac{33 \cdot 6.6 \cdot \left(2 \cdot 409.25 \cdot 14.53 + 2 \cdot 150.53 \cdot 24.13 + 32.93 \cdot 27.51\right)}{\dfrac{1.293 \cdot 20.97^3}{2} \cdot 33^2} = 0.673.$$

Note that the LDF obtained for this suction unit in [129] constitutes 0.71, which is sufficiently close to the value we have obtained (deviation of less than 6%).

The air velocity next to the bell-shaped suction unit (Figure 3.46) was measured using Testo 405-V1 electronic device at points on at AB interval (Figure 3.45).

Test and design values of axial velocity obtained using the developed program coincide quite well (Figure 3.48). The gradual increase in differences between the test and design velocity values

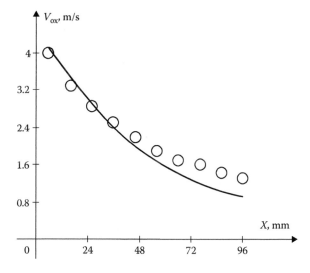

FIGURE 3.48   Comparison of test and design values of axial air velocity at a distance from two-gauge long bell-shaped suction unit with an opening angle of 45°.

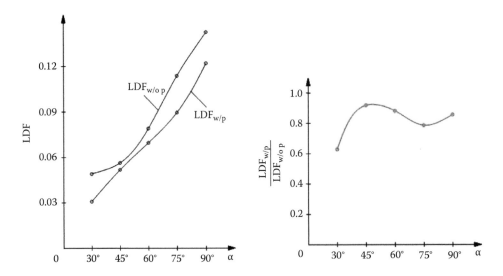

FIGURE 3.49 Two-gauge long bell-shaped suction unit LDF versus suction unit inclination angle α: $LDF_{w/o\,p}$ is LDF without a profile; $LDF_{w/p}$ is LDF with a profile.

proportional to the distance from the suction unit is related with the increase in 3D airflow effects observed at the approach to the glass edge.

The study conducted for bell-shaped suction units with and without profiles showed that the LDF value is mainly a function of the full pressure variation at points near the suction unit walls, that is, energy losses are mainly observed in the boundary layer.

The experimental outcome in Figure 3.49 shows that the most significant LDF decrease is observed at the bell inclination angle of 30°. On an average, the LDF of profiled suction units is decreased by about 30% relative to suction units without profiles.

3.5 SUPPLY AIR JET SUCTION UNITED EXHAUSTS

We will numerically analyze airflows at a local exhaust with a mechanically and aerodynamic suction unit based on the discrete vortex method. There is a direct flow annular jet used as an aerodynamic shield because when it flows on an impermeable surface (mechanical shield) a reverse airflow occurs that significantly increases the intake plume range.

Let us analyze the flow at a constant suction velocity v_0 and variable supply annular jet velocity v_s at various bell angles α and bell lengths d. We are interested in such α and d at which the reverse flow (Figure 3.50) has the highest axial velocity.

The problem was solved in the axially symmetric setup based on the discrete vortex method. The boundary is discretized by a system of continuously thin vortex rings with reference points (rings) right in between them where the boundary conditions are set: the condition of impermeability on solid walls and the appropriate normal velocity values at supply and exhaust outlets. There are three ring vortices ($L = 3$ in formulas (1) and (3)) separating from sharp edges (points A, B, and C) at each instant of time.

The discretization interval (the distance between the adjacent reference point and bound vortex) is $h = 0.01$ m in all calculations, time increment $\Delta t = h/v_s$ at $v_s \geq 1$ м/c and $\Delta t = h$ at $v_s < 1$, $R = 0.15$ m, and wall thickness: 0.02 m.

All kinematic and geometric parameters in the representation of the numerical computation results were converted into a dimensionless form $\left(\overline{x} = x/R, \quad \overline{d} = d/R, \quad \overline{H} = H/R, \quad \overline{z} = z/R, \quad \overline{v} = v/v_0 \right)$. Then we will operate dimensionless values only skipping hyphens above letters.

Since the problem is resolved in the nonstationary setup, the calculated values were averaged over time.

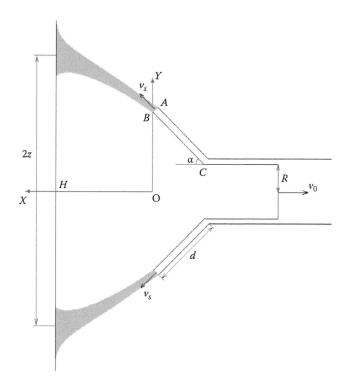

FIGURE 3.50 Round bell-shaped exhaust suction unit with an annular supply jet above an impermeable surface.

To begin with, let us analyze the flow around local exhaust with aerodynamic suction unit located in an unlimited space (Figure 3.51). The first jet, an effluent annular jet, is completely caught by the suction unit; in the second one, a part of the airflow is caught by the suction unit, another is rather quickly occluded and then is developed by the turbulent jet laws. Figure 3.52 shows the relation between the supply air jet's highest dimensionless velocity $v_s^{max} = v_s/v_0$ at which it is completely caught by the suction unit and the bell length and inclination angle. Here, we could not find geometric or kinematic parameters at which air inflow velocity would increase due to the shielding effect of the supply jet. In this case, the comparison was made between the exhaust with an aerodynamic suction unit and the exhaust with a *disabled* supply jet. The processed air rates were the same, that is, the suction airflow rate in the second case was equal to the sum of effluent and exhaust air rates for the exhaust with aerodynamic suction unit.

The case with a mechanical shield is of the greatest interest for practical tasks since the local exhaust is positioned above the components that emit pollutants from process equipment.

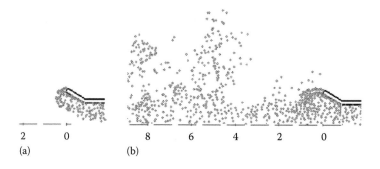

FIGURE 3.51 Vortex pattern of the flow around bell-shaped suction unit in an unlimited space: (a) the supply jet completely caught by the suction unit and (b) the supply jet divided into two jets.

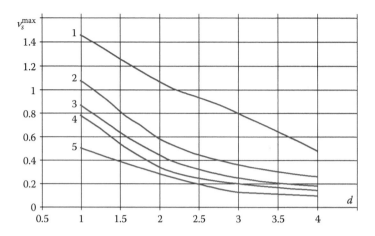

FIGURE 3.52   Relation between v_s^{max} and bell length d at various angles α: 1, 30°; 2, 45°; 3, 60°; 4, 75°; 5, 90°.

The specific flow patterns in the case of the exhaust positioned above a continuous surface are shown in Figure 3.53. When a supply jet is *disabled* (Figure 3.53a), the vortex region parameters are close to those studied earlier for bell-shaped suction units with continuously thin walls. Once a supply jet is enabled the vortex region will be expanding (Figure 3.53b) and at a definite velocity $v_п$ the jet will be divided into two parts: the one caught by the suction unit and the one flowing on an impermeable surface (Figure 3.53c). The greatest practical interest is represented by the case (Figure 3.53d) when the jet is divided into two at the surface onflow point. The axial flow velocity at certain points increases several dozens of times (Figure 3.54), which contributes to an efficient containment of pollutants occurring in this region. The best effect is achieved when the bell is inclined at 30°. The axial velocity measurement for this case at various bell lengths and distances from the impermeable surface is shown in Figure 3.55. The bell length and opening angle are proportional to the width of the air onflow region z (Figure 3.56). But in the last case the axial flow velocity is being reduced.

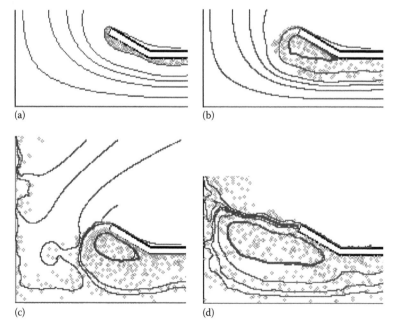

(a)

(b)

(c)

(d)

FIGURE 3.53   Vortex pattern of the flow and flow line around a local exhaust with suction unit at $\alpha = 30°$, $d = 1$, $H = 2$: (a) $v_s = 0$, (b) $v_s = 1$, (c) $v_s = 2$, and (d) $v_s = 4$.

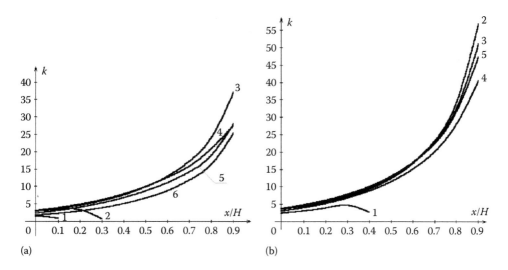

(a)

(b)

FIGURE 3.54 Relation between $k = v_{sx}/v_{cx}$ and the distance from the bell inlet (v_{sx} is an axial air velocity under the influence of supply air jets and exhaust, v_{cx} is an axial velocity without an aerodynamic shielding when the exhaust air rate is equal to the sum of supply jet and exhaust rates with a shielding jet enabled): 1, $\alpha = 0°$; 2, $\alpha = 15°$; 3, $\alpha = 30°$; 4, $\alpha = 45°$; 5, $\alpha = 60°$; 6, $\alpha = 75°$. (a) $v_s = 8$ and (b) $v_s = 4$.

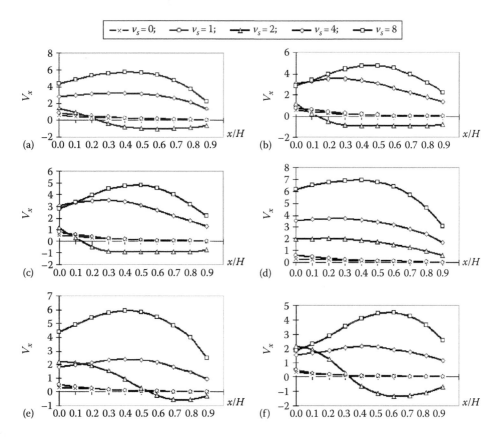

(a)

(b)

(c)

(d)

(e)

(f)

FIGURE 3.55 Relation between the axial velocity and distance from the bell inlet at an angle of $\alpha = 30°$: (a) $H = 2, d = 1$; (b) $H = 3, d = 1$; (c) $H = 4, d = 1$; (d) $H = 2, d = 2$; (e) $H = 3, d = 2$; (f) $H = 4, d = 2$.

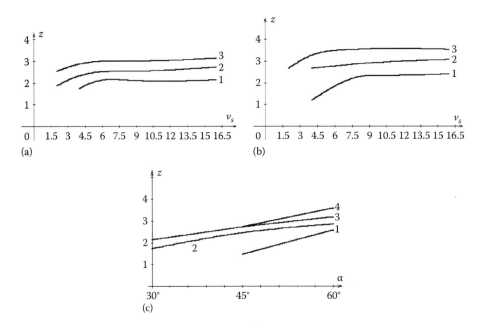

FIGURE 3.56   Variation of z: (a) $\alpha = 30°$, $H = 2$, 1: $d = 1$; 2: $d = 2$; 3: $d = 3$; (b) $\alpha = 30°$, $H = 3$, 1: $d = 1$; 2: $d = 2$; 3: $d = 3$; (c) $H = 2$, $d = 1$, 1: $v_n = 2$, 2: $v_n = 4$, 3: $v_n = 8$, 4: $v_n = 16$.

The jet axis equation is supposed to be obtained from the formula

$$y = \frac{z - 1 - d\sin\alpha - Htg\alpha}{H^2}(x - H)^2 + \frac{2z - 2 - 2d\sin\alpha - Htg\alpha}{H}(x - H) + z, \qquad (3.68)$$

where (x_0, y_0) are B point coordinates.

The jet axis plotted according to this formula with the accuracy adequate for practical purposes enables determining the pollution containment area (Figure 3.57).

In our opinion, the approach to the design of local exhausts with suction units demonstrated herein allows for obtaining the fullest information regarding the process air mechanics as compared to the previous calculations set forth in Sections 1.4.2 and 3.3.

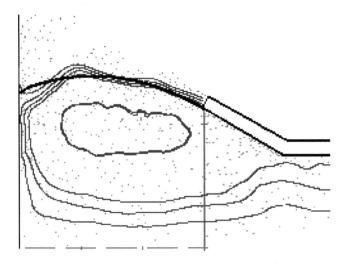

FIGURE 3.57   Jet axis plotted according to formula (11) at $\alpha = 30°$; $H = 2$; $d = 1$.

3.6 CLOSED REGION FLOWS: MODEL PROBLEMS

3.6.1 APPROACH FLOW PAST A THIN SUCTION UNIT

Figure 3.58 shows the vortex pattern and flow lines (Figure 3.58d) at the initial instants of time with the 2D parallel flow approaching a wall offset. In calculations, the approach flow velocity is 1 m/s, discretization interval $h = 0.02$, and time increment $\Delta t = 0.02$.

The photographic images of a plane plate at the initial instant of its sudden movement toward the normal [153] in water (Figure 3.59) demonstrate a fairly good coincidence of the vortex pattern development versus the computation data (Figure 3.58).

3.6.2 APPROACH FLOW PAST A RECTANGULAR OFFSET

The vortex region at flow past a rectangular offset (Figures 3.60 and 3.61) significantly depends on the approach flow velocity. At higher velocities, vortices behind the rectangle windward side approach

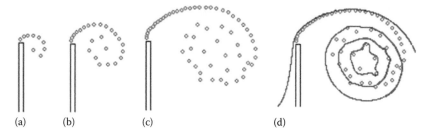

(a)   (b)   (c)   (d)

FIGURE 3.58 (a through d) shows the vortex pattern and flow lines (Figure 3.58d) at the initial instants of time with the 2D parallel flow approaching a wall offset. In calculations, the approach flow velocity is 1 m/s, discretization interval $h = 0.02$, time increment $\Delta t = 0.02$.

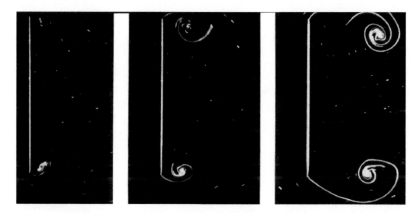

FIGURE 3.59 Sudden initial movement of a plane plate toward its normal. (From Dyke, V., Sketchbook of liquid and gas flows, M: Mir, 1986, 182p.)

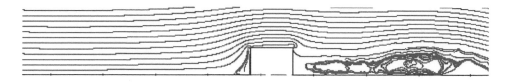

FIGURE 3.60 Flow lines at flow past a rectangular offset at the approach airflow velocity of 10 m/s at the instant of time $t = 2.74$ ($\Delta \tau = 0.005$, $h = 0.05$; the number of free vortices $N_s = 1216$), the flow separation from both edges of the rectangular offset.

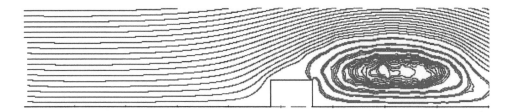

FIGURE 3.61 Flow lines at flow past a rectangular offset at the approach airflow velocity of 1 m/s at the instant of time $t = 13.9$ ($\Delta\tau = 0.05$, $h = 0.05$; $N_s = 820$), the flow separation from both edges of the rectangular offset.

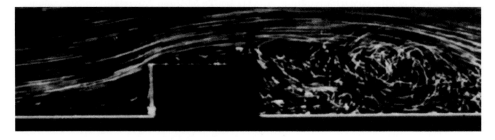

FIGURE 3.62 Turbulent separation at flow past a rectangular offset on a plate. (From Dyke, V., Sketchbook of liquid and gas flows, M: Mir, 1986, 182p.)

the floor (Figure 3.60). At lower velocities, the vortex region may be significantly higher. We also have a fairly good agreement between the experimental and design flow patterns (Figure 3.62).

3.6.3 TWO-DIMENSIONAL TURBULENT JET FLOWING OUT OF A PLANE WALL SLOT

As is clear from Figures 3.63 and 3.64, the jet is initially laminar, then turbulent. The flow patterns coincide both qualitatively and quantitatively as shown in [116]. Note that vortices forming on either side of the jet axis with the jet development are present in both the design and experimental vortex patterns.

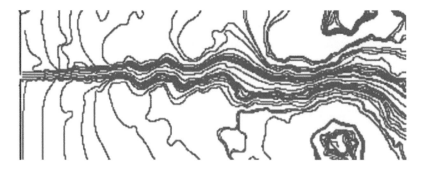

FIGURE 3.63 Flow lines at jet flowing out of a plane wall slot at the velocity of 33.2 m/s at the instant of time $t = 1.174$ ($\Delta\tau = 0.0032$, $h = 0.02$; the number of free vortices is 727).

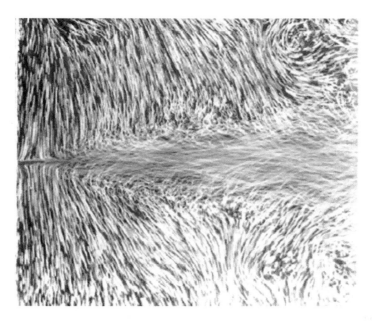

FIGURE 3.64 Two-dimensional turbulent jet of clean water flowing out of a long wall slot at the velocity of 30 m/s. (From Dyke, V., Sketchbook of liquid and gas flows, M: Mir, 1986, 182p.)

3.6.4 CIRCULATION FLOWS IN A RECTANGULAR REGION

Let us consider the flow area (Figure 3.65) within the closed rectangular region:

As is clear from the experiment, there is a vortex observed in the flow region nearly throughout the cavity. Note that the corner vortices in the picture (Figure 3.66b) is absent in the design images (Figures 3.66 and 3.67). The calculations made using the discrete vortex method (Figure 3.10) show many vortex formations in the central vortex covering the entire region. These vortex formations in the central vortex may eventually change positions (Figure 3.67). Note also that free vortices are circulating along the walls keeping away from the central part.

With the increase in the rectangular region length (Figure 3.68), the central vortex is being stretched along the region. In this case, according to the experiment, a vortex range occurs at the wall adjacent to the air inlet and starts growing with increase in length. As seen, this vortex region is observed in the DVM-based calculation too (Figure 3.68a).

Figures 3.69 through 3.72 show cases when supply and exhaust ports are located at the bottom of the model with the leftward inflow and different aspect ratios. The general nature of the flow remains approximately the same with the formation of a closed vortex region. In the experimental flow pattern with the model aspect ratio being equal to 1 the model is a circle that is stretching lengthwise with the increase in the rectangular region length.

FIGURE 3.65 Flow area diagram (circles denote free vortex separation points).

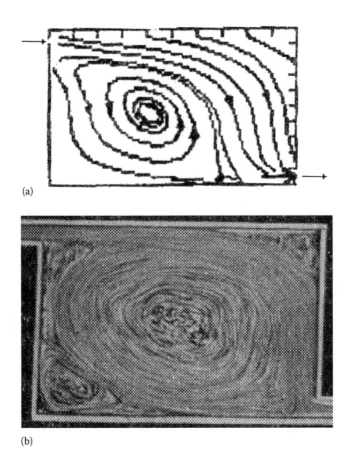

(a)

(b)

FIGURE 3.66 Vortex flow pattern: (a) estimated based on the numerical solution of continuity and Navier–Stokes equations in the Boussinesq approximation. (From Shaptala, V.G., *Mathematical Modeling in Applied Problems of Two-Phase Flow Mechanics*, Tutorial, BelGTASM Publishing, Belgorod, Russia, 1996, 103p.) (b) Experimental pattern. (From Baturin, V.V. and Khanzhonkov, V.I., Indoor air circulation based on arrangement of supply and exhaust ports, Heating and ventilation, 1939, No. 4–5, pp. 29–33.)

When the aspect ratio is less than 1 this region is being stretched across the model width; there are several oppositely turning vortices occurring rather than single one (Figure 3.14). No such vortices occur in the design flow patterns. Many vortices can be observed again in the central vortex.

The most specific aspect to the model with two supply inlets as shown in Figures 3.73 and 3.74 is the occurrence of a third jet moving in the opposite direction to the main jets. The design flow patterns show that at the initial instants of time, the flow pattern is asymmetric. Then the flow pattern eventually changes toward a symmetric one and a consistency between the design and the experiment is achieved.

Figure 3.74 shows the flow pattern with an aspect ratio equal to 3.5. The experimental flow pattern (Figure 3.74a) demonstrated a vortex behind the jet turn that was rotating opposite to the main jet. This vortex is absent in the design flow pattern.

In the experiment where a supply inlet is located in the center of the model, the flow-over direction is unstable (Figure 3.75). The jet is retained against a wall (Figure 3.75a). The supply jet departure from the center results in the jet being retained against the nearest wall. The design flow pattern shows a jet retained only against the wall adjacent to the exhaust outlet (Figure 3.75b).

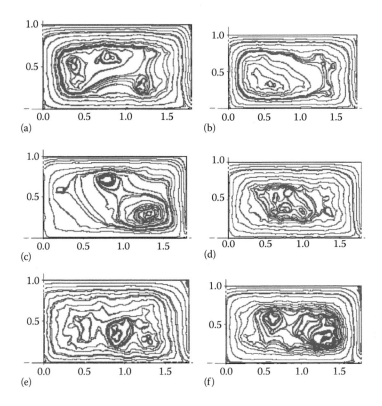

FIGURE 3.67 Flow lines in the rectangular region: (a) $\Delta\tau = 0.05$; $h = 0.02$; $v_0 = v_s = 0.4$ m/s; $t = 43.3$; $N_S = 1086$; (b) $\Delta\tau = 0.02$; $h = 0.02$; $v_0 = v_s = 1$ m/s; $t = 16.18$; $N_S = 1055$; (c) $\Delta\tau = 0.01$; $h = 0.02$; $v_0 = v_s = 1$ m/s; $t = 10.83$; $N_S = 1803$; (d) $\Delta\tau = 0.01$; $h = 0.02$; $v_0 = v_s = 2$ m/s; $t = 14.08$; $N_S = 1330$; (e) $\Delta\tau = 0.01$; $h = 0.02$; $v_0 = v_s = 4$ m/s; $t = 13.97$; $N_S = 1203$; (f) $\Delta\tau = 0.005$; $h = 0.02$; $v_0 = v_s = 4$ m/s; $t = 5.09$; $N_S = 1159$.

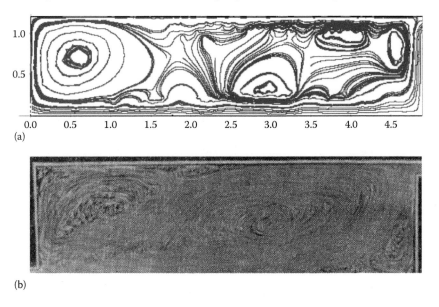

FIGURE 3.68 Vortex flow in a rectangular region: (a) design pattern at $\Delta\tau = 0.02$; $h = 0.02$; $v_0 = v_s = 1$ m/s; $t = 21.08$; $N_S = 647$, supply jet sharp edge separation and (b) experimental pattern. (From Baturin, V.V. and Khanzhonkov, V.I., Indoor air circulation based on arrangement of supply and exhaust ports, Heating and ventilation, 1939, No. 4–5, pp. 29–33.)

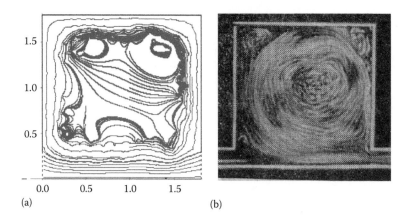

(a)   (b)

FIGURE 3.69   Vortex flow in a square region at (a) $\Delta\tau = 0.02$; $h = 0.02$; $v_0 = v_s = 1$ m/s; $t = 49.28$; $N_S = 375$, supply jet sharp edge separation and (b) experimental pattern. (From Baturin, V.V. and Khanzhonkov, V.I., Indoor air circulation based on arrangement of supply and exhaust ports, Heating and ventilation, 1939, No. 4–5, pp. 29–33.)

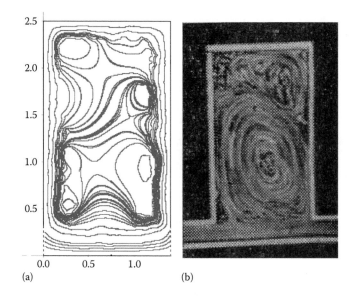

(a)   (b)

FIGURE 3.70   Vortex flow in a square region at (a) $\Delta\tau = 0.01$; $h = 0.02$; $v_0 = v_s = 1$ m/s; $t = 29.94$; $N_S = 610$, supply jet sharp edge separation and (b) experimental pattern. (From Baturin, V.V. and Khanzhonkov, V.I., Indoor air circulation based on arrangement of supply and exhaust ports, Heating and ventilation, 1939, No. 4–5, pp. 29–33.)

3.6.5   AIRFLOW DYNAMICS IN A ROOM WITH AN OFFSET

Let us consider the airflow area in a naturally ventilated room (Figure 3.76). The rectangular offset here is a table model (Figures 3.76 through 3.79).

As seen from the design flow patterns obtained using the software program developed based on the discrete vortex method, the vortex flow area between the table and the window is increased as

(a)

(b)

FIGURE 3.71 Vortex flow in a square region at (a) $\Delta\tau = 0.02$; $h = 0.05$; $v_0 = v_s = 1$ m/s; $t = 44.84$; $N_S = 796$, supply jet sharp edge separation and (b) experimental pattern. (From Baturin, V.V. and Khanzhonkov, V.I., Indoor air circulation based on arrangement of supply and exhaust ports, Heating and ventilation, 1939, No. 4–5, pp. 29–33.)

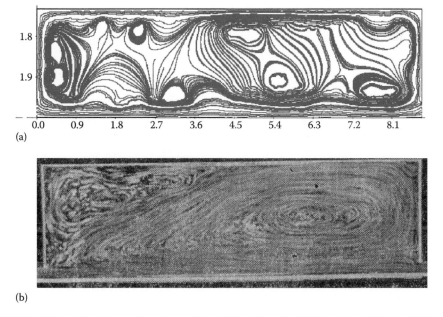

(a)

(b)

FIGURE 3.72 Vortex flow in a rectangular region: (a) based on the DVM at $\Delta\tau = 0.02$; $h = 0.05$; $v_0 = v_s = 1$ m/s; $t = 41.4$; $N_S = 913$, supply jet sharp edge separation and (b) experimental pattern. (From Baturin, V.V. and Khanzhonkov, V.I., Indoor air circulation based on arrangement of supply and exhaust ports, Heating and ventilation, 1939, No. 4–5, pp. 29–33.)

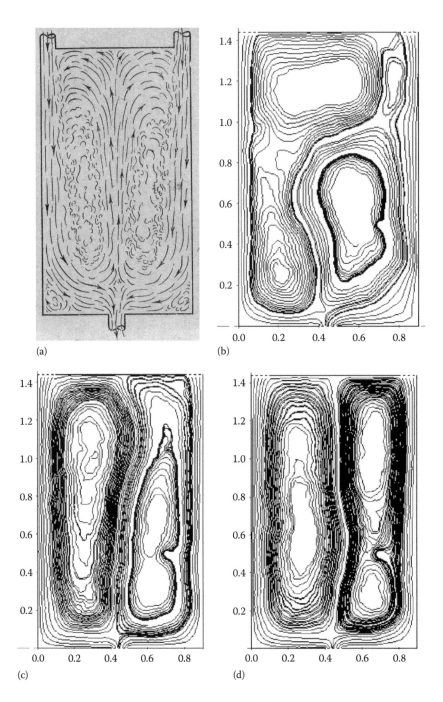

FIGURE 3.73 Vortex flow in a model with two supply jets: (a) Experimental flow pattern. (From Baturin, V.V. and Khanzhonkov, V.I., Indoor air circulation based on arrangement of supply and exhaust ports, Heating and ventilation, 1939, No. 4–5, pp. 29–33.) Design flow pattern at $\Delta\tau = 0.02$; $h = 0.02$; $v_0 = 1$ m/s; $v_s = 0.5$ m/s; at separation from non-wall-adjacent supply jet sharp edges; (b) $t = 13.92$; $N_S = 856$; (c) $t = 16.74$; $N_S = 933$; and (d) $t = 33.86$; $N_S = 1241$.

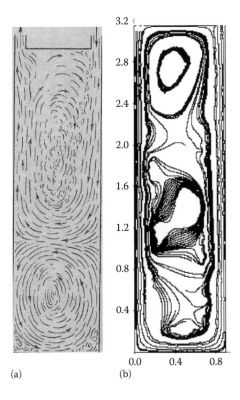

(a) (b)

FIGURE 3.74 Vortex flow in a region with supply and exhaust located on the same wall: (a) Experimental flow pattern. (From Baturin, V.V. and Khanzhonkov, V.I., Indoor air circulation based on arrangement of supply and exhaust ports, Heating and ventilation, 1939, No. 4–5, pp. 29–33.) (b) Design flow pattern at $\Delta\tau = 0.02$; $h = 0.02$; $v_0 = 1$ m/s; $v_s = 1$ m/s; supply jet sharp edge separation, $t = 13.8$; $N_S = 608$.

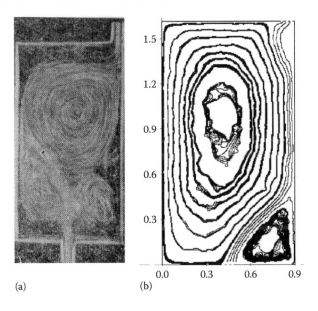

(a) (b)

FIGURE 3.75 Jet sticking to the wall with a supply inlet located in the center of the model at (a) experimental flow pattern. (From Baturin, V.V. and Khanzhonkov, V.I., Indoor air circulation based on arrangement of supply and exhaust ports, Heating and ventilation, 1939, No. 4–5, pp. 29–33.) (b) Design flow pattern at $\Delta\tau = 0.02$; $h = 0.02$; $v_0 = 1$ m/s; $v_s = 1$ m/s; $t = 15.7$; $N_S = 850$, separation from both sharp edges of the supply inlet.

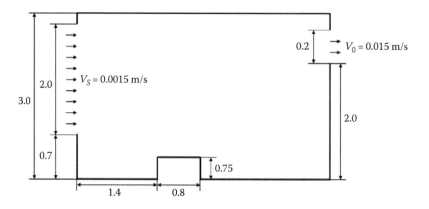

FIGURE 3.76   Diagram of natural ventilation with window infiltration and natural exhaust.

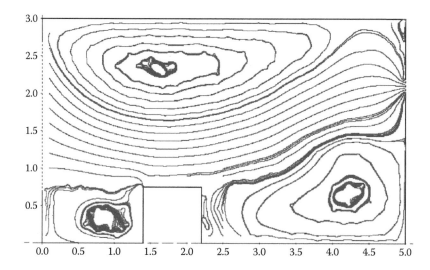

FIGURE 3.77   Indoor air circulation at the distance between the table and the window equal to 1.4 m at $\Delta\tau = 33.4$; $h = 0.05$; $v_0 = 0.015$ m/s; $v_s = 0.0015$ m/s; $t = 10020$; $N_S = 1793$; separation from both sharp edges of the supply inlet, sharp edges of the rectangle, and smooth ceiling and floor surface to the left and the right from the table.

the table is displaced from the window to the wall with the natural exhaust. Accordingly, the vortex at the right of the table is reduced slightly stretching along the wall. The vortex range near the ceiling remains unchanged. Note that the vortices near the ceiling and floor are rotating in opposite directions. The flow lines plotted can be used for choosing the indoor location that is most comfortable for the people inside the room (Figure 3.79).

Thus, the methodic research conducted allows in making the conclusion that the calculation results conform to the full-scale experimental data. The conformance of the calculation results to the experimental data may be violated when modeling a jet sticking to a wall or when modeling air exchange in a rectangular region with the aspect ratio above 3.5. The DVM does not *detect* some vortices that occur in corners. In general, the flow pattern in the central vortex range is consistent with the field studies.

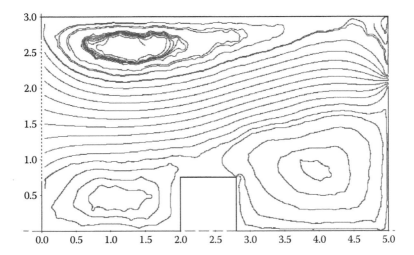

FIGURE 3.78 Indoor air circulation at the distance between the table and the window equal to 3 m at $\Delta\tau = 33.4$; $h = 0.05$; $v_0 = 0.015$ m/s; $v_s = 0.0015$ m/s; $t = 10921.8$; $N_S = 1984$; separation from both sharp edges of the supply inlet, sharp edges of the rectangle, and smooth ceiling and floor surface to the left and the right from the table.

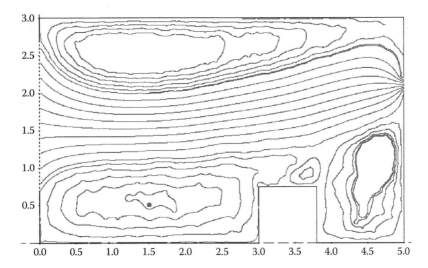

FIGURE 3.79 Indoor air circulation at the distance between the table and the window equal to 3 m at $\Delta\tau = 33.4$; $h = 0.05$; $v_0 = 0.015$ m/s; $v_s = 0.0015$ m/s; $t = 10420.8$; $N_S = 1488$; separation from both sharp edges of the supply inlet, sharp edges of the rectangle, and smooth ceiling and floor surface to the left and the right from the table.

3.7 MODELING FLOWS IN THIN SCREEN AREAS (SLOT-TYPE)

The objective of this chapter is to develop a mathematical model, computational algorithm, and software program for the calculation of nonstationary vortex flows inside a ventilated suction unit and upstream of slots equipped with thin screens arranged randomly in space.

3.7.1 PROBLEM SETUP

The physical setup of the problem: find the field of velocities in the suction unit and in the slot (Figure 3.80) at which inlet thin screens can be installed.

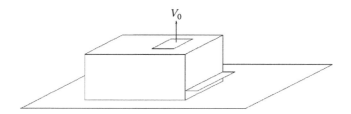

FIGURE 3.80   Diagram of a slot-type ventilated suction unit equipped with a screen.

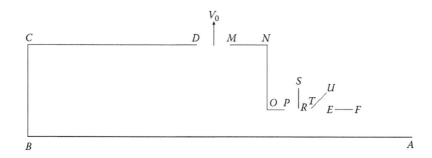

FIGURE 3.81   Regarding the problem setup.

There may not be just one (as shown in Figure 3.80) of such screens but many screens forming a labyrinth for the passage of air.

Since the slot length exceeds the slot width more than 10 times, there was a 2D problem considered (Figure 3.81).

The problem was solved using the discrete vortex method.

The mathematical setup of the problem consists of solving the Laplace equation for potential function ϕ at each design instant of time

$$\Delta\varphi = 0,$$

at the given normal boundary velocity values $\left.\dfrac{\partial\varphi}{\partial n}\right|_S = v_n(x) - U_n$ where x is S boundary point. Function U_n expresses the effect of free vortices converging into a flow from sharp edges at each design instant of time along the outward normal.

3.7.2   DERIVING THE BASIC DESIGN RELATIONS

Let us assume that the range boundary consist of z lines which we will discrete with a set of bound vortices and reference (design) points. There should be vortices located on the line jogs and ends. The reference points will be at the center between two bound vortices. Then, if we have N bound vortices there will be N-z reference points.

Let us consider the initial instant of time $t = +0$ when the suction port is enabled. There are only bound vortices in the region at this instant of time. The effect of all of these vortices on reference point x^p along the normal will be determined from the expression

$$v_n(x^p) = \sum_{k=1}^{N} G(x^p, \xi^k)\Gamma(\xi^k), \tag{3.69}$$

where

$$G(x^p, \xi^k) = \frac{(x_1 - \xi_1)n_2 - (x_2 - \xi_2)n_1}{2\pi \left[(x_1 - \xi_1)^2 + (x_2 - \xi_2)^2\right]}$$

(x_1, x_2) are x^p point coordinates

(ξ_1, ξ_2) are the coordinates of the bound vortex with circulation $\Gamma(\xi^k)$ located at point ξ^k

$\{n_1, n_2\}$ are the coordinates of the unitary vector of the normal $\vec{n}$ to the boundary of the range

$v_n(x^p)$ is the velocity at a point x^p along $\vec{n}$ already known at the problem setup

By measuring p from 1 to N-z in the expression given in Equation 3.69 we will obtain a system of N-z equations with N unknown circulations $\Gamma(\xi^k)$ where $k = \overline{1, N}$. We will supplement the system in question with z equations that are discrete countertypes of the Thompson condition of constant circulation around a liquid circuit covering a profile and trace (the sum of the circulations of bound vortices located on this line and free vortices separating from this line is equal to zero). Then we will obtain a closed-circuit system of linear algebraic equations:

$$\begin{cases} \sum_{k=1}^{N} G(x^p, \xi^k)\Gamma(\xi^k) = v_n(x^p); \quad p = \overline{1, N - z}, \\ \sum_{k=1+n_{c-1}}^{n_c} \Gamma(\xi^k) = 0, \quad c = \overline{1, z}. \end{cases}$$

Here, $\sum_{c=1}^{z} n_c = N$; $n_0 = 0$; n_c is the number of bound vortices on c-line.

As soon as the unknown circulations have been determined, the velocity at any point of the range along any given direction will be defined from the expression given in Equation 3.69 where x^p is substituted with the point in question.

At the instant of time $t = 1 \cdot \Delta t$, there are L free vortices separating from L sharp edges of the boundary. In the strict sense, vortices lying on these edges have already been free, since subject to Chaplygin–Zhukovski–Kutt hypothesis proven in [113] the bound vortex layer on the profile from which free vortex wake is separating will go to zero. Free vortices separate in the flow velocity direction. Their new position will be determined based on the old one from the following formulas:

$$x' = x + v_x\Delta t, \quad y' = y + v_y\Delta t, \tag{3.70}$$

where v_x, v_y are the velocity components calculated by the formulas given in Equation 3.69 at $\vec{n} = \{1, 0\}$ and $\vec{n} = \{0, 1\}$, respectively, while (x_1, x_2) are substituted with (x, y). Free vortex circulations remain unchanged over time. In view of separated free vortices, the system of equations for determining unknown circulations of bound vortices is given by

$$\begin{cases} \sum_{k=1}^{N} G(x^p, \xi^k)\Gamma(\xi^k) + \sum_{l=1}^{L} G(x^p, \varsigma^l)\gamma(\varsigma^l) = v_n(x^p); \quad p = \overline{1, N - z}, \\ \sum_{k=1+n_{c-1}}^{n_c} \Gamma(\xi^k) + \sum_{l=1+L_{c-1}}^{L_c} \gamma(\varsigma^l) = 0, \quad c = \overline{1, z}, \end{cases} \tag{3.71}$$

where

ς^l is a point where a free vortex separated from l-sharp edge is

L_c is the number of points of vortex wake separation from line c

$\sum_{c=1}^{z} L_c = L$; $L_0 = 0$

At the following instant of time L free vortices more will be separated, previous vortices will change their position determined from the formula given in Equation 3.70 where the velocity components are calculated considering the free vortices present in the flow:

$$v_n(x) = \sum_{k=1}^{N} G(x, \xi^k)\Gamma(\xi^k) + \sum_{l=1}^{L} G(x, \varsigma^l)\gamma(\varsigma^l).$$

At the instant of time $t = 2 \cdot \Delta t$, system in Equation 3.71 will be converted into the following form:

$$\begin{cases} \sum_{k=1}^{N} G(x^p, \xi^k)\Gamma(\xi^k) + \sum_{\tau=1}^{2}\sum_{l=1}^{L} G(x^p, \varsigma^{l\tau})\gamma(\varsigma^{l\tau}) = v_n(x^p), \quad p = \overline{1, N-z}, \\ \sum_{k=1+n_{c-1}}^{n_c} \Gamma(\xi^k) + \sum_{\tau=1}^{2}\sum_{l=1+L_{c-1}}^{L_c} \gamma(\varsigma^{l\tau}) = 0, \quad c = \overline{1, z}, \end{cases}$$

where

$\varsigma^l\tau$ is a point where a free vortex separated from l-sharp edge is at the instant of time τ
$\gamma(\varsigma^l\tau)$ is the vortex circulation

At a random instant of time $t = m \cdot \Delta t$, the system of equations for determining unknown circulations of bound vortices is given by

$$\begin{cases} \sum_{k=1}^{N} G(x^p, \xi^k)\Gamma(\xi^k) + \sum_{\tau=1}^{m}\sum_{l=1}^{L} G(x^p, \varsigma^{l\tau})\gamma(\varsigma^{l\tau}) = v_n(x^p), \quad p = \overline{1, N-z}, \\ \sum_{k=1+n_{c-1}}^{n_c} \Gamma(\xi^k) + \sum_{\tau=1}^{m}\sum_{l=1+L_{c-1}}^{L_c} \gamma(\varsigma^{l\tau}) = 0, \quad c = \overline{1, z}, \end{cases}$$

while the velocity at any given point is given by

$$v_n(x) = \sum_{k=1}^{N} G(x, \xi^k)\Gamma(\xi^k) + \sum_{\tau=1}^{m}\sum_{l=1}^{L} G(x, \varsigma^{l\tau})\gamma(\varsigma^{l\tau}).$$

If a free vortex was approaching an impermeable boundary for a distance of less than λ (the distance between the adjacent bound vortices and the reference point), it will be moved away normally from the boundary for a distance λ. If a free vortex was approaching a suction inlet for the same distance such vortex was disregarded.

In case the vortex was approached for distance $x < \lambda$, the velocity it induced was determined from the formula

$$v(x) = \frac{xv}{\lambda},$$

where v is the velocity induced by the vortex at distance λ.

There was a software program developed based on the described mathematical model to enable determining the field of velocities, plotting flow lines, and monitoring the vortex pattern behavior in a slot-type suction unit upstream of which there may be screens of various lengths installed randomly in space.

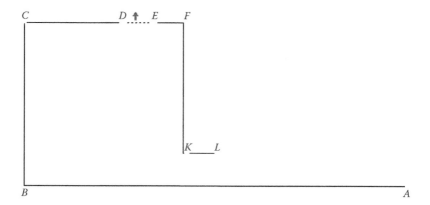

FIGURE 3.82 The boundary of a slot-type suction unit equipped with a screen the length of which is equal to its width.

In particular, the following flow boundary (Figure 3.82) was used for calculations set forth in Chapter 1 with the separation point at point L. The suction unit geometrical dimensions: $AB = 1.2$ m, $CB = 0.5$ m, $CD = 0.3$ m, $DE = 0.1$ m, $EF = 0.1$ m, $FK = 0.4$ m, and $KL = 0.1$ m. The suction velocity is 0.52 m/s.

A set of equations for determining unknown circulations of bound vortices at the instant of time $t = m \cdot \Delta t$ is given by

$$
\begin{cases}
\displaystyle\sum_{k=1}^{N} G(x^p, \xi^k)\Gamma(\xi^k) + \sum_{\tau=1}^{m} G(x^p, \varsigma^\tau)\gamma(\varsigma^\tau) = v_n(x^p), \quad p = \overline{1, N-1}, \\[2mm]
\displaystyle\sum_{k=1}^{N} \Gamma(\xi^k) + \sum_{\tau=1}^{m} \gamma(\varsigma^\tau) = 0,
\end{cases}
$$

where $\gamma(\varsigma^\tau)$ is a circulation of a free vortex separated from a sharp edge at the instant of time τ at point ς^τ.

The velocity at any given point is given by

$$
v_n(x) = \sum_{k=1}^{N} G(x, \xi^k)\Gamma(\xi^k) + \sum_{\tau=1}^{m} G(x, \varsigma^\tau)\gamma(\varsigma^\tau).
$$

3.7.3 Modeling Vortex Flows in a Slot-Type Suction Unit Equipped with Mechanical Screens

As an example there was a vortex flow inside a suction unit and upstream of the slot calculated for the following parameters: $CB = 0.35$ m, $CD = 0.93167$ m, $DM = 0.13333$ m, $MN = 0.29$ m, and $v_0 = 1$ m/s. These dimensions correspond to the dimensions the suction unit laboratory sample studied experimentally with the only exception that the suction cross section area rectangular in shape (see Figure 3.80) was stretched across the suction unit width.

There was a vortex flow calculated at various lengths of the screen installed upstream of the slot. With an increase in the screen length the suction cross section is quite quickly decreased. The highest drop is observed (Figure 3.83) at the variation in the screen length between 0 and 0.5 gauge ($d \approx 0.5h$).

The separated flow was calculated at $\Delta t = 0.0025$ s and free vortices (denoted with circles on Figure 3.83a) were separating from point P. Discretization interval $\lambda = 0.005$. Note that if the screen

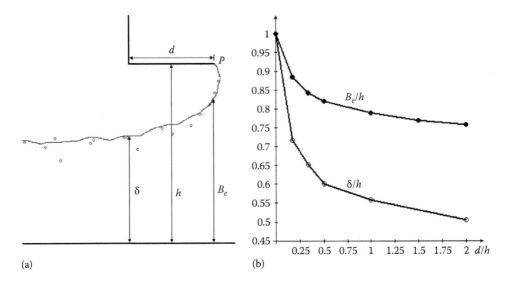

(a)　　　　　　　　　　　　　　　　　　(b)

FIGURE 3.83 Vortex flow characteristics upstream of a screened slot: (a) separated flow line and (b) dependency of the dimensionless suction width upstream of the slot B_e/h and upstream of the suction unit δ/h on the dimensionless length of the screen.

length exceeds two gauges the dimensionless effective suction width upstream of the slot B_e/h remains unchanged and is equal to 0.76. For a suction slot freely located in space, this value found using the Joukowski method is equal to 0.78. The Joukowski method implementation for the geometry of the range shown in Figure 3.84 allowed for determining $B_e/h = 0.81$ at the screen length of one gauge. With unlimited increase in the screen length this value is 0.775.

At a considerable distance from the suction unit inlet the dimensionless width of the jet δ_∞/h varies between 0.5 and 0.58 and does not depend on the screen width (Figure 3.84). In calculations using the Joukowski method, this value is 0.5.

The obtained design pattern of the separated flow upstream of the slot and the pattern observed in the full-scale experiment (Figure 3.85) satisfactorily agree with each other.

The flow lines plotted inside the suction unit using the developed software program (Figures 3.86 and 3.87) and the flow lines observed in the full-scale experiment (Figure 3.88) have a similar structure. There was a phenomenon of the central vortex displacement toward to the right wall of the suction unit observed with the increase in the screen length. The numerical experiment *detects* no such phenomenon. However, vortex formations eventually change their positions (Figure 3.87) and there are velocity pulsations observed.

According to the full-scale experiment conducted under the guidance of Yu.G. Ovsyannikov the jet separation line is broken approximately in 1–2 gauges at the suction unit inlet (Figure 3.85). With increase in the screen length, the separation line starts fluctuating in the numerical experiment too (Figure 3.89) but breaks in the full-scale experiment so that the intake flow completely fills the suction slot between the screen and the floor (Figure 3.88).

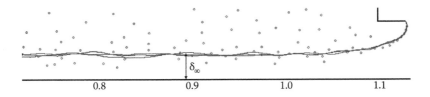

FIGURE 3.84 Separated flow lines at various instants of time.

(a)   (b)

FIGURE 3.85   Separated flows upstream of (a) screened slot and (b) unscreened slot.

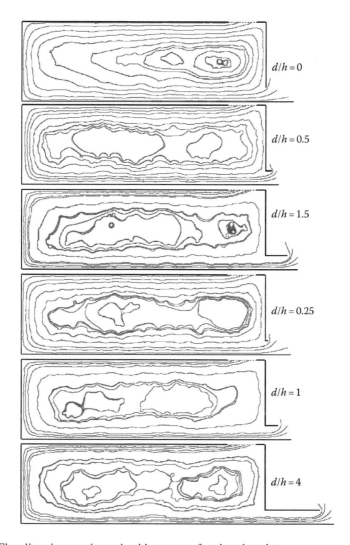

FIGURE 3.86   Flow lines in a suction unit with screens of various lengths.

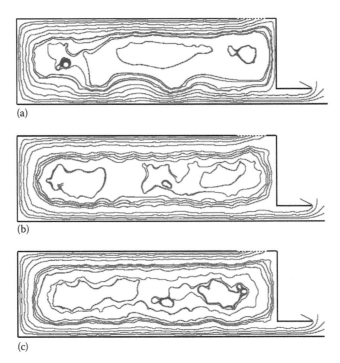

FIGURE 3.87   Flow lines at $d/h = 1.5$ and various instants of time: (a) 1.775 s, (b) 3.005 s, and (c) 5.025 s.

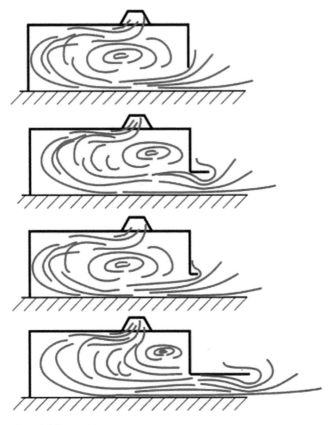

FIGURE 3.88   Experimental flow patterns.

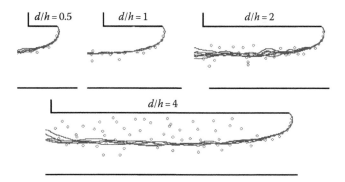

FIGURE 3.89 Separated flow line at various instants of time and various screen lengths.

The kinetic energy coefficient (Coriolis' coefficient) that expresses the irregularity of the field of velocities is calculated from the formula

$$\alpha = \frac{\int_S v_x^3 dS}{\left(v_{xav}\right)^3 S} \approx n^2 \frac{\sum_{i=1}^n v_i^3}{\left(\sum_{i=1}^n v_i\right)^3},$$

where

$v_{xav} = \int_S v_x dS/S$ is an average horizontal velocity component

v_x is a horizontal velocity component

S is the section in question

v_i is a horizontal velocity component at i-interval vertex

n is the number of intervals the given section is divided into

Coriolis' coefficient will be calculated in the suction unit slot inlet section. The number of partition intervals is $n = 11$ and horizontal velocity components were calculated at points located at the distance of 0.005; 0.01; …; 0.055 from the floor.

Since the flow is nonstationary, there were velocity pulsations observed in these points and accordingly Coriolis' coefficient pulsations are included in this section. Therefore, the coefficient value was averaged over time. After free vortices fill the whole suction unit area (the number of free vortices varied between 550 and 700), there were five instants of time randomly selected and α determined for each of them following which their arithmetic mean was found.

With the increase in the screen length Coriolis' coefficient increases significantly (Figure 3.90) to one gauge, then change insignificantly. Coriolis' coefficient is directly related to the LDF. For instance, at a sudden extension their values may be assumed to be the same. Note that the full-scale experiment showed that the vacuum-gauge pressure in the suction unit is the maximum when the screen length is equal to half of the gauge.

The flow vortex pattern and lines for the slot inlet with two screens is shown in Figure 3.91.

When slots are equipped with a combination of screens (Figure 3.92) the flow vortex pattern becomes significantly complicated. Energy losses due to the resulting local drags contribute to higher vacuum-gauge pressure in the suction unit that allows for decreasing the suction volumes and reducing the energy output of suction systems.

Thus, the following conclusions may be drawn:

1. Based on the discrete vortex method, there were a mathematical model, computation algorithm, and software program developed for the purpose of the design of nonstationary vortex flows in a slot-type suction unit equipped with thin screens.

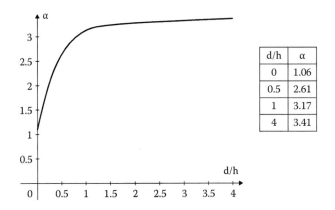

FIGURE 3.90   Coriolis' coefficient variation upstream of the suction unit with an increase in the screen length.

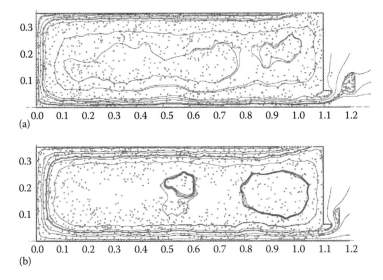

FIGURE 3.91   (a) Flow vortex pattern inside the suction unit and (b) upstream of the slots equipped with two screens.

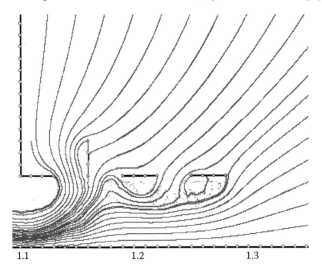

FIGURE 3.92   Flow vortex pattern upstream of the suction unit slot equipped with thin screens.

2. There were relations between the separated flow characteristics and the length of a screen installed at right angle to the slot inlet determined.

3. There was an effect of the position and presence of thin screens on the flow pattern shown as significantly increasing the flow turbulence as well as slot inlet flow resistance and, as a consequence, reducing air leaks and energy consumed from the operation of suction units.

3.7.4 Modeling Vortex Flows in a Trapezoidal Area

Modeling circulation flows in closed premises is required for the correct arrangement of ventilation. For instance, there are restrictions on air velocity applicable to commercial milk farm (CMF) facilities intended for cattle keeping. The maximum allowable velocity in cold season is 0.3–0.5 m/s. In summer, the air velocity may be increased to 0.6–1.0 m/s. In addition, air movement should be organized in the lower part of the facility. This section objective is to determine the most reasonable schemes of ventilation in such premises based on the mathematical model developed and the computational experiment.

The physical setup of the problem is to determine the field of velocities and vortex pattern of the flow in closed premises of nonrectangular shape with a suction inlet at the top and openings through which air is taken in from outside (Figure 3.93).

Since the flow is symmetrical relatively to the vertical axis we will consider only a single part of the design range (Figure 3.94) in which opening there also may be thin profiles.

The simply connected region (Figure 3.95) modeling was carried out for the following parameters: discretization interval $h = 0.05$ m, time increment $\Delta t = 0.06$ s, and the suction velocity $= 1$ m/s. The bound vortices are shown in the range boundary as circles. The following instants of time were considered: (a) $t = 77$ s (at this moment of time there are 2075 free vortices in the range), (b) $t = 118$ (2695 free vortices), (c) $t = 155$ (3151 free vortices), (d) $t = 178$ (3310 free vortices), (e) $t = 241$ (3696 free vortices), (f) $t = 423$ (4470 free vortices), (g) $t = 695$ (4980 free vortices), and (h) $t = 738$ (5118 free vortices).

The flow lines were plotted after free vortices (shown as dots in Figure 3.95) had completely filled the range. The vortex pattern is changing (Figure 3.95a through h) over time. Initially, the

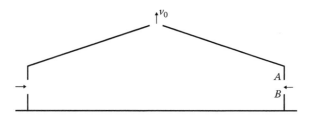

FIGURE 3.93 Regarding the problem setup.

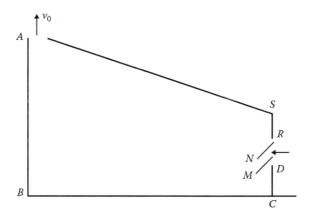

FIGURE 3.94 Design flow area.

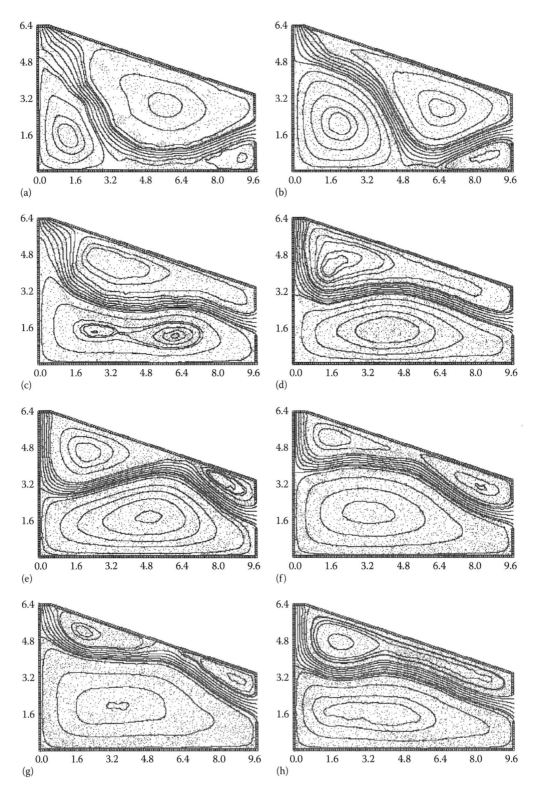

FIGURE 3.95 Flow vortex pattern at various instants of time: (a) $t = 77$s, (b) $t = 118$s, (c) $t = 155$s, (d) $t = 178$s, (e) $t = 241$s, (f) $t = 423$s, (g) $t = 695$s, and (h) $t = 738$s.

medium flow is being retained against the bottom (Figure 3.95a and b) and then it is moving to the center (Figure 3.95b through d) to be further positioned closer to the external part of the boundary (Figure 3.95e through h). However, it should be noted that there is the flow instability observed: dimensions of the top vortices are changing over time which can also be observed when modeling suction flows [156].

The dimensions of the design region correspond to the dimensions of the CMF facility having the capacity of 650 forage cows and located in the Krivtsovo village of the Yakovlevsk district of the Belgorod region (Figures 3.96 through 3.98). The modeling was carried out at various heights of the

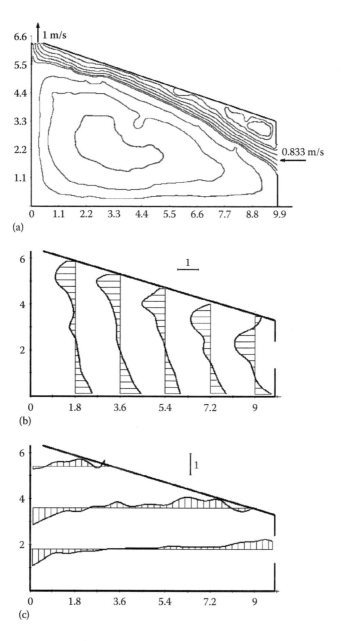

FIGURE 3.96 Flow pattern with 1.2 m high window opened completely: (a) flow lines, (b) horizontal velocity component profile (the unit segment shown corresponds to 1 m/s), and (c) vertical velocity component profile (model instant of time $t = 448.8$; the number of free vortices is 2321; time increment $\Delta t = 0.12$).

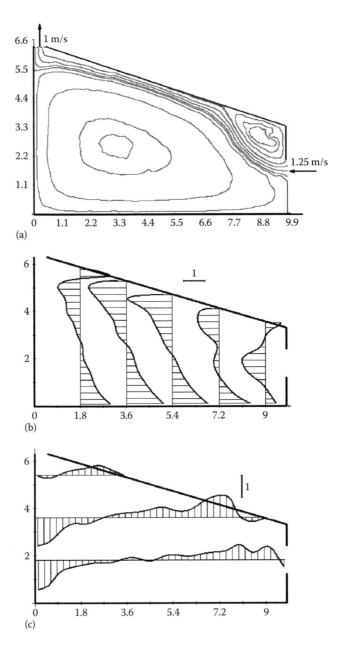

FIGURE 3.97   Flow pattern with 0.8 m high window opened partially: (a) flow lines, (b) horizontal velocity component profile (the unit segment shown corresponds to 1 m/s), and (c) vertical velocity component profile (model instant of time $t = 177$; the number of free vortices is 2005; time increment $\Delta t = 0.08$).

opening (window): 1.2 m (Figure 3.96), 0.8 m (Figure 3.97), and 0.4 m (Figure 3.98). The distance between two adjacent bound vortices (discretization interval) $h = 0.1$ m; time increment $\Delta t = h \cdot AB/ (v_0 \cdot 1 \text{ m})$. There were true-to-scale flow lines and profiles of the longitudinal and horizontal velocity components plotted.

If the window height is within 0.8–1.2 m the still (vortex) region will be in the facility lower part (Figures 3.96 and 3.97) which is undesirable for cattle keeping or occupational safety. If the window height is decreased to 0.4 m, the still region will be moved to the facility

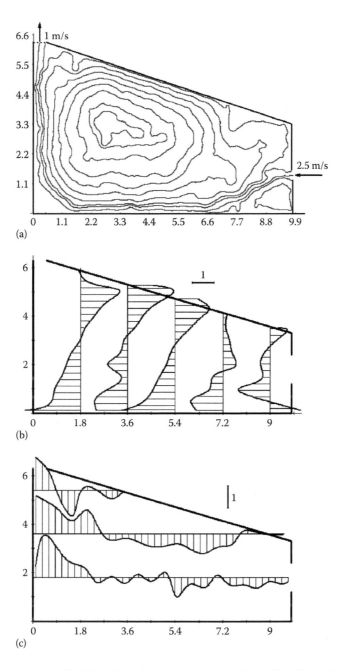

FIGURE 3.98 Flow pattern with 0.4 m high window opened partially: (a) flow lines, (b) horizontal velocity component profile (the unit segment shown corresponds to 1 m/s), and (c) vertical velocity component profile (model instant of time $t = 48.96$; the number of free vortices is 2017; time increment $\Delta t = 0.04$).

upper part (Figure 3.98). But in this case there are high velocity values observed in rather a narrow region adjacent to the facility lower part. This region may be extended by setting some profiles in the fully opened window aperture, for example, two profiles 20 cm long each at an angle of 45° (Figure 3.99). Using the resulting velocity profiles, it is possible to find the exhaust velocity v_0 required for animal safety. The calculations show that v_0 should be decreased twofold, that is, it should be equal to 0.5 m/s.

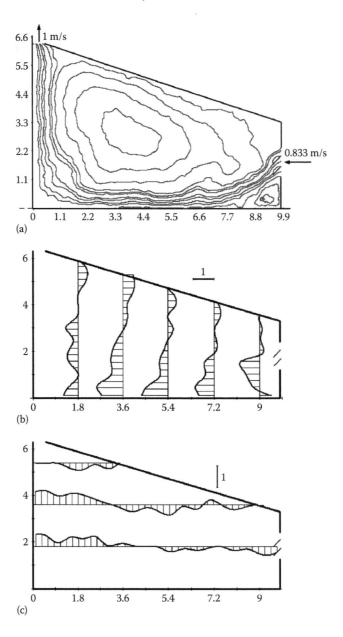

FIGURE 3.99 Flow pattern with completely opened window equipped with 1.2 m high air guides: (a) flow lines, (b) horizontal velocity component profile (the unit segment shown corresponds to 1 m/s), and (c) vertical velocity component profile (model instant of time $t = 56.16$; the number of free vortices is 2599; time increment $\Delta t = 0.12$).

The most effective method of ventilation is the installation of thin profiles in the window openings, which would retain the ventilating jet against the facility lower part. The flow lines and velocity profiles plotted may be used to determine the volumes of exhaust air required to ensure the correct velocity conditions for air jets.

Thus, based on the discrete vortex method and the Thompson condition of constant circulation around a liquid circuit covering a profile and trace, there was a mathematical flow separation model built, which is distinguished from the existing models by the consideration of multiple sections within the design range from which a vortex wake is separating.

There is a numerical method of this model implementation elaborated. It consists of plotting a recurrent computation scheme that solves sets of linear algebraic equations at each time increment where the right-hand side of such equations is determined using its values at the previous instant of time and adding discrete countertypes of the Thompson equations for each of the sections.

There was a software program developed for calculating nonstationary vortex flows in noncontinuous regions with numerous profiles that enable determining the field of velocities, plotting the flow lines, and visualizing the flow pattern behavior over time.

The computational experiment methods and algorithms developed are the basis for the systems of computer and simulation modeling in the area of ventilation air mechanics. The relevance of such conclusion is illustrated by the developed software support that enables the computational experiments described in the following.

3.8 MODELING SEPARATED FLOWS USING STATIONARY DISCRETE VORTICES

3.8.1 FLOW SEPARATION UPSTREAM OF A TWO-DIMENSIONAL PROTRUDING SUCTION DUCT

There is an ideal incompressible liquid analyzed. A vortex separates from the sharp boundary edge and the so-called free flow line originates from the edge in the position unknown in advance. Traditionally, this problem can be solved using the complex variable theory and ideal liquid jet theory methods [124]. It is assumed that the velocity on the free flow line is constant. Using such methods ensures the most accurate results. There is an analytical solution obtained in some of the simplest cases. The problem often resolves itself into quite lengthy nonlinear equations with coefficients of unknown variables in the form of Cauchy-type integrals. In the general case when a multiply connected flow region or, all the more, a 3D space is considered for solving the problems, using the aforementioned methods is impossible.

Using the discrete vortex method in nonstationary setup (NDVM) enables plotting free lines and streamline surfaces. However, according to numerical experiments for problems concerning suction spectra of exhaust slots, jet contraction δ_∞ at a considerable distance from the suction port remains virtually unchanged while some geometric dimensions are changing. In addition, this value pulsates in time and needs to be averaged over time. Therefore, using the NDVM to analyze δ_∞, determine the boundaries of separated flows and the relation of these values with air drag of suction inlets presents a certain problem.

The discrete vortex method application is known in the stationary setup (SDVM) to solve problems of a separated flow of ideal incompressible liquid upstream of a round continuous tube [137]. The problem was solved in the axially symmetric setup with discrete vortex rings used as discrete vortices. A clearer boundary of the free streamline surface would be obtained by placing a free vortex ring on the sharp edge.

This study's objective is to develop a method of mathematical modeling of separated flows upstream of suction-slot openings within the SDVM framework.

3.8.1.1 Basic Design Relations and Computation Algorithm

As an example, we considered a problem of a separated flow of ideal incompressible liquid upstream of a screened plane wall suction port (Figure 3.100) previously solved by various methods.

The following symbols were introduced: N is the number of bound vortices of the upper boundary (above OX-axis); N_S is the number of free vortices of the upper boundary. The vortex lying on sharp screen edge C is considered to be a free one that follows from the theorem set forth in [113]—the bound vortex intensity (circulation) at the flow separation point is equal to zero. Bound vortices were also positioned at the boundary fracture points. There were reference points positioned

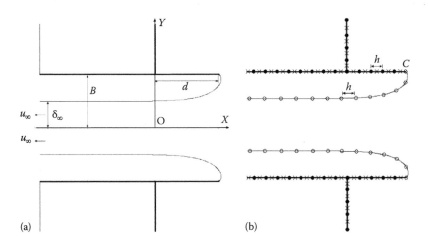

FIGURE 3.100 Regarding the problem setup: (a) physical flow area and (b) range boundary discretization: (● is for bound vortices; ○ is for free vortices; × is for reference points).

between the bound vortices. Point $\xi^k(\xi_1, \xi_2)$ is the bound vortex position point; $x^p(x_1, x_2)$ is the reference point. The action on point x^p (velocity at this point) by the vortex of a single circulation located at point ξ^k along the unit vector $\vec{n} = \{n_1, n_2\}$ is determined by the formula

$$G\left(x^p, \xi^k\right) = \frac{1}{2\pi} \cdot \frac{\left(x_1 - \xi_1\right)n_2 - \left(x_2 - \xi_2\right)n_1}{\left(x_1 - \xi_1\right)^2 + \left(x_2 - \xi_2\right)^2}. \tag{3.72}$$

This vortex action on point x^p is given by

$$v_n\left(x^p\right) = G\left(x^p, \xi^k\right)\Gamma\left(\xi^k\right), \tag{3.73}$$

where

$\Gamma(\xi^k)$ is a circulation of the vortex located at point ξ^k

$v_n(x^p)$ is the velocity at point x^p along the unit vector $\vec{n}$

With the boundary discretization method described, the number of bound vortices is equal to the number of reference points. Since the range boundary is symmetric around OX-axis and this axis is impermeable, the circulation of symmetric vortices will be inverse:

$$\Gamma\left(\xi^k\right) = -\Gamma\left(\xi^{k+N}\right) \quad \text{where } k = \overline{1, N}. \tag{3.74}$$

Fulfillment of this condition will automatically result in the noncircular flow condition (the sum of circulations of all vortices is equal to zero).

It was assumed that the circulation of vortices on the required free flow line is constant and is equal to γ. The distance between free vortices is a constant value equal to h. The first approximation for the free flow line was chosen as follows: The first three vortices were positioned parallel to OY starting from the sharp edge, while the rest of the vortices were positioned parallel to OX.

The resultant action of all vortices on the reference point x^p along the outward normal is given by

$$v_n\left(x^p\right) = \sum_{q=1}^{N} G\left(x^p, \xi^q\right)\Gamma\left(\xi^q\right) - \sum_{q=N+1}^{2N} G\left(x^p, \xi^q\right)\Gamma\left(\xi^{q-N}\right)$$

$$+ \gamma \sum_{k=1}^{N_S} G\left(x^p, \zeta^k\right) - \gamma \sum_{k=N_{S+1}}^{2N_S} G\left(x^p, \zeta^k\right)$$

$$= \sum_{q=1}^{N}\left(G\left(x^p, \xi^q\right) - G\left(x^p, \xi^{q+N}\right)\right)\Gamma\left(\xi^q\right) + \gamma \sum_{k=1}^{N_S}\left(G\left(x^p, \zeta^k\right) - G\left(x^p, \zeta^{k+N_S}\right)\right), \quad (3.75)$$

where ζ^k is a free vortex position point.

Since $v_n(x^p) = 0$ at all reference points, that is, the impermeability condition is fulfilled, expression Equation 3.75 will be converted as follows:

$$\sum_{q=1}^{N}\left(G\left(x^p, \xi^q\right) - G\left(x^p, \xi^{q+N}\right)\right)\Gamma\left(\xi^q\right) = -\gamma \sum_{k=1}^{N_S}\left(G\left(x^p, \zeta^k\right) - G\left(x^p, \zeta^{k+N_S}\right)\right). \quad (3.76)$$

When p is changed between 1 and N, the expression given in Equation 3.76 becomes a set of linear algebraic equations (SLAE) with N unknown variables $\Gamma(\xi^1)$, $\Gamma(\xi^2)$,..., $\Gamma(\xi^N)$. By solving this set using the Gaussian method with pivoting and determining the unknown circulations of bound vortices we will be able to compute the velocity at any point $x(x_1, x_2)$ of the range along any given direction from the formula

$$v_n\left(x\right) = \sum_{q=1}^{N}\left(G\left(x, \xi^q\right) - G\left(x, \xi^{q+N}\right)\right)\Gamma\left(\xi^q\right) + \gamma \sum_{k=1}^{N_S}\left(G\left(x, \zeta^k\right) - G\left(x, \zeta^{k+N_S}\right)\right). \quad (3.77)$$

If the distance from point x to the vortex at point ξ is less than $h/2$, then $G(x, \xi)$ is given by

$$G\left(x, \xi\right) = \frac{r}{\pi h} \cdot \frac{\left(S_1 - \xi_1\right)n_2 - \left(S_2 - \xi_2\right)n_1}{\left(S_1 - \xi_1\right)^2 + \left(S_2 - \xi_2\right)^2}, \quad (3.78)$$

where

$$S_1 = \frac{h}{2r}\left(x_1 - \xi_1\right) + \xi_1$$

$$S_2 = \frac{h}{2r}\left(x_2 - \xi_2\right) + \xi_2$$

$$r = \sqrt{\left(x_1 - \xi_1\right)^2 + \left(x_2 - \xi_2\right)^2}$$

The second approximation for the free flow line was chosen as follows: There is a flow line plotted from sharp edge C in the defined velocity field using the following formulas:

$$x = x' + v_x \Delta t, \quad y = y' + v_y \Delta t, \tag{3.79}$$

where

Δt is a quite small time increment (e.g., $\Delta t = 0.0001$)

(x', y') is the previous position of the point on the flow line

(x, y) is the previous point

v_x, v_y are found from Equation 3.77 at $\vec{n} = \{1;0\}$ for v_x and $\vec{n} = \{0;1\}$ for v_y

As soon as the distance between point (x, y) and the sharp edge becomes equal to h, a free vortex is placed at this point, that is, it will be the second approximation for this free flow line point. Then, a flow line is plotted again using Equation 3.79 until the distance between (x, y) and the previous position of the free vortex becomes equal to h. A free vortex is placed at this point and so on.

After defining the second approximation for the free flow line there should be a set of Equation 3.76 resolved again and bound vortex circulations determined. Then, the third approximation is made for the free flow line using the formulas given in Equations 3.77 through 3.79. This iteration process is repeated until the difference between the previous and next positions does not exceed the given accuracy.

Since the range boundaries *break* at some distance from the suction port, a situation occurs when the free flow line starts lifting up (for upper part of Figure 3.72), that is, its Y-coordinate starts increasing. That is why in this case y-coordinate is fixed for all other points and is equal to the maximum decrease point Y-coordinate.

The word description of the computational algorithm built using design relations in Equations 3.72 through 3.79 is as follows:

1. There are matrixes of the coordinates of reference points $kt[1...N, 1...2]$, coordinates of bound vortices $pv[1...N, 1...2]$, and coordinates of free vortices $sv[1...N, 1...2]$ formed.
   a. For the upper screen:

$$kt[i,1] = d - h\left(i - \frac{1}{2}\right); \quad kt[i,2] = B; \quad pv[i,1] = kt[i,1] - \frac{h}{2}; \quad pv[i,2] = B;$$

$$i = \overline{1, N_1}, \quad kt[N_1, 1] = \frac{h}{2}; \quad pv[N_1, 1] = 0.$$

   b. For the upper vertical wall:

$$kt[i,1] = 0; \quad kt[i,2] = B + h\left(k - \frac{1}{2}\right); \quad pv[i,1] = 0; \quad pv[i,2] = kt[i,2] + \frac{h}{2};$$

$$i = \overline{N_1, N}; \quad kt[N, 2] = kalibr - \frac{h}{2};$$

   kalibr is the vertical wall length (divisible by the suction port width B);

$$k = 1, 2, \ldots \text{ until } \left(kt[N, 2] - kalibr + \frac{h}{2}\right) < 0.000001.$$

c. For the range boundary symmetric around OX:

$$kt[i,1] = kt[i-n,1]; \quad kt[i,2] = -kt[i-n,2];$$
$$pv[i,1] = pv[i-n,1]; \quad pv[i,2] = -pv[i-n,2];$$

where $i = \overline{N+1,2N}$.

d. For the upper free flow line:

$$sv[i,1] = d; \quad sv[i,2] = b - h(i-1); \quad i = \overline{1,3};$$
$$sv[i,1] = d - h(i-3); \quad sv[i,2] = b - 2h; \quad i = \overline{4,k_2};$$

where k_2 is an integral part of $kalibr/h$; $N_S = i-1$.

e. For the lower free flow line:

$$sv[i,1] = sv[i-N_S,1]; \quad sv[i,2] = -sv[i-N_S,2]; \quad i = \overline{N_S+1,2N_S}.$$

2. Forming a matrix of the coefficients of unknown variables of SLAE Equation 3.76:

$$kof[p,q] = Skor\left(kt[p],pv[q],evn\right) - Skor\left(kt[p],pv[q+n],evn\right);$$

where $p = \overline{1,N}; \quad q = \overline{1,N}; \quad env[1...2]$ are outward normal coordinates;

$$evn[1] = 0; \quad evn[2] = -1 \text{ at } kt[p,2] = B; \quad evn[1] = 0; \quad evn[2] = 1 \text{ at}$$
$$kt[p,2] = -B; \quad evn[1] = -1; \quad evn[2] = 0 \text{ at } kt[p,1] = 0;$$
$$Skor\left(kt[p],pv[q],evn\right)$$

is a velocity along evn vector direction at point $kt[p]$ induced by the single circulation vortex located at point $pv[q]$. It is calculated using the formulas given in Equations 3.72 and 3.78.

3. Forming a column of free terms for SLAE, Equation 3.76:

$$pr[p] = -SvG \sum_{k=1}^{N_S} \left(Skor\left(kt[p],sv[k],evn\right) - Skor\left(kt[p],sv[k+N_S],evn\right)\right),$$

where $p = \overline{1,N}$; SvG is a circulation of vortices on the free flow line.

4. Solving SLAE $\sum_{q=1}^{N} kof[p,q]circ[q] = pr[p]$; $p = \overline{1,N}$ using the Gaussian method with pivoting and determination of the matrix of circulations of bound vortices $circ[1...N]$ at the upper boundary. Accordingly, for the lower boundary, $circ[i] = -circ[i-N]$, $i = \overline{N+1,2N}$.

5. Forming coordinates of the next approximation for the free flow line:
   a. There is an initial point of the flow line given: $x[1] = sv[1,1]$; $x[2] = sv[1,2]$.
   b. There is an increment in the flow velocity direction made, that is, there are coordinates calculated:

$$x[i] = x[i] + chag\left(\sum_{q=1}^{2N} circ[q]Skor(x, pv[q], evn)\right.$$

$$\left. + SvG\sum_{q=1}^{N_S}\left(Skor(x, sv[q], evn) - Skor(x, sv[q+N_S], evn)\right)\right),$$

where $i = \overline{1,2}$; $evn[1] = 1$; $evn[2] = 0$ at $i = 1$ and $evn[1] = 0$; $evn[2] = 1$ at $i = 2$; $chag = 0.0001$ is time increment.

   c. As soon as the distance from point x to the previous point of the free flow line is $\sqrt{\left(x[1] - sv[i-1,1]\right)^2 + \left(x[2] - sv[i-1,2]\right)^2} \geq h$, there are coordinates of the next approximation determined for the free flow line coordinates $psv[i] = x$.
   d. If the flow line point Y-coordinate exceeds the previous point Y-coordinate proceed to step (e), otherwise proceed to step (b). The count interruption condition: $x[1] < sv[N_S,1]$.
   e. There are free flow line points determined:

$$psv[i,1] = psv[i-1,1] - h; psv[i,2] = x[2]$$.

The count interruption condition: $psv[i,1] < sv[N_S,1]$.

   f. Old coordinates of the free flow line are changed for new ones:

$$sv[i] = psv[i]; \quad sv[i+N_S,1] = pvs[i,1]; \quad sv[i+N_S,2] = -psv[i,2],$$

where $i = \overline{1,N_S}$.

6. Then proceed to step 3, that is, determine the new approximation for the free flow line. This cycle is repeated as long as $|db - pv[N_S,2]| > 0.0001$, where $db = sv[N_S,2]$ is the Y-coordinate of the previous position of N_S point of the free flow line.

After the free flow line boundary has been defined its coordinates and circulations of bound vortices will be written to a file. Then these data can be used to determine the field of velocities. The velocity at the given point $x(x[1], x[2])$ along the given direction $evn(evn[1]; evn[2])$ will be determined from the formula

$$v = \sum_{q=1}^{2N} circ[q]Skor(x, pv[q], evn) + SvG\sum_{q=1}^{N_S}\left(Skor(x, sv[q], evn) - Skor(x, sv[q+N_S], evn)\right).$$

3.8.1.2 Calculation Data and Its Discussion

Using the developed software program, dimensionless velocity profiles (u_x/u_∞; u_y/u_∞) and free flow line coordinates were calculated. For convenient comparison of calculations made with various methods, they are represented in another system of coordinates (Figures 3.101 and 3.102).

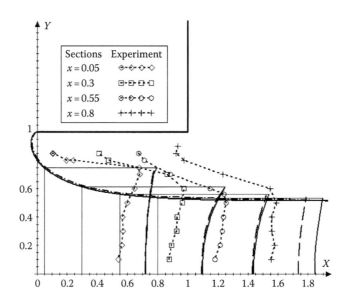

FIGURE 3.101   Free flow line and horizontal velocity component profiles.

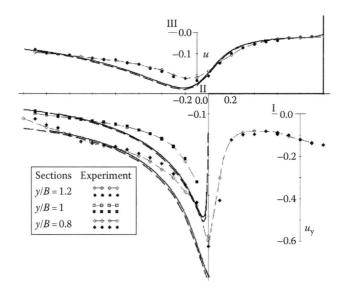

FIGURE 3.102   Vertical velocity component profiles.

Continuous lines on all figures denote calculations made using the Joukowski method while dotted and dash–dotted lines denote calculations made using the SDVM. As seen, the velocity and free flow line coordinates coincide. The biggest difference for the horizontal velocity component is observed in section $x = 0.8$ but it does not exceed 2% versus calculations made using the Joukowski method. The discretization interval in calculations made using the DVM is $h = 0.0125$, u_∞ was assumed to be the velocity on the suction axis two gauges away from the suction port.

The vertical wall length and the distance of the last free flow line point from the suction port are equal to eight gauges ($8C$).

The ratio of the difference of velocity u_∞ and average slot velocity $u_\infty \delta_\infty / B$ to the average velocity is $\Delta u = B/\delta_\infty - 1$, which was used as the LDF criterion in Chapter 4. Calculation of Δu using the

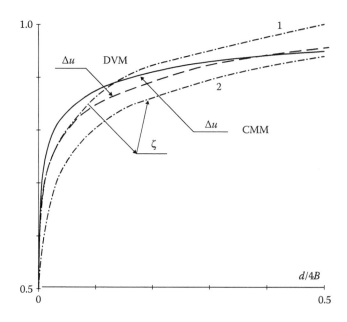

FIGURE 3.103 Variation in the LDF upstream of a flat pipe (ζ) and in the jet separation velocity (Δu) with an increase in the offset length ($d/4B$): dash–dotted lines denote experimental data by Idelchik [129] (Curve 1 is for the relative tube wall thickness $\delta/D_g = 0$; Curve 2 is for $\delta/D_g = 0.004$).

discrete vortex method demonstrates better conformance to the experimental data for the LDF ζ than in the case with the Joukowski method (Figure 3.103).

When the separation pattern is significantly enlarged, the initial separation section shows a noticeable difference between the coordinates of free flow lines obtained by various methods (Figure 3.104).

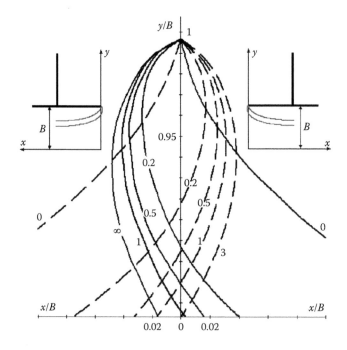

FIGURE 3.104 Curves of the jet flow separation next to the exhaust outlet with variation in the offset length d/B.

Thus, the mathematical model of separated flows elaborated based on the SDVM allows for obtaining an adequate flow pattern and its application will enable resolving a number of problems of separated flow air mechanics.

3.8.2 FLOW SEPARATION UPSTREAM OF A TWO-DIMENSIONAL SUCTION DUCT, COVERING A THIN PROFILE WITH A NONCIRCULATORY FLOW PAST IT

3.8.2.1 Deriving Basic Design Relations

The subject considered was a multiply connected flow region (Figure 3.105) upstream of a 2D suction duct the range of which covered numerous thin bodies. A vortex is separating from sharp edge C and a free flow line occurs. It is required to define the line position and flow velocity at any given point.

The discrete mathematical model is built as follows (Figure 3.106). Let the following symbols be introduced: N_1 is the number of bound vortices of the first connected boundary part equal to the number of reference points; N_2 is the number of bound vortices of the second connected boundary part being one unit more than the number of reference points; N_3 is the number of bound vortices of the third connected boundary part being one unit more than the number of reference points; N_l is the number of bound vortices of l connected boundary part being one unit more than the number of reference points. Thus, $N = \sum_{l=1}^{L} N_l$ is the number of bound vortices; $N - L + 1$ is the number of reference points.

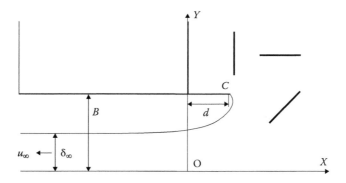

FIGURE 3.105 Regarding the problem setup.

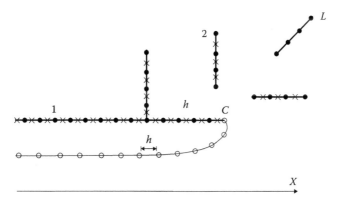

FIGURE 3.106 Range boundary discretization ($\bullet$ is for bound vortices; $\circ$ is for free vortices; $\times$ is for reference points).

The vortex lying on sharp screen edge C is considered to be a free one. Bound vortices were also positioned at the boundary fracture points. There were reference points positioned between the bound vortices. Point $\xi^k(\xi_1, \xi_2)$ is k bound vortex position point; $x^p(x_1, x_2)$ is p reference point. The action on point x^p (velocity at this point) by the vortex of a single circulation located at point ξ^k along the unit vector $\vec{n} = \{n_1, n_2\}$ is determined from the formula

$$G\left(x^p, \xi^k\right) = \frac{1}{2\pi} \cdot \frac{\left(x_1 - \xi_1\right)n_2 - \left(x_2 - \xi_2\right)n_1}{\left(x_1 - \xi_1\right)^2 + \left(x_2 - \xi_2\right)^2}.$$

Then the velocity at point x^p along the unit vector $\vec{n}$ induced by the action on point x^p by the vortex of circulation $\Gamma(\xi^k)$ located at point ξ^k is given by $v_n(x^p) = G(x^p, \xi^k)\Gamma(\xi^k)$.

In order to fulfill OX impermeability condition we will project all vortices symmetrically around OX so the circulations of symmetric vortices will be inverse: $\Gamma(\xi^k) = -\Gamma(\xi^{k+N})$ where $k = \overline{1, N}$. Fulfillment of this condition will automatically result in the noncircular flow condition (the sum of circulations of all vortices is equal to zero).

It was assumed that the circulation of vortices on the required free flow line is constant and equal to γ. The distance between free vortices is a constant value equal to h. The first approximation for the free flow line was chosen as follows: The first three vortices were positioned parallel to OY below the separation point starting from the sharp edge, while the rest of the vortices were positioned parallel to OX leftward the separation point.

The resultant action of all vortices on reference point x^p along the outward normal is given by

$$v_n\left(x^p\right) = \sum_{q=1}^{N} G\left(x^p, \xi^q\right)\Gamma\left(\xi^q\right) - \sum_{q=N+1}^{2N} G\left(x^p, \xi^q\right)\Gamma\left(\xi^{q-N}\right) + +\gamma\sum_{k=1}^{N_S} G\left(x^p, \zeta^k\right)$$

$$-\gamma\sum_{k=N_S+1}^{2N_S} G\left(x^p, \zeta^k\right) = \sum_{q=1}^{N}\left(G\left(x^p, \xi^q\right) - G\left(x^p, \xi^{q+N}\right)\right)\Gamma\left(\xi^q\right) + \gamma\sum_{k=1}^{N_S}\left(G\left(x^p, \zeta^k\right) - G\left(x^p, \zeta^{k+N_S}\right)\right),$$

$$(3.80)$$

where ζ^k is a free vortex position point.

Since $v_n(x^p) = 0$ at all reference points, that is, the impermeability condition is fulfilled, by changing $p = 1, 2, \ldots, N - L + 1$ and adding the noncirculatory flow condition for each of closed screens we will obtain the following set of equations for determining unknown circulations of bound vortices $\Gamma(\xi^1), \Gamma(\xi^2), \ldots, \Gamma(\xi^N)$:

$$\begin{cases} \sum_{q=1}^{N}\left(G\left(x^p, \xi^q\right) - G\left(x^p, \xi^{q+N}\right)\right)\Gamma\left(\xi^q\right) = -\gamma\sum_{k=1}^{N_S}\left(G\left(x^p, \zeta^k\right) - G\left(x^p, \zeta^{k+N_S}\right)\right), \\ \sum_{i=1}^{N_l}\Gamma\left(\xi^{N_1+N_2+\cdots+N_{l-1}+i}\right) = 0, \quad l = \overline{2, L}. \end{cases}$$

$$(3.81)$$

After unknown circulations of the bound vortices have been determined, the velocity at any point $x(x_1, x_2)$ of the range along any given direction will be determined from the formula

$$v_n\left(x\right) = \sum_{q=1}^{N}\left(G\left(x, \xi^q\right) - G\left(x, \xi^{q+N}\right)\right)\Gamma\left(\xi^q\right) + \gamma\sum_{k=1}^{N_S}\left(G\left(x, \zeta^k\right) - G\left(x, \zeta^{k+N_S}\right)\right) \qquad (3.82)$$

If the distance from point x to the vortex at point ξ is less than $h/2$, then $G(x, \xi)$ is given by

$$G(x,\xi) = \frac{r}{\pi h} \cdot \frac{(S_1 - \xi_1)n_2 - (S_2 - \xi_2)n_1}{(S_1 - \xi_1)^2 + (S_2 - \xi_2)^2}, \quad S_1 = \frac{h}{2r}(x_1 - \xi_1) + \xi_1, \quad S_2 = \frac{h}{2r}(x_2 - \xi_2) + \xi_2,$$

$$r = \sqrt{(x_1 - \xi_1)^2 + (x_2 - \xi_2)^2}.$$

The second approximation for the free flow line is made as follows. There is a flow line plotted from sharp edge C in the defined velocity field using the following formulas: $x = x' + v_x \Delta t$, $y = y' + v_y \Delta t$, where Δt is a quite small time increment (e.g., $\Delta t = 0.0001$), (x', y') is the previous position of the point on the flow line; (x,y) is the next point, and v_x, v_y are found from Equation 3.82 at $\vec{n} = \{1;0\}$ fort v_x and $\vec{n} = \{0;1\}$ for v_y. As soon as the distance between point (x,y) and the sharp edge becomes equal to h (e.g., to the accuracy of 0.000001), a free vortex is placed at this point, that is, it will be the second approximation for this free flow line point. Then, a flow line is plotted again until the distance between (x,y) and the previous position of the free vortex becomes equal to h. A free vortex is placed at this point and so on.

After defining the second approximation for the free flow line there should be a set of Equation 3.81 resolved again and bound vortex circulations determined. Then, the third approximation is made for the free flow line, etc. This iteration process is repeated until the difference between the next position of N_S free vortex and the previous positions does not exceed the given accuracy ε.

Since the range boundaries *break* at some distance from the suction port, a situation occurs when the free flow line starts lifting up (Figure 3.106), that is, its Y-coordinate starts increasing. That is why in this case Y-coordinate is fixed for all other points and is equal to the maximum decrease point Y-coordinate.

3.8.2.2 Analysis of a Separated Flow Upstream of a Two-Dimensional Duct with Two Screens

As it was previously determined analytically, numerically, and experimentally the highest increase in the LDF upstream of a 2D duct is observed when there is a screen with the length of $d \approx 0.5B$ (Figure 3.105) installed perpendicular to the suction plane. It is of interest to determine the effect of the second closed screen position and length on the separated flow parameters attributed to the suction duct width B: $(\bar{x}_{max}, \bar{y}_{max}) = (x_{max}/B, y_{max}/B)$ are the maximum jet offset point coordinates (*max* point on Figure 3.107), $\bar{B}_e = B_e/B$ is the efficient suction width upstream of the duct, and $\bar{\delta}_\infty = \delta_\infty/B$ is the jet contraction coefficient at infinity. Hereinafter, we will skip hyphens above values as dimensionless.

With a fixed length d_2 of the vertical screen and its distance from the horizontal one (Figure 3.107), there are extreme values observed for all parameters within $0.15 < r < 0.35$. This is also the range where the greatest variation of these parameters takes place while the further variation is insignificant. The value of $\Delta u = 1/\delta_\infty - 1$ will be the biggest at $r = 0.05$. When r distance is changed, the vertical screen flow direction is changed (Figure 3.107a through d). When $r \approx 0.2$, the flow line crosses the screen (Figure 3.107b).

With a fixed distance r and an increase in length d_2 of the vertical screen, there are also extreme values observed for the jet separation parameters (Figure 3.108). The lowest jet contraction coefficient is δ_∞, which means that the highest criterion $\Delta u = 1/\delta_\infty - 1$ is observed at $d_2 \approx 1$.

When such screen inclination angle is changed toward the positive direction of X-axis (Figure 3.109), jet separation parameter variation functions are steady. The lowest δ_∞ is observed at $\alpha = 0$. The increase in the second screen length at $\alpha = 0$ (Figure 3.110) also results in a steady variation in the jet separation parameters. The free flow line is displaced rightward (Figure 3.110a and b). The most considerable decrease in δ_∞ is within $0 < d < 1$ (Figure 3.110c).

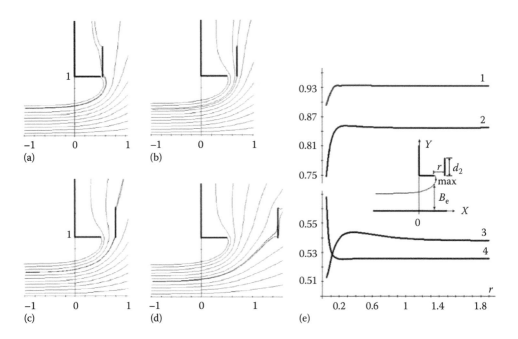

FIGURE 3.107 Flow pattern at various distances from the vertical screen: (a) $r = 0.05$, (b) $r = 0.2$, (c) $r = 0.3$, (d) $r = 1$, and (e) relation between the flow separation parameters at $d_2 = 0.5$ and distance r (1, y_{max}; 2, B_e; 3, δ_∞; 4, x_{max}).

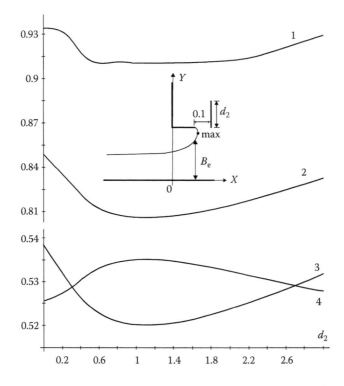

FIGURE 3.108 Relation between the flow separation parameters at the distance $r = 0.1$ between the screens and the vertical screen distance d_2: 1, y_{max}; 2, B_e; 3, δ_∞; 4, x_{max}.

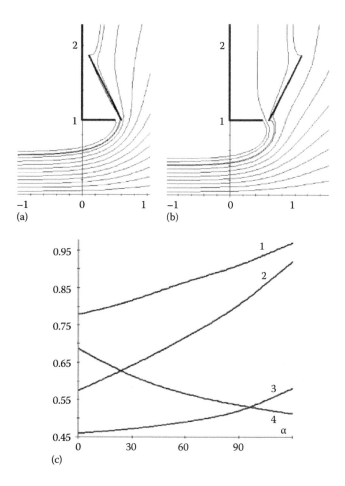

FIGURE 3.109 Flow pattern at $r = 0.1$; $d_2 = 1$: (a) $\alpha = 2\pi/3$, (b) $\alpha = \pi/3$, and (c) relation between the flow separation parameters and angle $\alpha°$: 1, y_{max}; 2, B_e; 3, δ_∞; 4, x_{max}.

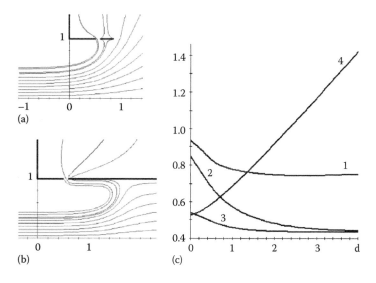

FIGURE 3.110 Flow pattern at $r = 0.1$: (a) $d_2 = 0.25$, (b) $d_2 = 4$, and (c) the second screen length dependence: 1, y_{max}; 2, B_e; 3, δ_∞; 4, x_{max}.

The following design parameters were used in the calculations mentioned: flow line plotting increment: 0.00001; discretization interval: $h = 0.025$; circulation of the free flow line vortices: $\gamma = -0.1$; free flow line determination accuracy: $\varepsilon = 0.0001$. The vertical wall distance from the origin of coordinates is 8; the last free vortex distance from the origin of coordinates is also 8. The total number of bound vortices in the calculations: $300 \div 461$; free vortices: about 340. The number of iterations for determining the free flow line coordinates constituted about 10–40.

Thus, according to the study results, in order to achieve higher air drag for a screened suction duct, it is advised to install the second horizontal 1 gauge long (the suction port width) screen on the same straight line with the first one and at the distance of 0.05–0.1 gauge. However, it is not consistent with the full-scale experimental data.

3.8.3 REGARDING CIRCULATORY AND SHOCK-FREE FLOW AROUND PROFILES WITHIN THE RANGE OF A TWO-DIMENSIONAL SUCTION DUCT

This section objective is to choose an adequate profile flow pattern.

3.8.3.1 Computational Approach Description

The subject considered was a multiply connected flow region (Figure 3.111) upstream of a 2D (or a round) suction duct, the range of which covered a thin profile (a disk with a round central hole). A vortex is separating from sharp edge C and a free flow line occurs. It is required to define the line position, the flow velocity at any given point, and the LDF upstream of the suction port.

The mathematical setup of the problem consists in solving the 2D Laplace's equation for potential function ϕ:

$$\Delta\varphi = 0,$$

at the given normal boundary velocity values $\left.\dfrac{\partial\varphi}{\partial n}\right|_S = v_n(x) - U_n$ where x is S boundary point.

Function U_n expresses the effect of free vortices located on the free flow line, the position of which is unknown in advance.

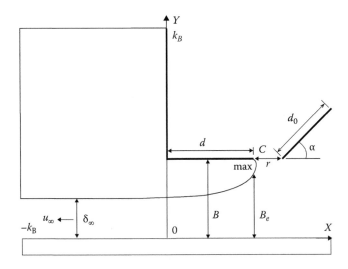

FIGURE 3.111 Regarding the problem setup.

This equation resolves oneself into the boundary singular integral equation:

$$\int_S G(x, \xi) m(\xi) ds(\xi) = v_n(x) - \mu \int_\sigma G(x, \xi) \, ds(\xi),$$

where

$m(\xi)$ is the bound vortex layer circulation (intensity)

$\mu = $ const is the intensity of the free vortex layer located continuously on σ line occurring at flow separation from the sharp edge

ξ is S boundary random point

From physical standpoint $G(x, \xi)$ is a velocity at point $x(x_1, x_2)$ along the unitary vector $\mathbf{n} = \{n_1, n_2\}$ induced by the vortex of a single circulation located at point $\xi(\xi_1, \xi_2)$. For the 2D problem in the rectangular coordinate system

$$G\left(x, \xi\right) = \frac{1}{2\pi} \frac{\left(x_1 - \xi_1\right) n_2 - \left(x_2 - \xi_2\right) n_1}{\left(x_1 - \xi_1\right)^2 + \left(x_2 - \xi_2\right)^2}. \tag{3.83}$$

For the axially symmetric problem in the cylindrical coordinate system

$$\begin{cases} G\left(x, \xi\right) = \dfrac{\left(A_1 b + A_2 a\right)}{b} \cdot \dfrac{4}{\left(a - b\right)\sqrt{a + b}} E\left(t\right) - \dfrac{A_2}{b} \cdot \dfrac{4}{\sqrt{a + b}} F\left(t\right) & \text{at } b \neq 0, \\[4mm] G\left(x, \xi\right) = \dfrac{\xi_2^2 \cdot n_1}{2a\sqrt{a}} & \text{at } b = 0, \end{cases} \tag{3.84}$$

$$2x_2\xi_2 = b > 0, \quad a = (x_1 - \xi_1)^2 + \xi_2^2 + x_2^2 > 0, \quad A_1 = \frac{\xi_2^2 n_1}{4\pi}, \quad A_2 = \frac{\xi_2}{4\pi}\left[(x_1 - \xi_1)n_2 - x_2 n_1\right],$$

$$F\left(t\right) = \int_0^{\pi/2} \frac{d\theta}{\sqrt{1 - t^2 \sin^2\theta}}, \quad E\left(t\right) = \int_0^{\pi/2} \sqrt{1 - t^2 \sin^2\theta}\, d\theta$$

are complete elliptic integrals of 1 and 2 kind, $t = 2b/(a + b)$,

$$F\left(t\right) = \sum_{i=0}^{4} c_i(1 - t)^i + \sum_{i=0}^{4} d_i(1 - t)^i \ln\frac{1}{1 - t}, \quad E\left(t\right) = 1 + \sum_{i=1}^{4} c_i(1 - t)^i + \sum_{i=1}^{4} d_i(1 - t)^i \ln\frac{1}{1 - t},$$

$$E\left(t\right) = 1 + \sum_{i=1}^{4} c_i(1 - t)^i + \sum_{i=1}^{4} d_i(1 - t)^i \ln\frac{1}{1 - t},$$

c_i, d_i were taken from tables [123].

The discrete mathematical model for the 2D problem is built as follows (Figure 3.112). Let the following symbols be introduced: N is the number of bound vortices; K is the number of reference points.

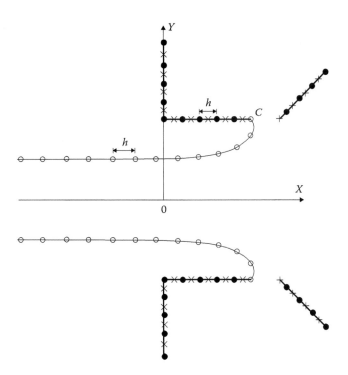

FIGURE 3.112   Range boundary discretization: ● is for bound vortices, ○ is for free vortices, × is for reference points.

Bound vortices were also positioned at the boundary fracture points. There were reference points positioned between the bound vortices. Point $\xi^k(\xi_1, \xi_2)$ is k bound vortex position point; $x^p(x_1, x_2)$ is p reference point. Then the velocity at point x^p along the unit vector $\boldsymbol{n}$ induced by the action on point x^p by the vortex of circulation $\Gamma(\xi^k)$ located at point ξ^k is given by $v_n(x^p) = G(x^p, \xi^k)\Gamma(\xi^k)$.

In order to fulfill OX impermeability condition we will project all vortices symmetrically around OX so the circulations of symmetric vortices will be inverse: $\Gamma(\xi^k) = -\Gamma(\xi^{k+N})$ where $k = \overline{1, N}$. Fulfillment of this condition will automatically result in the noncircular flow condition (the sum of circulations of all vortices is equal to zero).

It was assumed that the circulation of vortices on the required free flow line is constant and is equal to γ. The distance between free vortices is a constant value equal to h. The first approximation for the free flow line was chosen as follows. The first three vortices were positioned parallel to OY below the separation point starting from the sharp edge while the rest of the vortices were positioned parallel to OX leftward the separation point.

The resultant action of all vortices on reference point x^p along the outward normal is given by

$$v_n\left(x^p\right) = \sum_{q=1}^{N} G\left(x^p, \xi^q\right)\Gamma\left(\xi^q\right) - \sum_{q=N+1}^{2N} G\left(x^p, \xi^q\right)\Gamma\left(\xi^{q-N}\right) + \gamma\sum_{k=1}^{N_S} G\left(x^p, \zeta^k\right) - \gamma\sum_{k=N_S+1}^{2N_S} G\left(x^p, \zeta^k\right)$$

$$= \sum_{q=1}^{N}\left(G\left(x^p, \xi^q\right) - G\left(x^p, \xi^{q+N}\right)\right)\Gamma\left(\xi^q\right) + \gamma\sum_{k=1}^{N_S}\left(G\left(x^p, \zeta^k\right) - G\left(x^p, \zeta^{k+N_S}\right)\right) \qquad (3.85)$$

where ζ^k is a free vortex position point, G is given by Equation 3.83.

Since $v_n(x^p) = 0$ at all reference points, that is, the impermeability condition is fulfilled, by changing $p = 1, 2, ..., K$ expression (Equation 3.85) will be converted into a set of linear algebraic equations for determining unknown circulations $\Gamma(\xi^q)$ of bound vortices:

$$\sum_{q=1}^{N}\left(G\left(x^p, \xi^q\right) - G\left(x^p, \xi^{q+N}\right)\right)\Gamma\left(\xi^q\right) = -\gamma\sum_{k=1}^{N_S}\left(G\left(x^p, \zeta^k\right) - G\left(x^p, \zeta^{k+N_S}\right)\right). \quad (3.86)$$

K is the number of reference points. The case $K = N$ corresponds to the case of circulatory flow past profiles when the vortex layer intensity on one of the edges is considered to be finite.

When $K < N$, there are vortices on profile edges, that is, the vortex layer intensity is infinite. In this case, the noncirculatory flow condition is introduced for each profile: the sum of intensities of all vortices is equal to zero that allows for extending the set of equations.

When $K > N$ set of Equation 3.86 turns out to be superdefinite, regulation factors need to be introduced [113]. This variant corresponds to the case of shock-free flow past profiles implemented when reference points are placed at the ends to limit the intensity of the relevant vortex layer. For example, for a profile located in the flow area, the number of bound vortices is one unit less than the number of reference points and the expression from Equation 3.86 will be converted into a set of $N + 1$ equations with $N + 1$ unknown variables:

$$\sum_{q=1}^{N}\left(G\left(x^p, \xi^q\right) - G\left(x^p, \xi^{q+N}\right)\right)\Gamma\left(\xi^q\right) + \Lambda = -\gamma\sum_{k=1}^{N_S}\left(G\left(x^p, \zeta^k\right) - G\left(x^p, \zeta^{k+N_S}\right)\right), \quad (3.87)$$

where Λ is a regulating variable.

After unknown circulations of the bound vortices have been determined, the velocity at any point $x(x_1, x_2)$ of the range along any given direction will be determined from the formula

$$v_n(x) = \sum_{q=1}^{N}\left(G\left(x, \xi^q\right) - G\left(x, \xi^{q+N}\right)\right)\Gamma\left(\xi^q\right) + \gamma\sum_{k=1}^{N_S}\left(G\left(x, \zeta^k\right) - G\left(x, \zeta^{k+N_S}\right)\right). \quad (3.88)$$

Then, it is obvious that an attribute of a shock-free flow past profile existence is Λ vanishing. Otherwise, the boundary condition of impermeability is violated for velocity.

Velocity peaks are clipped as follows:

If the distance from point x to the vortex at point ξ is less than $h/2$, then $G(x, \xi)$ is given by

$$G\left(x, \xi\right) = \frac{r}{\pi h} \cdot \frac{\left(S_1 - \xi_1\right)n_2 - \left(S_2 - \xi_2\right)n_1}{\left(S_1 - \xi_1\right)^2 + \left(S_2 - \xi_2\right)^2},$$

$$S_1 = \frac{h}{2r}\left(x_1 - \xi_1\right) + \xi_1, \quad S_2 = \frac{h}{2r}\left(x_2 - \xi_2\right) + \xi_2, \quad r = \sqrt{\left(x_1 - \xi_1\right)^2 + \left(x_2 - \xi_2\right)^2}.$$

The second approximation for the free flow line is made using Runge–Kutta method of numerical solution of a set of ordinary differential equations:

$$\begin{cases} \dfrac{dx}{dt} = v_x, \\[2mm] \dfrac{dy}{dt} = v_y, \end{cases}$$

where v_x, v_y are given by Equation 3.88 at $n = \{1.0\}$ for v_x and $n = \{0.1\}$ for v_y.

There is a flow line plotted from sharp edge C. Time increment Δt is quite small (e.g., $\Delta t =$ 0.0001). As soon as the distance between point (x, y) and the sharp edge becomes equal to h, a free vortex is placed at this point, that is, it will be the second approximation for this free flow line point. Then, a flow line is plotted again until the distance between (x, y) and the previous position of the free vortex becomes equal to h. A free vortex is placed at this point and so on.

After defining the second approximation for the free flow line there should be a set of Equation 3.86 resolved again and bound vortex circulations determined. Then, the third approximation is made for the free flow line, etc. This iteration process is repeated until the difference between the next position of N_S free vortex and the previous positions does not exceed the given accuracy ε.

Since the range boundaries *break* at some distance from the suction port, a situation occurs when the free flow line starts lifting up (Figure 3.112), that is, its Y-coordinate starts increasing. That is why in this case Y-coordinate is fixed for all other points and is equal to the maximum decrease point Y-coordinate.

The LDF value ζ at the medium entry into the duct is expected to be determined using numerically defined jet thickness at infinity upstream δ_∞ by the following formula:

$$\zeta = \left(1 - 1/\delta_\infty\right)^2,$$

which results from Borda–Karnot formula [129].

For the axially symmetric problem the discrete mathematical model is built like that. There were continuously thin noninductive vortex rings used as discrete properties. The set of equations for determining unknown intensities of bound vortex rings will take the following form:

$$\sum_{q=1}^{N} G\left(x^p, \xi^q\right)\Gamma\left(\xi^q\right) = -\gamma \sum_{k=1}^{N_S} G\left(x^p, \zeta^k\right),$$

where G is given by Equation 3.84 and velocity is given by

$$v_n(x) = \sum_{q=1}^{N} G\left(x, \xi^q\right)\Gamma\left(\xi^q\right) + \gamma \sum_{k=1}^{N_S} G\left(x^p, \zeta^k\right).$$

If the distance between the point and the vortex is less than the discretization radius $h/2$, the velocity induced by such vortex will be given by

$$v_n(x) = 8\pi \frac{\left(x_1 - \xi_1\right)n_2 - \left(x_2 - \xi_2\right)n_1}{h^2}.$$

The LDF in such case is given by

$$\zeta = \left(1 - 1/\delta_\infty^2\right)^2.$$

3.8.3.2   Full-Scale Experiment Arrangement Description

For checking the resulting design values for validity there was an experimental arrangement developed and assembled (Figure 3.113).

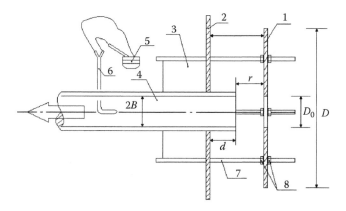

FIGURE 3.113 Diagram of the experimental arrangement for determining air drag upstream of a screened round hole: 1, screen with a central hole; 2, shield; 3, trihedral guide prism; 4, tube; 5, inclined-tube micromanometer; 6, pneumatic Pitot–Prandtl tube; 7, steel studs; 8, screen fixing nuts.

The LDF at air entry into a screened air duct will be determined in accordance with the universally accepted relation

$$\zeta = \frac{2\Delta P}{\rho u^2},$$

which is the relation of full differential pressure and dynamic (velocity) pressure. The experiments were conducted using the arrangement (Figure 3.113) the test section of which comprised MMN-2500 micromanometer and pneumatic Pitot–Prandtl tube. Pressure was measured in the section located 1.5 m away from the axis of the entry section of a vinyl tube featuring the inner diameter of 125 mm and the wall thickness of 1.7 mm. The screen system (a shield on a tube and a screen with a round hole) is fixed using Ø 4 and 400 mm long steel studs. In order to ensure the system perpendicularity to the tube axis and the system axial symmetry, the shield simulating a vertical impermeable wall is rigidly attached to 100 mm high trihedral regular prism sliding over the tube while the screen fixing studs are attached to the prism edges. The shield and screen are made of 4 mm thick pressed board in the shape of disks of $D_0 = 128$ mm, $D = 450$ mm and $D_0 = 128$ mm, $D = 360$ mm; accordingly, some tests featured the screen inner diameter $D_0 = 106.5, 102.5, 90,$ and 73 mm. Most of the tests were conducted with $D_0 = 128$ mm. For the purpose of rigid fixation and installation of the screen at the required distance r, there were nuts used for tightening from two sides.

The length of the projective part of the tube $d = 0$–100 mm and the gap between the entry section and the screen $r = 0$–150 mm were changed in tests. Average tube air velocity u was determined by means of measuring velocity pressure along the tube center line and introducing the factor of correction for irregularity of the field of velocities that was determined by measuring the velocity values at four points of equal-sized rings. The tube wall friction pressure losses from the entry to the gage section were disregarded (both due to the tube smoothness and due to the impossibility of proper accounting for such losses in flow stabilization conditions at stalled air intake). The average air velocity in the tube constituted 7–9 m/s.

3.8.3.3 Analysis of a Separated Flow Upstream of a Protruded Duct with a Vertical Profile

As it was previously determined analytically, numerically, and experimentally, the highest increase in the LDF upstream of a 2D duct is observed when there is a screen with the length of $d \approx 0.5B$ (Figure 3.111) installed perpendicular to the wall plane. It is of interest to determine the effect of the second closed screen position and length on the separated flow parameters attributed to the suction duct width (radius) B: $\left(\bar{x}_{max}, \bar{y}_{max}\right) = \left(x_{max}/B, y_{max}/B\right)$ are the maximum jet offset point coordinates (*max* point on Figure 3.111); $\bar{B}_e = B_e/B$ is the efficient suction width upstream of the duct; $\bar{\delta}_\infty = \delta_\infty/B$ is the jet contraction coefficient at infinity. Calculations were made at $B = 1$. Hereinafter, we will skip hyphens above values as dimensionless.

Let us analyze the effect of the following parameters: discretization step h, flow line time increment Δt, calculation accuracy ε, and distance gauge k_B (Figure 3.111) on design parameters x_{max}, y_{max}, δ_∞, B_e and on the number of iterations I for determining the free flow line and calculation time t.

As seen from Table 3.3, for calculation of the flow parameters it is sufficient to assume $k_B = 10$, and from Table 3.4, $h = 0.05$; $\Delta t = 0.0001$. The number of iterations does not exceed 15 in this case.

There were various regimes of a vertical profile flow within the duct suction range considered. It is of interest to determine the profile flow regime closest to the experimental data.

As it was noted, a shock-free regime does not always exist. By fixing the value of distance r and changing angle α we obtained the values at which $\Lambda \to 0$ in formula (Equation 3.85), which means that such flow regime did exist (Figure 3.114). That is to say, by fixing α and changing r we will obtain the parameters for a shock-free profile flow. For instance, at $\alpha = \pi/2$, $r = 0.1309235$, $\delta_\infty = 0.5433$, $x_{max} = 0.517$, $y_{max} = 0.9471$, $B_e = 0.8736$. δ_∞ is decreased and approximates to 0.518 with an increase in r, which means that the LDF is increased.

When the bound vortex is on the lower edge of the vertical profile and the reference point is on the upper edge, the calculated flow separation parameters have the same qualitative nature (Figure 3.115) as for a noncirculatory profile flow. A maximum δ_∞ is observed at $0.4 < r < 0.5$ (accordingly, the minimum for LDF ζ), which is not consistent with the experimental data. The difference between the flow patterns in various flow regimes is shown in Figure 3.116.

The closest to the experimental data are the results obtained when using the condition of velocity finiteness on the profile lower edge (Figure 3.117). Here, the unnumbered curves correspond to axially symmetric case and their positions are the same as in the case with the 2D problem. With a fixed length d_0 of the vertical profile and its various distances from the horizontal one (Figure 3.117), there are extreme values observed for all design parameters. In particular, δ_∞ for the 2D problem has its

TABLE 3.3
Design Flow Parameters at $d = 0.5$; $r = 0.6$; $d_0 = 0.525$

k_B	δ_∞	x_{max}	y_{max}	B_e	I	t
2	0.7337	0.5334	0.9091	0.7991	3	14s
5	0.5106	0.5292	0.9128	0.8124	14	3 min 31 s
10	0.5032	0.5287	0.9131	0.8139	15	8 min 43 s
20	0.5008	0.5286	0.9132	0.8146	16	23 min 24 s
40	0.4994	0.5286	0.9133	0.8146	12	39 min 53 s
80	0.4988	0.5286	0.9132	0.8147	11	1 h 34 min 23 s

TABLE 3.4
Design Flow Parameters at $d = 0.5$; $r = 0.6$; $d_0 = 0.525$; $k_B = 10$

H	Δt	δ_∞	x_{max}	y_{max}	B_e	I	t
0.05	0.0001	0.5032	0.5287	0.9131	0.8139	15	8 min 39 s
0.05	0.000005	0.5034	0.5288	0.9132	0.8138	15	2 h 4 min
0.025	0.0001	0.5011	0.5357	0.9120	0.7996	50	31 min 27 s
0.0125	0.00001	0.5018	0.5399	0.9133	0.7917	17	74 min 56 s
0.00625	0.000005	0.5007	0.5416	0.9147	0.7906	52	≈17 h

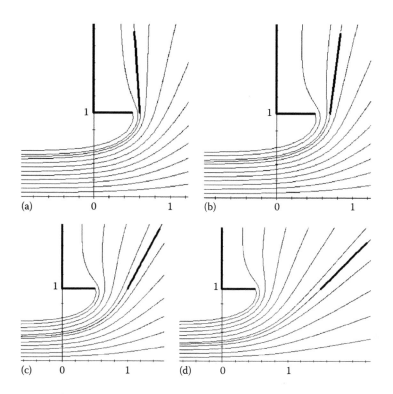

FIGURE 3.114 Flow lines for shock-free flow around a screen featuring length $d_0 = 1$: (a) $\delta_\infty = 0.5467$; $x_{max} = 0.5153$; $y_{max} = 0.9488$; $B_e = 0.8815$; $r = 0.1$; $\alpha = \pi/1.910331$. (b) $\delta_\infty = 0.542$; $x_{max} = 0.5174$; $y_{max} = 0.9467$; $B_e = 0.8708$; $r = 0.2$; $\alpha = \pi/2.19493$. (c) $\delta_\infty = 0.53962$; $x_{max} = 0.5183$; $y_{max} = 0.9456$; $B_e = 0.8662$; $r = 0.5$; $\alpha = \pi/3.00173$. (d) $\delta_\infty = 0.54$; $x_{max} = 0.5181$; $y_{max} = 0.9458$; $B_e = 0.8673$; $r = 1$; $\alpha = \pi/4.2994$.

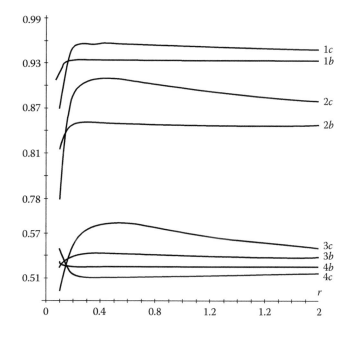

FIGURE 3.115 Variation in the separated flow parameters based on distance r from the vertical profile featuring length $d_0 = 0.5$ in the case of its noncirculatory flow (index b) and circulatory flow (index c) when the bound vortex is on the lower edge of the profile past: $1b$, $1c - y_{max}$; $2b$, $2c - B_e$; $3b$, $3c - \delta_\infty$; $4b$, $4c - x_{max}$.

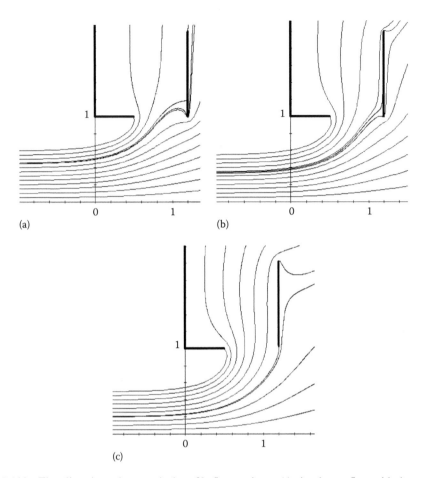

(a)   (b)

(c)

FIGURE 3.116 Flow lines in various vertical profile flow regimes: (a) circulatory flow with the condition of velocity finiteness on the upper edge ($\delta_\infty = 0.5985$; $x_{max} = 0.5087$; $y_{max} = 0.9594$; $B_e = 0.9222$; $r = 0.7$), (b) non-circulatory flow ($\delta_\infty = 0.5505$; $x_{max} = 0.5177$; $y_{max} = 0.947$; $B_e = 0.8727$; $r = 0.7$), and (c) circulatory flow with the condition of velocity finiteness on the lower edge ($\delta_\infty = 0.4957$; $x_{max} = 0.5322$; $y_{max} = 0.9102$; $B_e = 0.8012$; $r = 0.7$).

minimum within $0.55 < r < 0.75$, while for the axially symmetric problem it is within $0.3 < r < 0.35$. With a considerable increase in d_0 for the 2D problem, the extreme value is observed at $r = 0.75$, which corresponds to calculations made using the Joukowski method for $d_0 \rightarrow \infty$.

Note that the computational experiment for the axially symmetric problem was conducted with the same parameters of the discrete model as for the 2D problem. An extensive comparison of design and experimental values of ζ (Figures 3.118 through 3.121) demonstrates their satisfactory agreement.

A maximum of 15% upward bias of design values of ζ is observed at small r. The values of ζ closest to the experimental data are those found for the axially symmetric problem which is naturally as the full-scale experiment was conducted in the same setup.

The design and experimental extreme values of the LDF (Figures 3.118 and 3.62) coincide, which allows for making a conclusion that the mathematical modeling method elaborated and the study results are valid for the 2D problem (i.e., the problem of a separated flow upstream of a rectangular suction port featuring min 1:10 aspect ratio) as well as for axially symmetric one.

With a decrease in inner diameter D_0, the LDF is increased (Figure 3.121), but this case is of a low practical interest as a higher LDF is required without sealing the inlet section. Such sealing is impossible for technological reasons. Note a significant difference from experiments with a sleeved disk (Figure 3.122).

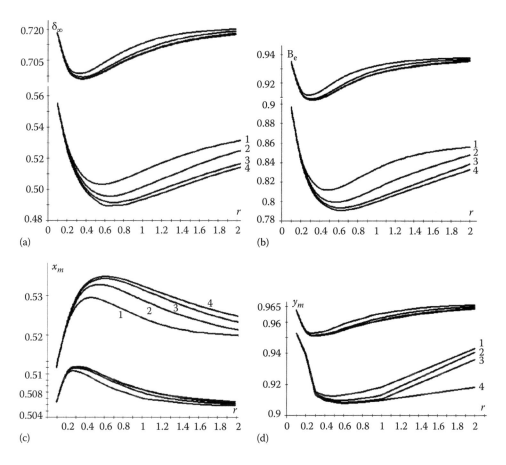

FIGURE 3.117 Variation in the separated flow parameters based on distance r of the vertical profile from the suction duct inlet at various lengths of its screen: (a) $d_0 = 0.5$, (b) $d_0 = 1$, (c) $d_0 = 2$, and (d) $d_0 = 10$.

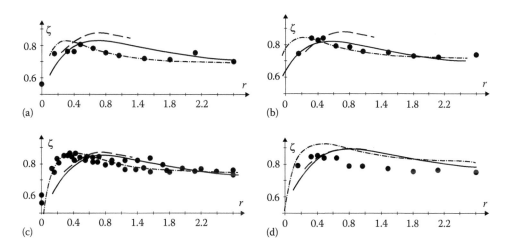

FIGURE 3.118 Design LDFs ζ versus r at $d_0 = 1.55$: (a) $d = 0.24$, (b) $d = 0.48$, (c) $d = 0.56$, and (d) $d = 1.2$ (the solid line denotes calculations for the 2D problem, the dash–dotted line denotes calculations for the axially symmetric problem, the dotted line denotes the Joukowski method, and the circles denote the experimental data).

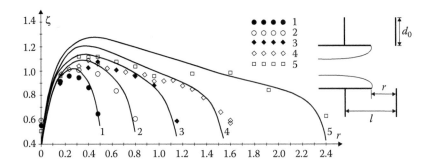

FIGURE 3.119 Design (solid lines) and experimental variation in LDF ζ for moving twin disks ($d_0 = 1.56$): 1, $l = 0.5$; 2, $l = 0.8$; 3, $l = 1.15$; 4, $l = 1.55$; 5, $l = 2.4$.

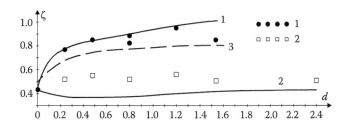

FIGURE 3.120 LDF ζ versus duct extension length d with constant $d_0 = 1.55$: 1, $r = \infty$; 2, $r = 0$; 3, Idelchik's experimental curve. (From Idelchik, I.E., *Hydraulic Resistance Guide Book*, M: Mashinostroeniye, 1977, 559p.)

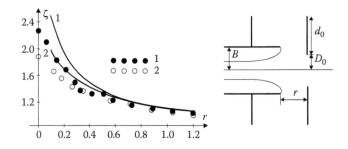

FIGURE 3.121 The outer disk inner diameter function at $D_0 < 1$ with an increase in the tube inlet air LDF at a distance to the disk r and $d = 0.8$: 1, at $D_0 = 0.82$; 2, $D_0 = 0.852$.

3.8.4 SEPARATED FLOW AROUND PROFILES WITHIN THE RANGE OF A SUCTION DUCT

This section objective is to develop a method of mathematical modeling of separated flows upstream of suction ducts within the profiled regions at separated flow around the profiles using stationary discrete vortices.

3.8.4.1 Computation Algorithm

The subject considered was a multiply connected flow region (Figure 3.123) upstream of a 2D suction duct, the range of which covered *MD* thin profile. A vortex is separating from sharp edges *C*, *M*, and *D* and free flow lines *CE*, *DS*, and *MS* occur. It is required to define the lines position, the flow velocity at any given point, and the LDF upstream of the suction port.

The discrete model is built in the same way as in the previous section. There are bound vortices positioned on the sharp edges and fractures (Figure 3.124).

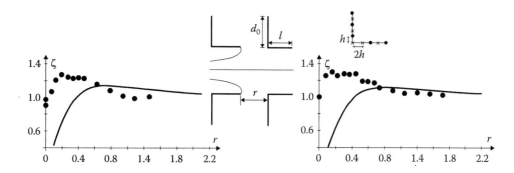

FIGURE 3.122 The effect of a sleeved disk on the tube inlet air LDF: (a) $l = 0.59$ and (b) $l = 0.96$.

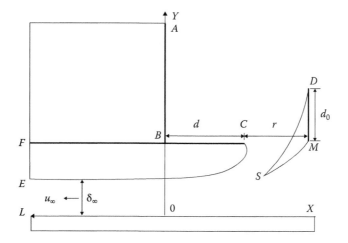

FIGURE 3.123 Regarding the problem setup.

The vortices lying on sharp edges C, M, and D and the symmetrical vortices are considered to be free ones that follow from the theorem set forth in I.K. Lifanov's book mentioned earlier: the bound vortex intensity (circulation) at the flow separation point is equal to zero. Bound vortices were also positioned at the boundary fracture points. There were reference points positioned between the bound vortices. Point $\xi^k(\xi_1, \xi_2)$ is k bound vortex position point; $x^p(x_1, x_2)$ is p reference point. Then, the velocity at point x^p along the unit vector $\boldsymbol{n}$ induced by the action on point x^p by the vortex of circulation $\Gamma(\xi^k)$ located at point ξ^k is given by $v_n(x^p) = G(x^p, \xi^k)\Gamma(\xi^k)$.

In order to fulfill OX impermeability condition we will project all vortices symmetrically around OX so the circulations of symmetric vortices will be inverse: $\Gamma(\xi^k) = -\Gamma(\xi^{k + N})$ where $k = \overline{1, N}$. Fulfillment of this condition will automatically result in the noncircular flow condition (the sum of circulations of all vortices is equal to zero).

It was assumed that the circulation of vortices on the required free flow line was known, constant and equal to γ. The distance between free vortices is a constant value equal to h. The first approximation for the free flow line was chosen as follows. The first three vortices were positioned parallel to OY below the separation point starting from the sharp edge while the rest of the vortices were positioned parallel to OX leftward the separation point.

No initial approximation was given for free flow lines DS and MS. The circulation on these lines is unknown in advance. We will assume that DS flow line circulation and MS flow line circulation are equal in absolute magnitude but different in sign, which is logical from physical standpoint. Particulates separated from sharp edges M and D should be rotating in opposite directions. When they converge at point S, they compensate each other and no rotation of particulates is further observed.

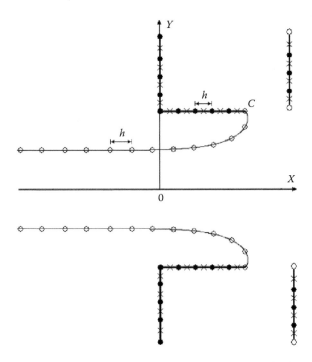

FIGURE 3.124 Range boundary discretization: ● is for bound vortices, ○ is for free vortices, × is for reference points.

Let us denote the number of bound vortices including two vortices on M and D edges as N. The number of vortices at points M and D will be N and N_1, respectively.

At the initial instant of time, the resultant action of all vortices on reference point x^p along the outward normal is given by

$$v_n(x^p) = \sum_{\substack{q=1, \\ q \neq N_1}}^{N-1} G(x^p, \xi^q) \Gamma(\xi^q) + G(x^p, \xi^{N_1}) \gamma_1 - G(x^p, \xi^N) \gamma_1$$

$$- \sum_{\substack{q=N+1, \\ q \neq N_1+N}}^{2N-1} G(x^p, \xi^q) \Gamma(\xi^{q-N}) - \left(G(x^p, \xi^{N_1+N}) \gamma_1 - G(x^p, \xi^{N_1+N}) \gamma_1 \right)$$

$$+ \gamma \sum_{k=1}^{N_{S1}} G(x^p, \zeta^k) - \gamma \sum_{k=N_{S1}+1}^{N_{S1}+N_S} G(x^p, \zeta^k)$$

$$= \sum_{\substack{q=1, \\ q \neq N_1}}^{N-1} \left(G(x^p, \xi^q) - G(x^p, \xi^{q+N}) \right) \Gamma(\xi^q) + \gamma \sum_{k=1}^{N_{S1}} \left(G(x^p, \zeta^k) - G(x^p, \zeta^{k+N_S}) \right)$$

$$+ \left(G(x^p, \xi^{N_1}) - G(x^p, \xi^N) - G(x^p, \xi^{N_1+N}) + G(x^p, \xi^{N_1+N}) \right) \gamma_1, \tag{3.89}$$

where
  ζ^k is a free vortex position point, G is given by (1)
  γ_1 are the circulations on free flow lines separating from sharp edges M and D
  N_{S1} is the number of free vortices located on CE free flow line
  N_S is the number of free vortices located on all free flow lines (at the first iteration $N_{S1} = N_S$)

Since $v_n(x^p) = 0$ at all reference points, that is, the impermeability condition is fulfilled, by changing $p = 1, 2, \ldots, N - 1$, expression (Equation 3.89) will be converted into a set of linear algebraic equations for determining unknown circulations $\Gamma(\xi^q)$ of bound vortices:

$$\sum_{\substack{q=1, \\ q \neq N_1}}^{N-1} \left(G\left(x^p, \xi^q\right) - G\left(x^p, \xi^{q+N}\right) \right) \Gamma\left(\xi^q\right) + \left(G\left(x^p, \xi^{N_1}\right) - G\left(x^p, \xi^N\right) - G\left(x^p, \xi^{N_1+N}\right) + G\left(x^p, \xi^{N+N}\right) \right) \gamma_1$$

$$= -\gamma \sum_{k=1}^{N_{S1}} \left(G\left(x^p, \zeta^k\right) - G\left(x^p, \zeta^{k+N_S}\right) \right). \tag{3.90}$$

By changing $p = 1, 2, \ldots, N - 1$ in expression (Equation 3.90) and solving the resulting set of linear algebraic equations we will find the unknown circulations $\Gamma(\xi^q)$, $q = 1, \ldots N_1 - 1, N_1 + 1, \ldots N - 1$ of bound vortices and the free vortex circulation γ_1. The velocity at any point $x(x_1, x_2)$ of the range along any given direction will be determined from formula (Equation 3.89).

Then, free flow lines are plotted using the Euler method of numerical solution of a set of ordinary differential equations:

$$\frac{dx}{dt} = v_x, \quad \frac{dy}{dt} = v_y,$$

where v_x, v_y are given by Equation 3.89 at $\boldsymbol{n} = \{1, 0\}$ at v_x and $\boldsymbol{n} = \{0, 1\}$ for v_y.

There are flow lines plotted from sharp edges C, M, and D. Time increment Δt is quite small (e.g., $\Delta t = 0.00005$). As soon as the distance between point (x, y) and the sharp edge becomes equal to h, a free vortex is placed at this point, that is, it will be the second approximation for this free flow line point.

After plotting the free flow lines, it is required to determine the circulations of bound vortices and free vortices located on DS and MS free flow lines. Set of Equation 3.90 at the next iteration will be converted as follows:

$$\sum_{\substack{q=1, \\ q \neq N_1}}^{N-1} \left(G\left(x^p, \xi^q\right) - G\left(x^p, \xi^{q+N}\right) \right) \Gamma\left(\xi^q\right)$$

$$+ \left(G\left(x^p, \xi^{N_1}\right) - G\left(x^p, \xi^N\right) - G\left(x^p, \xi^{N_1+N}\right) + G\left(x^p, \xi^{N+N}\right) \right) \gamma_1$$

$$= -\gamma \sum_{k=1}^{N_{S1}} \left(G\left(x^p, \zeta^k\right) - G\left(x^p, \zeta^{k+N_S}\right) \right) - \gamma_1' \sum_{k=N_{S1}+1}^{N_{S2}} \left(G\left(x^p, \zeta^k\right) - G\left(x^p, \zeta^{k+N_S}\right) \right)$$

$$+ \gamma_1' \sum_{k=N_{S2}+1}^{N_S} \left(G\left(x^p, \zeta^k\right) - G\left(x^p, \zeta^{k+N_S}\right) \right),$$

where
  N_{S1} is the number of free vortices on DS flow line
  N_{S2} is the number of free vortices on MS flow line
  γ_1' is the circulation of the free vortex on sharp edge D found at the previous iteration

The velocity at any point $x(x_1, x_2)$ of the range along any given direction will be determined from the formula

$$v_n(x) = \sum_{\substack{q=1, \\ q \neq N_1}}^{N-1} \left(G\left(x, \xi^q\right) - G\left(x, \xi^{q+N}\right) \right) \Gamma\left(\xi^q\right)$$

$$+ \left(G\left(x, \xi^{N_1}\right) - G\left(x, \xi^{N}\right) - G\left(x, \xi^{N_1+N}\right) + G\left(x, \xi^{N+N}\right) \right) \gamma_1$$

$$+ \gamma \sum_{k=1}^{N_{S1}} \left(G\left(x, \zeta^k\right) - G\left(x, \zeta^{k+N_S}\right) \right) + \gamma_1' \sum_{k=N_{S1}+1}^{N_{S2}} \left(G\left(x, \zeta^k\right) - G\left(x, \zeta^{k+N_S}\right) \right)$$

$$- \gamma_1' \sum_{k=N_{S2}+1}^{N_S} \left(G\left(x, \zeta^k\right) - G\left(x, \zeta^{k+N_S}\right) \right).$$

Then, flow lines are plotted again until the distance between (x,y) and the previous position of the free vortex becomes equal to h. A free vortex is placed at this point and so on.

3.8.4.2 Calculation Data and Its Discussion

The following discrete model parameters were used in calculations: integration step for the set of ordinary differential equations: $\Delta t = 0.00005$; discretization interval $h = 0.05$; suction duct width $b = 1$; $FB = AB = 10b$; CE flow line is plotted until E point x-coordinate reaches $-10b$; DS and MS flow lines are plotted until $d + r - d_0$ is reached; circulation of the free vortices on CE flow line $\gamma = -0.2$.

Note that the DS and MS flow lines do not actually converge but approach each other for a considerably close distance (shorter than the discretization interval). As soon as the x-coordinate of these lines does not exceed $d + r - d_0$, there will be no free vortices positioned thereon. But as soon as the iteration process becomes approximated, these lines were continued by the simple algorithm for plotting flow lines (Figure 3.125). The calculations show that the still region DM behind the vertical profile (Figure 3.125a) has small dimensions versus the horizontal one (Figure 3.125b).

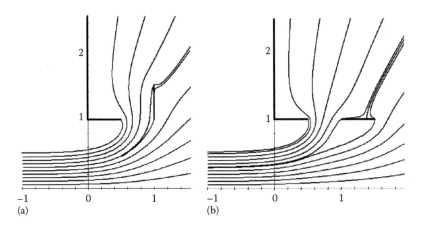

FIGURE 3.125 Flow lines at vertical and horizontal profile positions: (a) $r = 0.5$ and (b) $d_0 = 0.5$.

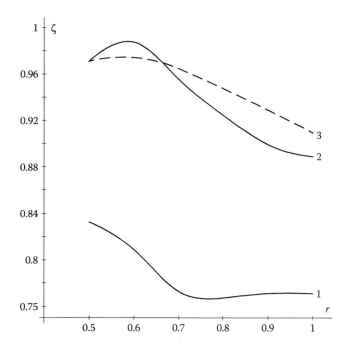

FIGURE 3.126 Relation between LDF ζ and distance r: 1, horizontal profile position; 2, vertical profile position for the separated model; 3, vertical profile position for the intact model.

When the profile is moving away from the suction duct LDF ζ has extreme values (Figure 3.126): the maximum is observed for the vertical profile and the minimum for the horizontal one. From the practical point of view, it is more desirable to use a vertical profile.

According to the computational experiments, an increase in the profile length for more than one gauge has no significant practical effect.

4 Behavior of Dust Particles in the Suction Range

4.1 MODELING THE BEHAVIOR OF SINGLE DUST PARTICLES IN THE RANGE OF SEMICLOSED AND CLOSED EXHAUST DUCTS

4.1.1 CALCULATING THE REQUIRED VOLUMES OF SUCTION IN SELF-CLEANING AND OVERHEAD DRILLING

The most efficient dust and slurry collector [103] is 0.08 m high circular cylinder fitting with a lateral opening equal to a drill rod in diameter 0.03 m and an exit branch to connect to the draft booster. The problem is to determine the volume of suction air that would enable efficient separation of dust and slurry generated in the course of drilling. We will disregard air leaks in a circular clearance around the drill rod that would reduce the problem of describing the field of velocities in the dust and slurry collector to the 2D form.

The dust and slurry collector airflow areas were discretized using a set of 429 boundary intervals. The numerical implementation of the boundary integral equations method allowed for plotting the flow lines and determining the distribution of velocities in the dust and slurry collector opening and at a distance from it.

The experimentally measured velocity of dust and slurry flow from Ø 40 mm upholes: $v_{exp} = 6$ m/s and from Ø 75 mm wells: $v_{exp} = 7$ m/s. The greatest lump size of bored out material: 0.3–0.5 m and density $\rho_1 = 3500$ kg/m³.

There was a 3D flight of a particle analyzed. The dimensional section of dust and slurry collectors is shown in Figure 4.1. OZ axis is directed vertically upward and crosses the drill rod symmetry axis. Due to a large particle size the Stokes flow law is not applicable.

The most appropriate is to describe the particle motion according to the law [81]:

$$\frac{\pi d_e^3}{6}\rho_1\frac{d\vec{v}_1}{dt} = \frac{\pi d_e^3}{6}\rho_1\vec{g} - \psi S_m\rho\frac{\left|\vec{v}_1 - \vec{v}\right|\cdot(\vec{v}_1 - \vec{v})}{2}, \tag{4.1}$$

where

$\vec{v}, \vec{v}_1$ is the medium velocity and particle velocity, respectively
d_e is the equivalent diameter of the particle
ρ, ρ_1 is the medium and particle density
$S_m = \pi d_e^2/4$ is the frontal area

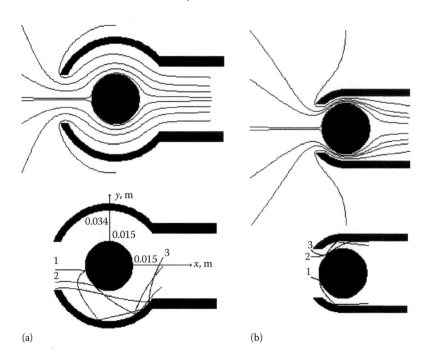

FIGURE 4.1 Flow lines and the path of Ø 0.5 mm dust particle at various positions of its escape points: 1,0.0025 m from the suction symmetry axis; 2,0.01 m; 3,0.014 m in dust and slurry collectors of the diameters (a) 68 mm and (b) 40 mm.

There were equal fields of air velocity components v_x, v_y assumed to be in any plane perpendicular to the drill rod and calculated using the method of boundary integral equations. The vertical air velocity component is determined by the formula

$$v_z = -v_{exp} + \sqrt{\frac{4\rho_1 d_e g}{3\psi\rho}},$$

which may be derived from Equation 4.1 on the assumption that the dust particle motion is uniform, that is, $d\vec{v_1}/dt = 0$.

The head drag coefficient is best represented in the following form:

$$\psi = \frac{24}{Re} + \psi_0, \tag{4.2}$$

where

$$Re = \frac{\rho d_e |\vec{v_1} - \vec{v}|}{\mu}$$

ψ_0 is a coefficient found experimentally for various particle shapes

In our calculations, $\psi_0 = 1.8$. Oseen-type formula given in Equation 4.2 is the closest to the experimental data for an isometric particle shape.

Equation 4.1 resolves oneself into the set of ordinary differential equations that was solved using Runge–Kutta numerical method:

$$\begin{cases} A\dfrac{dv_{1x}}{dt} = -B\left(v_{1x} - v_x\right), \\[2mm] \dfrac{dx}{dt} = v_{1x}, \\[2mm] A\dfrac{dv_{1y}}{dt} = -B\left(v_{1y} - v_y\right), \\[2mm] \dfrac{dy}{dt} = v_{1y}, \\[2mm] A\dfrac{dv_{1z}}{dt} = -Ag - B\left(v_{1z} - v_z\right), \\[2mm] \dfrac{dz}{dt} = v_{1z}, \end{cases} \tag{4.3}$$

where

$$A = \rho_1 \frac{\pi d_e^3}{6}$$

$$B = -\psi S_m \rho \left| \vec{v}_1 - \vec{v} \right| = -\psi S_m \rho \sqrt{(v_{1x} - v_x)^2 + (v_{1y} - v_y)^2 + (v_{1z} - v_z)^2}$$

After an impact with the dust particle, rigid surface tangential (v_2) and normal (v_{2n}) velocities were calculated from formulas [91]:

$$v_{2n} = -k \cdot v_{0n}, \quad v_{2\tau} = v_{0\tau} + \eta \cdot f \cdot (1+k) \cdot v_{0n}, \tag{4.4}$$

where

$$\eta = \min\left\{ -\frac{2v_{0\tau}}{7f(1+k)v_{0n}}, 1 \right\}$$

k is the coefficient of restitution (in our calculations $k = 0.5$)
f is the coefficient of sliding friction ($f = 0.5$)
v_{0n}, $v_{0\tau}$ are the normal and tangential velocity components of the particle prior to impact

Initial design data: dust particle density $\rho_1 = 3500$ kg/m^3, air density $\rho = 1.205$ kg/m^3, coefficient of dynamic viscosity $\mu = 1.809 \cdot 10^{-5}$ Pa $\cdot$ s, $d_e = 500$ mkm, $g = 9.81$ m/s^2.

Figure 4.1 shows the flow lines (upper part of the figure) in dust and slurry collectors and paths of 0.5 mm particles at various points of their possible escape. Air suction velocity v_0 was chosen so that a particle would get in the exit branch within its vertical fall distance 0.08 m. The least favorable point for catching dust is the point on the airflow symmetry axis. According to the design data such dust particles cannot be caught. However, in real conditions of dust and slurry collectors such case is unlikely. That is why the suction axis has 0.0025 m offset. Therefore, in order to contain exhaust dust emissions, the velocity of 80 m/s must be ensured for a dust and slurry collector with the inner diameter of 68 mm and 27 m/s for a dust and slurry collector with the inner diameter of 40 mm.

The dust particle path computation in the case of the perfectly elastic collision with the flow boundaries (that is often used in computations) had paradoxical results. For instance, at gas exhaust velocity of 80 m/s a dust particle gets in the suction inlet unlike at the velocity of 100 m/s when it misses the suction inlet tangled in numerous bounces between the walls and the drill rod.

There were three approaches to a 3D problem reduction to 2D one. The suction velocity was given by $v = L/S$, where L is the experimentally measured airflow rate and S is the suction port area.

In the first approach, S was assumed to be the real elliptic suction port area (*real* velocity in Figures 4.2 and 4.3): $S_{ell} = 0.001657$ m². In the second approach (*conventional* velocity in Figures 4.2 and 4.3), S is the area of a rectangle having dimensions $0.02 \cdot 0.08$ m² (0.02 m is the suction slot width, 0.08 m is the dust and slurry collector height). In the third approach, there was a separate computation made for the dimensionless field of velocities in the case with the suction slot width $S_{ell}/0.08$ m. The computations made in accordance with the third approach significantly differ from the second approach results.

The comparison of experimentally measured velocity values with the design velocities (Figures 4.2 and 4.3) allows for making a conclusion that experimental velocity values are between the analytical numerical data obtained using the first and in the second approach. That is why the airflow rate required for containing dust emissions is expected to be defined by the formula

$$L = \frac{v(S_{el} + S)}{2}. \tag{4.5}$$

For a dust and slurry collector of 68 mm diameter we will have $L = 0.13028$ m³/s and for a dust and slurry collector of 40 mm diameter we will have $L = 0.04397$ m³/s.

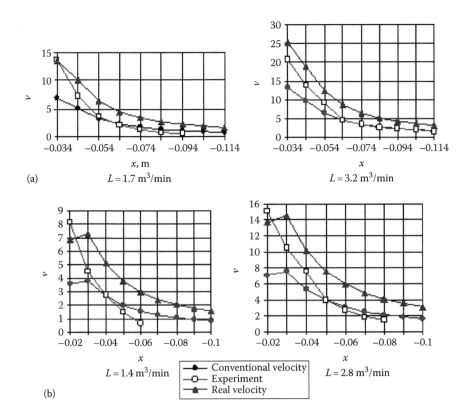

(a) $L = 1.7$ m³/min

$L = 3.2$ m³/min

(b) $L = 1.4$ m³/min

$L = 2.8$ m³/min

— Conventional velocity
—□— Experiment
— Real velocity

FIGURE 4.2 Air velocity attenuation at distancing from the dust collector lateral opening ($y = 0$): (a) dust collector inner diameter $D_{inner} = 68$ mm and (b) dust collector inner diameter $D_{inner} = 40$ mm.

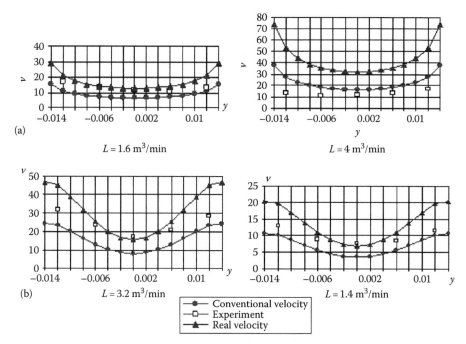

FIGURE 4.3 Air velocity distribution in the dust collector lateral opening section: (a) dust collector diameter = 68 mm ($x = -0.034$ m) and (b) dust collector diameter = 40 mm ($x = -0.02$ m).

4.1.2 COMPUTATION OF DUST PARTICLE PATHS IN SILO BIN CAVITIES

The path of a dust particle in the case with the Stokes airflow is described by equation [81]:

$$\frac{\pi d_e^3}{6}\rho_1 \frac{d\vec{v}_1}{dt} = \frac{\pi d_e^3}{6}\rho_1 \vec{g} - 3\pi v \chi \rho d_e (\vec{v}_1 - \vec{v}), \tag{4.6}$$

where
d_e is an equivalent diameter of the particle
ρ_1 is the particle density
v_1 is the particle velocity
ρ is the air density
v is the kinematic air viscosity
χ is the particle dynamic form factor
g is the gravity factor
v is the air velocity

At $A = \rho(\pi d_e^3/6)$ and $B = 3\pi v \rho \chi d_e$ Equation 4.6 will resolve oneself into the following set:

$$\begin{cases} A\dfrac{dv_{1x}}{dt} = -B\left(v_{1x} - v_x\right), \\[2mm] \dfrac{dx}{dt} = v_{1x}, \\[2mm] A\dfrac{dv_{1y}}{dt} = -Ag - B\left(v_{1y} - v_y\right), \\[2mm] \dfrac{dy}{dt} = v_{1y}. \end{cases} \tag{4.7}$$

By integrating the set of Equation 4.7 using the Runge–Kutta method and determining the air velocity at each step using the method of boundary integral equations, it is possible to plot the paths for dust particles.

In suction of bins being loaded with powdery material there was in increased ejection of dust particles by the exhaust air observed. The suspended materials concentration in suction air reaches 5...10 g/m³ that results both in noticeable losses of expensive fine-grained materials and to salvo emissions of dust into the atmosphere due to excessive dust loads on dust-collecting units.

Especially high dust concentration is observed in the suction of bins being loaded from pneumatic conveying systems. This is contributed by a more uniform distribution of powder conveying air as compared to the gravity chuting and recurrent whirling of dust particles at aerodynamic interaction between the air jet and the surface of a material in the bin. In this connection, the most preferable is the scheme of suction with a buffer bin, that is, when bin cavities are aerodynamically interconnected with dikes (bypass channels) and dust-laden air is exhausted from the bin adjacent to the one being loaded. This buffer bin is used as suction air precleaner dust bowl to reduce dust load on the exhaust system dust collectors.

For the purpose of the quantitative estimation of the rate of dust articles ejection from the bin cavity, there was particle motion by gravity with the Stokes air drag studied analytically.

The necessity of such study is also associated with the determination of the maximum diameter of dust particle d_{max} ejected from the bin by the exhaust air.

The field of air velocities was determined using the method of boundary integral equations. The paths of dust particles were plotted by means of numerical integration of motion equation (Equation 4.6).

Note that the problem can only be solved numerically. As an example, let us consider the calculation for 6 m high prismatic bin with the bottom of 4.4 m. The local exhaust outlet and inlet are 0.159 and 0.1 m slots, respectively. Airflow parameters: suction volume is 12,000 m³/h and dynamic viscosity is $1.776 \leq 10^{-5}$ Pa $\leq$ s. Dust particle density is 2500 kg/m³. The problem was reduced to the 2D setup.

The relation of d_{max} to the bin filling depth h was analyzed. d_{max} was calculated for h from 6 to 1 m (maximum allowable filling depth) in increments of $\Delta h = 1$ m. d_{max} remains virtually constant being approximately equal to 45 μm. Figures 4.4 and 4.5 show that as far as the bin is being filled, the horizontal air velocity component is increased (flow lines are adjusted), which contributes to a quick particle motion toward the local exhaust.

A screen that would prevent dust particles from being ejected from the bin is best to be installed at the outlet. Its displacement toward the local air exhaust results in increase in the maximum particle diameter d_{max} (Figure 4.6). A noticeable decrease in d_{max} also occurs with decrease in airflow rate (Figure 4.7).

The volume of air ejected from the bin cavity will be determined by the flow rate of air coming from bypass channels and through slots $Q_a = Q_b + Q_s$.

The highest suction volume is required when the most distant bin is loaded. The amount of air flowing through the dike to the adjacent bin is

$$Q_0 = Q_1 + Q_2 + Q_3,$$

where
  Q_0 is the flow rate of air in the dike of the bin being loaded, m³/s
  Q_1 is the flow rate of air coming in the bin from the pneumatic conveying system together with charging material, m³/s
  $Q_3 = G_m/\rho_p$ is the flow rate of air ejected from the bin by the transfer material particles, m³/s
  G_m is the material mass flow, kg/s
  ρ_p is the density of particles, kg/m³

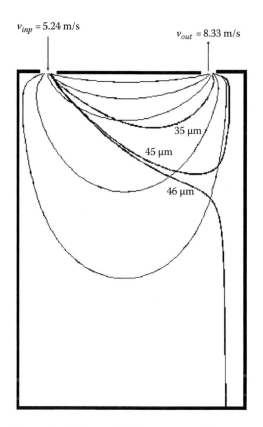

FIGURE 4.4   Flow lines and the paths of dust particles in an empty bin.

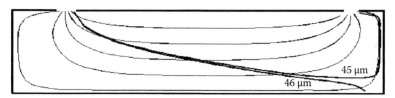

FIGURE 4.5   Flow lines and the paths of particles in the cavity of a fully filled bin.

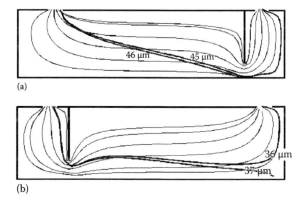

FIGURE 4.6   Flow lines and paths of dust particles in the case there have been a partition installed: (a) at the local exhaust and (b) at the outlet.

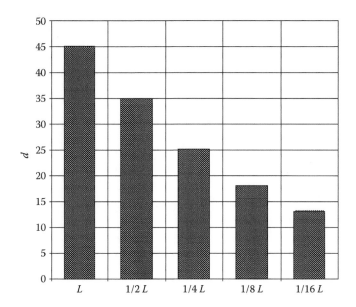

FIGURE 4.7 Chart of relation between maximum particle diameter d and the flow rate of air ejected by the local exhaust L (12,000 m³/h).

Q_2 is the flow rate of air coming in the bin cavity through slots (having area S, m²) under the vacuum-gauge pressure (P) maintained by the exhaust system given by

$$Q_2 = 0.65S\sqrt{\frac{2P_s}{\rho}}, \quad \text{m}^3/\text{s}$$

where
P_s is the standardized vacuum-gauge pressure in the bin being filled with material ($P_s \approx 5$ Pa)
ρ is the air density

If air is exhausted from the adjacent bin, the vacuum-gauge pressure in the bin cavity will exceed the standardized vacuum-gauge pressure for the dike pressure drop value

$$P = P_s + \zeta\frac{\rho}{2}\left(\frac{Q_0}{S_0}\right)^2,$$

where
ζ is the dike drag factor
S_0 is the dike cross-sectional area, m²

The suction volume will constitute

$$Q_a = Q_0 + 0.65S_a\sqrt{\frac{2P_d}{\rho}},$$

where S_a is the area of the adjacent bin slots.

If N bin is being exhausted rather than the adjacent one (the bin being loaded is the first one $Q_0 = Q_{o1}$) we will have

$$Q_{aN} = Q_{o1} + \sum_{i=2}^{N} Q_{si}, \quad Q_{si} = 0.65 S_{si} \sqrt{\frac{2 P_{si}}{\rho}},$$

where Q_{si} is the flow rate of air coming from slots of i bin (having area S_{si}, m^2) under the vacuum-gauge pressure,

$$P_{si} = P_s + \sum_{j=1}^{i} \zeta_j \frac{\rho}{2} \left(\frac{Q_{oj}}{S_{oj}} \right)^2 ; \quad i = 2,3,\ldots,N; \, j = 1,2,\ldots,i.$$

In this case, airflow rate Q_{oj} will be increased out of the flow rate of air coming through slots of the previous bins:

$$Q_{oj} = Q_{o1} + \sum_{k=2}^{j} Q_{sk}.$$

It is obvious that in order to decrease the suction volumes it is required to decrease the pneumatic conveying system capacity and the dike pressure losses. In such, there will be both dust material suction losses and the exhaust system energy losses reduced.

Thus, the maximum diameter of dust particles and, hence, the bin cavity suction air particle-size distribution can be increased with the decrease in the pneumatic conveying system capacity and the dike drag as well as due to a screen installed at the bin feed port. Dust losses are not a function of the bin filling depth.

4.1.3 FORECASTING PARTICLE-SIZE DISTRIBUTION AND CONCENTRATION OF COARSELY DISPERSED AEROSOLS

The study of intake plume is necessary not only for determining the capacity of ventilating plants but also for choosing optimum designs and efficient layouts of local exhausts [16]. When containing dust emissions, the hydrodynamic field of local exhaust has an active part in changing the structures of aerosol flows, that is, particle-size distribution and concentration of dust particles [104,120] determining the choice of a dust-collecting device for an exhaust system. Quantitative description of these changes is based either on the solution of dynamic equations (e.g., in the estimation of inertial separation of particles in the course of dust sampling [119]) or on the analysis of settling particles diffusion equations (when studying the dispersive regularities of dust emissions [121,122]). The discrete method analyzing the behavior of individual particles found its use in engineering applications, for example, in the design of dust extract systems.

In view of the approximate nature of various initial data (such as process material consumption rate, strength, particle-size distribution, and humidity), defining parameters of the initial state of dust flows this study sets forth an attempt to elaborate an approximate, substantially combined method of qualitative evaluation of compositional changes in the aerosol flow in the local exhaust range.

A dust particle behavior in the intake plume is mainly determined by the Earth's gravitational field and the local exhaust hydrodynamic field

$$\frac{d\vec{v}_1}{dt} = \vec{g} - \frac{3\pi\mu\delta}{m}\left(\vec{v}_1 - \vec{v}_2\right), \tag{4.8}$$

where

$\vec{v}_1$, $\vec{v}_2$ are the particle and air velocity vectors, m/s
δ is the equivalent diameter of an aerosol particle, m
m is the particle mass, kg
μ is the coefficient of air dynamic viscosity, Pa·s

By introducing the following values, velocity u_∞, length l_∞, time $t_\infty = l_\infty/u_\infty$ as basic, it is possible to make particle dynamic equation (Equation 4.8) dimensionless, which will reduce the multiparameter dependence of $\vec{v}_1$ to the two-parameter dependence of the Stokes and Froude numbers:

$$St \cdot \frac{d\vec{v}}{d\tau} + \vec{v} = \vec{u} + Fr \cdot \vec{e}_g \cdot St, \tag{4.9}$$

$$St = \frac{mu_\infty}{3\pi\mu\delta l_\infty}, \quad Fr = \frac{gl_\infty}{u_\infty^2},$$

where $\vec{e}_g$ is free-fall acceleration unitary vector $\vec{g}$

$$\vec{v} = \vec{v}_1/u_\infty$$

$$\vec{u} = \vec{v}_2/u_\infty$$

$$\tau = t/t_\infty = t \cdot u_\infty/l_\infty$$

These parameters can be eliminated if the airborne velocity and the particle inertial range length are taken for characteristic parameters:

$$u_\infty = c \equiv \frac{mg}{3\pi\mu\delta} \text{ is airborne velocity, m/s;}$$

$$l_\infty = \frac{c^2}{g} \text{ is length, m;} \quad t_\infty = \frac{c}{g} \text{ is time, s}$$

Then Equation 4.8 will take the following form:

$$\frac{d\vec{u}}{d\tau} + \vec{v} = \vec{e}_g + \vec{u}. \tag{4.10}$$

Note that Equation 4.8 can be reduced to the form of Equation 4.10 if the requirements are not that strict: $u_\infty = c$; $t_\infty = c/g$. However, is $l_\infty = H$ further integration of Equation 4.10 yields Froude number gH/c^2 (see, e.g., Equation 4.27).

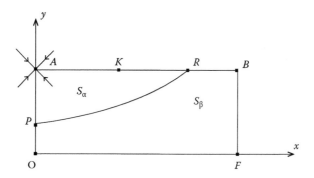

FIGURE 4.8 Region of dust emissions: L_{ARPA} is boundary S_α, L_{RBFOPR} is boundary S_β, and $S = S_\alpha + S_\beta$.

Despite its simplicity, in most of the practical cases this equation admits a numerical solution only. An exception is the case of hydrodynamic field constancy $\vec{u} = \vec{u}_k = \text{const}$ where Equation 4.10 becomes a linear one that admits the following solution:

$$\vec{v} = \vec{v}_0 e^{-\tau} + \left(\vec{e}_g + \vec{u}_k\right)\left(1 - e^{-\tau}\right), \tag{4.11}$$

which demonstrates a relaxational nature of particle motion. If $\tau \gg 1$, the effect of initial velocity vector $\vec{v}_0$ may be neglected. This allows for analyzing the behavior of particles in an intake plume studying a simpler flow of dust particles with initial zero velocities.

In order to simplify the presentation of the essence of the qualitative method proposed, we will consider a 2D flow. Let us assume that there are dust sources with the specific intensity of J_S [g/(s·m²)] distributed within plane region S (Figure 4.8) limited by boundary L, and dust sources with the specific intensity of J_L [g/(s·m²)] distributed along the boundary. These sources characterize the occurrence of dust particles of d size with zero initial velocity at the relevant points of region S (including the points of its boundary L). The occurring particle is exposed to the gravity and aerodynamic force resulting from the local exhaust intake plume located at point A. Since the gravity force is downward and the aerodynamic force is directed upward point A the general stream of particles with mass flow

$$G = G_S + G_L = \iint_S J_S dS + \int_L J_L dL \tag{4.12}$$

will be divided into two. The first one is the stream of particles featuring mass flow G_β and settling on the lower boundary of L_{OF}. The second one is the stream of particles carried away by the intake plume.

$$G_\alpha = G - G_\beta. \tag{4.13}$$

Let us assume that these streams are divided with line L_{RP}, that is, the stream of settling particles is moving toward S_β region while the stream of ejected particles is moving toward S_α region adjacent to suction port A. Then defining mass flows of particles at these points makes no significant difficulties. For instance, with uniform intensity of dust sources and convective particle transportation from the outside across KB interval

$$G_\beta = J_S \cdot S_\beta + J_L \cdot L_{RB}, \quad G_\alpha = J_S \cdot S_\alpha + J_L \cdot L_{KR}, \quad G = G_\alpha + G_\beta = J_S \cdot S + J_L \cdot L_{KB}.$$

The relation

$$\alpha = \frac{G_\alpha}{G} \tag{4.14}$$

is called [119] suction coefficient while

$$\beta = \frac{G_\beta}{G} \tag{4.15}$$

is called sedimentation coefficient interrelated by the apparent equation

$$\alpha + \beta = 1. \tag{4.16}$$

Then the concentration of dust particles in suction air with the known local exhaust capacity Q [m³/s · m] will be determined by the equation

$$\gamma = \alpha \gamma_{max}, \quad \gamma_{max} = \frac{G}{Q}, \tag{4.17}$$

as the suction coefficient determines reduction in suction air particle-size distribution against the maximum possible concentration (when all particles that occur are carried over by the intake plume, $G_\beta = 0$).

If there are dust particles of various sizes ($\delta_{min} < \delta < \delta_{max}$) in the region in question it is possible to find the suction air particle-size distribution. To this effect, it is necessary to know $S_\alpha(\delta_i)$ as narrow i fraction mean diameter function:

$$\delta_i = 0.5(\delta_{i\,min} + \delta_{i\,max}), \tag{4.18}$$

where $\delta_{i\,min}$, $\delta_{i\,max}$ are boundary dimensions of i fraction particles.

For instance, if there is no convective transportation of $(J_L(\delta_i) = 0)$, we have

$$G_\alpha(\delta_i) = J_S(\delta_i) \cdot S_\alpha(\delta_i), \tag{4.19}$$

$$\alpha(\delta_i) = \frac{G_\alpha(\delta_i)}{G(\delta_i)} = \frac{S_\alpha(\delta_i)}{S}. \tag{4.20}$$

The weight ratio of i fraction in exhaust air (m_i) will change against the weight ratio of these particles (M_i) in S region:

$$m_i = \frac{\alpha(\delta_i)}{\alpha \cdot M_i}, \tag{4.21}$$

where
$$m_i = G_\alpha(\delta_i)/G_\alpha$$
$$M_i = G(\delta_i)/G$$
$$\sum_{i=1}^{N} m_i = 1$$
$$\sum_{i=1}^{N} M_i = 1$$
$$G_\alpha = \sum_{i=1}^{N} G_\alpha(\delta_i)$$
$$G = \sum_{i=1}^{N} G(\delta)$$

N is the number of narrow fractions

As an example, let us consider two cases: a linear outflow in the upper semiplane that is the model of an open local exhaust and a linear outflow in the horizontal bar that is the model of a suction unit (to be exact, the model of a ventilated continuous tunnel).

Before we find $S_\alpha(\delta_i)$ we will introduce the following definition: let us refer to the locus of the flow area points at which the vertical velocity component is equal to the airborne velocity as the critical curve.

Conceptually, it is a locus of labile points. A particle that has separated therefrom will either be falling or ascending until it eventually gets into the local exhaust.

Let us plot critical curves for the simplest cases of airflow motion: a linear outflow in the upper semiplane and in the bar.

When two outflows of the same intensity Q [m³/(s·m)] are placed at points $\pm H$ of OY axis, the field of airflow velocities will be determined from the formulas

$$\left\{ \begin{aligned} v_{2x} &= -\frac{Q\bar{x}}{2\pi}\left[\frac{1}{\bar{x}^2 + (\bar{y}+H)^2} + \frac{1}{\bar{x}^2 + (\bar{y}-H)^2}\right], \\ v_{2y} &= -\frac{Q}{2\pi}\left[\frac{\bar{y}+H}{\bar{x}^2 + (\bar{y}+H)^2} + \frac{\bar{y}-H}{\bar{x}^2 + (\bar{y}-H)^2}\right], \end{aligned} \right. \tag{4.22}$$

where
$\bar{x}, \bar{y}$ are the dimensional coordinates
Q is the outflow intensity, m³/(s·m)

By reducing it to the dimensionless form we will obtain

$$\left\{ \begin{aligned} u_x &= -xq\left[\frac{1}{x^2 + (y+1)^2} + \frac{1}{x^2 + (y-1)^2}\right], \\ u_y &= -q\left[\frac{y+1}{x^2 + (y+1)^2} + \frac{y-1}{x^2 + (y-1)^2}\right], \end{aligned} \right. \tag{4.23}$$

where
$$q = \frac{Q}{2\pi H \cdot c} = \frac{u_H}{c}$$
$$u_H = Q/H$$

The set of critical curves is given by the equation

$$q\left[\frac{1-y}{x^2+(y-1)^2} - \frac{1+y}{x^2+(y+1)^2}\right] = 1. \tag{4.24}$$

Note that when $q \to \infty$, we will obtain an equation for the unit circle centered on the origin.

In case of a linear outflow located at point $(0, H)$ of a horizontal bar $0 \le y \le H$, the dimensionless field of velocities is given as follows:

$$\begin{cases} u_x = -2\pi q \dfrac{\operatorname{sh} \pi x}{\operatorname{ch} \pi x + \cos \pi y}, \\[3mm] u_y = 2\pi q \dfrac{\sin \pi y}{\operatorname{ch} \pi x + \cos \pi y}. \end{cases} \tag{4.25}$$

The set of critical curves in this case is better found from the equation

$$ch\pi x = 2\pi q \sin \pi y - \cos \pi y. \tag{4.26}$$

When $q \to \infty$, this curve will be converted into two straight lines: $y = 0$ and $y = 1$.

As seen from Figures 4.9 and 4.10 the critical curves plotted according to Equations 4.24 and 4.26 at $q = 1, 2, \ldots, 10, 850$ are closed lines limiting some region of uniformly distributed dust sources. Once aerosol particles have entered this region they will get into the linear outflow. Let this region be referred to as the active suction region.

The active suction region area is growing with an increase in q and is asymptotically approximated to $\pi/2$ (Figure 4.11) for the upper semiplane outflow. For the bar outflow, this region will be continuously growing with increase in q but its increase when $q > 90$ (Figure 4.12) will be small. Besides, for real physical problems $q < 8500$ ($q_{max} \approx 8500$ at $H_{min} = 0.01$ m; $c_{min} = 0.0075$ m/s for 10 μm particles; $Q_{max} = 4$ m³/s for a round fitting with the radius of 0.25 m and suction velocity of 20 m/s).

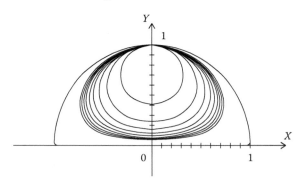

FIGURE 4.9   Critical curves of the outflow in the upper semiplane.

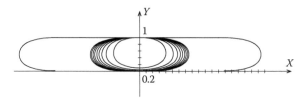

FIGURE 4.10   Critical curves of the outflow in the bar.

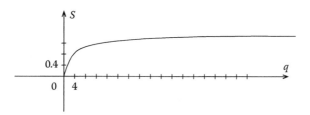

FIGURE 4.11 Active suction region area versus dimensionless velocity for the upper semiplane outflow.

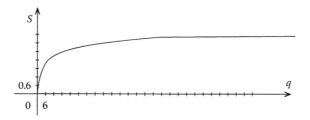

FIGURE 4.12 Active suction region area versus dimensionless velocity for the horizontal bar outflow.

There is a region where an aerosol particle will also be caught into the exhaust since it will eventually get into the active suction region. Let such region be referred to as the adjoined region. We will show that its boundary will actually be the borderline (L_R). To this effect, we will calculate the extreme path above which all aerosol particles will get into the exhaust unlike in the below part.

By reducing Equation 4.10 to the set of ordinary differential equations, we will obtain

$$\begin{cases} dx/d\tau = v_x c^2/(Hg), \\ dv_x/d\tau = -v_x + u_x, \\ dy/d\tau = v_y c^2/(Hg), \\ dv_y/d\tau = -1 - v_y + u_y. \end{cases} \tag{4.27}$$

We will consider the initial particle velocity values as equal to zero. The dimensionless y-coordinate of the particle escape $y = 1$ as a particle caught into the exhaust is most often below it in real-world problems. The critical path computation algorithm will include the following main steps:

1. Calculate interval $[a, b]$ where a particle gets into the exhaust at $x = a$ but does not at $x = b$.
2. Make the computation for $x = c = (a + b)/2$. If a particle falls we will assign $b = c$, otherwise $a = c$.
3. Check if the following conditions are fulfilled for each point of the aerosol particle path:

$$\left| x_p - x \right| < \varepsilon, \quad \left| y_p - y \right| < \varepsilon,$$

where
    (x_p, y_p) are the active suction region lower point coordinates
    (x, y) are the path point coordinates
    ε is the given computation accuracy

If the conditions are fulfilled the computation stops, otherwise step 2 is repeated. The computation process converges quite quickly (Figure 4.13).

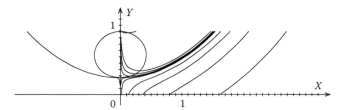

FIGURE 4.13   Regarding the extreme path definition at $a = 0.25$ and $b = 10$ for an aerosol particle featuring fineness $d = 50$ μm and the density of 2500 kg/m³ with the upper semiplane outflow intensity $Q = 2$ m³/s and the dynamical air viscosity of $1.8 \cdot 10^{-5}$ Pa·s.

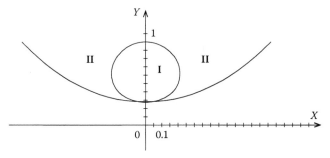

FIGURE 4.14   The active suction region (**I**) and the adjoined region (**II**) for an aerosol particle featuring fineness $d = 50$ μm and the density of 2500 kg/m³, with the upper semiplane outflow intensity $Q = 2$ m³/s, air density of 1.2 kg/m³, and dynamical viscosity of $1.5 \cdot 10^{-5}$ Pa·s.

As seen from the calculations (Figure 4.14) the adjoined region is significantly larger than the active suction region.

Note that the extreme path of an aerosol particle at initial velocities other than zero will coincide with the extreme path computed for a particle at zero initial velocities except for a small interval (Figure 4.15). Moreover, the extreme path is not affected by the *y*-coordinate from which it is found: only its length will be changing.

We will refer to the combined active suction region and the adjoined region as the suction region. Let us determine k equal to the relation of the active suction region area to the suction region area with variation in fineness of aerosol particles.

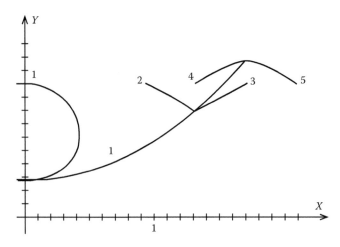

FIGURE 4.15   Extreme paths at various initial velocities: 1: (0, 0); 2: (20*c*, −10*c*); 3: (−20*c*, −10*c*); 4: (20*c*, 10*c*); 5: (−20*c*, 10*c*).

Suction region S_α includes both active suction region S_a and the adjoined region. S_α region area can be determined from the relation

$$k = \frac{S_a}{S_\alpha}$$

at various values of particle fineness (Table 4.1). As is clear from these data, the suction region grows with decrease in the size of particles and increase in the local exhaust capacity.

Thus, the suction quality may be adequately characterized by S_α region area, that is, by the suction coefficient. The latter value can be determined with known values of M_i from the formula

$$\alpha = \sum_{i=1}^{N} \left[\alpha(\delta_i) \cdot M_i \right] \tag{4.28}$$

TABLE 4.1
Relation of the Active Suction Region Absolute (S_a, m²) and Relative Area ($k = S_a/S_\alpha$) to a Size of Aerosol Particles (δ, μm) and Outflow Intensity (Q, m³/(s · m)) at $\rho = 2500$ kg/m³, $\mu = 1.8 \cdot 10^{-5}$ Pa · s

Q, m³/s·m	$S_a/2$ (Numerator) and k (Denominator) at δ, μm								
	5	10	15	20	30	45	60	80	100
Semiplane outflow									
0.5	0.700	0.57	0.443	0.332	0.173	0.062	0.025	0.009	0.003
	0.117	0.201	0.258	0.296	0.337	0.353	0.371	0.375	0.378
1.0	0.735	0.648	0.555	0.464	0.309	0.153	0.074	0.030	0.014
	0.086	0.157	0.212	0.250	0.303	0.345	0.354	0.373	0.425
1.5	0.748	0.682	0.606	0.530	0.389	0.228	0.126	0.058	0.028
	0.071	0.133	0.183	0.222	0.276	0.326	0.348	0.366	0.408
2	0.755	0.701	0.637	0.571	0.443	0.285	0.173	0.087	0.045
	0.063	0.117	0.162	0.202	0.258	0.318	0.338	0.364	0.389
2.5	0.760	0.714	0.658	0.599	0.482	0.330	0.215	0.116	0.064
	0.058	0.107	0.149	0.187	0.244	0.297	0.329	0.337	0.374
3.0	0.764	0.722	0.673	0.620	0.512	0.366	0.251	0.144	0.082
	0.052	0.098	0.139	0.175	0.230	0.285	0.320	0.353	0.377
Bar outflow									
0.5	1.774	1.333	1.075	0.892	0.637	0.390	0.235	0.116	0.058
	0.013	0.040	0.074	0.109	0.176	0.257	0.305	0.341	0.153
1.0	1.995	1.554	1.296	1.113	0.855	0.600	0.425	0.264	0.161
	0.008	0.023	0.045	0.067	0.117	0.188	0.245	0.300	0.336
1.5	1.124	1.683	1.425	1.242	0.984	0.727	0.547	0.375	0.254
	0.006	0.017	0.033	0.050	0.089	0.150	0.205	0.263	0.311
2	2.216	1.774	1.516	1.333	1.075	0.818	0.637	0.460	0.330
	0.0043	0.014	0.016	0.041	0.073	0.126	0.177	0.235	0.286
2.5	2.286	1.845	1.587	1.404	1.146	0.889	0.707	0.528	0.393
	0.0036	0.011	0.022	0.034	0.063	0.110	0.156	0.214	0.269
3.0	2.345	1.903	1.645	1.462	1.204	0.946	0.764	0.584	0.447
	0.003	0.0097	0.019	0.030	0.055	0.097	0.140	0.195	0.247

or with known values of m_i from the formula

$$\alpha = 1/\sum_{i=1}^{N} \frac{m_i}{\alpha(\delta_i)},$$

(4.29)

which makes it possible to determine the concentration and particle-size distribution of dust in the exhaust air or, knowing them, to find the true total and fraction-by-fraction intensity of dust emissions to be subsequently able to forecast dust flow patterns for other volumes.

Let us determine the aerosol flow pattern for 3000 × 6400 GSO [157] sieve hood of pellet plant No. 2 of the Northern Mining and Concentration Complex with known weight ratios of i narrow fractions in the exhaust air, material density ($\rho = 2600$ kg/m³), and rates of suction Q_a and induced air Q_{c1}, Q_{c2} through slots Q_n (Figures 4.16 and 4.17). There were extreme paths of medium diameter dust particles of narrow fractions plotted at given rates Q (5, 15, 30, 50, 80 μm) and their suction coefficients determined. The hood suction coefficient was determined from formula (Equation 4.29) while the intensity of dust emissions M_i inside the sieve was determined from Equation 4.21. It should be noted that the sieve mesh vibration and the material flow inside the hood cause the intensity of dust emissions *spread* throughout the volume; therefore, we will assume it to be constant throughout the region in question. The obtained values of M_i were used to forecast the particle-size distribution and the concentration of dust particles for various suction volumes and air leak schemes (Figures 4.16 and 4.17). In order to determine the concentration there as the initial mass flow of dust particles defined first $G = \alpha_Q \cdot Q_a \cdot \gamma_Q$, then, the required concentration γ_{Q_i} determined from the formula $\gamma_{Q_i} = \alpha_{Q_i} \cdot (G/Q_i) = \alpha_{Q_i} \cdot (\alpha_Q \cdot Q_a \cdot \gamma_Q/Q_i)$ at the given suction volumes Q_i.

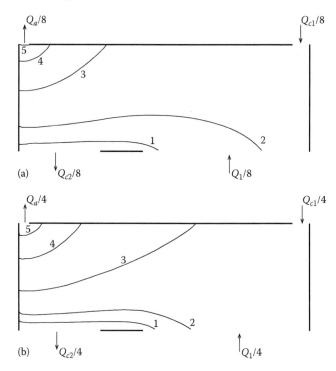

FIGURE 4.16 Extreme paths of dust particles with the density of 2600 kg/m³ and various sizes: 1: 5 μm, 2: 15 μm, 3: 30 μm, 4: 50 μm, 5: 80 μm at $Q_a = 1.26$ m³/(s·m), $u_a = 4.5$ m/s, $Q_{c1} = 0.05$ m³/(s·m), $u_{c1} = 0.125$ m/s, $Q_{c2} = 0.09$ m³/(s·m), $u_{c2} = 0.045$ m/s, $Q_1 = 1.30$ m³/(s·m), and $u_1 = 0.325$ m/s: (a) $Q/8$ and (b) $Q/4$. (*Continued*)

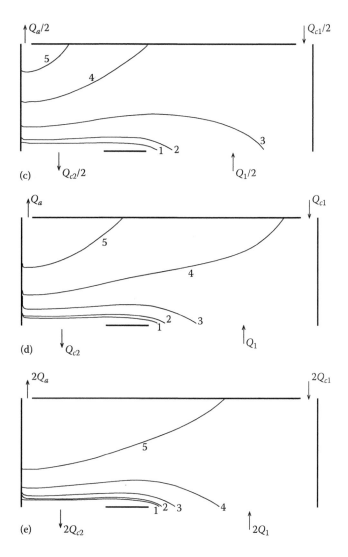

FIGURE 4.16 (*Continued*) Extreme paths of dust particles with the density of 2600 kg/m³ and various sizes: 1: 5 μm, 2: 15 μm, 3: 30 μm, 4: 50 μm, 5: 80 μm at $Q_a = 1.26$ m³/(s·m), $u_a = 4.5$ m/s, $Q_{c1} = 0.05$ m³/(s·m), $u_{c1} = 0.125$ m/s, $Q_{c2} = 0.09$ m³/(s·m), $u_{c2} = 0.045$ m/s, $Q_1 = 1.30$ m³/(s·m), and $u_1 = 0.325$ m/s: (c) $Q/2$, (d) Q, and (e) $2Q$.

The calculations (Tables 4.2 and 4.3) show that more preferable is the second air leaking scheme since suction coefficients are less and, accordingly, the rate of dust emissions into the suction network is decreased. Whence it follows that it is necessary to air proof primarily the lower sieve hood.

With increase in suction volumes, the suction air dust concentration is also increased. Note that there are various interpretations for this issue: some specialists believe that the concentration is proportional to the flow rate while others assert vice versa. There was a technical study conducted (Table 4.4). It was found that when the intensity of dust emissions is shifted toward coarse particles, the concentration will be proportional to the suction volumes. If dust emissions take place due to fine particles the concentration will be inversely related to the flow rate. When there are particles of medium size emitted, the concentration is not a monotone function of the suction volumes but has the maximum value that is shifted from fine to coarse aerosols if the intensity of dust emissions has the same direction.

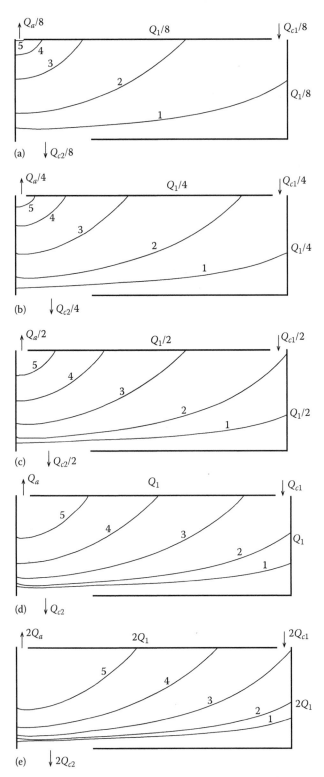

FIGURE 4.17 Extreme paths of dust particles with the density of 2600 kg/m³ and various sizes: 1: 5 μm, 2: 15 μm, 3: 30 μm, 4: 50 μm, 5: 80 μm at Q_a = 1.26 m³/(s·m), u_a = 4.5 m/s, Q_{c1} = 0.05 m³/(s·m), u_{c1} = 0.125 m/s, Q_{c2} = 0.09 m³/(s·m), u_{c2} = 0.045 m/s, Q_1 = 1.30 m³/(s·m), and u_1 = 0.325 m/s: (a) $Q/8$, (b) $Q/4$, (c) $Q/2$, (d) Q, and (e) $2Q$.

TABLE 4.2

Suction Air Dust Concentration against Suction Volumes and Dust Emission Intensity

	$M_1 = 1$		$M_2 = 1$		$M_3 = 1$		$M_4 = 1$		$M_5 = 1$		$M_i = 0.2$	
	α	γ	α	γ	α	γ	α	γ	α	γ	α	γ
$Q/8$	0.774	31.07	0.318	14.36	0.069	4.84	0.011	1.63	≈ 0	0.01	0.235	16.58
$Q/4$	0.824	16.54	0.450	10.16	0.169	5.93	0.041	2.91	0.006	1.36	0.298	10.54
$Q/2$	0.847	8.5	0.656	7.40	0.318	5.58	0.115	4.03	0.027	2.81	0.392	6.94
Q	0.857	4.3	0.762	4.3	0.49	4.3	0.243	4.3	0.081	4.3	0.487	4.3
$2Q$	0.865	2.17	0.815	2.3	0.656	2.88	0.405	3.58	0.187	4.96	0.586	2.59

TABLE 4.3

Aerosol Flow Parameters (Figure 4.16d) at Various Suction Volumes with $Q_a = 1.26$ m³/(s·m), $Q_{a1} = 0.05$ m³/(s·m), $Q_{c2} = 0.09$ m³/(s·m), $Q = 1.30$ m³/(s·m)

No.	$\delta_i^{min} < \delta < \delta_i^{max}$	δ_i	M_i	m_i	$\alpha(\delta_i)$
Suction Coefficient $\alpha = 0.194$ at $Q/8$, Concentration $\gamma = 0.815$ g/m³					
1	0...10	5	0.092	0.4578	0.963
2	10...20	15	0.105	0.4198	0.776
3	20...40	30	0.216	0.1000	0.090
4	40...60	50	0.328	0.0220	0.0126
5	60...100	80	0.259	0.0004	0.0003
Suction Coefficient $\alpha = 0.258$ at $Q/4$, Concentration $\gamma = 1.084$ g/m³					
1	0...10	5	0.092	0.3455	0.966
2	10...20	15	0.105	0.3695	0.908
3	20...40	30	0.216	0.2131	0.255
4	40...60	50	0.328	0.0649	0.051
5	60...100	80	0.259	0.0069	0.0069
Suction Coefficient $\alpha = 0.416$ at $Q/2$, Concentration $\gamma = 1.747$ g/m³					
1	0...10	5	0.092	0.2149	0.968
2	10...20	15	0.105	0.2385	0.944
3	20...40	30	0.216	0.4026	0.776
4	40...60	50	0.328	0.1248	0.158
5	60...100	80	0.259	0.0193	0.031
Suction Coefficient $\alpha = 0.559$ at Q, Concentration $\gamma = 2.348$ g/m³					
1	0...10	5	0.092	0.16	0.970
2	10...20	15	0.105	0.18	0.959
3	20...40	30	0.216	0.35	0.908
4	40...60	50	0.328	0.26	0.443
5	60...100	80	0.259	0.05	0.108
Suction Coefficient $\alpha = 0.758$ at $2Q$, Concentration $\gamma = 3.184$ g/m³					
1	0...10	5	0.092	0.1182	0.971
2	10...20	15	0.105	0.1336	0.9648
3	20...40	30	0.216	0.2684	0.944
4	40...60	50	0.328	0.378	0.873
5	60...100	80	0.259	0.1018	0.298

TABLE 4.4

Aerosol Flow Parameters (Figure 4.17d) at Various Suction Volumes with Q_a = 1.26 m³/(s·m), Q_{c1} = 0.05 m³/(s·m), Q_{c2} = 0.09 m³/(s·m), Q_1 = 1.30 m³/(s·m)

No.	$\delta_i^{min} < \delta < \delta_i^{max}$	δ_i	M_i	m_i	$\alpha(\delta_i)$
Suction Coefficient α = 0.1 at Q/8, Concentration γ = 9.61 g/m³					
1	0...10	5	0.066	0.5138	0.774
2	10...20	15	0.084	0.2671	0.318
3	20...40	30	0.253	0.1752	0.069
4	40...60	50	0.379	0.0438	0.0115
5	60...100	80	0.219	0.0001	0.00003
Suction Coefficient α = 0.152 at Q/4, Concentration γ = 7.37 g/m³					
1	0...10	5	0.066	0.3588	0.824
2	10...20	15	0.084	0.2479	0.450
3	20...40	30	0.253	0.2815	0.169
4	40...60	50	0.379	0.1026	0.0411
5	60...100	80	0.219	0.0092	0.0064
Suction Coefficient α = 0.241 at Q/2, Concentration γ = 5.84 g/m³					
1	0...10	5	0.066	0.2327	0.847
2	10...20	15	0.084	0.228	0.656
3	20...40	30	0.253	0.3342	0.318
4	40...60	50	0.379	0.181	0.115
5	60...100	80	0.219	0.0241	0.0265
Suction Coefficient α = 0.354 at Q, Concentration γ = 4.3 g/m³					
1	0...10	5	0.066	0.16	0.857
2	10...20	15	0.084	0.18	0.762
3	20...40	30	0.253	0.35	0.440
4	40...60	50	0.379	0.26	0.243
5	60...100	80	0.219	0.05	0.081
Suction Coefficient α = 0.486 at 2Q, Concentration γ = 2.95 g/m³					
1	0...10	5	0.066	0.1178	0.865
2	10...20	15	0.084	0.1404	0.815
3	20...40	30	0.253	0.3417	0.656
4	40...60	50	0.379	0.3160	0.405
5	60...100	80	0.219	0.0842	0.187

4.2   ANALYSIS OF DUST PARTICLE DYNAMICS INDUCED AT OPEN LOCAL EXHAUSTS BY ROTATING CYLINDERS

This section objective is to analyze dust and air dynamics at local exhausts as induced by a rotating cylinder given then process equipment elements and to determine the effect of airflow induced by the cylinder rotation on dust aerosol caught by a suction fitting.

The first design flow pattern is shown in Figure 4.18. The flow area boundaries are given in accordance with a roll-turning machine induction scheme.

The flow lines were plotted from the suction port in increment of 0.001 m and at the interval of 0.005 m between the lines. The number of boundary intervals was about 760.

In case of a resting cylinder the intake flow comprises three segments (Figure 4.19): I is above the cylinder, II is between the cylinder and the frame, and III is between the local exhaust and the support; the air rate relation in these flows is 14:5:11, respectively.

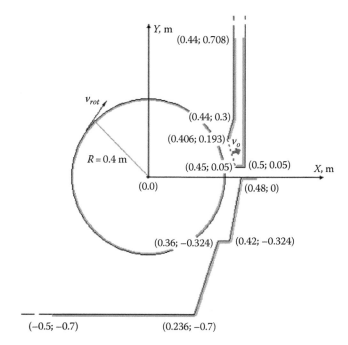

FIGURE 4.18 The flow pattern induced at a local exhaust by a roll-turning machine.

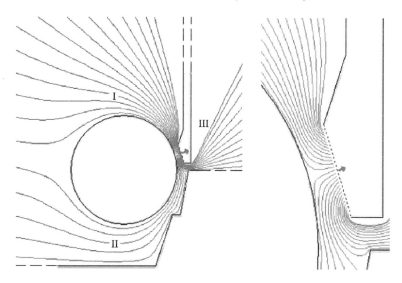

FIGURE 4.19 Flow lines for a resting cylinder at velocity $v_0 = 2$ m/s.

When the cylinder is rotating clockwise at the linear velocity of 0.5 m/s (these parameters conform to the machine roll processing workflow) and the suction velocity of 0.5 m/s, the flow pattern is changing significantly. Practically, the whole suction volume includes flow III, which is passing between the local exhaust and the support, circulating clockwise about the cylinder and getting into the suction port.

With an increase in the suction port velocity (Figures 4.20 through 4.23) the intake flow is divided into three parts again. The area of the circulatory flow around the cylinder is decreased with increase in the suction velocity. It is obvious that with a considerable increase in the suction velocity the flow pattern will be similar (Figure 4.19).

The paths of dust particles were plotted with the following parameters: integration step $h = 0.001$ m; dynamic form factor $\chi = 1.8$; air density $p = 1.205$ kg/m^3; dynamical air viscosity $\mu = 0.0000178$ Pa·s;

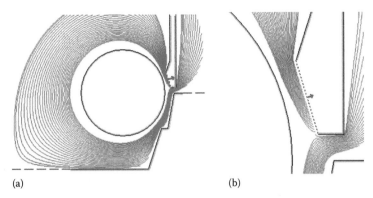

FIGURE 4.20   Flow lines for a cylinder rotating at linear velocity v_{rot} = 0.5 m/s and v_0 = 0.5 m/s: (a) general form and (b) near the suction opening.

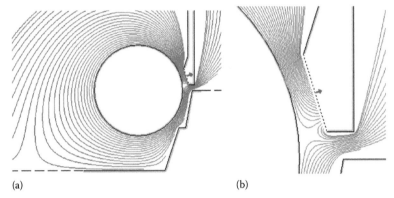

FIGURE 4.21   Flow lines for a cylinder rotating at linear velocity v_{rot} = 0.5 m/s and v_0 = 1 m/s: (a) general form and (b) near the suction opening.

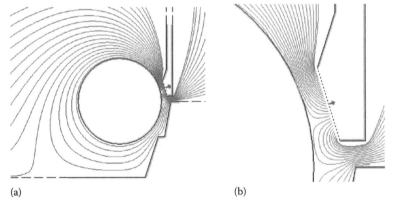

FIGURE 4.22   Flow lines for a cylinder rotating at linear velocity v_{rot} = 0.5 m/s and v_0 = 2 m/s: (a) general form and (b) near the suction opening.

coefficient of restitution k = 0.5; coefficient of sliding friction f = 0.5. Dust particle escape point x_0 = 0.41 m; y_0 = 0 (approximate location of dust emissions when machining the cylinder/roll). The initial velocity of dust particles was assumed to be zero. When v_0/v_{rot} = 2 (Figure 4.24), due to a great influence from the circulatory flow around the cylinder, dust particles of min 50 μm fly over the cylinder and are subsequently caught. Larger particles settle on the machine frame.

When v_0/v_{rot} = 4 (Figure 4.25), n particles are flying over the cylinder because of a great influence from the flow induced by the local exhaust.

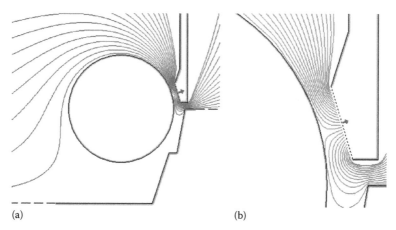

FIGURE 4.23 Flow lines for a cylinder rotating at linear velocity v_{rot} = 0.5 m/s and v_0 = 4 m/s: (a) general form and (b) near the suction opening.

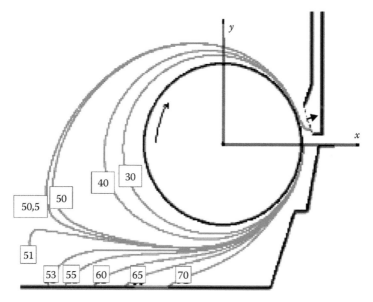

FIGURE 4.24 Paths of dust particles having the density of 1000 kg/m³ when a cylinder is rotating at linear velocity v_{rot} = 0.5 m/s and v_0 = 1 m/s.

It should be noted that the cylinder rotation has a significant impact on the maximum diameter d_{max} of dust particles caught by the local exhaust (Table 4.5).

Let us analyze the roll-turning machine suction pattern without air inflow between the local exhaust and the support (Figure 4.26) that corresponds to the case when a mechanical *screen* is installed to shut off air inflow from region III. In such cases, the exhaust airflow will be combined of two flows: above and below the cylinder (Figure 4.27). When v_0/v_{rot} = 2, a closed circulation region occurs around the cylinder (Figure 4.28).

With increase in v_0/v_{rot}, the circulation region area is reduced and the local exhaust starts having a greater effect.

The paths of dust particles also have ellipsoid form at certain parameters (Figure 4.29). A dust particle may fly around the cylinder several times and then settle on the machine frame. The circulation region prevents dust particles from getting into the suction port. With an increase in suction velocity, no particles fly around the cylinder and starting from a certain size they will be caught by the local exhaust (Table 4.6).

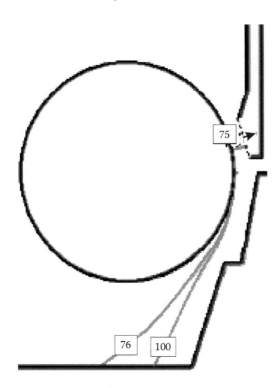

FIGURE 4.25 Paths of dust particles having the density of 1000 kg/m³ when a cylinder is rotating at linear velocity $v_{rot} = 0.5$ m/s and $v_0 = 2$ m/s.

TABLE 4.5
Maximum Diameters of Dust Particles Caught by the Local Exhaust as Shown in Figure 4.18

ρ_p, kg/m³	1000	2000	3000	4000	5000	6000	7000
$v_0 = 1$ m/s, $v_{rot} = 0$ m/s							
d_{max}, µm	271	180	143	121	107	97	89
$v_0 = 1$ m/s, $v_{rot} = 0.5$ m/s							
d_{max}, µm	50	36	29	25	23	20	19
$v_0 = 2$ m/s, $v_{rot} = 0.5$ m/s							
d_{max}, µm	75	51	41	36	31	28	26

4.3 MODELING DUST PARTICLE DYNAMICS IN SUCTION UNITS WITH ROTATING CYLINDERS

Maximum diameter d_{max} of dust particles carried over into the suction system from hoods that contain dust emissions from various processes is needed to be determined for the purpose of a scientifically grounded choice of dust-collecting equipment. According to the hypothesis by Professor V.A. Minko [104,105] using the maximum diameter value it is possible to forecast the particle-size distribution. This hypothesis was proven to be correct by pilot tests conducted for suction units of various designs in transfer of bulk materials [104].

There was the flight of round-shaped dust particles previously studied. As compared to the experimental data the design maximum diameter of dust particle was lower for 50–70 µm.

In addition, for practical purposes it is reasonable to decrease the suction system dust losses and use a suction unit as a settling chamber to reduce exhaust air treatment costs. For instance, there is

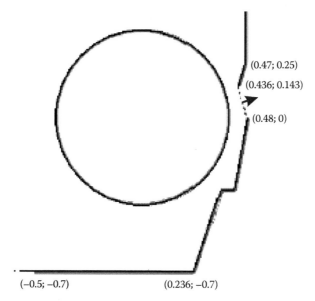

(0.47; 0.25)

(0.436; 0.143)

(0.48; 0)

(−0.5; −0.7)   (0.236; −0.7)

FIGURE 4.26 The flow pattern induced at a local exhaust by a roll-turning machine without air inflow between the local exhaust and the support.

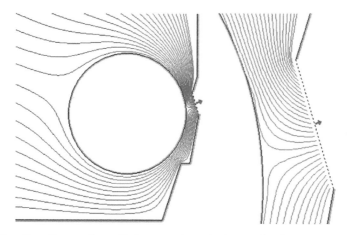

FIGURE 4.27 Flow lines for a resting cylinder at suction velocity $v_0 = 1$ m/s.

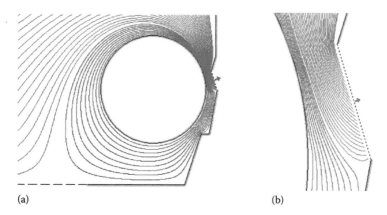

(a)   (b)

FIGURE 4.28 Flow lines for a cylinder rotating at linear velocity $v_{rot} = 0.5$ m/s and $v_0 = 1$ m/s: (a) general form and (b) near the suction opening.

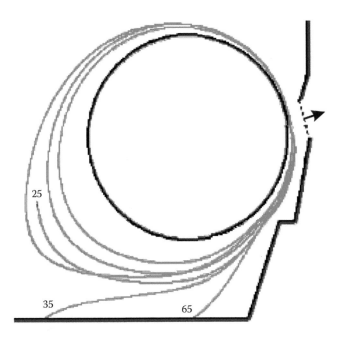

FIGURE 4.29 Paths of dust particles having the density of 1000 kg/m^3 when the cylinder is rotating at linear velocity $v_{rot} = 0.5$ m/s and $v_0 = 1$ m/s.

TABLE 4.6
Maximum Diameters of Dust Particles Caught by the Local Exhaust as Shown in Figure 4.26

ρ_p, kg/m^3	1000	2000	3000	4000	5000	6000	7000
$v_0 = 1$ m/s, $v_{rot} = 0$ m/s							
d_{max}, μm	283	188	148	126	111	101	92
$v_0 = 1$ m/s, $v_{rot} = 0.5$ m/s							
d_{max}, μm	0	0	0	0	0	0	0
$v_0 = 2$ m/s, $v_{rot} = 0.5$ m/s							
d_{max}, μm	144	99	79	68	60	55	51

a partition installed inside a hood along with chain curtains to reduce dust losses. Or, as we have already suggested, a rotating suction cylinder can be installed inside a hood.

Therefore, this section objective is to determine d_{max} of finely dispersed aerosols of various physical properties carried over into the suction system and to elaborate constructive solutions as to the design of suction units with settling chamber functions.

4.3.1 MODEL PROBLEMS

Let us analyze a hood used in transfers of bulk materials (Figure 4.30) studied in industrial conditions [104]. The hood dimensions, inlet and outlet velocities are specified in Table 4.7. We will determine the maximum diameter of dust particles carried over into the suction system.

Note that the coincidence of design and experimental maximum diameter values of dust particles is determined by the choice of form factor χ for these particles. Therefore, it may be concluded that the form factor of particles in transfers of bulk materials studied in [104] was within 2 ÷ 6.4.

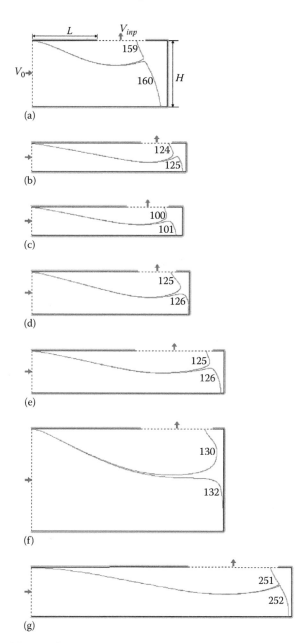

FIGURE 4.30 Paths of dust particles in the suction units of various geometric dimensions (Table 4.7) at the initial particle velocity of V_o; air density $\rho = 1.205$ kg/m³, dynamical air viscosity $\mu = 0.0000178$ Pa·s, impact slide factor of 0.5, coefficient of sliding friction of 0.5, integration step of 0.005 s: (a) experiment 1, (b) experiment 2, (c) experiment 3, (d) experiment 4, (e) experiment 5, (f) experiment 6, and (g) experiment 7.

4.3.2 CALCULATING THE MAXIMUM DIAMETER OF DUST PARTICLES FROM SINGLE-WALL HOODS

The hood under analysis (Figure 4.31) is used for the containment of dust emissions from a conveyor feed unit [159] at ore preparation plants.

With an increase in the initial vertical escape velocity of a dust particle (downward) having the density of 3500 kg/m³ the maximum diameter is decreased for 10–25 µm (Figure 4.32).

We will assume, hereinafter, that a dust particle escape velocity is equal to the inlet airflow velocity (2.25 m/s).

TABLE 4.7
Design d_{max} versus Experimental d_{max}

Test No.	Particle Density, ρ_1, kg/m³	V_{inp}, m/s	V_o, m/s	Hood Length, L, m	Hood Height, H, m	Particle Diameter, Test, d_{max}, μm	Particle Diameter, Design, d_{max}, μm	Form Factor, χ
1	2600	1.44	1.11	0.3	0.3	160	159	3.25
2	2600	1.07	1.2	0.8	0.2	125	124	6.4
3	2600	0.84	1.25	0.7	0.2	100	100	3.9
4	2600	1.56	1.41	0.78	0.3	125	125	3.5
5	2600	1.18	2.2	0.76	0.3	125	129	2
6	3500	1.47	1.03	0.8	0.73	130	130	2.5
7	1400	1.97	3.73	1.15	0.35	250	251	2.15

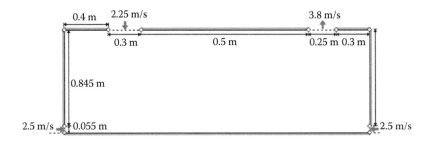

FIGURE 4.31   Single-wall hood.

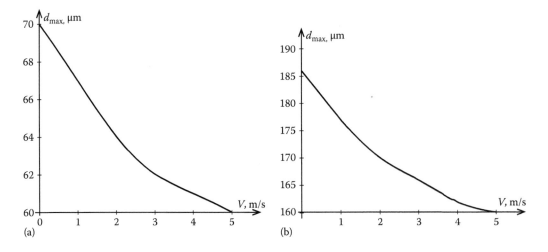

FIGURE 4.32   Maximum diameter versus initial escape velocity of dust particle: (a) at $\chi = 1$ and (b) at $\chi = 6$.

With an increase in the dust particle dynamic form factor and a decrease in the dust particle density, the maximum diameter of such particle significantly increases (Figure 4.33).

4.3.3   CALCULATING THE MAXIMUM DIAMETER OF DUST PARTICLES FROM DOUBLE-WALL HOODS

Let us analyze the relation of d_{max} from the screen location (Figure 4.34) and length. Geometrical and aerodynamic characteristics of the hood are the same as in Figure 4.31.

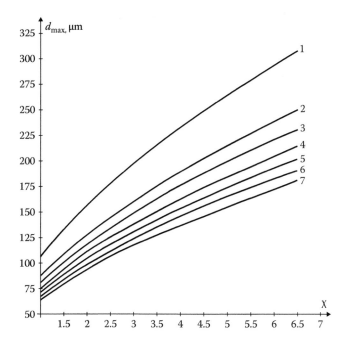

FIGURE 4.33 Maximum diameter versus form factor and density of dust particle: 1: 1400 kg/m³, 2: 2000 kg/m³, 3: 2300 kg/m³, 4: 2600 kg/m³, 5: 2900 kg/m³, 6: 3200 kg/m³, 7: 3500 kg/m³.

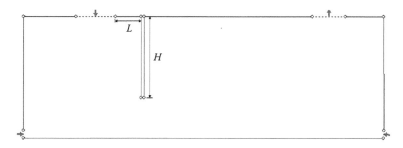

FIGURE 4.34 Double-wall hood.

If the screen is displaced from the inlet to the outlet, d_{max} will be virtually constant. A very slight decrease in this value can be observed if the screen is at a distance of less than 0.5 m (Figure 4.35).

The analysis of calculations shown in Figures 4.36 and 4.37 demonstrates that the greatest decrease in the dust particle maximum diameter for 10–15 µm is observed when the screen is positioned at a distance $L = 0.25$ m from the inlet. The numeric characters shown in the figures describe the density of dust particles shown in Figure 4.33.

If a dust particle escape velocity is changed, d_{max} remains the same. For instance, if there is a spherical dust particle having the density of 3500 kg/m³ and the velocity variation is between 0 and 5 m/s, screen length $H = 0.55$ m and the distance between the screen and the inlet $L = 0.25$ m, then $d_{max} = 54$ µm.

The screen length has a great impact on d_{max} (Figures 4.38 and 4.39). If H is increased from 0.3 to 0.5 m, d_{max} will be decreased for 8–21 µm based on the dynamic form factor.

In all calculations hereinafter the initial point of dust particle escape will be in the rightmost inlet position, which is due to the fact that this point is the closest one to the outlet and the maximum diameter value will be the greatest in such cases.

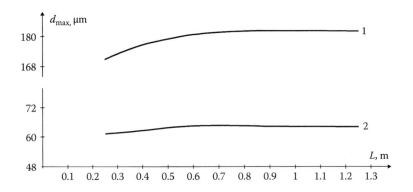

FIGURE 4.35   Relation of d_{max} to the screen distance from the inlet if its length $H = 0.4$ m and dust particle density of 3500 kg/m³: 1: $\chi = 6.5$; 2: $\chi = 1$.

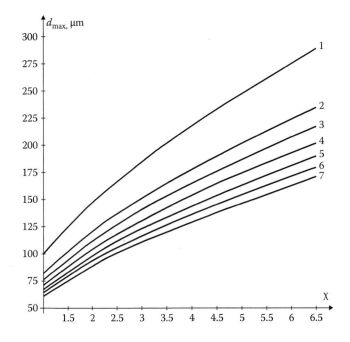

FIGURE 4.36   Relation of d_{max} to dust particle form factor and density if $H = 0.4$ m and $L = 0.25$ m.

4.3.4   CALCULATING THE MAXIMUM DIAMETER OF DUST PARTICLES FROM DOUBLE-WALL HOODS EQUIPPED WITH A ROTATING CYLINDER

We will analyze dust dynamics in a double-wall suction unit with a rotating cylinder installed inside. All other geometric and kinematic parameters are the same as for the hood shown in Figure 4.31. The screen distance L and length H were chosen based on the conditions most favored for reduction in the maximum diameter of a dust particle carried over into the suction system and based on the technological conditions (Figure 4.40).

As it was shown by the numerical experiment, the best cylinder position is below the screen. With a decrease in the cylinder radius, d_{max} is increased. The radius chosen for the study is 0.1 m, which is stipulated by the technological conditions: there should be a distance between the cylinder and the belt conveying a bulk material sufficient to pass the material. The airflow induced by the cylinder rotation should be expected to contribute to settlement of dust particles on the hood bottom.

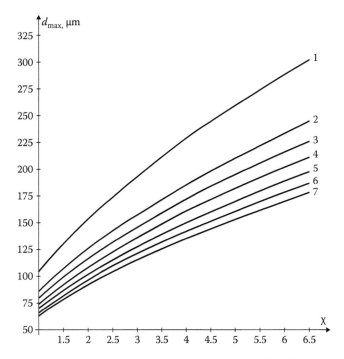

FIGURE 4.37 Relation of d_{max} to dust particle form factor and density if $H = 0.4$ m and $L = 0.45$ m.

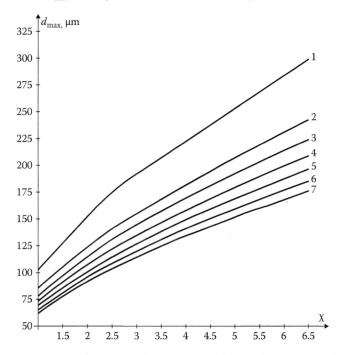

FIGURE 4.38 Relation of d_{max} to dust particle form factor and density if $H = 0.3$ m and $L = 0.25$ m.

During the clockwise rotation of the cylinder the maximum diameter is slightly greater than in the case with counterclockwise rotation; however, there is a number of interesting phenomena observed. The airflow induced by the cylinder affects the flight of fine particles most. Therefore, fine particles entrained at the right of the cylinder are settling within some range (Figure 4.41). With further decrease in the preset dust particle diameter, particles fly around the cylinder, escaping into

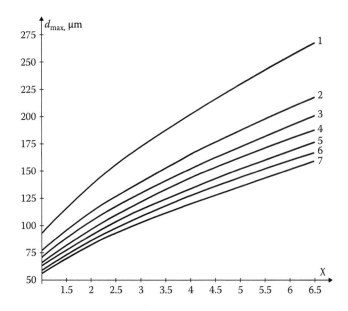

FIGURE 4.39 Relation of d_{max} to dust particle form factor and density if $H = 0.5$ m and $L = 0.25$ m.

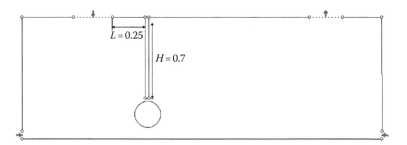

FIGURE 4.40 The model of a double-wall hood equipped with a rotating cylinder.

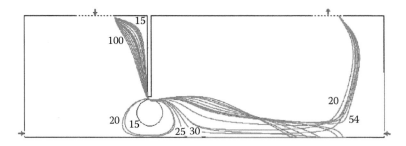

FIGURE 4.41 Paths of dust particles having the density of 3500 kg/m³ and $\chi = 1.83$ when the cylinder is rotating clockwise at the linear velocity of 8 m/s.

the slot between the cylinder and the screen to be further carried over into the suction port. It should be also noted that all dust particles are escaping into the space beyond the screen only through the slot between the cylinder and the screen where the velocity is rather high (e.g., about 35 m/s for Figure 4.41).

The pattern of paths for plate-like dust articles is somewhat different (Figure 4.42). There is no intermediate fraction that would settle on the hood bottom. The maximum diameter here (Figure 4.43) is lower than in the case with counterclockwise rotation of the cylinder.

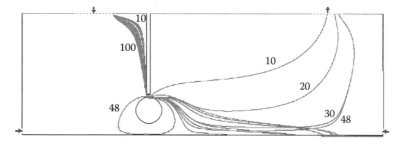

FIGURE 4.42 Paths of dust particles having the density of 3500 kg/m³ and $\chi = 6.5$ when the cylinder is rotating clockwise at the linear velocity of 8 m/s.

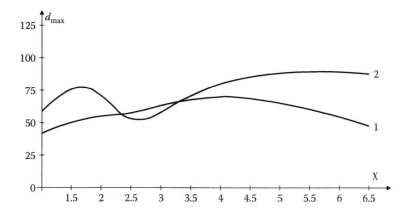

FIGURE 4.43 Relation of d_{max} (μm) to form factor χ when the cylinder is rotating clockwise at the linear velocity of 8 m/s with the dust particle density of 1: 3500 kg/m³; 2: 1400 kg/m³.

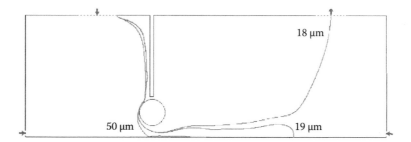

FIGURE 4.44 Paths of ball-shaped dust particles having the density of 3500 kg/m³ when the cylinder is rotating counterclockwise at the linear velocity of 5 m/s.

However, in our view, no particle-size distribution can be forecasted by d_{max} when the cylinder is rotating clockwise. There will be no logarithmic normal distribution of dust particles here since some intermediate fractions are settling on the hood bottom.

During counterclockwise rotation of the cylinder (Figure 4.44), the air flow induced by the cylinder contributes to the precipitation of particles. At low rotation velocities (up to 2 m/s), the point where particles collide with the cylinder is near the screen and a bounced dust particle takes the direction that contributes to the particle being carried over by the airflow between the cylinder

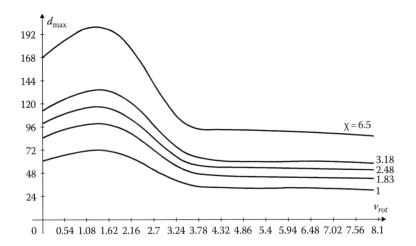

FIGURE 4.45 Relation of d_{max} to the cylinder counterclockwise rotation velocity with the dust particle density of 3500 kg/m³.

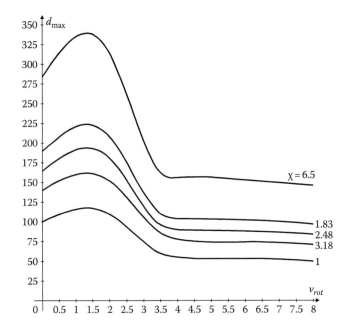

FIGURE 4.46 Relation of d_{max} to the cylinder counterclockwise rotation velocity with the dust particle density of 1400 kg/m³.

and the screen toward the exhaust outlet. This explains an increase in d_{max} as respects the case with a resting cylinder (Figures 4.45 and 4.46). With an increase in the cylinder rotation velocity, the bouncing point is displaced counterclockwise and at definite rotation velocities a dust particle settles on the hood bottom of the cylinder or is flying between the cylinder and the bottom. Note that the flow between the cylinder and the screen also changes its direction. In this case, d_{max} will be decreased. Starting from the cylinder rotation velocity of over 4 m/s, such decrease is insignificant (Figures 4.45 and 4.46).

Note that when the cylinder is lowered the maximum diameter is decreased (Figure 4.47). The best position of the cylinder is (1.97; −1.05). Here, the relation of d_{max} to χ is actually linear.

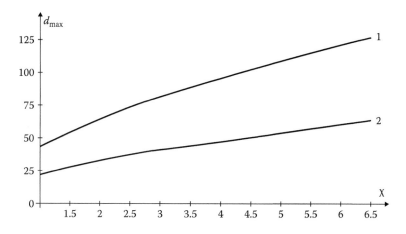

FIGURE 4.47 Relation of d_{max} (µm) to form factor χ during the counterclockwise rotation of the cylinder centered on (1.97; −1.05) at the velocity of 8 m/s and with dust particle density of 1: 1400 kg/m³; 2: 3500 kg/m³.

4.3.5 CALCULATING THE MAXIMUM DIAMETER OF DUST PARTICLES FROM DOUBLE-WALL HOODS EQUIPPED WITH TWO ROTATING CYLINDERS

When the second cylinder rotating counterclockwise is installed in the corner between the screen and the inflow wall (Figure 4.48), the maximum diameter decrease effect is even clearer. It should be noted that the cylinder in the corner is also the model of vortex (Figure 4.48a) that can be observed in reality. Dust particles are being displaced leftward at the initial flight segment already following which the second cylinder installed under the screen has its effect. Relation of d_{max} to χ in such cases is shown in Figure 4.49.

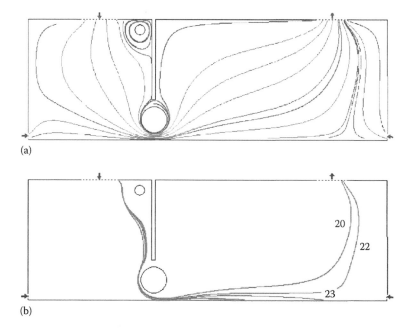

FIGURE 4.48 The hood equipped with two cylinders rotating counterclockwise at the velocity of 8 m/s: (a) flow lines and (b) paths of ball-shaped dust particles having the density of 3500 kg/m³.

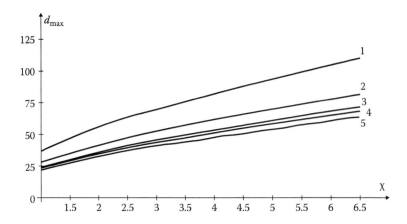

FIGURE 4.49 Relation of d_{max} (µm) to form factor χ during the counterclockwise rotation of the cylinder at the velocity of 8 m/s and with dust particle density of 1: 1400 kg/m³; 2: 2300 kg/m³; 3: 2900 kg/m³; 4: 3200 kg/m³; 5: 3500 kg/m³.

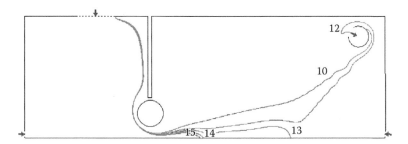

FIGURE 4.50 Flight of dust particles when the cylinder (1.97; −1.02) is rotating counterclockwise at the velocity of 8 m/s and the suction cylinder (3.55; −0.45) is rotating counterclockwise at the velocity of 4 m/s.

4.3.6 CALCULATING THE MAXIMUM DIAMETER OF DUST PARTICLES FROM DOUBLE-WALL HOODS EQUIPPED WITH A ROTATING CYLINDER AND A SUCTION CYLINDER

When a rotating suction cylinder is installed in the upper right corner (Figure 4.50) of the suction unit the maximum diameter is significantly decreased. For instance, there is a 10 µm decrease observed for spherical particles having the density of 3200 kg/m³ and 32 µm decrease observed for particles having the form factor of 6.5 and the same density.

A sinuous nature of the paths of dust particles caught by the suction cylinder should be noted, which is due to the pulsating field of velocities in the suction unit: the velocity at the given point is changing regularly at the time interval equal to the suction cylinder turnaround time.

4.4 MODELING THE BEHAVIOR OF DUST PARTICLES IN PULSATING FLOWS

Studying the behavior of dust particles in the aerodynamic field of a rotating suction cylinder installed in a suction unit is both of scientific and of practical interest (Figure 4.51). Such problem set up per se is novel in the industrial ventilation area and, thus, is poorly studied.

The study objective was to determine such a position of the suction cylinder and its rotation velocity at which dust losses in the suction system would be minimum, which is directly connected with a reduction in the maximum diameter (d_{max}) of a dust particle caught by the suction unit. Besides, it is necessary to determine to what extent d_{max} is decreased if decreased at all as compared to the traditional configurations of suction units.

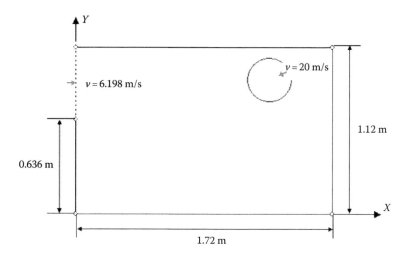

FIGURE 4.51 The scheme of a suction unit equipped with a suction cylinder.

The solution for the scientific and technical problem set is based on the elaboration of a mathematical model of dust and air mechanics in multiply connected regions with rotating suction cylinders, algorithms for its numerical implementation, and a software program that could become an efficient research tool for a wide class of industrial ventilation problems.

4.4.1 DERIVING BASIC DESIGN RELATIONS

Let a multiply connected flow region be limited by boundary S on which the normal velocity is given as a function of coordinates and time $v_n(x_0, t)$ where $x_0 \in S$. There may be impermeable cylinders (denoted by circles) within the region rotating at linear velocities v_i. We will assume that there are sources (outflows) of $q(\xi, t)$ intensity unknown in advance distributed continuously along the boundary. Let linear vortexes be located in the cylinder centers $a_i(a_{i1}, a_{i2})$ with circulations

$$\Gamma_i = 2\pi r_i \cdot v_i, \tag{4.30}$$

where r_i is i cylinder radius. The impact of all these sources (outflows) and vortexes on inner point x of the flow area will be given by the integral equation

$$v_n(x, t) = \int_S F(x, \xi) q(\xi, t) dS(\xi) + \sum_{i=1}^{m} \Gamma(a_i) G(x, a_i),$$

where
$v_n(x, t)$ is the velocity at point $x(x_1, x_2)$ along $\vec{n} = \{n_1, n_2\}$ at the instant of time t
m is the number of rotating cylinders
$$F(x, \xi) = \frac{n_1(x_1 - \xi_1) + n_2(x_2 - \xi_2)}{2\pi\left[(x_1 - \xi_1)^2 + (x_2 - \xi_2)^2\right]}$$

$$G(x, a_i) = \frac{n_2(x_1 - a_{i1}) - n_1(x_2 - a_{i2})}{2\pi\left[(x_1 - a_{i1})^2 + (x_2 - a_{i2})^2\right]}$$

$dS(\xi)$ means that the integration variable is ξ

Letting inner point to boundary point x_0 along the outward normal, we have the boundary integral equation

$$v_n(x_0, t) = -\frac{1}{2}q(x_0, t) + \int_S F(x_0, \xi)q(\xi, t)dS(\xi) + \sum_{i=1}^{m}\Gamma(a_i)G(x_0, a_i),$$

(4.31)

where the first term results from the calculation of the integral singularity at $x_0 = \xi$; therefore, the integral itself does not contain this point.

By discretizing the region boundary into N boundary intervals at each of which intensity $q(\xi, t)$ will be assumed to be constant, we will obtain a discrete countertype of Equation 4.31:

$$v_n^p = -\frac{1}{2}q^p + \sum_{\substack{k=1, \\ k \neq p}}^{N} q^k F^{pk} + \sum_{i=1}^{m}\Gamma_i G_i^p,$$

(4.32)

where
$v_n^p = v_n(x_0^p, t)$

x_0^p is the middle of p interval

$q^p = q(x_0^p; t)$

$q^k = q(\xi^k, t)$

ξ^k is a random point of k interval

$F^{pk} = \int_{\Delta S^k} F(x_0^p, \xi^k)dS(\xi^k)$ is an integral over k interval

$\Gamma_i = \Gamma(a_i)$

$G_i^p = G(x_0^p, a_i)$

Searching through p values from 1 to N, we have a set of N linear algebraic equations with N unknown variables solving which we will find intensities of sources (outflows) $q^1, q^2, \ldots, q^N$ at the given instant of time t. Accordingly, the required velocity at inner point x will be determined from the formula

$$v_n(x) = \sum_{k=1}^{N} q^k F^k + \sum_{i=1}^{m}\Gamma_i G_i,$$

(4.33)

where
$F^k = \int_{\Delta S^k} F(x, \xi^k)dS(\xi^k)$
$G_i = G(x, a_i)$

In order to plot the flow line it is required to set an initial point and calculate horizontal (v_x) and vertical (v_y) air velocity components at this point, thus having defined flow direction $\vec{v}$; take a step in this direction and repeat the calculation procedure set forth earlier. The reverse calculation procedure is also possible when the flow line is plotted from the suction port, that is, a step is taken in the direction opposite to vector $\vec{v}$. The calculation stops as soon as the air exhaust line is reached or as the flow line length exceeds the given value.

A dust particle path is plotted based on the following motion equation integration using the Runge–Kutta method:

$$\rho_1 \frac{\pi d_e^3}{6} \cdot \frac{d\vec{v_1}}{dt} = -\psi \cdot \frac{|\vec{v_1} - \vec{v}|(\vec{v_1} - \vec{v})}{2} \rho \chi S_m + \rho_1 \frac{\pi d_e^3}{6} \vec{g}, \qquad (4.34)$$

where

ρ_1, ρ are dust particle and medium densities, respectively
$\vec{v_1}$ is the particle velocity vector
$\vec{v}$ is the air velocity calculated from formula (Equation 4.33)
d_e is an equivalent diameter
$S_m = \pi d_e^2/4$ is the transparent frontal area
χ is the particle dynamic form factor
$\vec{g}$ is the gravity factor
ψ is the drag factor

$$\psi = \begin{cases} 24/\text{Re} & \text{at Re} < 1 \text{ (Stokes formula)}, \\ 24\left(1 + 1/6 \cdot \text{Re}^{2/3}\right)/\text{Re} & \text{at } 1 \le \text{Re} < 10^3 \text{ (Klyachko formula)}, \\ 24/\text{Re} \cdot \left(1 + 0.065\,\text{Re}^{2/3}\right)^{1.5} & \text{at Re} \ge 10^3 \text{ (Adamov formula)}. \end{cases} \qquad (4.35)$$

When a particle is hitting a solid wall, tangential ($v_2\tau$) and normal (v_{2n}) velocities are calculated from the formula [91]

$$v_{2n} = -k \cdot v_{0n}, \quad v_{2\tau} = v_{0\tau} + \eta \cdot f \cdot (1+k) \cdot v_{0n}, \qquad (4.36)$$

where

$$\eta = \min\left\{-\frac{2v_{0\tau}}{7f(1+k)v_{0n}}, 1\right\}$$

k is the coefficient of restitution
f is the coefficient of sliding friction

When dust particle paths are calculated in a region with time-variant boundary conditions in order to determine the air velocity, the intensity of sources (outflows) distributed along the flow boundary must be recalculated at each instant of time by solving the set given in Equation 4.32.

Spectrum software program was developed based on the algorithms described to determine the field of velocities, plotting flow lines and paths of dust particles in multiply connected regions with complex boundaries where the normal velocity component may vary in time and where the given number of rotating cylinders can be present.

4.4.2 CALCULATION DATA AND DISCUSSION

There are a number of difficulties in the problem with a rotating suction cylinder related to determining whether a particle will get into the suction or not.

The first difficulty is that particles escaping from the same point and having the same diameter may either get into the suction or not. This depends on the suction cylinder rotation angle at the moment of the particle escape. Figure 4.52 shows an example of such a situation. According to Figure 4.52a, the particle does not get into the suction because at the instant of time when it was at the cylinder, the suction port was turned in another direction. Flying around the cylinder the particle is moving away from it, eventually settling on the suction unit bottom.

Therefore, we should apparently speak of a percent of particles caught in the suction unit. It is clear that all the permissible initial positions of the suction cylinder cannot be checked. There should be the maximum diameter of particles caught by the suction cylinder determined within the framework of the set problem rather than an exact percentage of hits. Herewith, if at any diameter of particles the percentage of particles getting in the suction cylinder is low for analysis simplicity's sake we can assume that no particles of this diameter are getting into the suction cylinder. For instance, there were four variants of the suction cylinder initial position analyzed in the course of the study (Figure 4.53). The difference between these positions is 90°. If no particle is getting into the suction cylinder with either of these positions, we can conclude that a particle of this diameter does not reach the suction cylinder at all.

Another difficulty consists of the fact that particles on having approached to the suction cylinder can be moving close to the cylinder along a closed path (Figure 4.54a). It is impossible to unambiguously conclude whether a pulsating particle gets into the suction cylinder or not. The particle may collide with another one and having changed its path get into the suction cylinder. However, the most probable assumption is that a pulsating particle will not reach the suction cylinder because it will coagulate with another one and settle on the suction unit bottom or following the collision will change its path, again, to settle on the bottom. Therefore, if the situation occurs where a particle is moving along a closed path, it is assumed that the particle does not get into the suction cylinder.

The suction unit configuration shown in Figure 4.51 is used in experiments.

The initial data common for all experiments: the particle dynamic form factor is 1; the particle density is 3500 kg/m³; the medium density is 1.205 kg/m³; dynamic air viscosity is 0.0000178 Pa · s;

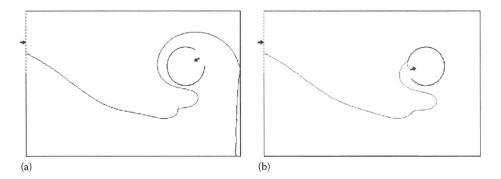

(a)   (b)

FIGURE 4.52 Various paths of similar particles escaping from the same point at different instants of time: (a) particles are deposited and (b) particles are trapped.

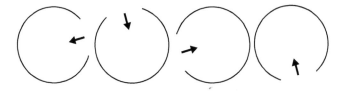

FIGURE 4.53 Suction cylinder initial positions.

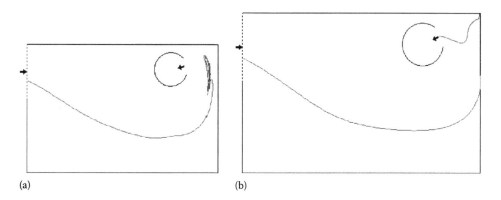

(a)   (b)

FIGURE 4.54 Paths of dust particles for the cylinder is rotating: (a) clockwise and (b) counterclockwise.

particle escape velocity is equal to zero; the cylinder outlet air velocity is 20 m/s; the inlet air velocity is 6.198 m/s.

Some experimental data (shown in Figure 4.55) are quite obvious. With an increase in the difference between the particles, escape point and the suction cylinder center point ordinates (Figure 4.55a and b) as well as in the case of its displacement toward the inlet (Figure 4.55c), d_{max} of a caught particle increases.

Relation of d_{max} to the suction cylinder linear rotation velocity is not that obvious (Figure 4.56).

The airflow in the problem under analysis is induced by the exhaust and the cylinder rotation. At the beginning of a particle motion, the suction has the dominant impact on its path. As seen from Figure 4.54a and b, at low suction cylinder rotation velocities the particle path at the initial segment of its motion remains virtually constant versus the rotation direction. d_{max} is continuously decreasing with the increase in the counterclockwise rotation velocity (Figure 4.56b). Besides the fact that the circulatory flow from the cylinder makes particle miss the exhaust outlet near it, it is also explained by the fact that at the initial segment of a particle motion at significant rotation velocities the circulatory flow holds the particle down to the suction unit bottom together with the gravity force. If the suction cylinder is rotating clockwise, there is an oscillating particle motion observed near the suction cylinder (Figures 4.54a and 4.57b). This is due to the circulatory flow dragging the particle at the right side of the suction cylinder down together with the gravity force. The suction cylinder, on the contrary, induces a flow that is dragging a particle upward. When the suction cylinder is turned to the particle, its effect is greater and the particle is being lifted up. When the suction cylinder is turned in another direction, its effect is lesser so the particle is going down under the gravity force and the downward airflow from the cylinder rotation. As seen from Figure 4.56a, d_{max} at first decreases with the increase in velocity, then starts going up being the minimal at the rotation velocity close to 1. Indeed, if at low rotation velocities the circulatory flow around the cylinder has no significant effect on the behavior of dust particles at the initial motion segment, its increase results in a higher ascension of particles above the suction unit bottom and, accordingly, makes particles approach the exhaust and be caught (Figure 4.57). Therefore, d_{max} increases in this case.

With increase in the radius of the suction cylinder rotating clockwise, d_{max} is changing at first insignificantly (Figure 4.56c), then it is going down due to an increase in circulation Γ (Equation 4.30) and, accordingly, growing effect from the circulatory flow. In such an event there is also an oscillating motion of particles observed near the suction cylinder. Note that with the cylinder rotating counterclockwise no oscillating motion of particles could be detected with these geometric parameters.

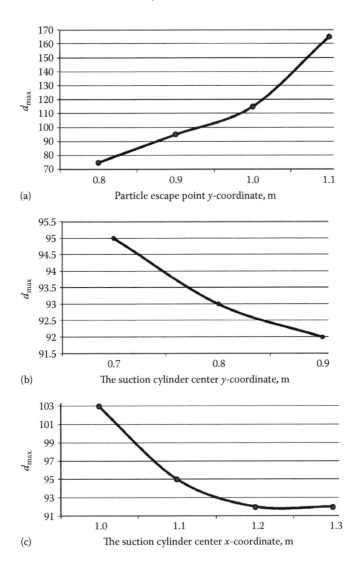

FIGURE 4.55 Relation of d_{max} to (a) the cylinder centered on point (1.3; 0.9) is rotating clockwise at the velocity of 1.5 m/s, (b) a particle is escaping from point (0.01; 0.8 and the suction cylinder (the center x-coordinate is 1.3) is rotating clockwise at the velocity of 3 m/s, (c) a particle is escaping from point (0.01; 0.8). The suction cylinder (the center x-coordinate is 0.8) is rotating clockwise at the velocity of 3 m/s.

Since the main practical task is d_{max} reduction the values of d_{max} were compared for various traditional configurations of suction units (Figure 4.58) and hoods where the exhaust outlet is replaced with a rotating suction cylinder. For all configurations, the exhaust outlet width was the same as the cylinder suction width. The particle escape point was chosen based on the condition most favorable for ingress of dust in the suction unit. For example, on Figure 4.58a, this point is in the upper corner of the inlet. The diameter was decreased approximately for 30–40 or even for 70 μm (Table 4.8).

Thus, the numerical experiment showed that it is practicable to install rotating suction cylinders inside suction units and that the developed software program enables selecting the best geometric and kinematic parameters for suction units with settling chamber functions.

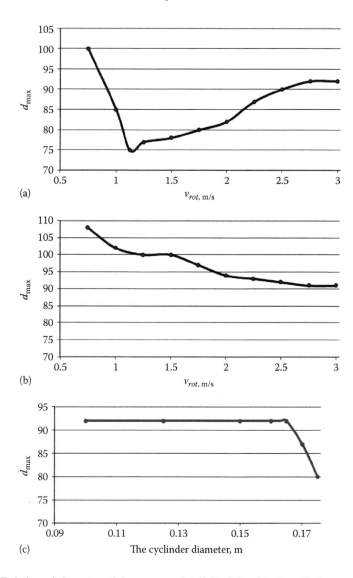

FIGURE 4.56 Relation of d_{max} at particle escape point (0.01; 0.8) with the cylinder centered on (1.3; 0.9) to (a) the cylinder clockwise rotation velocity, (b) the cylinder counterclockwise rotation velocity, and (c) the cylinder diameter.

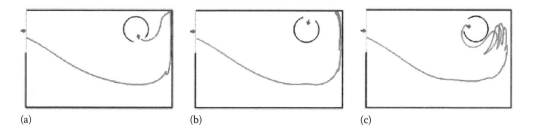

FIGURE 4.57 Flight paths of particles of 85 μm diameter at various clockwise rotation velocities of the suction cylinder: (a) 0.75 m/s, (b) 1.5 m/s, and (c) 2.5 m/s.

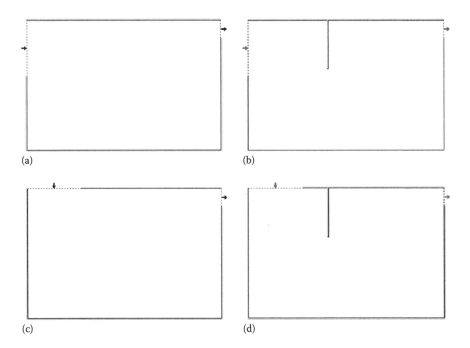

FIGURE 4.58   Traditional configurations of suction units: (a) scheme 1, (b) scheme 2, (c) scheme 3, and (d) scheme 4.

TABLE 4.8

Comparison of d_{max} Values for the Suction Units of Traditional Configurations and for Hoods Equipped with a Suction Cylinder

Traditional hood configuration figure number	Figure 4.58a	Figure 4.58b	Figure 4.58c	Figure 4.58d
Dust particle escape point in a hood of traditional configuration	(0.01; 1.1)	(0.01; 1.04)	(0.48; 1.11)	(0.4; 1.11)
Dust particle escape point in a hood with a suction cylinder	(0.01; 1.1)	(0.01; 1.1)	(0.48; 1.11)	(0.4; 1.11)
d_{max} for traditional hood configuration	205	171	238	129
Coordinates of the cylinder center with the radius of 0.15 m rotating clockwise at the velocity of 1.5 m/s	(1.3; 0.9)	(1.3; 0.9)	(1.3; 0.9)	(1.3; 0.9)
d_{max} for hoods with a suction cylinder	165	100	210	100

4.5 MODELING THE BEHAVIOR OF POLYFRACTIONAL AEROSOL IN A SUCTION UNIT

4.5.1 ALGORITHMS FOR COMPUTATION OF FLOWS IN REGIONS WITH MULTIPLE SUCTION CYLINDERS

This section objective is to develop a mathematical model, a computational algorithm for its numerical implementation and a software program for the calculation of nonstationary vortex gas flows in pulsating aerodynamic fields with time-variant boundary conditions, behavior of dust aerosols in such fields, and a method of forecasting the concentration and particle-size distribution of dust in suction systems.

Let us consider a multiply connected region of ideal incompressible liquid flow containing L rotating cylinders wherefrom gas can be exhausted. There is the normal time-variant velocity given along the region boundary. It is required to define the field of velocities within the region at any given instant of time.

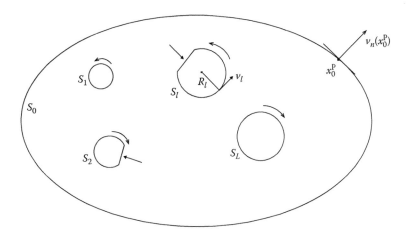

FIGURE 4.59 Flow region containing rotating cylinders wherefrom gas can be exhausted.

Let us consider the initial instant of time $t = 0$. We will discretize boundary S_0 (Figure 4.59) by a set of reference points with vortexes situated centrally in between. Cylinder boundaries S_1, S_2, ..., S_L will be partitioned with a set of straight line intervals on which there are sources (outflows) continuously distributed. The intensity of sources (outflows) on each of the intervals is assumed to be constant. The cylinders with radii R_l are rotating at velocities v_l where $l = 1, 2, ..., L$. In order to consider an impact on a random inner point x by gas flow induced by rotation of these cylinders, there will be vortexes with circulations $\Gamma_0^l = 2\pi R_l v_l$ placed there centrally.

By integrating the effects from all sources (outflows) and summarizing the effect of the bound vortexes (located on boundary S_0) and the vortexes located on the cylinders centers we will obtain the following expression for the flow velocity $v_n(x)$ at a random inner point of range $x(x_1, x_2)$ along the given unit vector $\vec{n} = \{n_1, n_2\}$:

$$v_n(x) = \sum_{i=1}^{N} F(x, \xi^i)q(\xi^i) + \sum_{k=1}^{M} G(x, \xi^k)\Gamma(\xi^k) + \sum_{l=1}^{L} G(x, \xi^l)\Gamma(\xi^l), \qquad (4.37)$$

where

$$F(x, \xi^i) = \frac{1}{2\pi} \int_{\Delta S_i} \frac{(x_1 - \xi_1)n_1 + (x_2 - \xi_2)n_2}{(x_1 - \xi_1)^2 + (x_2 - \xi_2)^2} \, dS(\xi) \qquad (4.38)$$

$\xi^i(\xi_1, \xi_2)$ is a random point of ΔS_i interval on which there are sources (outflows) continuously distributed

$G(x, \xi)$ is given by Equation 2.1

ξ is the k bound vortex position point (ξ^k) or cylinder center point (ξ^l)

Letting x to boundary point x_0^p along the outward normal we have a discrete countertype of the boundary integral equation for determining unknown intensities of sources (outflows) $q(\xi^i)$ and circulations $\Gamma(\xi^k)$:

$$v_n(x_0^p) = \sum_{i=1}^{N} F(x_0^p, \xi^i)q(\xi^i) + \sum_{k=1}^{M} G(x_0^p, \xi^k)\Gamma(\xi^k) + \sum_{l=1}^{L} G(x_0^p, \xi^l)\Gamma_0(\xi^l),$$

where x_0^p is the middle of p interval of cylinder boundaries $S_1, S_2, \ldots, S_n$ or p reference point of S_0 boundary. If $x_0^p = \xi^i$, $F(x_0^p, \xi^i) = -1/2$. If $x_0^p = \xi^k$, matrix element $G(x_0^p, \xi^k) = 0$ (the vortex has no impact on itself).

Denoting $F(x_0^p, \xi^i) = F^{pi}$, $v_n(x_0^p) = v^p$, $q(\xi^k) = q^k$, $G(x_0^p, \xi^k) = G^{pk}$, $\Gamma(\xi^k) = \Gamma^k$, introducing the constant circulation condition; introducing Lifanov's regulating variable Λ [113], and changing p between 1 and $N + M$, we have $N + M + 1$ linear algebraic equations with $N + M + 1$ unknown variables $q^1, q^2, \ldots q^N, \Gamma^1, \Gamma^2, \ldots \Gamma^M, \Lambda$:

$$\begin{cases} \sum_{i=1}^{N} F^{pi} q^i + \sum_{k=1}^{M} G^{pk} \Gamma^k + \Lambda = v^p - \sum_{l=1}^{L} G^{pl} \Gamma_0^l; \\ \sum_{k=1}^{M} \Gamma^k = 0. \end{cases} \qquad (4.39)$$

Having resolved Equation 4.39, we will determine the distribution of sources (outflows) intensities q^i and bound vortex circulations Γ^k at the initial instant of time $t = 0$.

At the next instant of time, there are free vortexes separating from the sharp edges and smooth surface. At the instant of time $t = \tau\Delta t$, the set of equations for determining circulations of bound vortexes and sources (outflows) will look as follows:

$$\begin{cases} \sum_{i=1}^{N} F^{pi} q^i + \sum_{k=1}^{M} G^{pk} \Gamma^k + \Lambda = v^p - \sum_{l=1}^{L} G^{pl} \Gamma_0^l - \sum_{b=1}^{B} \sum_{a=1}^{\tau} G^{pab} \Gamma^{ab}; \\ \sum_{k=1}^{M} \Gamma^k + \sum_{a=1}^{\tau} \sum_{b=1}^{B} \Gamma^{ab} = 0, \end{cases} \qquad (4.40)$$

where

Γ^{ub} is the circulation of the free vortex separated from b point of vortex wake separation at the instant of time a

G^{pab} is given by Equation 2.1 where there is x_0^p point coordinate at the current instant of time used as (x_1, x_2), (ξ_1, ξ_2) is the point where there is a free vortex having circulation Γ^{ub}

B is the number of vortex wake separation points

At the next instant of time, free vortexes are separating from all separation points B. There is a new position determined for all free vortexes in the flow using the formula given in Equation 2.6. The velocity components as well as the velocity at any given point along $\vec{n} = \{n_1, n_2\}$ are given by

$$v_n(x) = \sum_{i=1}^{N} F^i q^i + \sum_{k=1}^{M} G^k \Gamma^k + \sum_{b=1}^{B} \sum_{a=1}^{\tau r} G^{ab} \Gamma^{ab} + \sum_{l=1}^{L} G^l \Gamma_0^l, \qquad (4.41)$$

where (x_1, x_2) is substituted with these point coordinates in F and G formulas.

Note that since the suction cylinders are rotating, the boundary conditions are changing in time. Therefore, even if there are no free vortexes separating the intensities of sources (outflows) and the circulations of bound vortexes will change at each design step. There is a pulsating aerodynamic field observed.

4.5.2 Forecasting Particle-Size Distribution and Concentration of Coarsely Dispersed Aerosols in Local Exhaust Air

Let us assume that we know the concentration of dust in gas (C [kg/m³]) coming from *a* wide and 1 m deep air inlet, the particle-size distribution given by the lower and upper fractional boundaries (Table 4.9), and the inlet gas velocity v_n.

Determine the particle-size distribution and concentration of dust in the exhaust gas flow.

P dust particles diameter distribution function is drafted based on the particle-size distribution. This function has a step form (Figure 4.60). Each rectangle area is equal to the corresponding fraction ratio.

We will divide the suction inlet into k equal parts and at each instant of time generate k random numbers (diameters) distributed by the law given by P function. Thus, there will be k dust particles getting into the hood at each separate instant of time. It is required to find such time increment Δt that the concentration of dust particles coming into the hood was equal to the given concentration C.

The mass of dust coming in the region in question during Δt

$$C_{\Delta t} = km_{av} = Cv_s a\Delta t,$$

where the average mass of dust particle according to P distribution function

$$m_{av} = \sum_{i=1}^{n} \int_{d_{il}}^{d_{iu}} \frac{\pi x^3 l_i}{6(d_{iu} - d_{il})} \rho dx = \frac{\pi \rho}{24} \sum_{i=1}^{n} l_i \left(d_{iu} + d_{il}\right)\left(d_{iu}^2 + d_{il}^2\right). \tag{4.42}$$

TABLE 4.9

Particle-Size Distribution Given by the Lower and Upper Fraction Boundaries

Fraction boundaries	$d_{1l} - d_{1u}$	$d_{2l} - d_{2u}$	...	$d_{il} - d_{iu}$	...	$d_{nl} - d_{nu}$
Fractions	l_1	l_2	...	l_i	...	l_n

Note: $\sum_{i=1}^{n} l_i = 1.$

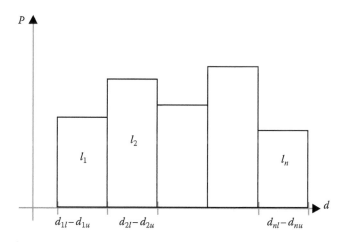

FIGURE 4.60 Particles diameter distribution function.

From that the time increment is

$$\Delta t = \frac{\pi k \rho}{24 C v_n a} \sum_{i=1}^{n} l_i \left(d_{il} + d_{iu} \right)\left(d_{il}^2 + d_{iu}^2 \right). \tag{4.43}$$

In order to calculate the concentration of dust in the exhaust outlet we will take n number for the instant of time at which a set of k particles is getting into the region. Then, the motion of $n \cdot k$ particles will be modeled until they all settle down or are caught by the suction unit. Total weight of particles (m_o) will be caught by the suction unit calculated during the modeling process. The outlet concentration here is $C_o = m_o / V$, where $V = v_s \cdot a \cdot \Delta t \cdot n$.

There are parameters of dust particles caught by the suction unit stored and the percentage co position of dust fractions in the exhaust air determined in the course of modeling.

The actual resulting concentration of dust in the exhaust air differs from the one given due to the model discreteness. In order to determine the actual inlet concentration C_r we will calculate the total weight of $n \cdot k$ particles (m_r) entering the suction unit from the suction inlet and, accordingly, $C_r = m_r / V$. With an increase in the quantity of $n \cdot k$ particles, concentration C_r is approximating to C with any given accuracy.

Dust particle motion is modeled based on the motion equation integration using the Runge–Kutta method by using the formulas in Equations 4.34 through 4.36.

4.5.3 ANALYSIS OF VARIATION IN THE PARTICLE-SIZE DISTRIBUTION AND CONCENTRATION OF DUST PARTICLES IN A ROTATING SUCTION CYLINDER LOCATED IN A SUCTION UNIT

Let us assume that the suction unit contains a rotating suction cylinder (Figure 4.61).

The input design data: suction velocity $v_0 = 3$ m/s; inlet velocity $v_s = 1$ m/s; suction unit height $h = 0.7$ m; suction unit width $H = 1$ m; suction unit depth is 1 m; the distance between the inlet and the left wall $e = 0.05$ m; cylinder radius $R = 0.1$ m; inlet width $a = 0.3$ m; suction outlet width $b = 0.1$ m; dust particles density $\rho_1 = 3500$ kg/m³; dynamic air viscosity air coefficient $\mu = 0.0000178$ Pa·s; dust particles dynamic form factor $\chi = 1$; coefficient of dust particle restitution $k = 0.5$; coefficient of sliding friction dust particle $f = 0.5$. The number of dust particles getting into the hood at each of 200 instants of time is 20. Particle fractions are shown in Table 4.10. The given concentration of dust particles in the exhaust air: $30 \cdot 10^{-6}$ kg/m³.

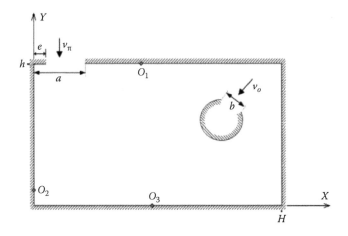

FIGURE 4.61 Flow area with a rotating suction cylinder.

TABLE 4.10

Particle-Size Distribution of Inlet Air Dust

Fraction boundaries, µm	10–30	30–50	50–70	70–90	90–110
Fractions	0.2	0.2	0.2	0.2	0.2

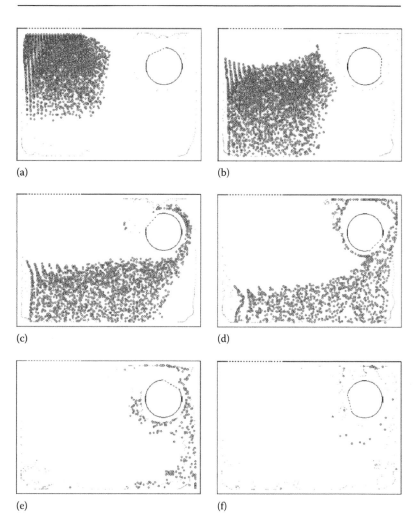

(a) (b)

(c) (d)

(e) (f)

FIGURE 4.62 Dust cloud behavior in time with the suction cylinder centered on (0.8; 0.5) (dots are free vortexes, circles are dust particles) and rotating at the velocity of 2 m/s: (a) first time, (b) second time, (c) third time, (d) fourth time, (e) fifth time, and (f) sixth time.

There were three vortex wake separation points O_1, O_2, and O_3 (Figure 4.61) considered in the course of modeling, distance between bound vortexes $h = 0.05$, and the aerodynamic design time increment $\Delta t = h$, the suction cylinder rotation is counterclockwise.

Initially, dust particles are moving down in the total flow (Figure 4.62) and then start moving toward the suction cylinder, coarser particles are settling down, and finer particles are going around the cylinder being gradually caught by it. Note that at the rotation velocities of 4–8 m/s, dust particles are rotating for a long time with the same conditions (neither being caught nor settling). In such cases, the calculation stops and it was assumed that these dust particles are coagulating and are finally settling on the hood bottom.

TABLE 4.11

Particle-Size Distribution and Concentration of Exhaust Dust Aerosol Based on the Suction Cylinder Rotation and Suction Velocities

Parameters			Velocity, m/s			Fractions, μm					Outlet Concentration, mg/m³
$C(x, y)$	b	R	V_{rot}	v_o	v_n	10–30	30–50	50–70	70–90	90–110	
A hood of standard configuration (w/o suction cylinder)											
—	0.1	—	—	1.5	0.5	0.78	0.22	0	0	0	0.1981
—	0.1	—	—	3	1	0.591	0.367	0.042	0	0	0.6765
—	0.1	—	—	6	2	0.438	0.384	0.177	0.001	0	1.8290
—	0.1	—	—	12	4	0.390	0.403	0.203	0.005	0	2.5316
Cylinder rotation velocity increase											
(0.8; 0.5)	0.1	0.1	0	3	1	0.567	0.398	0.037	0	0	0.6849
(0.8; 0.5)	0.1	0.1	1	3	1	0.580	0.388	0.032	0	0	0.5795
(0.8; 0.5)	0.1	0.1	2	3	1	0.586	0.379	0.035	0	0	0.5045
(0.8; 0.5)	0.1	0.1	4	3	1	0.742	0.255	0.003	0	0	0.1770
(0.8; 0.5)	0.1	0.1	8	3	1	0.972	0.028	0	0	0	0.0163
(0.8; 0.5)	0.1	0.1	16	3	1	0.987	0	0.013	0	0	0.0057
(0.63; 0.5)	0.1	0.1	1	3	1	0.575	0.361	0.064	0	0	0.7161
(0.63; 0.5)	0.1	0.1	2	3	1	0.542	0.424	0.034	0	0	0.4485
(0.63; 0.5)	0.1	0.1	4	3	1	0.994	0.006	0	0	0	0.0600
(0.63; 0.5)	0.1	0.1	8	3	1	1	0	0	0	0	0.0008

Note: The actual inlet concentration in all cases is 29.821 mg/m³, the suction cylinder rotation is counterclockwise; the suction outlet velocity is v_0; the inlet velocity is v_n; the rotation velocity is v_{rot}; the cylinder radius is R; the suction outlet width is b.

When the suction cylinder is not rotating, dust cloud inflow is from various directions, which is the effect of the field of velocities. The concentration of exhaust aerosol is even somewhat higher than in a standard hood (Table 4.11).

When the cylinder is rotating, the flow lines are spiral shaped, which prevents dust particles from getting into the suction unit. With an increase in the suction cylinder rotation velocity in any direction, the concentration of exhaust aerosol is decreased (Table 4.11). A favorable case for reduction in dust losses is the counterclockwise rotation since the direction of the airflow induced by the cylinder rotation is the same as the direction of dust particles gravity force at the initial segment of their motion (at the left of the suction cylinder). With a decrease in the suction cylinder height, the concentration of dust aerosol in the exhaust air is decreased. The smallest dust losses in the suction system can be observed when the suction cylinder is centered on point C (0.45; 0.35) (Table 4.12). A decrease in the concentration and the particle-size distribution shift toward finer particles can also be observed with reduction in cylinder radius R and suction outlet width b.

4.5.4 DUST AEROSOL BEHAVIOR IN A SUCTION UNIT OF STANDARD CONFIGURATION

Let us consider a suction unit in which the air mechanics of airflows was studied experimentally in [147] (Figure 4.63).

TABLE 4.12

Particle-Size Distribution and Concentration of Exhaust Dust Aerosol Based on Geometric Properties of the Hood

Parameters			Velocity, m/s			Fractions, µm					Outlet Concentration, mg/m³
$C(x, y)$	b	R	V_{rot}	v_o	v_n	10–30	30–50	50–70	70–90	90–110	
Cylinder position variation at the constant rotation velocity											
(0.45; 0.5)	0.1	0.1	4	3	1	0.875	0.125	0	0	0	0.1182
(0.45; 0.43)	0.1	0.1	4	3	1	0.974	0.026	0	0	0	0.0563
(0.45; 0.35)	0.1	0.1	4	3	1	0.967	0.033	0	0	0	0.0558
(0.45; 0.28)	0.1	0.1	4	3	1	0.935	0.065	0	0	0	0.0628
(0.45; 0.2)	0.1	0.1	4	3	1	0.913	0.087	0	0	0	0.0690
Suction inlet width variation											
(0.45; 0.35)	0.05	0.1	4	3	0.5	1	0	0	0	0	0.0033
(0.45; 0.35)	0.075	0.1	4	3	0.75	1	0	0	0	0	0.0170
(0.45; 0.35)	0.15	0.1	4	3	1.5	1	0	0	0	0	0.24
Cylinder radius variation											
(0.45; 0.35)	0.075	0.05	8	3	0.75	0	0	0	0	0	0
(0.45; 0.35)	0.075	0.075	5.34	3	0.75	1	0	0	0	0	0.0035
(0.45; 0.35)	0.075	0.1	4	3	0.75	1	0	0	0	0	0.0160
(0.45; 0.35)	0.075	0.125	3.2	3	0.75	0.997	0.03	0	0	0	0.0433
(0.45; 0.35)	0.075	0.15	2.67	3	0.75	0.837	0.163	0	0	0	0.1374

FIGURE 4.63 Laboratory bench for the analysis of air mechanics of hoods: 1, hood model; 2, chute; 3, local exhaust; 4, fan.

According to the experiments, the pattern of airflows distribution inside the hood is determined by the interaction between a supply air jet and a nozzle suction range. The field of velocities plotted using an electric thermal anemometer and the air motion pattern are shown in Figure 4.64a. The geometric and kinematic parameters of the hood are shown in Figure 4.65a.

We will reduce the problem to 2D setup as follows. Let us store the areas of inlet and outlet ports stretching them across the hood width (Figure 4.65b).

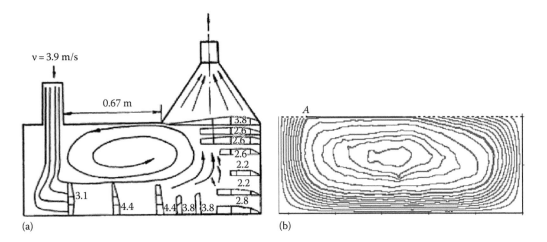

FIGURE 4.64 The motion pattern and fields of airflow velocities inside the hoods: (a) experimental pattern and (b) design flow lines.

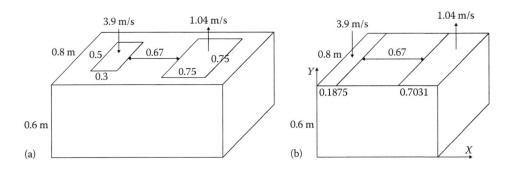

FIGURE 4.65 Suction unit diagram: (a) test bench model and (b) reduction to 2D problem.

Liquid was assumed to be ideal and incompressible. The condition of impermeability was given for the solid wall: normal velocity is equal to zero. Velocities along the outward normal in suction ports and inlets were known and, in addition, the intake air volume is equal to the exhaust air volume. The air mechanics inside the hood was modeled using the discrete vortex method or its combination with the method of boundary integral equations. The discretization interval, that is, the distance between the adjacent bound vortexes or reference points, is approximately the same and equal to $h = 0.02$ m. There are bound vortexes on the region boundary fractures. At the initial instant of time, air is still throughout the space. At the following instant of time all inlets and outlets are *enabled*.

The difference of the velocities along the hood bottom (at the distance of 0.05 m from it) and at the right wall (at the distance of 0.01 m from it) calculated using the software program from the experimental velocities (Figure 4.66) is 27% in average (4%–45%). In view of the fact that the aerodynamic experiment has the accuracy of about 25%, it is possible to have a satisfactory agreement between the design and the experimental data. Note that the calculation was conducted 5 times at each point, and then averaged, which is associated with the model nonstationarity. The fixed point velocity is pulsating in time. For example, the velocity at point (0.1; 0.5) varied between 2 and 3.1 m/s (the coordinate system is shown in Figure 4.65b). The flow vortex pattern plotted using the program at the fixed instant of time when free vortexes have filled the region in full (Figure 4.64b) is identical to the experimental flow pattern.

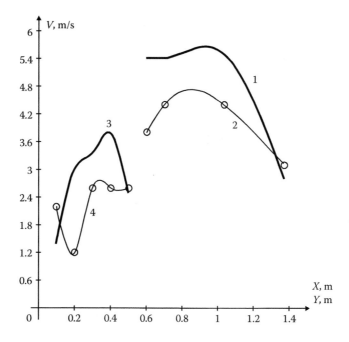

FIGURE 4.66 Comparison of design and experimental (circles) velocities: 1–2, across x along the hood bottom; 3–4, along y at the right hood wall.

There were four models for plotting the field of velocities inside a hood analyzed:

1. Free vortexes completely fill the flow region but keep moving and a definite instant of time a polyfractional assembly of dust particles starts coming out from the inlet.
2. Free vortexes completely fill the flow region and their positions are fixed at a definite instant of time, that is, we have a stationary flow (Figure 4.64b), then dust particles start coming into the flow region.
3. A dust cloud is entering the hood while a free vortex is separating, that is, the vortex pattern development and the dust particles distribution are concurrent.
4. There are no free vortexes in the flow region, that is, there is a potential flow observed in which the behavior of dust aerosol is analyzed.

We analyzed the motion of 20,000 dust particles ($l = 1000$, $s = 20$) having the density of 3, 500 kg/m^3 and the dynamic form factor 3.6 (sharp-grained particles). The initial dust particle-size distribution is given in Table 4.13. The physical and mechanical properties of dust aerosol correspond to the process of crushed iron ore transfer in a suction unit of the South mining-and-processing integrated works.

In model 1, the dust cloud is entering the hood moving at first translationally downward, then, due to the flow nonstationarity, some dust particles sliding along the vortex region boundary start being dragged by free vortexes into the central vortex region and is distributed throughout the flow area (Figure 4.67a) eventually getting into the suction inlet. Most part of the dust aerosol settles on the hood bottom.

In model 2, dust is moving translationally downward within the vortex-free flow area clearing the central vortex region: along the hood bottom and then upward at the right lateral wall (Figure 4.67b). Note that the *tail* following the main stream of particles is sliding along the central vortex boundary. This *tail* is *diffused* in the previous model.

In model 3, dust particles being twirled by separating vortexes are forming a spiral and then start *diffusing* throughout the flow area (Figure 4.68a).

TABLE 4.13
Particle-Size Distribution of Dust

10–16 μm	16–25 μm	25–40 μm	40–63 μm	63–250 μm
Initial composition				
0.236	0.456	0.229	0.067	0.012
Model 1 ($C_{in}/C_{out} = 5.67$)				
0.26	0.481	0.221	0.039	0
Model 2 ($C_{in}/C_{out} = 4.85$)				
0.25	0.475	0.227	0.048	0
Model 3 ($C_{in}/C_{out} = 3.58$)				
0.234	0.463	0.235	0.067	0.002
Model 4 ($C_{in}/C_{out} = 4.04$)				
0.244	0.467	0.229	0.059	0.001

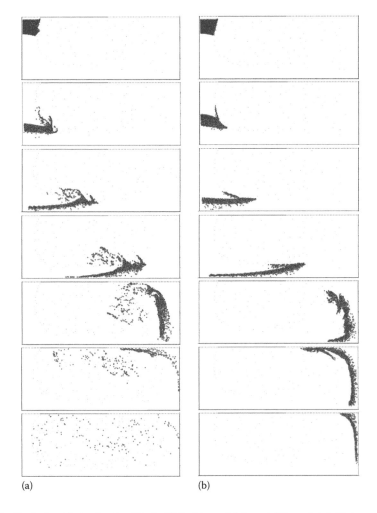

(a)   (b)

FIGURE 4.67 Dust cloud behavior ($s = 20$, $l = 100$): (a) model 1 and (b) model 2 (dots are free vortexes, circles are dust particles).

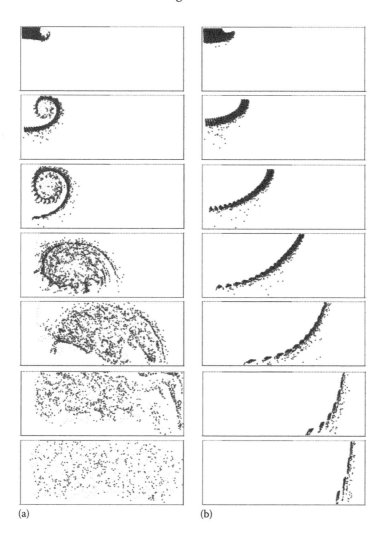

(a)   (b)

FIGURE 4.68 Dust cloud behavior ($s = 20$, $l = 100$): (a) model 3 and (b) model 4.

In model 4, aerosol dust particles are moving from the inlet forming an increasing-radius arc of circle to the suction inlet lining up vertically at last instants of their motion until being completely caught (Figure 4.68b).

It also should be noted that the coarse fractions of particle are settling on the hood bottom in all four models.

Following the modeling it was found that the particle-size distribution of dust particles is close to the initial one for all four models (Table 4.13) with implicitly confirms Minko's hypothesis that the particle-size distribution can be determined by the maximum diameter of dust particles getting into the suction inlet even if the model is significantly simplified based on the assumption that the flow is potential. However, a separating effect for coarse particles should be noted.

The concentration of dust aerosol is significantly decreased against the initial value. The greatest increase can be observed in model 1 that, in our view, is closest to the real air mechanics inside the hood.

Thus, various approaches to the description of air mechanics inside the hood have no significant effect on the design values of particle-size distribution in exhaust air, at least for iron ore transfer units. A suction unit of standard configuration cannot separate dust aerosol in which particle-size distribution in the exhaust air is close to the particle-size distribution of dust in a loading chute.

When the particle-size distribution of dust aerosols in a loading chute is given by the lognormal law, the particle-size distribution of dust in a suction duct will be subject to the same law (Table 4.13). A departure from this regularity can be observed for the initial and final fractions in the case of polydisperse materials ($\alpha = 30°–40°$). Here, α is an angle between the straight lines of dust particle distribution plotted in a logarithmically probable coordinate grid. This line slope ratio is equal to [158]

$$tg\alpha = \frac{1}{\lg\sigma},$$

where $\lg\sigma$ is the standard (root mean square) deviation of diameter logarithms from their mean values.

The more the uniform dust aerosol is ($\alpha > 45°$), the more this regularity manifests itself. Dust aerosol in a suction duct is more uniform than in a loading chute. However, with an increase in α this difference becomes less and the composition of inlet and exhaust aerosol is actually the same. With an increase in dust particle dynamic form factor χ, these compositions become even more similar.

In the case of variation in the hood geometric parameters and inlet and suction air velocities the particle-size distribution of dust aerosols changes insignificantly (Table 4.14), which allows for applying the findings also to the hoods of other standard sizes.

When a rotating cylinder of suction cylinder is placed inside a hood, the particle-size distribution of intake dust aerosol is not subject to a lognormal distribution. A suction cylinder commonly exhibits a significant displacement the particle size distribution of dust particles toward finer fractions. Table 4.15 contains only the modified parameters and all the other parameters remain as shown in Figure 4.63. When a suction cylinder is placed inside a hood AB suction port is *disabled*.

When a uniform particle-size distribution is given for dust particles (Table 4.16), the particle-size distribution of dust in the exhaust air is not subject to a lognormal distribution.

The findings can be used for a scientifically grounded selection of dust-collecting devices of suction systems in the transfer of bulk materials.

4.6 SOME GAS–DUST FLOW CALCULATION DATA OBTAINED USING RANS AND LES METHODS

There were two methods used in the solution of problems herein: RANS (Reynolds-averaged Navier–Stokes) and LES (large eddy simulation).

Modeling using RANS method utilized a standard $k–\varepsilon$ turbulence model, Reynolds averaged Navier–Stokes, and continuity equations:

$$\frac{\partial\rho}{\partial t} + \frac{\partial(\rho u_i)}{\partial x_i} = 0,$$

$$\frac{\partial(\rho u_i)}{\partial t} + \frac{\partial(\rho u_i u_j)}{\partial x_j} = -\frac{\partial p}{\partial x_i} + \frac{\partial}{\partial x_j}\left[\mu\left(\frac{\partial u_i}{\partial x_j} + \frac{\partial u_j}{\partial x_i} - \frac{2}{3}\delta_{ij}\frac{\partial u_l}{\partial x_l}\right)\right] + \frac{\partial}{\partial x_j}\left(-\rho\overline{u_i'u_j'}\right),$$

where all values of velocity u, pressure p, and density ρ are time averaged. Reynolds stresses $-\rho\overline{u_i'u_j'}$ are determined within the scope of a standard $k–\varepsilon$ turbulence model [159,160]:

$$\frac{\partial(\rho k)}{\partial t} + \frac{\partial(\rho k u_i)}{\partial x_i} = \frac{\partial}{\partial x_j}\left[\left(\mu + \frac{\mu_t}{\sigma_k}\right)\frac{\partial k}{\partial x_j}\right] - \rho\varepsilon + \tau_{ij}\frac{\partial(u_i)}{\partial x_j},$$

TABLE 4.14
Dust Particle-Size Distribution with Various Polydispersity of the Material

Dust particle-size distribution in a loading chute at $\alpha = 30°$, µm, and fractions

2–5	5–10	10–20	20–40	40–60	60–100	100–300	>300
0.0425	0.0575	0.1	0.16	0.14	0.13	0.24	0.13

Dust particle-size distribution in a suction duct at $\chi = 1$, $\alpha = 47.7°$, tg$\alpha = 1.1$

| 0.176 | 0.189 | 0.354 | 0.281 | 0 | 0 | 0 | 0 |

$\chi = 3.5$, $\alpha = 42°$, tg$\alpha = 0.9$

| 0.106 | 0.114 | 0.239 | 0.321 | 0.211 | 0.009 | 0 | 0 |

$\chi = 6.5$, $\alpha = 39.7°$, tg$\alpha = 0.83$

| 0.092 | 0.098 | 0.207 | 0.288 | 0.238 | 0.076 | 0 | 0 |

Dust particle-size distribution in a loading chute at $\alpha = 40°$

2–5	5–10	10–20	20–40	40–60	60–100	100–200	>200
0.07	0.15	0.23	0.25	0.12	0.1	0.06	0.02

Dust particle-size distribution in a suction duct at $\chi = 1$, $\alpha = 50.2°$, tg$\alpha = 0.84$

| 0.128 | 0.287 | 0.383 | 0.203 | 0 | 0 | 0 | 0 |

$\chi = 3.5$, $\alpha = 45.8°$, tg$\alpha = 1.029$

| 0.096 | 0.212 | 0.301 | 0.299 | 0.09 | 0.001 | 0 | 0 |

$\chi = 6.5$, $\alpha = 44.8°$, tg$\alpha = 0.992$

| 0.088 | 0.196 | 0.288 | 0.284 | 0.114 | 0.031 | 0 | 0 |

Dust particle-size distribution in a loading chute at $\alpha = 50°$

2–5	5–10	10–20	20–40	40–60	60–600
0.13	0.29	0.34	0.18	0.03	0.03

$\chi = 1$, $\alpha = 53.6°$, tg$\alpha = 1.357$

| 0.164 | 0.372 | 0.379 | 0.086 | 0 | 0 |

$\chi = 3.5$, $\alpha = 51.6°$, tg$\alpha = 1.26$

| 0.141 | 0.33 | 0.35 | 0.162 | 0.017 | 0 |

$\chi = 6.5$, $\alpha = 50.9°$, tg$\alpha = 1.23$

Dust particle-size distribution in a loading chute at $\alpha = 60°$

2–5	5–10	10–20	20–200
0.3	0.5	0.19	0.01

$\chi = 1$, $\alpha = 61.14°$, tg$\alpha = 1.814$

| 0.326 | 0.505 | 0.167 | 0.002 |

$\chi = 3.5$, $\alpha = 60.15°$, tg$\alpha = 1.743$

| 0.315 | 0.499 | 0.183 | 0.003 |

$\chi = 6.5$, $\alpha = 60.15°$, tg$\alpha = 1.743$

| 0.313 | 0.497 | 0.185 | 0.005 |

$$\frac{\partial(\rho\varepsilon)}{\partial t} + \frac{\partial(\rho\varepsilon u_i)}{\partial x_i} = \frac{\partial}{\partial x_j}\left[\left(\mu + \frac{\mu_t}{\sigma_\varepsilon}\right)\frac{\partial\varepsilon}{\partial x_j}\right] + C_{1\varepsilon}\frac{\varepsilon}{k}\tau_{ij}\frac{\partial u_i}{\partial x_j} - C_{2\varepsilon}\rho\frac{\varepsilon^2}{k},$$

$$\tau_{ij} = -\rho\overline{u_i'u_j'} = \rho\mu_t\left(\frac{\partial u_i}{\partial x_j} - \frac{\partial u_j}{\partial x_i}\right) - \frac{2}{3}\rho k\delta_{ij},$$

$$\mu_t = \rho C_\mu\frac{k^2}{\varepsilon}, \quad C_{1\varepsilon} = 1.44; \ C_{2\varepsilon} = 1.92; \ C_\mu = 0,09; \ \sigma_k = 1.0; \ \sigma_\varepsilon = 1.3.$$

TABLE 4.15

Dust Particle-Size Distribution with Variation in the Suction Unit Geometric and Kinematic Parameters

Dust particle-size distribution in a loading chute at $\alpha = 45°$

2–5	5–10	10–20	20–40	40–60	60–100	100–600
0.1	0.23	0.31	0.24	0.07	0.04	0.01

Dust particle-size distribution in a suction duct at $\chi = 1$, $\alpha = 52°$, $tg\alpha = 1.28$

0.137	0.333	0.373	0.157	0	0	0

$BC = 0.2$

0.133	0.327	0.364	0.176	0	0	0

$BC = 0.4$

0.134	0.328	0.366	0.172	0	0	0

$h = 0.4$

0.135	0.329	0.369	0.166	0	0	0

$h = 0.2$

0.132	0.327	0.394	0.147	0	0	0

$v_0 = 2.08$, $v_\pi = 7.8$

0.151	0.354	0.399	0.095	0	0	0

$v_0 = 0.52$, $v_\pi = 1.95$

0.142	0.331	0.377	0.15	0	0	0

In the case with a cylinder having radius of 0.1 m centered on (0.34; 0.27) and rotating counterclockwise at the velocity of 5 m/s

0.141	0.35	0.409	0.1	0	0	0

In the case with a suction cylinder having radius of 0.1 m and inlet width of 0.1 m centered on (0.3; 0.31) and rotating counterclockwise at the velocity of 5 m/s

0.514	0.243	0.135	0.108	0	0	0

$\chi = 3.5$, $\alpha = 46°$, $tg\alpha = 1.043$

0.111	0.272	0.325	0.238	0.053	0.001	0

$\chi = 6.5$, $\alpha = 46°$, $tg\alpha = 1.043$

0.107	0.262	0.321	0.233	0.065	0.012	0

TABLE 4.16

Dust Particle-Size Distribution with Uniform Distribution of Dust Particles over Fractions

Dust particle-size distribution in a loading chute at $\chi = 1$

2–12 μm	12–22 μm	22–32 μm	32–42 μm	42–52 μm	52–62 μm	62–72 μm	72–82 μm	82–92 μm	92–102 μm
0.1	0.1	0.1	0.1	0.1	0.1	0.1	0.1	0.1	0.1

Dust particle-size distribution in a suction duct

0.368	0.358	0.255	0.019	0	0	0	0	0	0

Equations forming the basis of LES method were obtained by filtering unsteady-state Navier–Stokes and continuity equations. The filtering process consists of disregarding of vortexes with size less than the size of differential grid cells. The filtered variable is given by

$$\overline{\phi}(x,y,z) = \frac{1}{V} \iiint_V \phi(x',y',z')\,dx'dy'dz',$$

where $(x', y', z') \in V$, V is a computation cell volume or, alternatively, the scale of turbulence tolerable by the filter.

The continuity and Navier–Stokes equations will take the following form in such cases:

$$\frac{\partial \rho}{\partial t} + \frac{\partial}{\partial x_i}\left(\rho \overline{u}_i\right) = 0,$$

$$\frac{\partial}{\partial t}\left(\rho \overline{u}_i\right) + \frac{\partial}{\partial x_j}\left(\rho \overline{u}_i \overline{u}_j\right) = -\frac{\partial \overline{p}}{\partial x_i} + \frac{\partial}{\partial x_j}\left[\mu\left(\frac{\partial \overline{u}_i}{\partial x_j}\right)\right] - \frac{\partial}{\partial x_j}\left[\rho\left(\overline{u_i u_j} - \overline{u}_i \overline{u}_j\right)\right].$$

The paths of dust particles were determined using the equations of their motion in the rectangular coordinate system

$$\frac{d\vec{u}_p}{dt} = F_D\left(\vec{u} - \vec{u}_p\right) + \frac{\vec{g}(\rho_p - \rho)}{\rho_p},$$

where

$$F_D = \frac{18\mu}{\rho_p d_p^2}\,\frac{C_D\,\mathrm{Re}}{24}$$

$$\mathrm{Re} = \frac{\rho d_p \left|\vec{u} - \vec{u}_p\right|}{\mu}$$

$$C_D = \frac{24}{\mathrm{Re}}\left(1 + b_1\,\mathrm{Re}^{b_2}\right) + \frac{b_3\,\mathrm{Re}}{b_4 + \mathrm{Re}}$$

$$b_1 = \exp(2.3288 - 6.4581\phi + 2.4486\phi^2); \quad b_2 = 0.0964 + 0.5565\phi;$$

$$b_3 = \exp(4.905 - 13.8944\phi + 18.4222\phi^2 - 10.2599\phi^3);$$

$$b_4 = \exp(1.4681 - 12.2584\phi + 20.7322\phi^2 - 15.8855\phi^3); \quad \phi = \frac{s}{S},$$

ϕ is the form factor
s is a surface of a sphere having the same volume as the particle
S is the actual particle surface area
$\vec{u}_p, \vec{u}$ is the particle and the medium velocity, respectively
μ is the medium molecular viscosity
ρ_p, ρ is the particle and the medium density, respectively
d_p is the particle diameter

4.6.1 MODELING THREE-DIMENSIONAL GAS–DUST FLOWS

The physical problem setup consists of the determination of the field of velocities, pressure, and paths of dust particles of various fractions for the hood shown in Figure 4.69. Two rectangular openings are given in the hood cover. The first opening intakes air coming from 0.5×0.3 m chute to the hood at the velocity of 3.9 m/s, while the second opening is the suction fitting inlet section through which air is exhausted at the average velocity of 1.04 m/s. Thus, the chuted air rate of 0.585 m³/s is equal to the rate of air exhausted by the suction fitting.

The calculations were made for the hood shown in Figure 4.69 with the following parameters: turbulent kinetic energy is 1 m²/s², turbulent dissipation rate is 1 m²/s², air density is 1.225 kg/m³, particle density is 3500 kg/m³, and air viscosity is 1.7894 kg/(m·s).

The computational grid was generated using Gambit software. The 3D grid plotted (Figure 4.70) contains 2,600,000 nodes. For the purpose of a correct description of the flow behavior in the boundary layers, the density of the grid nodes on the structure boundaries was increased several times.

Qualitatively, the pattern in both RANS and LES modeling methods is similar (Figure 4.71). There are two large-scale vortexes observed in the hood: 1—the central vortex between the inlet, outlet, and the upper wall; 2—at the left wall. In view of the fact that each flow line corresponds to a definite color, it is possible to conclude that the color scale for various flow areas is the same in both methods.

However, it may be noted that dimensions and location of vortex regions differ. For LES (at the instant of time 1.26 s), vortex formations are located at the inlet, while in the case with RANS, vortexes are more diffused in space. In addition, for a constant number of flow lines, the highest density of such lines for LES can be observed at the top of the structure, while in the case with RANS,

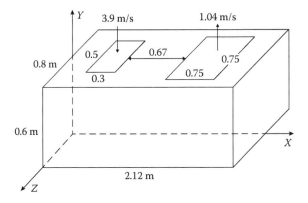

FIGURE 4.69 Regarding problem setup.

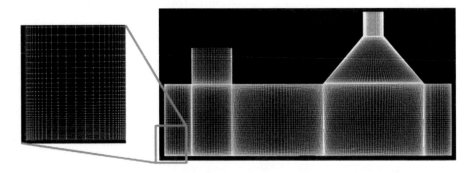

FIGURE 4.70 Three-dimensional grid scheme.

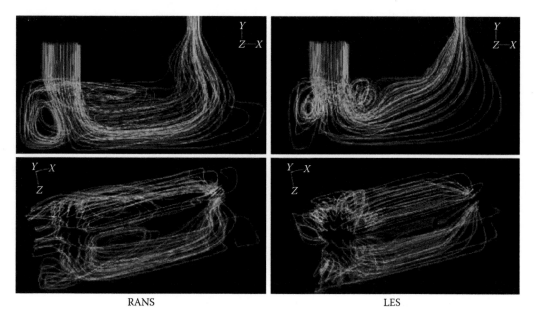

FIGURE 4.71   Suction unit flow lines plotted with various methods.

the flow lines are relatively uniformly distributed across the structure. It is evident that the density and pattern of the flow lines depend on the structure of vortex formations at the inlet. With respect to the spatial pattern of flow lines, it may be noted that in case with LES method, vortexes occur all over the inlet perimeter, while in the case with RANS, vortexes occur only along the main flow line direction toward the outlet.

The static pressure field in the structure (Figure 4.72) was analyzed against the field of velocities (Figure 4.73). Note that the flow is laminar in the most part of the suction unit structure volume.

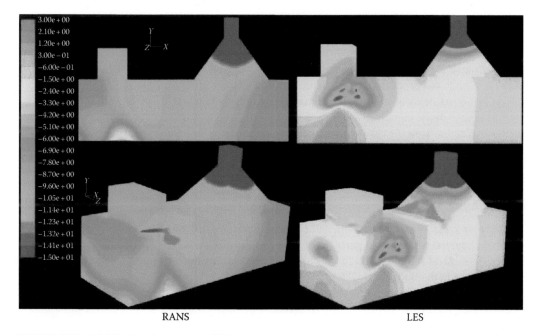

FIGURE 4.72   Fields of static pressures [Pa].

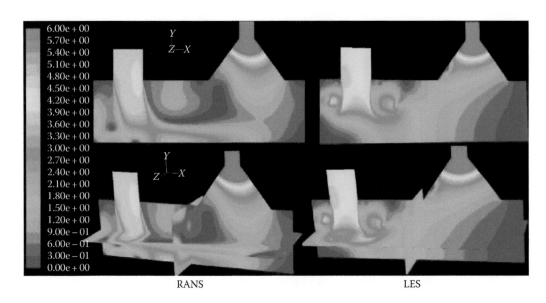

FIGURE 4.73 Fields of absolute velocities [m/s].

Hence, qualitatively static pressure and velocity may be related based on the Bernoulli law (the sum of static and dynamic pressures is a constant value).

The patterns of the field of velocities plotted using the LES method show that starting from the inlet the absolute velocity values (which means the static pressure values as well) are decreasing along the flow lines. As far as the outlet is being approached, the absolute velocity value is growing to reach the maximum at the outlet (the dynamic pressure at this point is maximum). In this connection, the static pressure variation will be inversely related, that is, the static pressure is growing in low flow velocity regions and vice versa. The static pressure is growing from the inlet to the outlet along the flow lines and starts going down to reach the minimum value at the outlet. It should also be separately noted that there is an excessive velocity and, hence, low static pressure observed in vortex regions. Similarly, the absolute velocity value at the structure corners is small that makes the static pressure increase.

Results obtained using the RANS method feature similar relation between the velocity and the static pressure. Regarding the static pressure value for both methods, it in average is higher for LES than for RANS except for the exhaust outlet.

The resulting fields of absolute velocities also feature a noticeable decrease in the velocities in vortex regions in the case with the RANS method and increased values in the case with the LES method. In addition, in the case with the LES method there is a vacuum zone observed immediately downstream of the inlet and absent in the case with the RANS method.

It should be noted that the RANS method has more benefits as compared with the experimental flow pattern (Figure 4.74) obtained by Kolesnik. The central vortex of significant dimensions is also observed in calculations made using the discrete vortex method. The suction nonuniformity confirmed by both methods and experimental data should also be mentioned. Here, the modeling results were visualized (Figure 4.74a and b) using Fieldview software.

There were the paths of dust particles of various sizes plotted in the field of velocities found using the RANS method (Figure 4.75). In the case with fine fractions of dust particles their paths are sufficiently close to the flow lines.

Starting from 40 μm particles, vortexes cease dragging dust particles in motion along closed paths (Figure 4.75f) and starting from the diameter of 60 μm they are settling on the hood bottom.

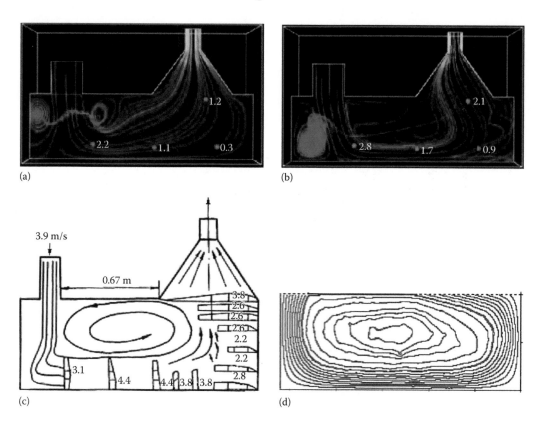

FIGURE 4.74 Suction unit flow patterns obtained: (a) with LES method, (b) with RANS method, (c) experimentally, and (d) with discrete vortex method.

The maximum diameter of an entrained dust particle in the calculations of the field of velocities made using the discrete vortex method was also 60 μm.

4.6.1.1 Parallelization of Computations

Parallel computations in Fluent software were carried out by running several processes executed on one or more computers in the network. Parallelization using Fluent includes an interaction of the software with the basic process and a number of subprocesses executed on computers in a cluster. Fluent interacts with the basic process and a number of subprocesses by means of Cortex utility that controls Fluent user interface and basic graphic functions. Fluent utilizes MPI (message passing interface) parallelization both on the cluster computers and on the multiprocessor common memory computers.

The computations were carried out on IBM cluster comprising 340 dual core Intel processors of which only 8 are available for Fluent applications. Figure 4.76 shows the flow lines that qualitatively correspond to the pattern implemented in the experiment (Figure 4.74c).

Airflow passing through a suction unit was modeled using 1, 2, and 4 processors. The results are given in Table 4.17; Figure 4.77 shows the relevant speedups.

It is obvious that the computation parallelization efficiency starts noticeably deteriorating already when there are four processors in use: the result is identical to the one obtained by McDonough and Endean [161].

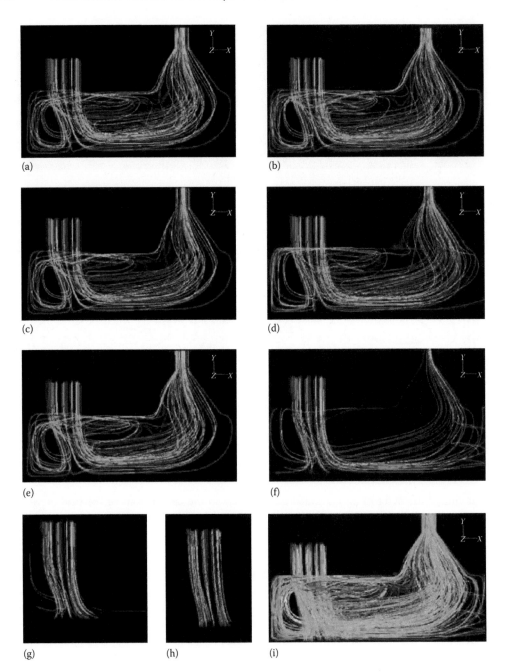

FIGURE 4.75 Paths of dust particles of various diameters: (a) 1 μm, (b) 3–5 μm, (c) 5–10 μm, (d) 10–20 μm, (e) 20–40 μm, (f) 40–60 μm, (g) 60–100 μm, (h) 100–300 μm, and (i) all sizes.

4.6.2 MODELING TWO-DIMENSIONAL FLOW IN A SUCTION UNIT WITH AIR INFLOW THROUGH SLOTS

It is of interest to analyze the flow vortex structure based on the hood dimensions and air inflow through the hood slots. The following parameters were recorded (Figure 4.78): $CD = RS = 0.1$ m; $DM = NR = EF = OP = 0.2$ m; $CB = ST = 0.63$ m; $\delta = 0.03$ m; $AB = TK = 0.03$ m; $WE = PL = 0.8$ m; $ED = FM = ON = PR = 0.14$ m; $WQ = LH = 0.8$ m; $v_1 = 2$ m/s; v_3 varied from 0 to 0.075 m/s in an

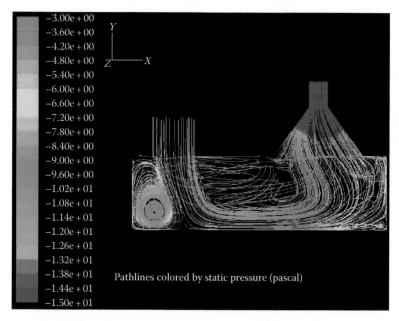

FIGURE 4.76 Airflow lines inside the hood.

TABLE 4.17
Modeling Time Based on the Number of Processors

Number of Processors	Modeling Time
1	45 h, 01 min
2	26 h, 16 min
4	17 h, 41 min

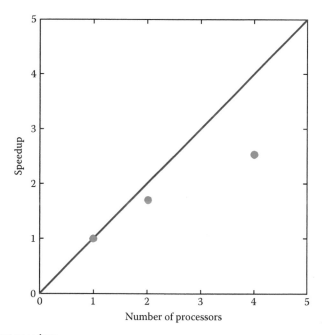

FIGURE 4.77 Fluent speedups.

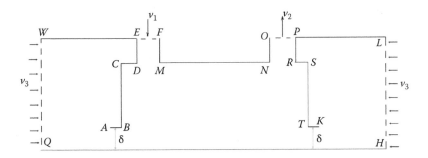

FIGURE 4.78   Design flow area.

increment of 0.015 m/s. *MN* varied from 0.2 to 1.2 m in an increment of 0.2 m. The screen length was 0; 0.02 m, 0.04 m, and 0.06 m.

There are two large-scale vortex patterns observed in a suction unit disregarding air inflows through slots (Figure 4.79): to the left of the inlet (first vortex) and to the right of the inlet (second vortex). Both vortexes occur in the result of the flow separation from the inlet sharp edges. The first vortex is sharply growing in size as soon as air starts leaking through the slots ($v_3 = 0.015$ m/s) and its center is being displaced downward. With an increase in air inflow velocity the separation area that occurs when a flow is separating from sharp edges of the slots is expanding. Thus, the third vortex occurs to the right of the left slot. As far as it is growing the first vortex is becoming smaller and being displaced upward. The fourth vortex that occurs to the left of the right slot is much bigger than the third one. Let us also note that if without slots the airflow between the inlet and the outlet is observed virtually all over the hood from the upper wall to the lower wall with increase in the volume of air passing through the slots this flow is moving away from the floor, and, for example, at $v_3 = 0.075$ m/s, is occupying approximately the half of the hood height. The lowest static pressure can be observed in the second vortex region that is becoming even smaller with the increase in the air inflow velocity. The static pressure drop can be observed in the rest region of the suction unit too, which is due to the increase in air velocity and is in line with the relation that follows from the Bernoulli equation: a sum of the static pressure and the dynamic head is a constant value along the flow lines or vortex lines.

The described flow pattern can also be observed in the case with a hood of other dimensions. The difference is only in the main air stream position between the inlet and the outlet. This stream is being retained against the lower hood wall. The highest velocity can be observed at the hood bottom right in between the inlet and the outlet. With an increase in the hood length, the second vortex region dimensions are growing significantly. If there are screens installed upstream of the slot at right angle to the slot (Figure 4.80) the flow vortex structure remains unchanged but the separation point changes its position. In this case it is positioned at the end of the screen. With the increase in the slot air inflow velocity, a vacuum zone occurs beneath the screen. Meanwhile, the velocity is increasing because the efficient suction area is growing smaller in size. A pressure drop upstream and downstream of the slot is also noticeable, which is indicative of a higher local drag factor. The greatest pressure drop can be observed when the screen length is about half the slot height, which is in line with the experimental observations of the assistant professor Yu.G. Ovsyannikov. Note that such pressure drop in the right slot is greater than in the left one. Thus, air inflows through the slot adjacent to the suction unit port are greater than through the slot adjacent to the air inlet. It follows thence that the slots of the first type must be sealed at first instance.

Note that in all figures where the static pressure fields are depicted, such fields correspond to the fields of velocities shown in the first half of each figure.

The flow pattern in slots is shown in Figures 4.81 through 4.83. As seen from the figures, with an increase in the slot height the velocity is decreased. The vortex region in the right slot is decreased and at $\delta = 0.12$ m the airflow in both slots has the similar pattern.

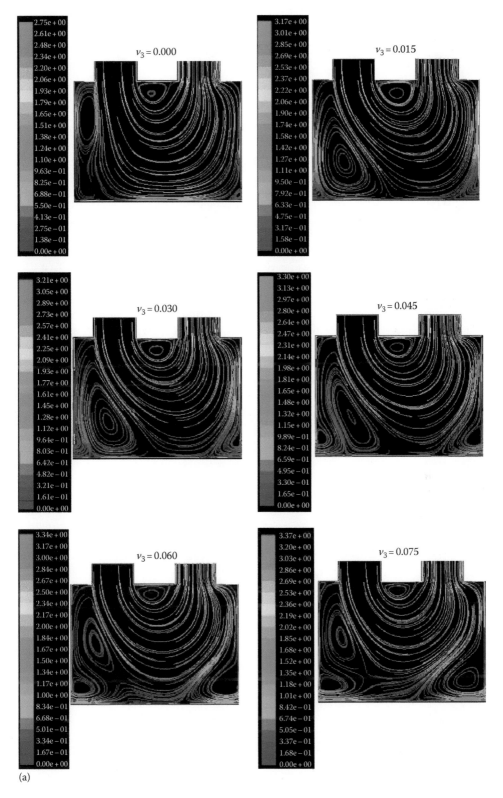

(a)

FIGURE 4.79 (a) Flow lines (the color scale corresponds to the velocity in m/s). (*Continued*)

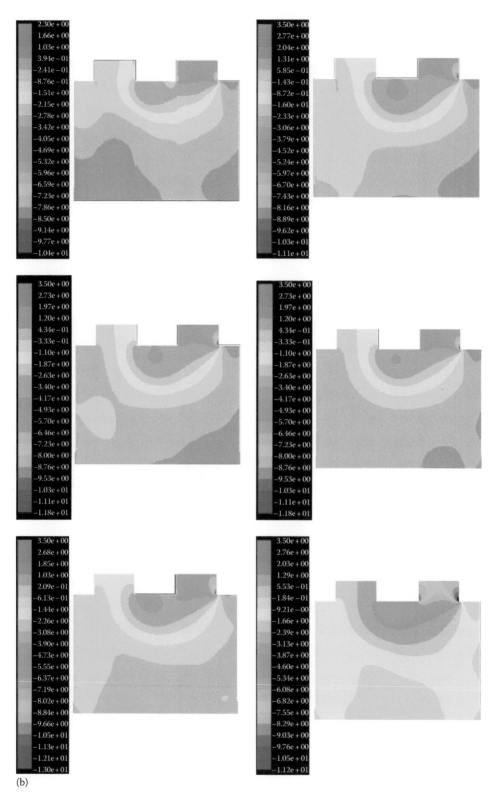

(b)

FIGURE 4.79 (*Continued*)   (b) The static pressure field (Pa) in a suction unit at $MN = 0.2$ m (beginning).

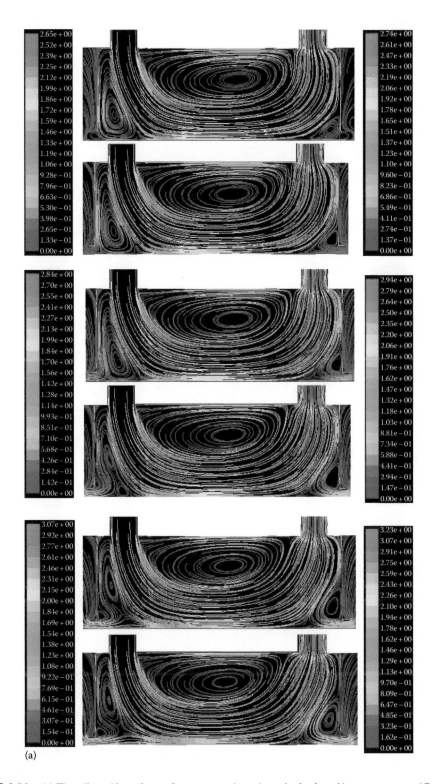

FIGURE 4.80 (a) Flow lines (the color scale corresponds to the velocity in m/s). *(Continued)*

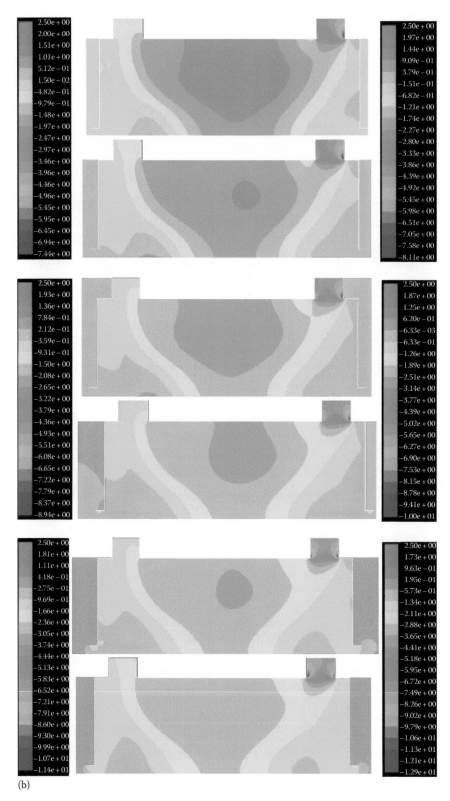

(b)

FIGURE 4.80 (*Continued*)   (b) The static pressure field (Pa) in a suction unit at $MN = 1.2$ m, $AB = TK = 0.06$ m (beginning).

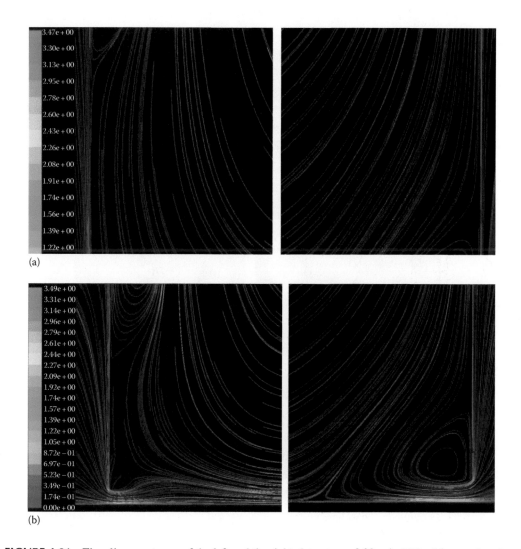

(a)

(b)

FIGURE 4.81 Flow lines upstream of the left and the right slots at $v_3 = 0.03$ m/s, $MN = 1.2$ m, and various heights: (a) $\delta = 0.02$ m, (b) $\delta = 0.03$ m.   (*Continued*)

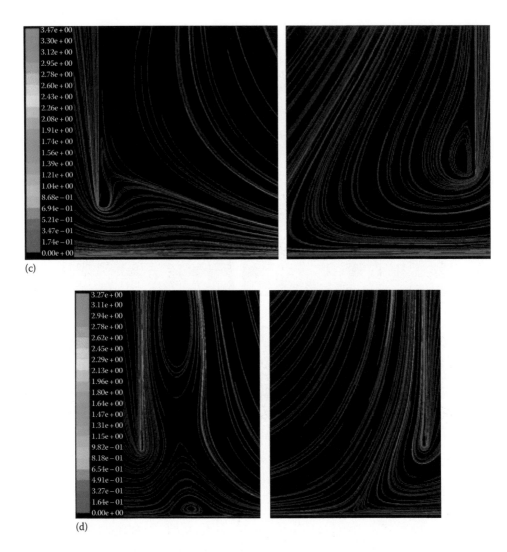

FIGURE 4.81 (*Continued*)   Flow lines upstream of the left and the right slots at $v_3 = 0.03$ m/s, $MN = 1.2$ m, and various heights: (c) $\delta = 0.06$ m, and (d) $\delta = 0.12$ m.

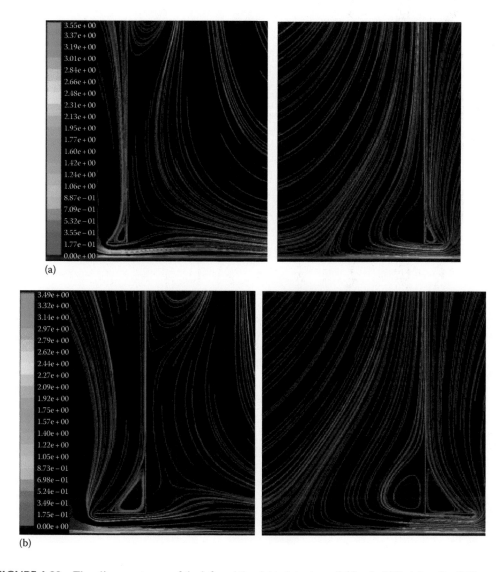

(a)

(b)

FIGURE 4.82 Flow lines upstream of the left and the right slots at $v_3 = 0.03$ m/s, $MN = 1.2$ m; $\delta = 0.03$ m and various lengths of the screen: (a) $AB = TK = 0.04$ m, (b) $AB = TK = 0.08$ m. *(Continued)*

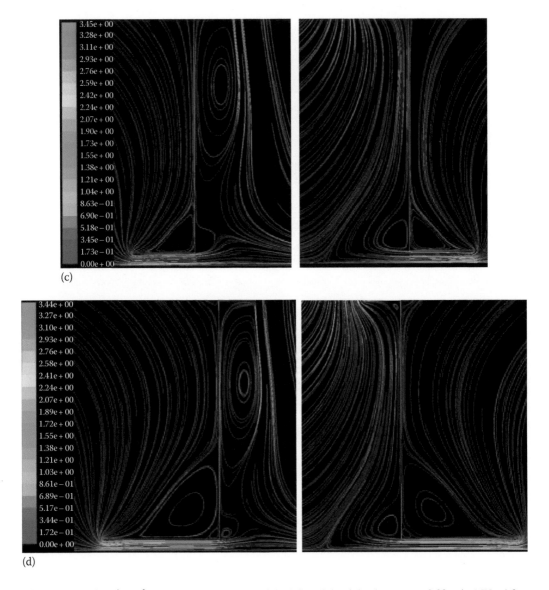

(c)

(d)

FIGURE 4.82 (*Continued*)   Flow lines upstream of the left and the right slots at $v_3 = 0.03$ m/s, $MN = 1.2$ m; $\delta = 0.03$ m and various lengths of the screen: (c) $AB = TK = 0.16$ m, and (d) $AB = TK = 0.32$ m.

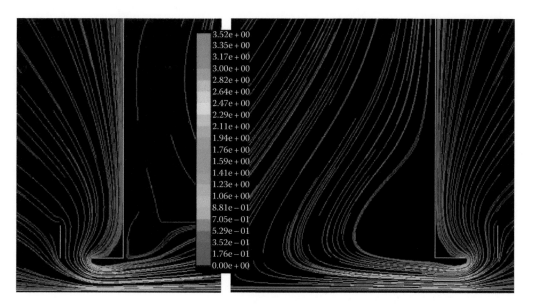

FIGURE 4.83 Flow lines upstream of the left and the right slots with an additional 0.03 m long vertical screen installed, $v_3 = 0.03$ m/s, $MN = 1.2$ m; $\delta = 0.03$ m and the distance between the screens is 0.03 m.

When a screen is installed (Figure 4.82), another vortex can be observed (above the screen), which is caused by the flow separation from a smooth surface of the hood vertical wall. With the increase in the screen length, the flow pattern slightly changes. Also note the absence of any vortex region beneath the screen while such region can be observed in calculations made using the discrete vortex method and in experiments set forth in Chapter 2.

Installation of a vertical screen (Figure 4.83) enables expanding this separation area, which is indicative of an increase in the slot inlet drag.

Section II

Recirculation of Ejected Air in Chutes

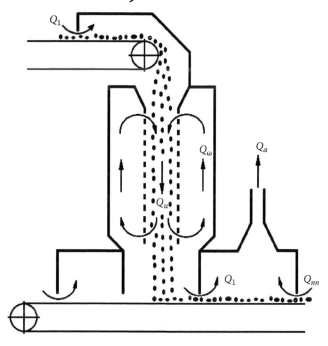

II.1   GENERAL NOTATION CONVENTIONS*

II.1.1   LATIN CHARACTERS

A

a, b, c are the coefficients for the linear equation Equation 5.44 as determined with the relation of Equation 5.45

A, B are the parameters of the linearized equation describing ejection in a perforated duct as determined with relations shown in Equation 5.167

a_1, a_2 are the roots of a characteristic equation as determined with the relation in Equation 5.170

---

* Some of the items listed may differ from the notation used in the text. In such cases, the latter should have priority.

a_3, b_3 are the auxiliary quantities as determined with the relations of Equation 6.29; same, as determined with the relations of Equation 2.68

a_τ is the acceleration of particle flow in an inclined chute (see Equation 7.10a) (m/s^2)

A_k is the extension of the elevator bucket (m)

a, b is the cross-sectional dimension of bucket elevator enclosure (m)

A_n is an auxiliary dimensionless quantity accounting for the sum total of aerodynamic coefficients for particles of spilled grain falling within the enclosure of the return run of the conveyor

a_{t0}, a_{t3} are the respective accelerations of loose matter flowing through inclined loading and discharge chutes of the bucket elevator (m/s^2)

B

b_0, b_1, b_2 are the constants of square trinomial Equation 5.32 determining the velocity of a putative particle flow in a pipe

B, B_l are the conveyor belt widths (m)

b is a constant determined with the relation of Equation 5.179

B is the ratio of the average coefficient of drag encountered by a falling particle to the drag coefficient of an individual particle in the self-similarity area (see Equation 7.19)

B_k is the elevator bucket width (m)

b is the bucket elevator enclosure width (m)

C

c_1, c_2, C_1, C_2 are the integration constants for second-order linear equations

c_k is a dimensionless coefficient of aerodynamic (front) drag of an empty bucket

c_{kz} is a dimensionless coefficient of aerodynamic (front) drag of a grain-laden bucket

D

d_e is an equivalent diameter of a particle of loose matter (m)

D_w is the hydraulic diameter of bucket elevator conveyor belt enclosure (m)

D_l is the same for bucket elevator conveyor belt (see Equation 7.80) (m)

E

E is the perforation parameter $E = \dfrac{\hat{S}_t}{\sqrt{\zeta_0}} = \dfrac{\tilde{\Pi}\tilde{l}\varepsilon}{\hat{s}_t\sqrt{\zeta_0}}$ (see Equation 5.23)

E_k is the ejection head created by the return run of the conveyor as it moves with empty buckets (see Equation 7.48) (Pa)

E_0, E_3 are the respective ejection properties for flows of material handled in loading and discharge chutes (Pa/(m^3/s)^2)

E_1, E_2 are the same for buckets moving within enclosures of carrying and return runs of the bucket elevator conveyor ((Pa/(m^3/s)^2) (see Equations 8.49 and 8.50)

F

$\tilde{F}_m$, F_m are midsectional areas of particles (m^2)

f_{nw} is the total leakage area of the upper cowl (drive drum cowl) (m^2)

f_w is the coefficient of friction of a particle against chute bottom

F_k is the midsectional area of a bucket (m^2)

f_u is a function, specific to the enclosure of the return run of belt conveyor, that accounts for both the ejective properties of a belt traveling with empty buckets and spillages of grain, as well as the frictional resistance to airflow along enclosure walls (see Equations 7.72 and Equation 7.75)

f_v is a function, specific to the enclosure of the carrying run of belt conveyor, that accounts for ejective properties of a belt traveling with loaded buckets combined with frictional resistance to airflow along enclosure walls (see Equations 7.73 and 7.76)

F_d, F_b, F_s, F_c are the respective cowl leakage areas for the driving drum of the lower conveyor/feeder, bucket elevator boot, and driving sprocket and top conveyor loading location (m²)

G

$\tilde{g}, g$ are the gravitational acceleration (m/s²)

G_p, G_m are the mass flow rate of particles in a pipe/chute (kg/s)

G_{a1} is the mass flow rate of air aspirated from the upper cowl (kg/s)

G_{a2} is the same for the lower cowl (kg/s)

G_{n1} is the flow rate of air entering into the upper cowl through leaky spots (kg/s)

G_{n2} is the same for leaky spots of the lower cowl (kg/s)

G_g is the mass flow rate of air entering (or being ejected) through a chute (kg/s)

G_p is the mass flow rate of grain spillages (during the discharge of elevator belt) (kg/s)

g_1 is the ratio of the sum total of local resistance coefficients (LRCs) for the enclosure of the carrying run of bucket elevator conveyor to a dimensionless ejection parameter M_1 (see Equation 7.137)

g_2 is the same for the enclosure of the return line of bucket elevator conveyor (see Equation 7.116)

H

h_d, h_s are the respective negative pressures in unaspirated cowls for the driving drum of the lower conveyor/feeder and the driving sprocket of the bucket elevator (Pa)

$\tilde{\omega}$ are the respective negative pressures maintained in aspirated enclosures of elevator boot and top (receiving) conveyor loading location (Pa)

h_{bc} is the negative pressure in the buffer (inner chamber) of the cowl at conveyor loading location (see Equation 8.88) (Pa)

h_y is the negative pressure in the outer chamber of the cowl at conveyor loading location (see Equation 8.89) (Pa)

K

k is the ratio of relative velocities of particles and air to the velocity of particles to air velocity (see Equation 5.34)

Ke is the ejection volume reduction coefficient $\mathrm{Ke} = u_{I3}/u_k$; same for the case of a porous pipe without a bypass chamber (see Equation 6.40)

k_1, k_2 are the auxiliary constants from the solution of second-order linear equation (Equation 5.168) defined with the ratios expressed by Equation 5.169; same for Equation 5.47 although with a different formula (Equation 5.50); same for Equation 6.50 although with a different formula (Equation 6.55)

K_2 is the maximum coefficient of reduction in the volume of ejected air in a porous pipe ($K_2 = u_2/u_0^{\min}$)

K_m is the ratio of the midsection of a particle to its volume (see Equation 7.9) (1/m)

K is the ratio of the true value of ejection head in an inclined chute to the mean value of this pressure (see Equation 7.21)

K_0 is the ratio K absent ejection airflow ($u = 0$)

k is the difference between air velocity in the return run enclosure to air velocity in the carrying run enclosure of bucket elevator conveyor (see Equation 7.57) (m/s)

L

$\tilde{l}$ is the total length of pipe/chute (m)

$L(x)$ is an auxiliary function defined by Equation 5.108

l_k is the distance between neighboring elevator buckets (m)

L is a dimensionless quantity describing the ratio of the area of the gap between end sides of elevator belt to the area of elevator conveyor enclosure walls (see Equation 7.87)

l is the overall length (height) of bucket elevator conveyor enclosure (m)

M

M, N are the coefficients of the linear equation (Equation 6.18) describing changes in the velocity of ejected air in a porous pipe without bypass chamber (these coefficients are defined by the relations of Equation 6.19)

$M_r(N_r)$ are the coefficients of the linear equation (Equation 6.50) describing changes in the velocity of ejected air in a porous pipe with a bypass chamber (these coefficients are defined by the formulas of Equation 6.51)

$M_2(M_u)$ is a dimensionless parameter accounting for ejection forces created by conveyor belt, buckets, and spilled grain inside the enclosure of the return run of bucket elevator conveyor (see Equation 7.78)

$M_1(M_v)$ is a dimensionless parameter accounting for ejection forces created by a conveyor belt with grain-laden buckets inside the enclosure of the carrying run of bucket elevator conveyor (see Equation 7.79)

m^* is a dimensionless constant describing a dimensionless difference between the air velocity inside the return run housing and air velocity inside the carrying run of the bucket elevator conveyor (see Equation 7.93)

N

n is the ratio of particle velocity at pipe/chute inlet to its velocity at pipe/chute outlet

N is the fan motor power (kW)

n_1, n_2 are roots of a characteristic equation (corresponding to the nonuniform linear equation of Equation 6.18) determined with the relations of Equation 6.22; same for the linear equation of Equation 6.50 although with different relation (Equation 6.52)

P

Δp_c are the pressure losses in the trunk line of aspiration network (Pa)

$\tilde{p}_a$ is the excess static pressure at bypass chamber inlet/outlet (Pa)

p_a is the same, dimensionless

$\tilde{p}$ is the excess static pressure of air inside pipe/chute (Pa)

p is the same, dimensionless

$\tilde{p}_\omega$ is the excess static pressure in the bypass chamber (Pa)

p_ω is the same, dimensionless

p_n is the negative pressure maintained inside the aspiration cowl (Pa)

$\tilde{P}_{atm}$ is the absolute static pressure in the room (Pa)

$\tilde{P}_n$ is the same for the upper cowl (Pa)

$\tilde{P}_f$ is the same for the flow-shaping chamber (inner chamber) of the lower cowl (Pa)

$\tilde{P}_a$ is the same for bypass chamber inlet/outlet (Pa)

$\tilde{P}_0$ is the same for the location downstream of inlet section of the pipe (Pa)

$\tilde{P}_k$ is the same for the location upstream of inlet section of the pipe (Pa)

p_t is the thermal head inside the chute (Pa)

p_e is the ejection head inside the chute (see (7.5)) (Pa)

p_f is the excess pressure in the flow-shaping chamber of aspiration cowl at belt conveyor loading location (Pa)

p_1 is the rated negative pressure inside the upper cowl (inside the cowl of the belt conveyor drive drum) (Pa)

p_2 is the rated negative pressure inside the lower cowl (inside the cowl of the belt conveyor loading location) (Pa)

p_{un} is a dimensionless static excess pressure in the upper cowl/chamber abutting on the loading chute

p_{uf} is a dimensionless static excess pressure in the lower cowl (inside the flow-shaping chamber)

p_{u0} is a dimensionless excess pressure in the inlet section of the pipe (at $x = 0$)

p_{uk} is a dimensionless excess pressure in the outlet section of the pipe (at $x = 1$)

p_{ey} is the ejection head inside an inclined chute at mean particle drag coefficient (see Equation 7.11) (Pa)
p_{eu} is the ejection head inside an inclined chute with a varying particle drag coefficient (see Equation 7.20) (Pa)
p is the pressure inside elevator enclosure
p_n is the pressure inside the cowl of the driving drum of bucket elevator conveyor belt (at elevator head) (Pa)
p_k is the pressure inside the cowl of bucket elevator boot (Pa)
p_1, p_v is the pressure inside the cover of the carrying run of bucket elevator conveyor (Pa)
p_2, p_u is the pressure inside the cover of the return run of bucket elevator conveyor (Pa)
P_A, P_a is the absolute static pressure of air in the room (Pa)
P_s is the same for the cowl of belt conveyor drive sprocket (Pa)
P_b is the same for the cowl of bucket elevator boot (Pa)
P_c is the same for the cowl of the upper conveyor loading location (Pa)
P_d is the same for the cowl of the lower conveyor/feeder driving drum (Pa)
$P_0(Q_0)$, $P_1(Q_1)$, $P_2(Q_2)$, $P_3(Q_3)$ are the ejection pressures as functions of airflows in the loading chute, inside enclosures of carrying and return runs of a bucket elevator, and in the discharge chute, respectively (Pa)
P_b, P_y are the absolute pressures in the buffer (inner) and outer chambers of the conveyor loading location cowl (Pa)
$P_e(Q_c)$ is the ejection head as a function of airflow in the chute of handling facility (Pa) (see the formulas of Equations 8.86 and 8.94)
P_c is the absolute pressure in the outer chamber of the upper (receiving) cowl (see Figure 8.16) (Pa)

Q

Q_t is the volumetric flow of air ejected in the perforated pipe/chute (m³/s)
Q_a is the volumetric flow of aspirated air (m³/s)
Q_1 is the volumetric flow of air coming in through a sealed enclosure of the carrying run of bucket elevator conveyor (m³/s)
Q_2 is the same for the enclosure of the return line of bucket elevator conveyor (m³/s)
Q_0 is the volumetric flow rate of air ejected through the loading chute of the elevator (m³/s)
Q_3 is the same for the discharge chute of the elevator (m³/s)
Q_d, Q_b, Q_s, Q_c are the flow rates of air Q_d coming in through leaky areas in the cowl of feeder drive drum Q_B, bucket elevator drive sprocket Q_S, and the cowl of upper (receiving) conveyor loading location Q_C (m³/s)
Q_e is the volumetric flow rate of air sucked away from the cowl of the bucket elevator boot (m³/s);
Q_a is the same for the upper conveyor loading location (m³/s)
Q_c, Q_{bp}, Q_y are the respective airflows in the chute, in the bypass air duct, and in the gap between vertical walls of the inner chamber and the carrying run of the conveyor (see Equation 8.85) (m³/s)
Q_4, Q_6, Q_7 are the flow rates of air circulating through bypass ducts of the bucket elevator (see Figure 8.16) (m³/s)
Q_5, Q_8 are the flow rates of air coming in, respectively, from the buffer chamber into the bucket elevator boot enclosure and from the inner chamber into the outer chamber of the cowl at upper conveyor loading location (see Figure 8.16) (m³/s)

R

r is the ratio of the cross-sectional area of bypass chamber to that of pipe/chute
R_z is the relative flow rate of recirculated air (see Equation 6.45)
r_{min} is the cross-sectional dimension of the chamber corresponding to the minimum speed of ejected air in a porous pipe
R is the aerodynamic force of a falling particle (N)

R_b is the aerodynamic force of a bucket in a bucket elevator installation (N)

R_z is the aerodynamic force of a single falling particle/grain (N)

R is a dimensionless difference between air pressures in enclosures of the carrying run and the return run (see Equation 7.93)

R_0, R_1, R_2, R_3 are the respective aerodynamic indices of the loading chute, enclosures of the carrying and return runs of elevator conveyor, and discharge chute $(Pa/(m^3/s)^2)$ that refer to the resistance posed by ducts to cross flows of air (see Equations 8.25 through 8.28)

R_d, R_b, R_s, R_c are the respective aerodynamic properties of leaky areas in the feeder drive drum cowl, elevator boot and head cowls, and upper conveyor loading location cowl $(Pa/(m^3/s)^2)$ that determine the resistance to ingress of air into cowls through leaky areas (see Equations 8.29 through 8.32)

R_u is the aerodynamic property of enclosures in the upper chamber of the cowl at belt conveyor loading location $(Pa/(m^3/s)^2)$ (see Equation 8.89)

R_y is the aerodynamic property of the gap between buffer chamber walls and carrying run of the conveyor $(Pa/(m^3/s)^2)$ (see Equation 8.88)

R_c is the aerodynamic property of the chute of handling facility $(Pa/(m^3/s)^2)$ (see Equation 8.86)

R_{bd} is the aerodynamic property of the bypass duct of handling facility $(Pa/(m^3/s)^2)$ (see Equation 8.87)

R_4, R_6, R_7 are the aerodynamic properties of the respective bypass channels of the bucket elevator $(Pa/(m^3/s)^2)$ (see Equation 8.120)

S

S_g is the cross-sectional area of a chute (m^2)

S_n is the total leakage area for a cowl (m^2)

$\tilde{s}_t$ is the cross-sectional area of a pipe/chute (m^2)

$\tilde{s}_b$ is the cross-sectional area of a bypass chamber (m^2)

$\hat{S}_t$ is the ratio of the total area of perforation holes in pipe/chute walls to the cross-sectional area of the pipe/chute itself

$\hat{S}_b$ is the ration of the total area of perforation holes in pipe/chute walls to the cross-sectional area of a bypass chamber

S_0, S_1, S_2, S_3 are the respective cross-sectional areas of the loading chute, enclosures of carrying and return runs of elevator conveyor, and discharge chute (m^2)

T

$\tilde{t}$ is the time

T is a dimensionless parameter accounting for frictional forces of air acting against enclosure walls (see Equation 7.77)

t_1 is the ratio of a dimensionless difference in excess pressures at ends of an enclosure housing the carrying run of bucket elevator conveyor to a dimensionless parameter M_1 determining ejection forces in this enclosure (see Equation 7.105)

t_2 is the same for the enclosure of the return line of bucket elevator conveyor (see Equation 7.115)

U

u_2 is mean velocity of ejected air in chute (m/s)

u_1 is a dimensionless velocity of air ejected (through a pipe with impermeable walls $E = 0$) by a flow of uniformly accelerated falling particles

u_1 is the same for a putative (accelerated) particle flow

$\bar{u}$ is a dimensionless velocity averaged over pipe height

$\tilde{u}$ is the air velocity in pipe/chute (m/s)

u is the same, dimensionless

$\tilde{u}_n$ is a projection of air velocity vector onto the outer normal line of a surface element (m/s)

$\tilde{u}_0$ is the air velocity at pipe/chute inlet (m/s)

u_0 is the same, dimensionless

u_{l3}, u_{Lu} are the dimensionless velocities of air ejected by a putative (accelerated) flow assuming that volumetric forces of interaction between components are linearized

u_{l4}, u_{lq} is the same for a uniformly accelerated flow with velocity defined by the ratio of Equation 5.24

u_p is a dimensionless velocity of air ejected by a putative flow of particles falling through a perforated chute with flow velocity varying according to a parabolic law (see Equation 5.131)

u_k is a dimensionless velocity of ejected air in the outlet section of a porous pipe (at $x = 1$)

u_m is the maximum dimensionless velocity of ejected air in a porous pipe

u_0^{min} is the minimum u_0 at r_{min}

u_1, u_2 are the respective velocities of air accordingly inside enclosures of the carrying and return runs of bucket elevator conveyor (m/s)

u^* is the dimensionless velocity of air in the enclosure of the return run of the conveyor (see Equation 7.93)

V

V_p, $\tilde{V}_p$ are the particle volumes (m³)

$\tilde{v}$ is the velocity of particles (m/s)

v is a dimensionless velocity of particles

$\overline{v}$ is a dimensionless velocity of particles averaged over pipe height

$\tilde{v}_k$ is the velocity of particles at pipe/chute outlet (m/s)

$\tilde{v}_n$ is the velocity of particles at pipe/chute inlet (m/s)

$\overline{v}_q$ is a dimensionless velocity of uniformly accelerated particle flow, averaged over pipe height

$\overline{v}_y$ is the same for a putative (accelerated) particle flow

$\overline{v}$ is the averaged (within the range $0 < x < 1$) dimensionless fall velocity of particles in a pipe/chute

v_a is the mean arithmetic fall velocity of particles in a chute (see Equation 7.19) (m/s)

v_e is the bucket elevator conveyor traveling velocity (m/s)

v_z is the fall velocity of grain particles (m/s)

v is the velocity of air inside the enclosure of the carrying run of bucket elevator (m/s)

W

$\tilde{w}$ is the velocity of air flowing through the perforated holes of a pipe/chute (m/s)

w is the same, dimensionless

$|\overline{w}|$ is a dimensionless modulus of velocity w, averaged over pipe height

w_0 is a dimensionless velocity of air flowing through the perforated holes of pipe inlet section (at $x = 0$)

w_k is a dimensionless velocity of air flowing through the perforated holes of pipe outlet section (at $x = 1$)

w is the velocity of cross flow of air from the enclosure of conveyor carrying run to that of its return run as long as the conveyor is installed inside a common elevator enclosure (see Equation 7.53) (m/s)

X

x_r is a dimensionless circumference of the upper circulation ring

x_{min} is the distance from porous duct inlet to the cross section with the minimum velocity of ejected air

x_{max} is the distance from porous duct inlet to the cross section with the maximum velocity of ejected air

Y

y_1, y_2 are the auxiliary functions that determine changes in the relative velocity of ejected air and differential pressure (see Equations 5.145 and 6.32)

Z

z is a relative velocity of particles in a pipe, defined as the difference $v - u$

II.1.2   GREEK CHARACTERS

α

α_1 is a constant used in the linearized equation of Equations 5.47 for relative velocity z

α is the ratio of the velocity of ejected air at the inlet of porous pipe to its velocity at the outlet of the pipe (see Equation 6.36)

α_m is the same for the maximum velocity of ejected air (see Equation 6.42)

β

β is the volumetric concentration of particles in any cross section of a pipe/chute (m^3/m^3)

β_k is the volumetric concentration of particles at the end of a pipe/chute (m^3/m^3)

β_1 is a constant used in the linearized equation of Equation 5.47 for relative velocity z

β_e is a volumetric concentration of spilled grain particles in the enclosure of bucket elevator return run (see Equation 7.81)

γ

γ is the sign of differential pressure ($\tilde{p}_\omega - \tilde{p}$) or ($p_\omega - p$)

γ_u is the sign of difference ($v_e - u$) (see (7.83)); same ($1 - u^*$) (see (7.98))

γ_v is the sign of difference ($v_e - v$) (see (7.83)); same ($1 - v^*$) (see (7.98))

δ

δ is the sign Δp (see Equation 7.85) or R (see Equation 7.97)

ε

ε is the extent of pipe wall perforation (total area of the perforation holes divided by the total side surface area of the pipe)

ε is the ratio of air density to particle density (see Equation 7.10)

ζ

ζ_0 is the local resistance coefficient (LRC) for a perforation hole

ζ_n is the LRC for resistance to air entering a pipe/chute

ζ_n is the LRC for leakage areas in the cowl

ζ_k is the LRC for resistance to air exiting a pipe/chute

$\Sigma\zeta$ is the sum total of LRCs for air moving through a pipe/chute, $\Sigma\zeta = \zeta_n + \zeta_k$

ζ_{nw} is the LRC for air entering through the leaky areas of the upper cowl (drive drum cowl) referred to the dynamic head of ejected air at the outlet section of the chute

ζ_z is the total LRC for two gaps between chute walls and the end sides part of carrying and return runs of bucket elevator conveyor

ζ_u is the total aerodynamic coefficient accounting for ejective properties of a bucket elevator with empty buckets inside the housing of elevator conveyor return run (see Equation 7.59)

ζ_v is the total aerodynamic coefficient accounting for ejective properties of a bucket elevator with grain-laden buckets inside the housing of elevator conveyor carrying run (see Equation 7.66)

ζ_1 is the LRC for the enclosure of the carrying run of bucket elevator conveyor (see Equation 7.104)

ζ_{be} is the LRC for a motionless bucket inside the enclosure of bucket elevator conveyor (see Equation 8.3)

ζ_d, ζ_b, ζ_s, ζ_c are LRCs of leakage areas in the driving drum of the lower conveyor/feeder, bucket conveyor boot and driving sprocket, and top conveyor loading location, respectively, within the cowl

ζ_0, ζ_3 are the sum totals of LRCs for the loading and discharge chutes of bucket elevator

η

η_f is the efficiency factor of a fan

η_g is the efficiency factor of a gear

λ

λ_1, λ_2 are the roots of the characteristic equation for the linearized equation (Equation 5.47) for relative velocity z

λ_w is the hydraulic friction coefficient for bucket elevator enclosure walls ($\lambda_w = 0.02$ was assumed for calculations)

λ_l is the hydraulic friction coefficient for conveyor surface ($\lambda_l = 0.03$ was assumed for calculations)

ξ

ξ_0 is a coefficient determining the viscous resistance of a porous pipe (see Equation 6.6)

ξ_u is the sign u (see Equation 7.92) or u^* (see Equation 7.97)

ξ_v is the sign v (see Equation 7.92) or v^* (see Equation 7.97)

π

$\tilde{\Pi}$ is the perimeter of pipe/chute section

ρ

ρ_0 is the ambient air density (kg/m³)

$\tilde{\rho}$ is the air density (kg/m³)

ρ_p, ρ_m are the densities of loose matter particles (kg/m³)

ρ_z is the density of grain particles (kg/m³)

τ

τ_w is the frictional stress at bucket elevator enclosure walls (see Equation 7.36) (Pa)

τ_l is the frictional stress at bucket elevator conveyor belt (see Equation 7.35) (Pa)

φ

φ_1, φ_2 are the respective ejection coefficients inside the enclosures of the carrying and return runs of bucket elevator conveyor ($\varphi_1 = u_1/v_e$; $\varphi_2 = u_2/v_e$)

φ is the ejection coefficient ($\varphi_1 = u_2/v_k$)

Φ is an auxiliary quantity (see Equation 7.15)

ψ

ψ is the frontal aerodynamic resistance (frontal drag) coefficient for a single particle

ψ_0 is the frontal aerodynamic resistance (frontal drag) coefficient for a single particle in turbulent surface flow mode (within self-similarity area)

ψ_y is the frontal drag coefficient of a falling particle averaged over fall height (see Equations 7.2 and 7.3)

ω

$\tilde{\omega}$ is the velocity of upward airflow in the bypass chamber (m/s)

ω is a dimensionless velocity of upward airflow in the bypass chamber

$\bar{\omega}$ is a velocity averaged over pipe height ω

II.1.3 Criteria and Complexes

Le is the ejection parameter: $\text{Le} = \psi \beta_k \tilde{l} \dfrac{\tilde{F}_M}{\tilde{V}_p} = \dfrac{1.5\psi v_k G_m (1-n^2)}{d_e \rho_m S_g g}$

Bu is the Butakov–Neykov criterion associated with ejection parameter:

$$\text{Bu} = \frac{2\text{Le}}{(1-n)\displaystyle\sum \zeta} = \frac{1.5\psi v_k G_m}{d_e \rho_m S_g g \displaystyle\sum \zeta}$$

Le_0, Bu_0 are the transition ejection constants determined using the formulas of Equations 5.90 and 5.91

$$\text{Euler's criterion :} \quad Eu = 2\frac{p_2 - p_1 - p_t}{\tilde{v}_k^2 \tilde{\rho} \sum \zeta}$$

II.2 INTRODUCTION

Thanks to their supreme efficiency in terms of containing dust releases and cleaning dust-contaminated air, aspiration systems have been used for more than 100 years as a reliable technology to keep atmospheres of manufacturing rooms and industrial sites free of contaminants introduced by mechanical processing and transportation of loose matter. Many enterprises in the building materials industry (plants turning out cement, lime, and wall construction materials), ferrous and nonferrous metallurgy (iron ore crushing and sorting sections, agglomeration/pelletization sections, and blast-furnace shops) have aspiration systems drawing more power than the process equipment. Nevertheless, their power consumption can be reduced dramatically with the skilled design of aspiration plant components and, above all, with well-informed choice of optimum-capacity local suction units, rational structural design, and layout planning.

Primary consumers of electric power in aspiration systems include forced-draft units (fans and smoke exhausters). The power requirement of electric motors of these forced-draft units is determined under the formula

$$N = \frac{Q_a \Delta P_n}{1000 \eta_f \eta_t} \text{ (kW)}, \tag{I.1}$$

where

Q_a is the total volumetric flow rate of air evacuated by the fan (also known as aspiration volume) (m^3/s)

ΔP_n is the resistance of mainline aspiration network (air ducts, dust traps, and vent pipe) (Pa)

η_f, η_t is the efficiency factor of the fan and transmission

Formula (I.1) clearly reveals the possible options for power saving in aspiration systems: (1) Minimizing aspiration volumes, (2) minimizing pressure losses in aspiration network components, and (3) improving fan efficiency.

II.3 PHILOSOPHY OF MINIMIZING ASPIRATION VOLUMES IN LOOSE-MATTER HANDLING FACILITIES

We will be considering various techniques (means and methods) of reducing aspiration volumes in the classical case of loose matter transferred from an upper conveyor to a lower conveyor. Let us begin with an air balance diagram for two aspirated cowls: the cowl of upper conveyor drive drum and that of lower conveyor loading location. We will assume that heated loose matter is being transferred over the handling chute (Figure II.1). As a result of dynamic interaction (owing to aerodynamic forces of falling particles and air), the directed flow of ejected air arises within the chute.

We will assume ejected air to flow in a forward pattern (from top to bottom) with a mass flow rate of G_g (kg/s). In addition, intercomponent heat exchange within the chute produces an upward thermal head p_t (Pa) that counteracts the downward-acting ejection pressure p_e (Pa). A running fan

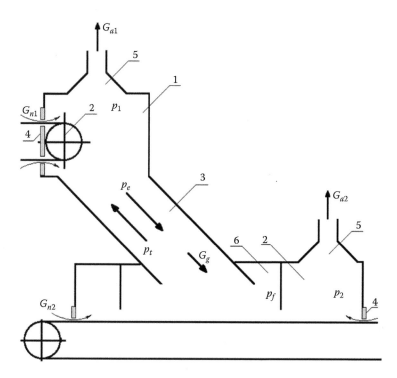

FIGURE II.1 Air balance diagram for aspiration cowls of a loose-matter handling facility: 1, upper conveyor drive drum cowl; 2, lower conveyor drive drum cowl; 3, transfer chute; 4, cowl seals; 5, aspiration flanges; 6, internal (flow-shaping) chamber of the upper cowl.

evacuates air from the respective aspirated cowls at rates of G_{a1} (kg/s) and G_{a2} (kg/s), while specified negative pressures of p_1 (Pa) and p_2 (Pa) are maintained inside to ensure suction of outside air through leaking gaps in cowls at a rate of G_{n1} (kg/s) for the upper cowl and G_{n2} (kg/s) for the lower cowl. The following balance equation may be put forward for this airflow layout, based on the law of mass conservation (assuming no intercomponent mass exchange):

$$G_{a1} = G_{n1} - G_g; \quad G_{a2} = G_{n2} + G_g, \tag{I.2}$$

which translates into an obvious equation:

$$G_{a1} + G_{a2} = G_{n1} + G_{n2}. \tag{I.3}$$

The total flow rate of air evacuated from cowls is equal to the sum total of air leaking in through gaps in cowls. This is a convincing testimony of the impact the sealing of aspiration cowls (equipment enclosures) has on minimizing aspiration volumes.

Besides, the first equation (I.2) may result in $G_{a1} < 0$ (this can be the case at $G_g > G_{n1}$, i.e., when the flow rate of ejected air exceeds the flow rate of air sucked into the cowl through the leaking gaps with a specified negative pressure p_1 inside the cowl). Such outcome will obviate the need for an upper suction unit because the cross flow of air through the chute into the lower cowl will be enough to maintain negative pressure within the upper cowl above its specified value. Thus, the solutions of combined equations for air balance are analyzed with dual goals of quantifying required aspiration volumes and optimizing aspiration flange layouts.

To ensure closure of the combined equations (I.2), we will introduce a ratio, known from engineering practice, for the flow rate of air entering the cowl through leaky locations:

$$G_n = S_n \sqrt{\frac{2 p_n \rho_o}{\zeta_n}}, \tag{I.4}$$

where
S_n is the total leakage area of the cowl (m²)
p_n is the negative pressure maintained within the cowl (Pa)
ζ_n are the LRCs of leakage areas
ρ_o is the density of ambient environment (kg/m³)

The values of S_n and p_n are specified in the engineering practice depending on the type of process equipment, cowl design, and the type of material handled/transported.

The mechanism behind cross flows of air in chutes is much more complex than would be the case for trivial leakages through holes. A common engineering practice involves the use of ejection factors $\varphi = u_2/v_k$ where u_2 is the mean velocity of ejected air inside chute (m/s) and v_k is the velocity of falling particles leaving the chute. Mass flow rate in this case will appear as

$$G_g = \varphi v_k S_g \rho \text{ (kg/s)} \tag{I.5}$$

where
S_g is the cross-sectional area of the chute (m²)
ρ is the density of air inside the chute (kg/m³)

For the case of a uniformly accelerated particle flow in a chute, ejection coefficient is computed using the criterial equation

$$\varphi |\varphi| = \text{Eu} + \frac{\text{Bu}}{3} \left[\left| 1 - \varphi \right|^3 - \left| n - \varphi \right|^3 \right], \tag{I.6}$$

where
Bu is Butakov–Neykov number characterizing the ejection pressure created by a uniformly accelerated flow of particles with a mean equivalent diameter d_e (m)

$$\text{Bu} = \frac{1.5 \psi G_m v_k}{d_e \rho_m S_g g \sum \zeta}. \tag{I.7}$$

ψ is the frontal drag coefficient for particles (one should keep in mind that $\psi = f(G_m, v_k, d_e)$)
G_m is the flow rate of loose matter (kg/s)
ρ_m is the density of loose matter (kg/m³)
g is the gravitational acceleration (m/s²)
$\sum \zeta$ is the sum total of LRCs for the chute

Eu is Euler's number referring to the thermal head and the impact that negative pressures in cowls have on ejected airflow rate:

$$\text{Eu} = \frac{p_2 - p_1 - p_t}{\sum \zeta \frac{v_k^2}{2} \rho}. \tag{I.8}$$

To ensure the clarity of classification schemes for minimizing aspiration volumes, separate treatment will be given to means and techniques designed to minimize the ingress of air entering through leakage areas G_n (Figure II.2) and to those designed to minimize the ingress of air into the cowl over the chute G_g (Figure II.3).

Cowls are normally sealed with flexible strips of used conveyor belt laid over holes so as to ensure maximum overlap (mainly within the areas where conveyor belts run close to fixed cowl walls). It is less important to overlap these leaks as much as possible as to maximize the aerodynamic drag of the seal, for example, by installing double aprons, screens, or labyrinth seals on the wall (Figure II.2b, layout 1).

Another important way of minimizing G_n involves reducing the mean negative pressure within the cowl by ensuring uniform distribution of air along the cowl (p_y). This objective can be met, first, by increasing cowl capacity (height h; a height of $h = 0.7B$, where B is conveyor belt width, would be enough for a considerable leveling of negative pressure, see the graph on Figure II.2b, layout 2); second, by installing a cowl equipped with an internal chamber with vertical walls posing additional resistance to moving air, thereby not only reducing the negative pressure along the outer chamber but bringing down the flow rate G_g (Figure II.2b, layout 3); and third, by using two horizontal partitions to split the internal space of the cowl into halves, profiling the partitions so as to ensure a variable aperture for the uniform suction of ejected air away from the bottom space of the cowl (Figure II.2b, layout 4).

Equalization of negative pressure along the cowl makes the reduction of the specified negative pressure p_n (averaged over length) possible. So, for a normal (single-walled) cowl, $p_n = 10 \div 12$ Pa, while for a two chamber cowl, $p_n = 6$ Pa, that is, 1.5 times higher. An important point with respect to reducing negative pressure is to relocate any possible openings (where cowl walls meet with running conveyor belt) as far away as possible from the free-fall zone of the material being handled. This would also prevent compression-driven carryover of dust particles with air as falling particles land on the conveyor belt. The particle free-fall zone can be removed from potential locations of

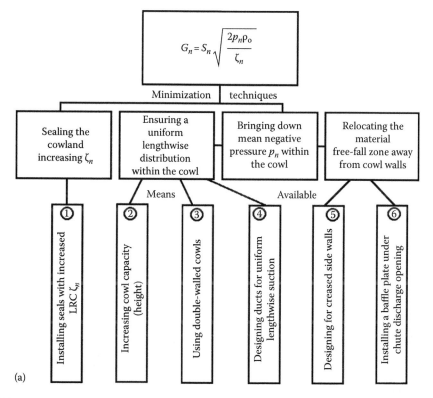

(a)

FIGURE II.2 (a) Classification of means and techniques for minimizing the flow rate of air leaking into the lower cowl (G_n). *(Continued)*

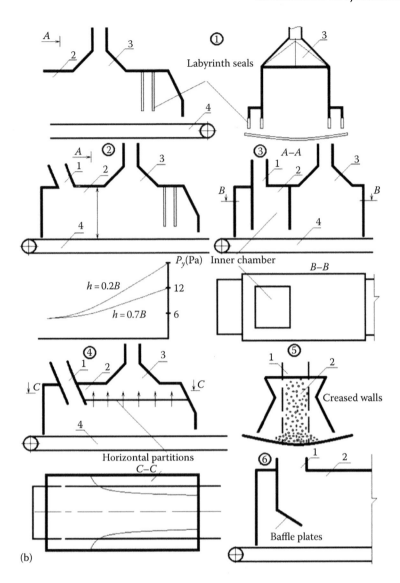

(b)

FIGURE II.2 (*Continued*) (b) Design layout options for minimizing G_n for the lower cowl (lower conveyor loading location): 1, transfer chutes; 2, aspiration cowls; 3, aspiration flanges; 4, belt conveyors.

concentrated leakages by erecting creased lengthwise walls or by designing baffle plates or pockets within the boot of the chute (Figure II.2, layouts 5 and 6). This confers an added benefit of preventing abrasive wear of the conveyor belt.

The following approaches are available for minimizing ejected air volumes G_g (Figure II.3a):

- Reducing free-fall velocity of loose-matter particles to weaken the forces of interaction between components
- Promoting drag along the path of ejected air toward aspiration flange, thereby lowering Eu and Bu and ultimately bringing down the ejection coefficient φ
- Designing for closed-loop circulation (recycling) of air by providing bypass chambers and, thereby, reducing the volume of ejected air that enters the aspiration flange

The following technical means are available for implementing both approaches (Figure II.3b).

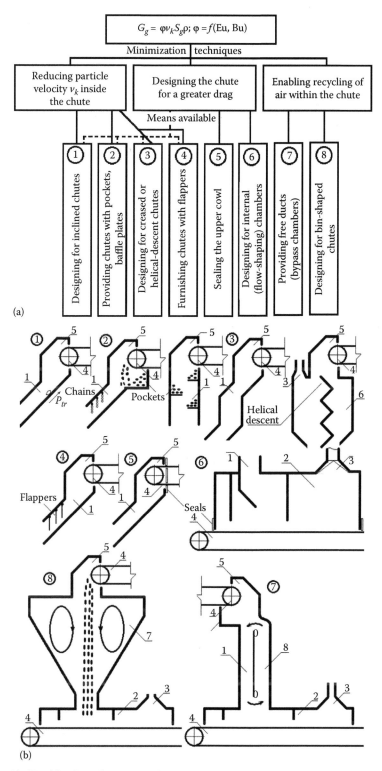

FIGURE II.3 (a) Classification of means and techniques for minimizing the flow rate of air entering the lower cowl through the chute (G_g). (b) Design layout options for minimizing G_g for the lower cowl: 1, transfer chutes; 2, aspiration cowls; 3, aspiration funnels; 4, belt conveyors; 5, cowl of the upper-cowl drive drum; 6, bin; 7, bin-shaped chute.

Reducing particle free-fall velocity by multiple means: designing for inclined chutes (layout 1) with increased particle friction against the inclined chute bottom (p_{tr}), installing (suspending) chain strings or pockets (layout 2), installing flappers (layout 4), and designing the chute to have a creased or helical path (layout 3). In addition to decreasing particle fall velocity by promoting friction forces, these devices also increase the aerodynamic resistance to the flow of ejected air, that is, they are dual purpose. Moreover, it is possible to increase the sum total of LRCs and, as a consequence, decrease Euler's and Butakov–Neykov numbers by using special devices to seal the upper cowl (layout 5).

This promotes increased resistance against the ingress of air (ζ_{nw}) through the remaining holes (f_{nw}) due to a known relation

$$\zeta_{nw} = 2.4\left(\frac{S_g}{f_{nw}}\right)^2.$$ (I.9)

Another option involves setting up internal chambers (flow-shaping chambers, layout 6) with walls that increase the drag of air exiting the chute.

Air can be recycled either by setting up bypass chambers (layout 7) or by designing special chambers for the natural recirculation of air within bin-shaped chutes (layout 8). The latter option brings about a sharp reduction in the amount of ejected air entering into the aspiration flange (see Chapters 5 through 8).

II.4 IMPROVING DUST RELEASE CONTAINMENT DESIGNS IN BELT-CONVEYOR HANDLING FACILITIES TO MINIMIZE POWER CONSUMPTION

A prerequisite for the effective containment of dust releases with aspiration is enclosing dust release locations into ventilated cowls. Release of airborne dust into process rooms through leaks in these cowls is prevented by maintaining a negative pressure within cowls by sucking away specific amounts of dust-laden air.

Industrial hygiene design requirements for aspiration cowls are well detailed both in literature [163,164] and in regulations [162,165].

In recent years, however, designers faced a new concern of optimizing containment devices with the overall goal of an energy-efficient aspiration system.

Increased unit power of process equipment and more sophisticated process technology with attendant multiplication of conveyor-based transfer units call for huge aspiration assemblies to ensure proper dust containment. At the same time, large amounts of dustlike material escaping into aspiration network in many cases mean losses of scarce and valuable commodities.

A significant portion of escaping material is lost permanently due to poor efficiency of dust-containment devices and difficulties surrounding the reuse of products trapped in them.

Therefore, aspiration cowl design improvements involve various means and techniques with dual goals of reducing aspiration volumes needed to maintain dust-containment efficiency and mitigating escape of airborne material into aspiration network [167].

Conveyor handling facilities are known to have the most number of categories of dust sources [168]. Therefore, as a fact that agrees comfortably with analytical findings, today's design improvements are mainly focused on these facilities [192,198–209].

II.4.1 DESIGNS OF POWER-SAVING DEVICES INSTALLED INSIDE CHUTES

A review of Soviet/Russian patent filings and scientific and technical publications reveals two primary fields of research with regard to minimization of ejection volumes: promoting aerodynamic drag within the chute and enabling closed-loop circulation of dust-laden air inside it.

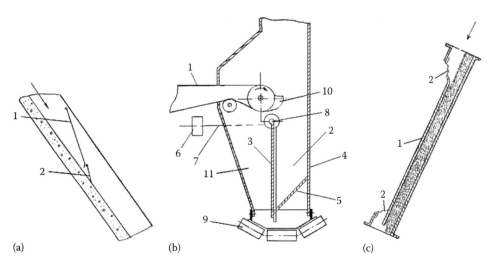

(a) (b) (c)

FIGURE II.4 Devices proposed for increasing drag in chutes: (a) Certificate of Authorship (CoA) #662462, (b) CoA #745817, and (c) CoA #615001.

Promoting aerodynamic drag within the chute: Means and techniques proposed to increase the drag of the chute have long become traditional, for example, the suggestion to use elastic aprons overlapping the cross section of the chute [169].

This involves modifying known designs to improve the overall gain from using a particular device. This can be illustrated by considering an improved design of a rigid valve [170] mounted inside the chute as an example. The shut-off valve (Figure II.4a) uses a combined design, comprising a solid curtain (1) with an articulated mount to the upper chute wall that has a swiveling multiple-part shutter (2) at its lower edge. Loose matter moving over the chute displaces the solid curtain (1) and pours further below along its path, while individual parts of the shutter (2) slide over the particle stream. This valve design is tolerant to inclusions of oversized material within the stream, while only insignificantly increasing the leakage area between the duct and the surface of flowing material. An oversized piece will be deflected by one or more members of the shutter (2). Other members will occlude the opening between the lower edge of the curtain and the surface of flowing loose matter as they come in contact with the stream.

It should be noted that elastic and rigid valves have a significant drawback: they abrade and wear out quickly, ultimately defeating their purpose of increasing resistance to air within the chute.

Ejection volumes can be reduced significantly if not eliminated by gating the chute with the material being handled. A loaded gate would be the simplest means of maintaining the necessary level of material in the chute. However, loaded gates often turn out to be technically infeasible to implement: material may become stuck within the chute, causing a risk of failure.

This risk is mitigated significantly in the design of a chute with a transfer cavity (Figure II.4b) [171]. Loose matter from conveyor (1) enters working vacuity (2) between shut-off piece (3), enclosure (4), and overflow ramp (5). Lever (7) carrying load (6) prevents the shut-off piece (3) from yielding until the working cavity fills with material to the required level. Once that level is reached, the head of the material forces shut-off piece (3) to swivel on axle (8), opening a gap to discharge the material. The position of load (6) on layer (7) may be adjusted to ensure continuous uniform unloading of loose matter onto conveyor (9) with a minimum free-fall height. The shut-off piece moves vertically in a reciprocal fashion with a jolt in the bottom position, actuated by cam (10) acting on axle (8).

If larger pieces of material become stuck in the unloading opening, working cavity (2) overflows and material escapes through a gap between axle (8) and conveyor drive drum (1) into the transfer cavity (11) to land on conveyer (9). The accompanying increase in dust emissions serves as a signal to force-open shut-off piece (3).

In order to maintain the operation of inclined chutes as they become *buried* even with varying flow rates of loose matter, a movable-wall chute has been designed (Figure II.4c) [172]. Upper wall (1) of the chute is joined in a leak-tight manner with its side walls using flexible enclosures, enabling it to move across the longitudinal axis of the chute. This minimizes the gap between the surface of the flowing material and the upper movable wall of the chute.

Buried operation of chutes poses strict requirements to the material being handled: it should not be prone to caking, must not be excessively moist, and its particles must have a rounded shape.

Closed-loop circulation of air: The most common chute design relies on the recirculation of air without any additional forced-draft devices, merely due to the pressure gradient arising within the chute as air is ejected by a flow of loose matter. Special ducts are commonly used to connect excess-pressure and negative-pressure zones with each other. Ducts may be installed in several layouts: inside the chute, coaxially with its centerline [173], or near side wall [174], as well as outside of chute [175–177].

It should be emphasized specifically that the mere reliance on air recirculation without aspiration will fail to address the issue of containing dust emission in full. Recirculation will only be able to reduce the aspiration volume necessary for effective dust containment. This problem owes itself to the impossibility of completely eliminating leaks within the cowl of loading conveyor drive drum.

Therefore, it is essential to locate the recirculation duct outlet properly when designing the duct as this design choice will determine the aspiration capacity for a particular transfer assembly. The main precondition for enabling aspiration is the existence of drag between the recirculation duct inlet opening and the dust-receiving section of aspiration column. Increasing drag promotes recirculation effect.

Failure to take this factor into account may inhibit recirculation, as air inside the duct will flow in the same direction as the material being handled inside the chute. An example of unsuccessful solution is provided by the inclined chute design proposed by Karaganda (Qaraghandy, Kazakhstan) Polytechnic Institute [174]. The bottom of the chute (Figure II.5a) carries plates (1) that form recirculation zones with side walls (2). The aspiration chamber (4) is installed immediately above chute discharge opening (3). This design requires excess pressure to be maintained inside the aspiration chamber (4) to enable recirculation. However, this condition would make aspiration redundant: excess air from the chamber (4) will be displaced into recirculation zones and will also leak into the industrial room through gaps in the cowl.

Another design with a similar effect was proposed by the Occupational Safety Research Institute (Yekaterinburg, Russia) for a double-walled cowl (Figure II.5b) [175]. Discharge opening (1) of recirculation duct (2) located outside the chute abuts on the upper part of chute (3). The duct is

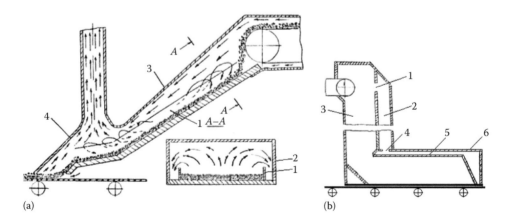

(a)   (b)

FIGURE II.5 Devices based on natural air circulation (no aerodynamic drag between the recirculation duct and the local suction unit): (a) CoA #998269 and (b) CoA #1030563.

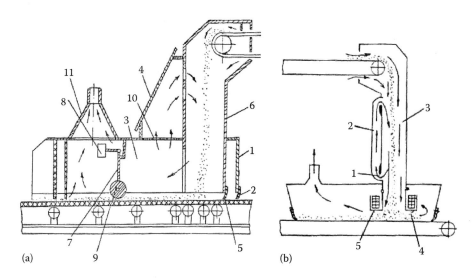

FIGURE II.6 Devices based on natural air circulation (with aerodynamic drag between the recirculation duct and the local suction unit): (a) CoA #445599 and (b) CoA #1105406.

routed below to the horizontal upper wall of the outer cowl enclosure that has an opening (4) in this location. The outer enclosure houses input enclosure (5) as a hollow triangular prism without a bottom face and sloping end and side faces, installed with a gap between it and outer enclosure walls.

A stream of dust-laden air ejected by falling material passes through loading chute (3) into the inner enclosure (5) of the cowl where excess pressure mounts. This pressure forces dust-laden air into the other enclosure (6) of the cowl, whence it passes through the opening (4) into duct (2) and returns through another opening (1) into the loading chute (3). One should keep in mind that only a fraction of the volume of ejected air is spent for recirculation whereas the remaining air will escape into the room through leakage areas so that the outer enclosure will have an excess pressure somewhat below that in the inner enclosure.

This device uses a local suction unit, usually installed on the external enclosure in two-walled cowl designs, and, thereby, rules out the recirculation of air.

Drawbacks inherent in the design described have been addressed by a design proposed by the Alma-Ata (Almaty, Kazakhstan) branch of the Santekhproyekt Design Institute [176], which is also intended for double-walled cowls (Figure II.6a). The cowl features outer box (1) with rubber seals (2) along its perimeter. The outer box houses inner box (3) with a recirculation duct (4) and a perimeter seal (5) along the lower edge next to chute (6).

Opposite the chute, mobile rigid partition (7) is installed inside the side walls of the inner box, equipped with counterweights (8) and sealing roller (9) fitted to the lower side of the partition with free turning and vertical travel as it touches the moving conveyor belt or the material atop it. It is easy to see that, unlike device [175], this design has inlet openings (10) of the recirculation duct on the inner box rather than the outer (aspirated) box. This enables recirculation of air in addition to aspiration of the handling facility. Some of the air entering the inner box (3) through the chute passes over to the outer one (1) whence it is removed by local suction unit (2) while the rest circulates in a closed-loop pattern over chute (6) and duct (4).

It is possible to combine aspiration with recirculation of air in the case of single-walled cowls as well (Figure II.6b) [177]. Inlet opening (1) of recirculation duct (2) should be located inside the lower part of chute (3) in this case to ensure sufficient sealing of the discharge opening of the chute. Sealing options include loaded gates and baffle plates. When handling loose ferromagnetic materials, it would be wise to use magnetic sealing unit (4) with magnetic system (5) manipulating particles of the material being handled.

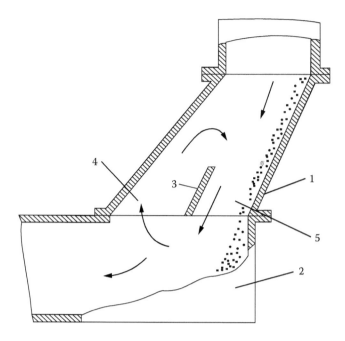

FIGURE II.7 Gravity-flow chute with two cavities. (From CoA #485928, IPC B65 G19/28, A gravity-flow chute for granular materials/O.D. Neykov, Ya.I. Zilberberg, E.D. Nesterov et al., No. 1688112/27-11, field September 08, 1971//Otkrytiya, Izobreteniya, 1975, No. 36, 45p.)

A magnetic sealing unit will achieve an added benefit of reducing dust content in air evacuated by suction as dust particles will be retained within the porous medium of the sealing unit.

When evaluating proposed designs, it should be noted that a transfer chute design that expands in the same direction with flowing material [178] serves as the most straightforward concept of closed-circuit air circulation. In addition to its expanding profile, the lower part of the chute (Figure II.7) has longitudinal partition wall (3) on unloading section (1) connected with loose-matter receiver (2). This partition wall splits the bottom of the chute into two cavities (4 and 5) abutting on the front and back walls of the gravity-flow chute, respectively. Cavity (4) serves as a recirculation duct by returning a part of ejected air into cavity (5).

Unfortunately, air circulation processes in such chutes have not been described analytically to this day, therefore, the impact of chute capacity, among others, on recirculation volumes cannot be ascertained. Informed choice of design parameters for handling facilities is, thereby, hindered.

Ejection of air by a flow of transferred material gives rise to differential pressure between end sections of gravity-flow chutes. Experience shows that this pressure usually measures 10–50 Pa.

Some authors dismiss the option of harnessing this pressure gradient to enable natural circulation and propose using additional external forced-draft devices.

For example, it is proposed to recirculate air by relying on the ejecting capacity of a jet of dispersed water emanating from a nozzle installed within the cowl of a handling facility (Figure II.8) [179–181]. Water jets also aid in decreasing airborne dust content inside the cowl.

However, using dispersed water alone may be insufficient to contain dust emissions when dust formation is intensive enough and the material falls from a great height.

This is because dispensing with aspiration will inevitably give rise to excess-pressure zones inside the cowl, thereby forcing the air–dust mixture out of it. Given that water droplets in the jet perform poorly as a cleaning medium, dust content in air escaping from the cowl into the room will be above exposure limits.

A noteworthy proposal is based on using a fan to recirculate with its primary function of ensuring the suction of air at cowls in conveyor loading locations [182,183].

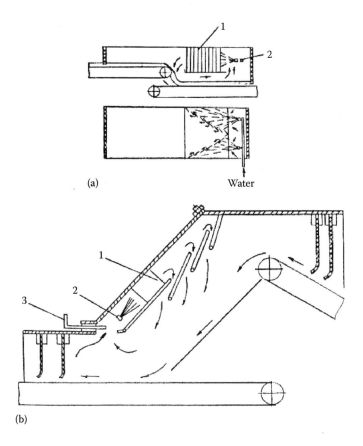

(a)

Water

(b)

FIGURE II.8 Devices based on forced recirculation of air (ejection by water droplets): (a) CoA #777238 and (b) CoA #939347: 1, recirculation duct/dust trap; 2, water spraying nozzle; 3, gate valve for varying the volume of circulating air.

For example, the Krivyy Rih Ore Mining Institute (Ukraine) proposes [182] supplementing the standard handling facility layout with a recirculation air duct (1, Figure II.9a). To partially compensate for the amount of ejected air, air sucked from cowl (4) of destination conveyor (5) is discharged by aspiration fan (2) through the recirculation duct (1) into the cowl (3) of the source conveyor drive drum. Air discharge rate is adjusted using damper (6) installed on the main air duct (7). The rest of air, depending on its dust content, is either directed onto the second cleaning stage or discharged into atmosphere.

A dedusting device designed by the Ural branch of the All-Union Thermal Engineering Research Institute [183] operates in a similar way. The only relevant difference of this device (Figure II.9b) is the use of foam filter (8) for containing dust releases along the free-fall path of the handled material as it lands on the conveyor belt. Fan (2) is sized so as to maintain the specified negative pressure in the cowl of the conveyor loading location. The volume of air delivered into foam generator (9) is determined by the capacity of the latter. The volume of air routed over the recirculation air duct (1) into the maximum negative pressure zone of the chute depends on the amount of the material handled and must be so as to keep the pressure in this zone below atmospheric pressure.

The use of a fan for forced recirculation of air only addresses the issue of reducing aspiration releases into atmosphere. In this case its required performance will be lower than the absence of a recirculation air duct.

Another option achieves recirculation by jointly using the pressure gradient inside the chute and enabling additional suction of air by mechanical ventilation [183]. This approach is suitable for double-walled cowls (Figure II.10) that have a perforated section of upper wall (3) of inner box (4) between the loading chute (1) and the duct (2) with making space for additional suction chamber (5)

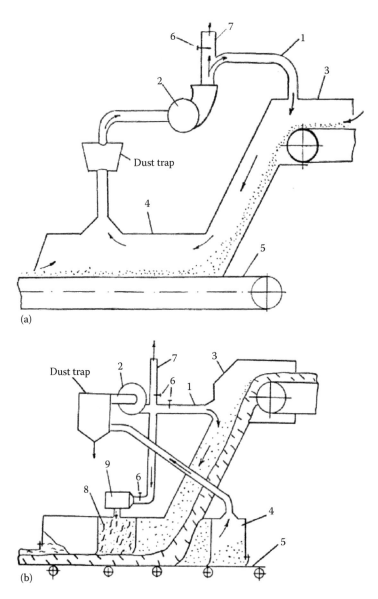

FIGURE II.9 Devices with forced circulation of air (using an aspiration fan): (a) CoA #962127 and (b) CoA #1190064.

provided with aspiration flange (6). Air evacuated through the chamber (5) (amounting to 5% of air entering the cowl through the loading chute) forces the greater part of the ejected air to adhere to perforated wall (3) and become routed into the recirculation duct (2), then passing through the holes (7) to return into the loading chute (1) for repeated ejection.

Thus, in any case, closed-loop circulation of air fails would never adequately replace aspiration of the handling facility, meaning that some energy will have to be expended on maintaining aspiration.

II.4.2 SEALING SYSTEMS OF ASPIRATION COWLS

A significant fraction of the total volume of air aspirated from a cowl (~30%–50%) is comprised of air sucked into the cowl through leakage areas and openings in the exterior enclosures of the cowl.

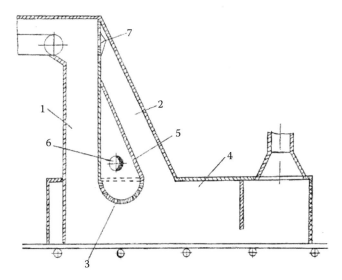

FIGURE II.10 A design involving the circulation of air (equipped with an additional local suction unit). (From CoA #950925, IPC E21 F5/00, Cowl design for loose-matter handling locations/V.D. Olifer, V.B. Rabinovich, S.A. Kozinets, No. 3233960/22-03, filed September 1, 1981//Otkrytiya, Izobreteniya, 1979, No. 30, 124p.)

Therefore, the key industrial hygiene requirement for aspirated cowls mandates maintaining certain minimum areas of leakages and process openings as long as permitted by process conditions. The volume of air introduced by suction also depends on the negative pressure maintained within the cowl with a local suction unit operating. In this case the required negative pressure is specified depending on the impact that an incoming jet of air ejected by a flow of loose matter has on leakage areas, mainly those of longitudinal shape. Mitigating this impact would enable the cowl to operate at a lower negative pressure without detriment to dust release containment.

To sum up, two primary directions are available and pursued presently in aspiration cowl design improvements for reducing the volume of in-leaking air: devising means enabling cowls to operate at lower required negative pressures and improving the leaktightness of cowls.

Necessary negative pressure inside the cowl may be reduced by mitigating the impact of the ejection jet on lengthwise seals. In addition to well-known means [166], an option is to control the jet. The device [186] implementing this approach (Figure II.11) has cowl enclosure (2) equipped with a lid (1) as a continuation of the upper wall (3) of the loading chute.

This lid bulges into the enclosure in the area between the aspiration flange (4) and upper wall (3). In accordance with the Coanda effect, a flow of material ejected by air passes through a constriction between the convex cowl enclosure lid and baffle plate (5) and then, without reaching the conveyor belt or longitudinal seals, completely adheres to the convex lid and follows its curve into the aspiration flange. This enables negative pressure inside the cowl to be reduced significantly as dust-laden air is no longer knocked out through leaking openings.

Cowl sealing improvement: All sources of leaks within the handling facility are concentrated in locations where external enclosures meet the running conveyor belt. Two approaches are available for ensuring the necessary tightness of lengthwise seals in conveyor loading locations: eliminating vertical swing of a running conveyor belt or expanding the surface area of contact between it and sealing band. The latter approach is applied, in particular, in the cowl [187] featuring a lengthwise seal that comprises two flexible cylinders, with the internal cylinder bent toward conveyor centerline and the external cylinder bent in opposite direction.

Insights from the experience of operating various aspiration cowl types indicate that the most expensive sources of leakage in cowls (except for loading locations of long-stretching assembly-line

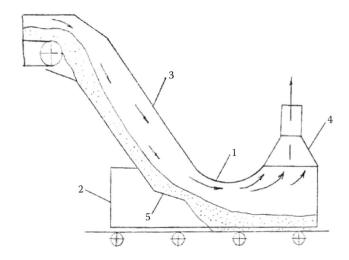

FIGURE II.11   A cowl with a device for controlling the ejection jet. (From CoA #921994, IPC B65 G21/00, Aspiration cowl design for belt conveyor loading location/S.A. Kozinets, V.D. Olifer, G.Yu. Khvostov, No. 2972381/27-03, filed August 15, 1980//Otkrytiya, Izobreteniya, 1982, No. 15, 98p.)

conveyors) occur at front/back enclosures of cowls. End openings are usually sealed with elastic aprons that have their bottom side cut out to approximate the cross-sectional shape of a layer of loose matter transported by the conveyor. Minimizing the gap between the apron and the surface of transported material, however, is a problem insurmountable in practice, as the material layer will vary both in its cross-sectional area and its shape as a result of process load fluctuations. The resulting leakage may become too large.

   To a certain extent this drawback is addressed in the design [188] where the layer of carried loose material is forcibly shaped to the required cross-section prior to its contact with the end seal. The proposed design (Figure II.12) has a distributor installed at the outlet of loose matter from chute (1) comprised of fixed lower (2) and movable upper (3) articulated partition plates with the lower plate loaded with a spring against the upper one. The lower edge of the mobile

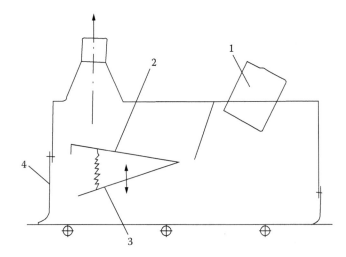

FIGURE II.12   A cowl with a distributor for shaping the material layer. (From CoA #882854, IPC B65 G21/00, Belt conveyor loading location cowl, I.F. Yufit, No. 2897490/27-03, filed March 19, 1980//Otkrytiya, Izobreteniya, 1981, No. 43, 75p.)

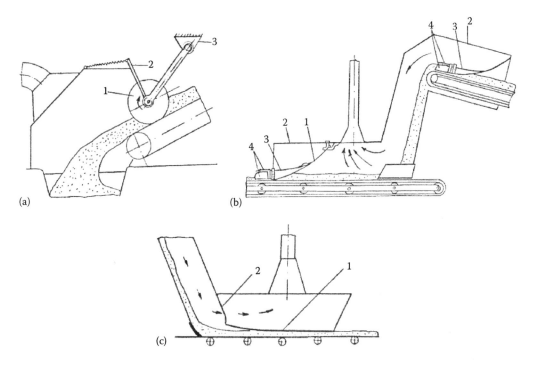

FIGURE II.13 Cowls featuring flow-shaping sealing arrangements: (a) CoA #589431, (b) CoA #1148814, and (c) CoA #947015.

plate (3) is shaped similar to the cutout in end apron (4) making it possible to minimize gaps as loose matter travels under aprons.

A more efficient solution is implemented in designs [189–191] where cross-sectional shaping of the material layer and sealing of end openings happen simultaneously.

While similar in function, flow-shaping sealing arrangements may significantly differ in their structural design. For example, one of the possible designs [189] (Figure II.13) is based on a roller with concave groove (1) that, while lying freely atop the material transported by the conveyor, rotates with the moving belt due to frictional forces. Scrapper plate (2) is attached directly to the concave surface of the roller in its upper part. This arrangement may be used both for sealing the entry of the carrying run into conveyor head cowl and for sealing the end opening in conveyor loading location from where laden conveyor belt emerges.

A design proposed by Novocherkassk Polytechnic Institute (Figure II.13) [190] has end openings of both upper and lower cowls sealed by means of curvilinear mobile wall (1) attached with an articulate suspension to the upper wall of the enclosure (2). Rigid plates (3) are attached to the sides of the wall while elastic plates (4) are attached to the ends of wall (1) and plates (3). Material transported by a running conveyor passes freely under a curved mobile wall (1) sitting atop it firmly. Sealing is also improved by plates (4) that are elastic enough to remain in contact with the material even when a large piece occurs, dislodging the rigid wall (1).

A similar principle is used in the device [191] where the standard enclosure is additionally provided (Figure II.13c) with mobile plate (1) made of sheet metal fixed to cloth-filter apron (2).

Aspiration causes flows of dust-laden air arising in the loading chute to be sucked through cloth filter (2), which retains dust particles. Mobile plate (1) rests continually on the surface of the transported material, thereby, ensuring a high degree of sealing. Dust particles are removed from the cloth filter surface by shaking as an uneven flow of the material being handled causes the plate (1) to move from one position to another.

The proposed cowl design ensures improved sealing at the same time with reducing carryover of dustlike material into aspiration network.

It should be noted that cowl end-opening seal designs [189–191] all share a single common deficiency. These designs fail to address the issue of sealing out locations where sealing arrangements come close to the side walls of cowls.

It was shown that the main contemporary approach for reducing the volume of air leaking in is based on increasing the contact area in locations where seals meet with moving surfaces. This approach ultimately faces a dead end as it is impossible to increase the area of this contact to an unlimited degree and because the area of frictional contact between two surfaces will face increased wear. A more rational path to take is devising techniques for reducing the amounts of air leaking in while minimizing or completely avoiding the said contact, for example, by increasing enclosure wall thickness around the leaking gap, maintaining a uniform field of velocities in uncovered openings, optimizing aspiration funnel layout with respect to leaking gaps, or promoting drag for air entering the cowl (see Sections 1.1 and 3.1).

II.4.3   Reducing the Carryover of Dustlike Material from Cowls

Dustlike fractions of material are carried over into the aspiration network mainly as a result of reduction in the volume of aspirated air. At the same time, specific moisture content in aspirated air remains high, failing to arrest the rapid build-up of material on evacuation air ducts. Depending on the specific factor being targeted, all existing means for reducing concentrations of dust in evacuated air can be referred to either of the two subgroups: measures for reducing dust formation intensity and devices for cleaning aspiration air immediately in dust release locations. The first case targets the flow of the material being handled, while the second case targets dust-laden air inside cowls.

A detailed review of existing means and techniques for minimizing carryover of dustlike material into aspiration network enables the following requirements to be put forward with respect to these means.

First of all, means for optimizing aspiration in terms of material carryover must draw as little power as possible from external sources or do away with external power sources altogether. Second, the dust-trapping process should ensure the return of dust into the process using the simplest of devices. Third, the means used for optimization must be efficient both in dust trapping and in maintainability. Finally, an essential requirement is the comprehensive effect of the facilities being designed, that is, their use must lead to multiple positive effects both in terms of mitigating carryover and in terms of reducing aspiration volume.

To the greatest extent these requirements are met by process-design measures that should be preferred. Industrial hygiene measures are of secondary importance and should be used when process-design measures are not feasible or to augment their effect. Process-design measures ensuring a comprehensive industrial hygiene effect may be illustrated with screw [168] and telescopic [193] chutes, as well as loaded gates mentioned earlier.

Application of these devices fulfills the goal of mitigating dust releases while reducing the volume of ejected air. The option of returning trapped dust directly into the process makes it advantageous to equip cowls and local suction units with dry dedusting devices. Gravity [194], electric [195], and magnetic [177] forces together with the precoalescence of dust particles are preferable approaches from the standpoint of minimizing energy consumption. Coalesced dust particles can be trapped rather effectively by the simplest of devices with only a minor drag, including trapping immediately inside the cowl cavity as a result of gravitational forces [196].

Considerations detailed here have been taken into account in the design of a transfer facility (Figure II.14) that seeks to address aspiration optimization problems comprehensively.

Material is discharged from the conveyor (1) onto an articulated flow-shaping piece (2) producing a cylindrically profiled compacted flow. The position of this piece may be adjusted using counterweights (3) depending on the material flow rate. The material then passes into a hollow spherical

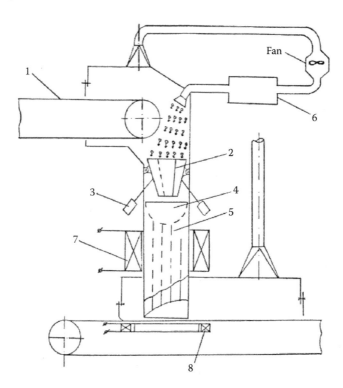

FIGURE II.14 Handling facility optimized comprehensively in terms of both aspiration volume and material carryover rates. (From An inquiry into airborne dust control and occupational safety improvement in iron ore agglomeration shops, Final Research Study Report/All-Union Institute for Occupational Safety in Ore Mining (VNIIBTG), Director: I.N. Logachev, No. GR 73052312, Inv. No. B452597, Krivyy Rih, Ukraine, 1975, 158pp.)

cutout (4) of guiding cage (5). The spherical cutout serves as a receiving hopper from which the material is transferred into the cells of guiding cage (5). At this point, loose material separates into several flows, maintaining a linked mode of motion as it fills the cells.

Loose material is treated with ionized air and multiple instances of magnetic field as it passes through the handling system. Ionizer (6) and electromagnets (7, 8) are provided for that purpose. This design proposal has numerous benefits. Enabling the linked motion of material makes it possible to mitigate air ejection, reduce aspiration volume, and dust formation intensity. As the material is handled with ionized air and magnetic field, the efficiency of dust settlement within the cowl increases.

II.5 CONCLUSIONS

1. Relevant problems in the design of aspiration cowl for loose-matter handling facilities include devising means and techniques for reducing aspiration volumes necessary to maintain proper dedusting efficiency, as well as reducing the carryover of dustlike material into the aspiration network.
2. In order to reduce the volume of air ejected by a flow of loose matter, the most rational and rewarding approach would be to ensure closed-loop circulation of dust-laden air streams in transfer chutes without using any additional forced-draft devices. Air circulation processes in such chutes have not been described analytically to this day.
3. Options for reducing the volume of air leaking into the cowl by increasing the area of contact between the seal and moving material are virtually exhausted by now. Further

development should focus on means and techniques that would ensure the necessary sealing effect while minimizing the area of this contact or avoiding it altogether.

4. The goal of reducing carryover of dust into the aspiration network calls for minimizing both the total carryover rate and specific dust content in aspiration system air. It would be preferable to address that goal with process–design measures and by equipping enclosures with dry dedusting devices.

5. When it comes to the means and techniques for optimizing aspiration systems of transfer assemblies, it is imperative that solutions be comprehensive, achieving wide-ranging positive effects both in terms of reducing material carryover rates and bringing down the volume of air evacuated by suction.

5 Specific Features of Air Ejection in a Perforated Duct with a Bypass Chamber

A common solution for reducing the required aspiration volume comprises bypass ducts that enable internal air recycling while reducing ingress of air from transfer chutes into aspirated cowls. Some of the known solutions have the hollow space of the chute connected with the bypass chamber not only in the upper and lower parts but also along the entire height of the chute by perforating its walls. In order to measure the efficiency of this solution, let us consider aerodynamic processes of air ejection for a case of a uniformly distributed flow of particles in a vertical duct, with particles falling at a constant acceleration (understandably, such a model is applicable for particle flows with a small height of free fall as long as the drag force is still easily overcome by the weight of a falling particle)

$$\frac{d\tilde{v}}{d\tilde{t}} = \tilde{g}. \tag{5.1}$$

In the following text, a tilde above a symbol will be used to distinguish dimensional quantities from their dimensionless counterparts: $\tilde{v}$ is particle velocity (m/s), $\tilde{t}$ is time (s), and $\tilde{g}$ is gravitational acceleration (m/s²).

5.1 ONE-DIMENSIONAL DYNAMICS EQUATIONS FOR EJECTED AND RECYCLED AIR

Let us consider an axially symmetrical flow of particles in a round pipe with a cross-sectional area of $\tilde{s}_t$, m² (Figure 5.1). Inside the pipe there is a cylindrical bypass chamber (with a cross-sectional area of $\tilde{s}_b$ (m²)) aerodynamically coupled with the pipe by a perforated wall. Both the velocity of ejected air in the pipe ($\tilde{u}$ (m/s)) and the velocity of upward airflow in the bypass chamber ($\tilde{\omega}$, m/s) vary along the duct as a result of air escaping from the pipe into the bypass chamber. These velocities will be determined using the following equation for the conservation of the amount of air motion in an immobile volume $\tilde{V}$ (m³) confined within surface $\tilde{S}$ (m²):

$$\int_{\tilde{S}} \tilde{\rho}\vec{\tilde{u}}\tilde{u}_n d\tilde{S} = \int_{\tilde{V}} \vec{\tilde{M}} d\tilde{V} + \int_{\tilde{S}} \vec{\tilde{p}}_n d\tilde{S}, \tag{5.2}$$

where
$\tilde{u}_n$ is the projection of air velocity vector onto external normal $\vec{n}$ of the surface $\tilde{S}$ (m/s)
$\vec{\tilde{u}}$ is the air velocity vector (m/s)
$\tilde{\rho}$ is the air density (kg/m³)
$\vec{\tilde{M}}$ is the vector of mass forces (N/m³)
$\vec{\tilde{p}}_n$ is the stress vector for a surface force applied to the elementary area $d\tilde{S}$ with an external normal $\vec{n}$ (Pa)

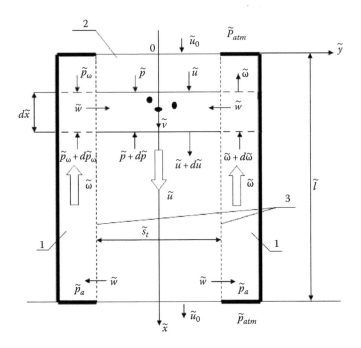

FIGURE 5.1   Design layout of a perforated chute with a bypass chamber: 1, bypass chamber; 2, chute (pipe); 3, perforated chute walls.

Expanding this integral relation for elementary volume $d\tilde{x} \cdot \tilde{s}_t$ in view of the mass forces of interaction between components yields

$$\widetilde{M} = \psi\beta \frac{\tilde{F}_m}{\tilde{V}_p} \frac{(\tilde{v}-\tilde{u})\left|\tilde{v}-\tilde{u}\right|}{2} \tilde{\rho}, \tag{5.3}$$

where

ψ is the aerodynamic drag coefficient for a single particle
$\tilde{F}_m$ is the midsection area of a particle (m²)
$\tilde{V}_p$ is the volume of the particle (m³)
β is the volumetric concentration of particles inside the pipe

$$\beta = \frac{\tilde{G}_p}{\tilde{\rho}_p \tilde{v}\tilde{s}_t}, \tag{5.4}$$

where

$\tilde{G}_p$ is the maximum flow rate of particles (kg/s)
$\tilde{\rho}_p$ is the particle density (kg/m³)

If aerodynamic resistance forces created by duct walls were left out as negligible, a projection onto axis *OX* would take the form

$$\tilde{\rho}\tilde{u}(-\tilde{u})\tilde{s}_t + \tilde{\rho}(\tilde{u}+d\tilde{u})(\tilde{u}+d\tilde{u})\tilde{s}_t = d\tilde{x}\tilde{s}_t\psi\beta \frac{\tilde{F}_m}{\tilde{V}_p} \frac{(\tilde{v}-\tilde{u})\left|\tilde{v}-\tilde{u}\right|}{2} \tilde{\rho} + \tilde{p}\tilde{s}_t - (\tilde{p}+d\tilde{p})\tilde{s}_t. \tag{5.5}$$

Whence, ignoring minor higher-order quantities, the basic equation for the 1D problem of ejected air dynamics in a pipe could be written as follows:

$$d\tilde{p} + 2\tilde{\rho}\tilde{u}d\tilde{u} = \tilde{E}\frac{|\tilde{v} - \tilde{u}|(\tilde{v} - \tilde{u})}{2}\tilde{\rho}d\tilde{x}; \quad \tilde{E} = \psi\beta\frac{\tilde{F}_m}{\tilde{V}_p}, \tag{5.6}$$

or, translating into dimensionless form (dividing both parts of the equation by $\tilde{\rho}\tilde{v}_k^2/2$),

$$dp + 4udu = \frac{\mathrm{Le}(v - u)|v - u|}{v} \cdot dx, \tag{5.7}$$

where

$$\mathrm{Le} = \psi\beta_k\tilde{l}\frac{\tilde{F}_m}{\tilde{V}_p}, \quad \beta_k = \frac{\tilde{G}_p}{\tilde{\rho}_p\tilde{v}_k\tilde{s}_t} \tag{5.8}$$

$$p = 2\tilde{p}/(\tilde{\rho}\tilde{v}_k^2), \quad u = \tilde{u}/\tilde{v}_k, \quad v = \tilde{v}/\tilde{v}_k, \quad x = \tilde{x}/\tilde{l} \tag{5.9}$$

$\tilde{p}$ is the static pressure inside the pipe (Pa) (from now on, usual static pressure will be considered)
$\tilde{v}_k$ is the velocity of particles at pipe outlet (m/s)
$\tilde{l}$ is the total length of the pipe (m)

Note that the dimensionless number Le (ejection parameter) represents the ratio of maximum ejection pressure forces (at $\tilde{v} - \tilde{u} = \tilde{v}_k$) to the dynamic pressure of ejected air:

$$\mathrm{Le} = \tilde{s}_t \cdot \tilde{l} \cdot \beta_k\psi\frac{\tilde{F}_m}{\tilde{V}_p} \cdot \frac{\tilde{v}_k^2}{2}\tilde{\rho}/\left(\tilde{s}_t \cdot \tilde{\rho}\frac{\tilde{v}_k^2}{2}\right). \tag{5.10}$$

In a similar way, a dynamics equation can be derived for an upward flow of air in a bypass duct with a cross-sectional area of $\tilde{s}_b$ (assuming that the duct walls create no aerodynamic resistance forces):

$$-\tilde{\rho}\tilde{\omega} \cdot (\tilde{\omega})\tilde{s}_b - \tilde{\rho}(\tilde{\omega} + d\tilde{\omega})(-\tilde{\omega} - d\tilde{\omega})\tilde{s}_b = \tilde{p}_\omega \cdot \tilde{s}_b - (\tilde{p}_\omega + d\tilde{p}_\omega)\tilde{s}_b. \tag{5.11}$$

This leads to a dimensionless equation for the dynamics of recirculated air:

$$dp_\omega + 4\omega d\omega = 0; \quad p_\omega = 2\tilde{p}_\omega/(\tilde{\rho}\tilde{v}_k^2); \quad \omega = \tilde{\omega}/\tilde{v}_k, \tag{5.12}$$

where p_ω is a dimensionless pressure in cross sections of the bypass duct.

The magnitude of this pressure is determined with an obvious equation in our case:

$$p_\omega = p_a - 2\omega^2, \quad p_a = 2\tilde{p}_a/(\tilde{\rho}\tilde{v}_k^2), \tag{5.13}$$

where $\tilde{p}_a$ is the excess static pressure at the duct inlet/outlet (Pa).

Elementary volumes $d\tilde{x} \cdot \tilde{s}_t$ and $d\tilde{x} \cdot \tilde{s}_b$ with the following integral mass conservation relation,

$$\int_{\tilde{S}} \tilde{\rho}\tilde{u}_n d\tilde{S} = 0, \tag{5.14}$$

applied, would result in the following relation between air velocities in the upstream section of the perforated pipe and bypass chamber*:

$$\frac{du}{dx} = \hat{S}_t w, \quad \frac{d\omega}{dx} = \hat{S}_b w, \quad \frac{d(u - \omega \cdot r)}{dx} = 0, \tag{5.15}$$

$$\omega = \frac{(u - u_0)}{r}, \quad r = \frac{\hat{S}_t}{\hat{S}_b} = \frac{\tilde{s}_b}{\tilde{s}_t}, \tag{5.16}$$

where

$$\hat{S}_t = \frac{\tilde{\Pi}\tilde{l}}{\tilde{s}_t}\varepsilon$$

$$\hat{S}_b = \frac{\tilde{\Pi}\tilde{l}}{\tilde{s}_b}\varepsilon$$

$\tilde{\Pi}$ is the pipe diameter (m)
$\tilde{s}_b$ is the cross-sectional area of the bypass duct (m²)
ε is the degree of the pipe wall perforation ($\varepsilon = 0$ absent perforation, $\varepsilon = 1$ absent pipe walls)

$$w = \frac{\tilde{w}}{\tilde{v}_k}, \quad u_0 = \frac{\tilde{u}_0}{\tilde{v}_k}, \tag{5.17}$$

$\tilde{w}$ is the velocity of air in the perforation holes (m/s); $\tilde{u}_0$ is the velocity of ejected air at the pipe inlet (m/s) (considering that $\omega(0) = \omega(1) = 0$, it is evident from Equation 5.16 that $\tilde{u}_k = \tilde{u}_0$, where $\tilde{u}_k$ is the velocity at the pipe outlet (m/s)).

The velocity of the cross flow of air from the pipe into the bypass chamber via perforation holes is determined with the relation

$$\tilde{w} = \gamma_1 \cdot \sqrt{\frac{2|\tilde{p}_\omega - \tilde{p}|}{\zeta_0 \tilde{p}}}, \quad \gamma_1 = \text{signum}\left(\tilde{p}_\omega - \tilde{p}\right) \tag{5.18}$$

or, in dimensionless form

$$w = \frac{\gamma\sqrt{|p_\omega - p|}}{\sqrt{\zeta_0}}, \quad \gamma = \text{signum}\left(p_\omega - p\right), \tag{5.19}$$

where ζ_0 is the local resistance coefficient (LRC) of a perforation hole.

Thus, in view of the relations of Equations 5.13 and 5.16, 5.7 and 5.15, and 5.19, it follows that

$$\frac{du}{dx} = E \cdot \gamma\sqrt{|\Delta p|}; \tag{5.20}$$

$$\frac{dp}{dx} = \frac{-4uE\gamma\sqrt{|\Delta p|} + \text{Le}(v - u)|v - u|}{v}, \tag{5.21}$$

where

$$\Delta p = p_\omega - p = p_a - 2\left(\frac{u - u_0}{r}\right)^2 - p \tag{5.22}$$

---

* It is assumed that air escapes through perforation holes from the bypass chamber into the pipe at inlet stage, reversing its direction from the pipe into the chamber at the outlet stage.

$$E = \hat{S}_t / \sqrt{\zeta_0} = \tilde{\Pi}\tilde{l}\varepsilon / (\tilde{s}_t \sqrt{\zeta_0}) \tag{5.23}$$

$$v = \sqrt{(1-n^2)x + n^2}, \quad n = \tilde{v}_0 / \tilde{v}_k \tag{5.24}$$

where $\tilde{v}_0$ is the velocity of particle flow at the pipe inlet (m/s).

In order to solve two equations, Equations 5.20 and 5.21 numerically, the following boundary conditions will be introduced:

$$u(0) = u_0, \quad p(0) = -\zeta_n u_0^2, \tag{5.25}$$

$$u(1) = u_0, \quad p(1) = \zeta_k u_0^2. \tag{5.26}$$

where ζ_n, ζ_k are LRCs for the pipe air inlet and outlet. Certain complications arise due to the fact that air velocity at the pipe inlet/outlet u_0 and excess pressure in upstream/downstream sections of the bypass duct p_a are at the same time unknown variables.

Therefore, the boundary-value problem is solved using the false position method: a Cauchy problem is solved with u_0 as the solve-for variable and initial conditions from Equation 5.25. At the same time, p_a is solved for assuming the condition

$$p(1) > p_a > p(0)\theta,* \tag{5.27}$$

that is,

$$p_a \approx \frac{(\zeta_k - \zeta_n)}{2} u_0^2 \tag{5.28}$$

and then checking for conditions from Equation 5.26. Search for u_0 and p_a can be facilitated using the bisection method (i.e., by recurrently dividing in halves).

In the case of a flow of light particles (or particles falling from a great height), their velocities will be reduced due to counteracting drag from air. The motion equation for a vertical particle flow can be described with the following equation for change in momentum:

$$\tilde{v}\frac{d\tilde{v}}{dx} = \tilde{g} - \psi \frac{\tilde{F}_m}{\tilde{V}_p} \cdot \frac{|\tilde{v} - \tilde{u}|(\tilde{v} - \tilde{u})}{2} \cdot \frac{\tilde{\rho}}{\tilde{\rho}_p}. \tag{5.29}$$

Or, translating into dimensionless form (dividing both parts of the equation by $\tilde{v}_k^2 / \tilde{l}$),

$$v\frac{dv}{dx} = \frac{1-n^2}{2} - \mathrm{Le} \cdot G_b \frac{(v-u)|v-u|}{2}, \tag{5.30}$$

where G_b is the ratio of maximum airflow rate (at $\tilde{u} = \tilde{v}_k$) to the flow rate of particles:

$$G_b = \frac{\tilde{\rho}\tilde{v}_k \tilde{s}_t}{\tilde{G}_p} = \frac{\tilde{\rho}}{\beta_k \tilde{\rho}_p}, \tag{5.31}$$

---

* It could be also helpful to use the results produced by approximate solution (e.g., solving a linearized equation that will be derived later on).

$\tilde{v}_k$ is the velocity of particle flow at the duct outlet assuming a constantly accelerated fall (m/s)

$$\tilde{v}_k = \sqrt{2\tilde{g}\tilde{l} + \tilde{v}_n^2}, \tag{5.31a}$$

where $\tilde{l}$ is the height of the duct/pipe (m).

Boundary conditions for Equation 5.30 appear as $v(0) = n$; $v(1) = 1$. Solving the problem in this case requires integrating three combined equations, Equations 5.20, 5.21, and 5.30, with the respective boundary conditions.

5.2 LINEARIZATION OF RELATIVE MOTION EQUATIONS FOR AIR EJECTED IN A PERFORATED PIPE

Before moving on to the quantitative study of ejected air dynamics described by a set of nonlinear equations, Equations 5.20 and 5.21, with boundary conditions from Equations 5.25 and 5.26, let us consider air ejection induced in a perforated pipe with a bypass chamber by a certain artificially constructed flow of particulate matter. The change to this putative flow is motivated by the intent to obtain an analytical solution for the asymptotic behavior of the equations of type (Equations 5.20 and 5.21) in areas of $\varepsilon = 0 (E = 0)$, $\varepsilon = 1$, and $\zeta_0 \rightarrow 0 (E \rightarrow \infty)$, as well as at $r \rightarrow 0$ (at $\tilde{s}_b \rightarrow 0$) and $r \rightarrow \infty$ (at $\tilde{s}_b \rightarrow \infty$).

Unlike real conditions, it will be assumed that the acceleration of this flow varies over the height of the pipe while the velocity is described by a quadratic trinomial

$$v = b_0 + b_1 x + \frac{b_2}{2} x^2, \tag{5.32}$$

the coefficients of which,

$$b_0 = n, \quad b_1 = \sqrt{8(1+n^2)} - 3n - 1, \quad b_2 = 4(1 + n - \sqrt{2(1+n^2)}), \tag{5.33}$$

result from the condition that the velocity putative flow is equal to that of the real flow (determined with Equation 5.24) in reference points, namely, the pipe inlet, center, and outlet. This binding serves to minimize deviations between putative and real flows in terms of kinematics.

In addition, the force of aerodynamic interaction between particles and air will be linearized by positing

$$\text{Le} \frac{|v - u|(v - u)}{v} \approx \text{Le} \cdot k(v - u), \quad k = \overline{\left|1 - \frac{u}{v}\right|} \approx 1 - \frac{\overline{u}}{\overline{v}}, \tag{5.34}$$

assuming drag forces acting against cross flows of air through perforation holes to be equal:

$$p_\omega - p \approx \zeta_0 \overline{|w|} w \tag{5.35}$$

with the same drag forces for flows through the pipe inlet and outlet sections

$$p(0) = -\zeta_n \overline{u} \cdot u_0, \tag{5.36}$$

$$P(1) = \zeta_k \overline{u} \cdot u_0, \tag{5.37}$$

and additionally allowing for

$$udu \approx \bar{u}du, \tag{5.38}$$

$$\omega d\omega \approx \bar{\omega}d\omega, \tag{5.39}$$

where a bar above a variable denotes averaging over the interval $x = 0\dots1$.

Taking into account condition, Equation 5.39, based on Equation 5.12, the *exact* Equation 5.13 will be replaced with

$$p_\omega = p_a - 4\bar{\omega}\omega \tag{5.40}$$

and, in view of Equation 5.16,

$$p_\omega = \frac{p_a - 4\bar{\omega}(u - u_0)}{r}. \tag{5.41}$$

Whereby, condition given in Equation 5.35 in view of Equation 5.15 will become

$$p = p_a - 4\bar{\omega}\frac{(u - u_0)}{r} - \zeta_0\overline{|w|} \cdot \frac{1}{\hat{S}_t} \cdot \frac{du}{dx}. \tag{5.42}$$

Then, original Equation 5.7 in view of Equations 5.38 and 5.34 can be written as

$$\frac{dp}{dx} = -4\bar{u}\frac{du}{dx} + k\mathrm{Le}(v - u) \tag{5.43}$$

or, in view of Equation 5.42, it can be expressed as a linear nonuniform second-order equation

$$a\frac{d^2u}{dx^2} - b\frac{du}{dx} - c(u - v) = 0, \tag{5.44}$$

where the constants a, b, and c are, respectively, equal to

$$a = \frac{\zeta_0}{\hat{S}_t}\overline{|w|} = \frac{\sqrt{\zeta_0}}{E}\overline{|w|}, \quad b = 4\bar{u} - 4\frac{\bar{\omega}}{r}, \quad c = k \cdot \mathrm{Le}. \tag{5.45}$$

Let us introduce a new function to represent the difference between dimensionless velocities of aerodynamic interaction components:

$$z = v - u, \tag{5.46}$$

The formula given in Equation 5.32 will be used to determine the velocity of particles in order to simplify the right-hand part of the equation). Then Equation 5.44 could be rewritten as

$$z'' - 2\alpha_1 z' - \beta_1 z = k_1 + k_2 x, \tag{5.47}$$

where z is a new function of x

$$z'' = \frac{d^2z}{dx^2}, \quad z' = \frac{dz}{dx}, \tag{5.48}$$

$$\alpha_1 = \frac{2}{\sqrt{\zeta_0}}E\frac{\bar{u} - \bar{\omega}/r}{\overline{|w|}}, \quad \beta_1 = \frac{k\mathrm{Le}}{\sqrt{\zeta_0}} \cdot \frac{E}{\overline{|w|}}, \tag{5.49}$$

$$k_1 = -2\alpha_1 b_1 + b_2, \quad k_2 = -2\alpha_1 b_2. \tag{5.50}$$

Boundary conditions for this equation can be expressed as follows. In view of Equations 5.42, 5.32, 5.36, and 5.37 we end up with

$$z(0) = n - u_0, \quad z(1) = 1 - u_0, \tag{5.51}$$

$$z'(0) = b_1 + \frac{\hat{S}_t}{\zeta_0 |w|} \left[-\zeta_n \bar{u} \cdot u_0 - p_a \right] = b_1 + \frac{E}{\sqrt{\zeta_0 |w|}} \left[-\zeta_n \bar{u} \cdot u_0 - p_a \right], \tag{5.52}$$

$$z'(1) = b_1 + b_2 + \frac{\hat{S}_t}{\zeta_0 |w|} \left[\zeta_k \bar{u} \cdot u_0 - p_a \right] = b_1 + b_2 + \frac{E}{\sqrt{\zeta_0 |w|}} \left[\zeta_k \bar{u} \cdot u_0 - p_a \right]. \tag{5.53}$$

The common solution of nonuniform equation (Equation 5.47) with constant coefficients would have the form

$$z = c_1 e^{\lambda_1 x} + c_2 e^{\lambda_2 x} + \frac{1}{\beta_1} \left[2 \frac{\alpha_1}{\beta_1} k_2 - k_1 - k_2 x \right], \tag{5.54}$$

whence we can find

$$z' = c_1 \lambda_1 e^{\lambda_1 x} + c_2 \lambda_2 e^{\lambda_2 x} - \frac{k_2}{\beta_1}, \tag{5.55}$$

$$z'' = c_1 \lambda_1^2 e^{\lambda_1 x} + c_2 \lambda_2^2 e^{\lambda_2 x}, \tag{5.56}$$

where

$$\lambda_1 = \alpha_1 + \sqrt{\alpha_1^2 + \beta_1}, \quad \lambda_2 = \alpha_1 - \sqrt{\alpha_1^2 + \beta_1} \tag{5.57}$$

are the roots of a characteristic equation corresponding to the homogeneous differential equation (Equation 5.47). By substituting Equations 5.54 through 5.56 into Equation 5.47, it is easy to verify that the relation given by Equation 5.54 is indeed the common solution of this equation in view of Equation 5.57.

Application of boundary conditions given by Equations 5.51 through 5.53 produces the following set of four algebraic equations for unknown constants in the problem—c_1, c_2, u_0, and p_a:

$$n - u_0 = c_1 + c_2 + \frac{1}{\beta_1} \left[2 \frac{\alpha_1}{\beta_1} k_2 - k_1 \right], \tag{5.58}$$

$$1 - u_0 = c_1 e^{\lambda_1} + c_2 e^{\lambda_2} + \frac{1}{\beta_1} \left[2 \frac{\alpha_1}{\beta_1} k_2 - k_1 - k_2 \right], \tag{5.59}$$

$$b_1 + \frac{E}{\sqrt{\zeta_0 |w|}} \left[-\zeta_n \bar{u} \cdot u_0 - p_a \right] = c_1 \lambda_1 + c_2 \lambda_2 - \frac{k_2}{\beta_1}, \tag{5.60}$$

$$b_1 + b_2 + \frac{E}{\sqrt{\zeta_0 |w|}} \left[\zeta_k \bar{u} \cdot u_0 - p_a \right] = c_1 \lambda_1 e^{\lambda_1} + c_2 \lambda_2 e^{\lambda_2} - \frac{k_2}{\beta_1}. \tag{5.61}$$

5.3 NUMERICAL STUDIES

5.3.1 ESTIMATING BOUNDARY CONDITIONS

Before solving combined Equations 5.20 and 5.21 numerically (in view of the relations given in Equations 5.22 through 5.26), certain analytical transformations would be apt in order to draw a tentative estimate of the parameters appearing in the problem of source equations.

First of all, let us demonstrate that the boundary conditions given in Equations 5.25 and 5.26 can be expanded to include the case of a perforated chute abutting on aspirated cowls (Figure 5.2).

The following absolute pressures are indicated in this figure:

$\tilde{P}_{atm}$—Inside the room (Pa)
$\tilde{P}_n$—In the upper cowl (Pa)
$\tilde{P}_f$—In the flow-shaping chamber (inner chamber) of the lower cowl (Pa)
$\tilde{P}_a$—At the inlet and outlet of the bypass duct (Pa)
$\tilde{P}_0$—Immediately downstream of the inlet section of the pipe (Pa)
$\tilde{P}_k$—Immediately upstream of the outlet section of the pipe (Pa)

These pressures are linked with absolute pressures at inlet/outlet cross sections by obvious equations:

$$\tilde{P}_n - \tilde{P}_0 = \zeta_n \frac{\tilde{u}_0^2}{2} \tilde{\rho}, \tag{5.62}$$

$$\tilde{P}_k - \tilde{P}_f = \zeta_k \frac{\tilde{u}_0^2}{2} \tilde{\rho}, \tag{5.63}$$

$$\tilde{P}_a - \tilde{P}_0 = \zeta_0 \frac{|\tilde{w}_0| \tilde{w}_0}{2} \tilde{\rho}, \tag{5.64}$$

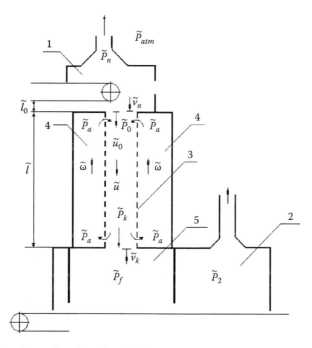

FIGURE 5.2 Aspiration layout for a handling facility equipped with upper (1) and lower (2) aspirated cowls, a perforated chute (3) with a bypass chamber (4), and a flow-shaping chamber (5) in the lower cowl.

$$\tilde{P}_k - \tilde{P}_a = \zeta_k \frac{|\tilde{w}_k| \tilde{w}_k}{2} \tilde{\rho} \tag{5.65}$$

or, translating into dimensionless form (dividing both parts of equations by $\tilde{v}_k^2 \tilde{\rho} / 2$)

$$P_n - P_0 = \zeta_n u_0^2, \tag{5.66}$$

$$P_k - P_f = \zeta_k u_0^2, \tag{5.67}$$

$$P_a - P_0 = \zeta_0 |w_0| w_0, \tag{5.68}$$

$$P_k - P_a = \zeta_k |w_k| w_k. \tag{5.69}$$

Boundary conditions of concern to us, Equations 5.66 and 5.67, can be expressed using excess pressures

$$(P_n - P_{atm}) - (P_0 - P_{atm}) = \zeta_n u_0^2, \tag{5.70}$$

$$(P_k - P_{atm}) - (P_f - P_{atm}) = \zeta_k u_0^2 \tag{5.71}$$

or

$$p_{un} - p_{uo} = \zeta_n u_0^2, \tag{5.72}$$

$$p_{uk} - p_{uf} = \zeta_k u_0^2, \tag{5.73}$$

where p_{un}, p_{u0} are dimensionless excess pressures in the upper cowl and at the inlet section of the chute, while p_{uk}, p_{uf} are dimensionless excess pressures in the outlet section of the chute and in the flow-shaping chamber.

Thus, boundary conditions for pressure will change from Equations 5.25 and 5.26 as follows:

$$p(0) = p_{un} - \zeta_n u_0^2, \tag{5.74}$$

$$p(1) = p_{uf} + \zeta_k u_0^2, \tag{5.75}$$

with the difference between them being equal to the value of constants p_{un} and p_{uf}.

The volumetric airflow at perforated chute inlet or outlet is defined by formula

$$\tilde{Q}_t = \tilde{u}_0 \tilde{s}_t \ (\text{m}^3/\text{s}), \tag{5.76}$$

where
$\tilde{u}_0$ is the velocity of air at pipe/chute inlet section (m/s)
$\tilde{s}_t$ is the cross-sectional area of perforated pipe/chute (m²)

If a dimensionless quantity u_0 is used for air velocity at the inlet/outlet section, then the computational formula would appear in a different form:

$$\tilde{Q}_t = u_0 \tilde{v}_k \tilde{s}_t \ (\text{m}^3/\text{s}), \tag{5.77}$$

where $\tilde{v}_k$ is the velocity of particles at the pipe/chute outlet:

$$\tilde{v}_k = \sqrt{2\tilde{g}(\tilde{l} + \tilde{l}_0)} = \sqrt{\tilde{v}_n^2 + 2\tilde{g}\tilde{l}}, \tag{5.78}$$

where

$$\tilde{v}_n = \sqrt{2\tilde{g}\tilde{l}_0} \tag{5.79}$$

$\tilde{l}$ is the height of the perforated pipe/chute (m)
$\tilde{l}_0$ is the height of the free-fall path of particles prior to their entry into the perforated pipe/chute (m)

In this case, the particle velocity ratio would appear as

$$n = \frac{\tilde{v}_n}{\tilde{v}_k} = \sqrt{\frac{\tilde{l}_0}{\tilde{l}_0 + \tilde{l}}}. \tag{5.80}$$

5.3.2 SPECIAL CASE FOR AIR EJECTION IN A PIPE WITH IMPERMEABLE WALLS ($E = 0$)

The energy efficiency of pipe wall perforation as a means of enabling the recirculation of air in the bypass chamber will be evaluated by comparing the velocity u_0 with velocity of air in a pipe with impermeable walls u_2. This case was examined in sufficient detail by Russian and foreign researchers. Therefore, we will be focusing exclusively on the classical results of the dynamic ejection theory, which rests upon applying an equation for momentum change law to a two-velocity continuum comprised by solid particles and air.

Let us integrate Equation 5.21 along the duct (in view of Equation 5.20)

$$\int_{P(0)}^{P(1)} dp = -4\int_{u(0)}^{u(1)} u\,du + Le\int_0^1 \frac{|v-u|}{v}(v-u)dx. \tag{5.81}$$

Transformation of this equation in view of boundary conditions given in Equations 5.25 and 5.26 results in the following integral relation for determining the velocity of ejected air u_0:

$$u_0^2 = \frac{Le}{\sum \zeta}\int_0^1 \frac{|v-u|}{v}(v-u)dx, \quad \sum \zeta = \zeta_n + \zeta_k \tag{5.82}$$

given a known function

$$u = F_u(x, Le, E, r, \zeta_n, \zeta_k, \zeta_0). \tag{5.83}$$

It should be noted that, in the case of impermeable walls ($\varepsilon = 0$; $E = 0$), this function

$$u = u_0 \approx \text{const} \tag{5.84}$$

and the integral relation shown by Equation 5.82 in view of Equation 5.24 allows in putting forward the following algebraic equation for determining the velocity of ejected air in a regular (nonperforated) duct of a constant cross section:

$$u_0^2 = \text{Bu} \int_n^1 |v - u_0|(v - u_0)dv = \text{Bu} \cdot \begin{cases} \dfrac{(1-u_0)^3 - (n-u_0)^3}{3} & \text{at } u_0 \le n, \\[4mm] \dfrac{(1-u_0)^3 - (u_0-n)^3}{3} & \text{at } 1 \ge u_0 \ge n, \end{cases} \qquad (5.85, 5.86)$$

where

$$\text{Bu} = \frac{2\text{Le}}{(1-n^2)\sum \zeta}, \qquad (5.87)$$

that is, at $0 \le \text{Le} \le \text{Le}_0$ $(0 \le \text{Bu} \le \text{Bu}_0)$

$$\frac{3u_0^2}{(1-u_0)^3 - (n-u_0)^3} = \text{Bu}, \qquad (5.88)$$

while, at

$$\infty \ge \text{Le} \ge \text{Le}_0 \ (\infty \ge \text{Bu} \ge \text{Bu}_0)$$

$$\frac{3u_0^2}{(1-u_0)^3 - (u_0-n)^3} = \text{Bu}, \qquad (5.89)$$

where

$$\text{Le}_0 = \frac{3n^2(1-n^2)\sum \zeta}{2(1-n)^3}, \qquad (5.90)$$

$$\text{Bu}_0 = \frac{3n^2}{(1-n)^3}, \qquad (5.91)$$

are transition ejection constants.

The latter equation can be used to determine the maximum velocity of ejected air (discovered for the first time in our earlier work [210])

$$\lim_{\text{Bu} \to \infty} u_0 \equiv u_0^{\max} = \frac{1+n}{2}. \qquad (5.92)$$

A problem of no less importance is studying the change in excess pressure along the pipe as a function of excess pressures in the upper and lower chambers abutting to the inlet and outlet sections of this pipe. While still considering a pipe with impermeable walls and assuming that the velocity of ejected air is constant along the pipe ($u = $ const at low constant volumetric concentrations of particles ($\beta_k < 0.01$), a case characteristic of transfer chutes at industrial facilities handling loose matters), we will integrate a differential equation (Equation 5.7) at boundary conditions Equations 5.74 and 5.75. The latter can be rewritten (assuming a direct flow with $u_0 > 0$) in the customary form by introducing putative LRCs that not only make allowance for air pressure losses at inlet and outlet but also account for the magnitude of excess pressures in adjacent chambers (cowls):

$$p(0) = -\zeta_n^* u_0^2, \quad p(1) = \zeta_k^* u_0^2, \tag{5.93}$$

where

$$\zeta_n^* = \zeta_n - \frac{p_{un}}{u_0^2}, \quad \zeta_k^* = \zeta_k + \frac{p_{uf}}{u_0^2}. \tag{5.94}$$

If the analysis of pressure changes in a pipe was to be extended to include the case of inverse motion of air (i.e., the counterflow case $u_0 < 0$), initial boundary conditions (Equations 5.74 and 5.75) should be specified as

$$p(0) = p_{un} - \zeta_n^{\uparrow} u_0 |u_0|, \quad p(1) = p_{uf} + \zeta_k^{\uparrow} u_0 |u_0|$$

and then

$$p(0) = -\zeta_n^{*\uparrow} u_0 |u_0|, \quad p(1) = \zeta_k^{*\uparrow} u_0 |u_0|, \tag{5.95}$$

where putative LRCs would appear as

$$\zeta_n^{*\uparrow} = \frac{p_{un}}{u_0^2} - \zeta_n^{\uparrow} \frac{|u_0|}{u_0}, \quad \zeta_k^{*\uparrow} = \frac{p_{uf}}{u_0^2} - \zeta_k^{\uparrow} \frac{|u_0|}{u_0}, \tag{5.96}$$

and the $\uparrow$ symbol would express LRCs of inlet and outlet pipe sections in counterflow conditions (generally, for $\zeta_n \neq \zeta_n^{\uparrow}$ and $\zeta_k \neq \zeta_k^{\uparrow}$).

We will assume a negative pressure, that is, $p_{un} < 0$ for the upper chamber/cowl, along with $p_{uf} > 0$ and, similarly, directed flows of air and particles (direct flow case) for the lower (flow-shaping) chamber.

Before moving on to studying the behavior of pressure along the pipe, we will determine u_0 from integral relation given in Equation 5.82 using the accepted convention for conventional LRCs, Equation 5.94, ending with

$$u_0^2 = \frac{\text{Le}}{\sum \zeta^*} \int_0^1 \frac{|v - u_0|}{v} (v - u_0) \, dx, \tag{5.97}$$

where

$$\sum \zeta^* = \zeta_n^* + \zeta_k^*. \tag{5.98}$$

This case poses a certain inconvenience due to the presence of an unknown variable, u_0, in $\sum \zeta^*$. However, that is not relevant in the case of computer processing, especially considering that u_0 can be quantified easily and without introducing ζ_n^* and ζ_k^* by writing the differential equation given in Equation 5.82 in view of *old* boundary conditions from Equations 5.74 and 5.75 as

$$u_0^2 = \frac{Le}{\sum \zeta} \int_0^1 \frac{|v-u|(v-u)}{v} dx + \frac{p_{un} - p_{uf}}{\sum \zeta}, \tag{5.99}$$

with the difference being the presence of an additional summand in the right-hand part of the equation comprising Euler's criterion

$$Eu = \frac{p_{un} - p_{uk}}{\sum \zeta} = \frac{\tilde{p}_{un} - \tilde{p}_{uf}}{\sum \zeta \frac{\tilde{v}_k^2}{2} \tilde{\rho}}, $$

where $\tilde{p}_{un}$ and $\tilde{p}_{uf}$ are the respective excess pressures in the upper and lower chambers/cowls (Pa).

It is supposed that values of these pressures are known (e.g., from specifications) when connecting local suctions to these chambers.

Therefore, the value u_0 for a constantly accelerated flow is determined using the algebraic equation

$$|u_0| u_0 = Bu \frac{|1 - u_0|^3 - |n - u_0|^3}{3} + Eu, \tag{5.100}$$

while in the case of putative (accelerated flow), it is determined using the integral relation

$$|u_0| u_0 = Eu + \frac{Le}{\sum \zeta} \cdot \begin{cases} \int_0^1 \frac{(v - u_0)^2}{v} dx & \text{at } u_0 < n, \\ -\int_0^{x_0} \frac{(v - u_0)^2}{2} dx + \int_{x_0}^1 \frac{(v - u_0)^2}{2} dx & \text{at } n < u_0 < 1, \\ -\int_0^1 \frac{(v - u_0)^2}{v} dx & \text{at } u_0 > 1. \end{cases} \tag{5.101}$$

The left-hand part of these equations is written in a universal form relevant to all the cases including the one of upward counterflow of air in the pipe ($u_0 < 0$). That said, it should be noted that in this case the sum

$$\sum \zeta \uparrow = \zeta_n \uparrow + \zeta_k \uparrow \tag{5.102}$$

may not be equal to the similar sum in the direct flow case. However, allowing for a modest margin of error, it can be posited that

$$\zeta_k \uparrow \approx \zeta_n, \quad \zeta_n \uparrow \approx \zeta_k \tag{5.103}$$

and then

$$\sum \zeta \uparrow = \sum \zeta. \tag{5.104}$$

Assuming that the value of u_0 is known, let us examine changes in pressure along an unperforated pipe. Integration of Equation 5.7 produces

$$p(x) - p(0) = \text{Le} \int_0^x \frac{|v - u_0|(v - u_0)}{v} dx. \qquad (5.105)$$

It would be the easiest to solve this equation for the case of a constantly accelerated flow with a velocity determined by the radical

$$v = \sqrt{(1 - n^2)x + n^2}, \qquad (5.106)$$

which enables the transition to a new independent variable v. From

$$2v\,dv = (1 - n^2)dx, \qquad (5.107)$$

it follows that the right-hand part can be easily expressed using quadrature, that is,

$$L(x) = \frac{2\text{Le}}{1 - n^2} \int_n^v |v - u_0|(v - u_0)\,dv = \frac{2\text{Le}}{1 - n^2} \frac{(v - u_0)^3 - (n - u_0)^3}{3} \quad \text{at } u_0 < n. \qquad (5.108)$$

If $1 > u_0 > n$, then

$$L(v) = \frac{2\text{Le}}{1 - n^2} \begin{cases} \dfrac{(v - u_0)^3 - (n - u_0)^3}{3} & \text{within } u_0 \geq v > n, \\[4mm] \dfrac{(v - u_0)^3 - (u_0 - n)^3}{3} & \text{within } 1 \geq v > u_0. \end{cases} \qquad (5.109)$$

Thus, Equation 5.105 in view of Equation 5.74 results in

$$p(x) = p_{un} - \zeta_n u_0^2 + L(v), \qquad (5.110)$$

showing that the excess pressure is rising. At $x = 0$ $(v = n)$, the function is $L(n) = 0$ and, therefore,

$$p(0) = p_{un} - \zeta_n u_0^2, \qquad (5.111)$$

at $x = 1$

$$L(1) = \frac{2\text{Le}}{1 - n^2} \begin{cases} \dfrac{(1 - u_0)^3 - (n - u_0)^3}{3} & \text{at } u_0 < n, \\[4mm] \dfrac{(1 - u_0)^3 - (u_0 - n)^3}{3} & \text{at } 1 > u_0 > n \end{cases} \qquad (5.112)$$

and, consequently,

$$p(1) = p_{un} - \zeta_n u_0^2 + L(1). \qquad (5.113)$$

It can be proven easily that it is identical to

$$p(1) = p_{uf} + \zeta_k u_0^2 \tag{5.114}$$

due to integral relation from Equation 5.101 for quantifying u_0.

And, at a certain distance x_0 of the inlet section, the excess pressure in the pipe is zero. This distance can be discovered from equation

$$p_{un} - \zeta_n u_0^2 + L(v_0) = 0, \tag{5.115}$$

where

$$v_0 = \sqrt{(1-n^2)x_0 + n^2}. \tag{5.116}$$

For the case of small velocities of ejected air ($u_0 < n$), the following computational relation can be written taking this equation as the base:

$$x_0 = \frac{m_2^2 - n^2}{1 - n^2}, \quad m_2 = u_0 + \sqrt[3]{\frac{3(1-n^2)m_1}{2Le} + (n-u_0)^3}, \tag{5.117}$$

$$m_1 = \zeta_n u_0^2 - p_{in}. \tag{5.118}$$

Moreover, the minimum of the function $p(x)$ exists at $x = x_m$. The integral relation Equation 5.105 would yield

$$\left.\frac{dp(x)}{dx}\right|_{x=x_m} = Le \frac{|v(x_m) - u_0|(v(x_m) - u_0)}{v(x_m)} = 0, \tag{5.119}$$

whence an obvious equation follows:

$$v(x_m) = u_0$$

or

$$x_m = \frac{u_0^2 - n^2}{1 - n^2}. \tag{5.120}$$

The condition of the absence of moving air in the pipe ($u_0 = 0$) when loose matter is transferred over it Equation 5.100 is determined with the following relation

$$Eu_0 = -Bu \frac{1 - n^3}{3}. \tag{5.121}$$

Thus, at

$$Eu \leq Eu_0 \tag{5.122}$$

there is the case of a counterflow ($u_0 < 0$) or, in an opposing situation, a direct flow.

Figure 5.3 illustrates plots of pressure changes in the pipe at different excess pressures in the lower chamber. A common property of these changes is the presence of a positive pressure gradient

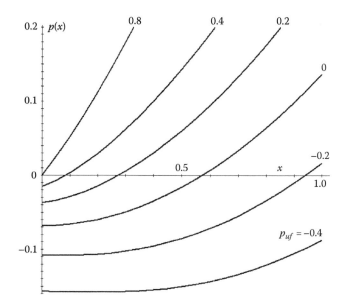

FIGURE 5.3 Variation of pressure along a nonperforated pipe as a function of counterpressure p_{uf} (at $\sum \zeta = 1.5$; $\zeta_n = 0.5$; $\zeta_n^\uparrow = 1.0$; $\zeta_k = 1.0$; $\zeta_k^\uparrow = 0.5$; Le $= 1.0$; $p_{un} = 0$).

(due to Equation 5.105). At zero pressures in adjacent chambers ($p_{un} = p_{uf} = 0$), large externs of vacuum can be noted in the upper part of the pipe and in the positive pressure area ($p(x) > 0$) in its lower part.

These areas diminish when the sign is reversed or the magnitude of excess pressure p_{uk} is changed. Thus, positive pressures result in a contraction of the vacuum zone while negative pressures rather reduce the excess pressure zone, additionally reversing the sign of the pressure gradient (at $p_{uf} \le 0.4$) and giving rise to a minimum $p(x)$. The features indicated will affect the air recycling behavior if pipe walls are perforated.

5.3.3 Averaging Functions and Simplifying Equations

The problem of air ejection in a perforated pipe with a bypass chamber can be substantially simplified by linearizing the mass forces of interaction between components. As demonstrated by the analytical examination of air ejection by a free jet of particles, the following approximation yields an adequate result:

$$\text{Le} \frac{|v-u|}{v}(v-u) \approx k\text{Le} \cdot (v-u), \tag{5.123}$$

where

$$k = 1 - \overline{\left(\frac{u}{v}\right)} \approx 1 - \frac{\overline{u}}{\overline{v}}. \tag{5.124}$$

Hereinafter, a bar above a variable will be used to denote its averaging over duct length, for example,

$$\overline{u} = \int_0^1 u\,dx, \quad \overline{\omega} = \int_0^1 \omega\,dx, \quad \overline{v} = \int_0^1 v\,dx = \frac{2}{3}\frac{1-n^3}{1-n^2} \tag{5.125}$$

for a constantly accelerated flow. The same for an accelerated (putative) flow would appear as

$$\bar{v} = b_0 + 0.5b_1 + \frac{1}{6}b_2^2. \tag{5.126}$$

In this case, integral relation Equation 5.82 will become

$$u_0^2 = k\frac{Le}{\sum \zeta}(\bar{v} - \bar{u}), \tag{5.127}$$

Whence,

$$\bar{u} = \bar{v} - u_0^2\frac{\sum \zeta}{kLe}, \tag{5.128}$$

or, in view of Equation 5.124,

$$u_0^2 = \frac{Le}{\bar{v}\sum \zeta}(\bar{v} - \bar{u})^2, \tag{5.129}$$

then

$$\bar{u} = \bar{v} - u_0\sqrt{\frac{\bar{v}\sum \zeta}{Le}}. \tag{5.130}$$

In order to simplify the problem further, we will supplement our study of an equally accelerated flow of particles described with radical Equation 5.24 by examining a putative flow of particles with a velocity described by a quadratic trinomial

$$v = \frac{b_0 + b_1x + b_2x^2}{2}. \tag{5.131}$$

The coefficients b_0, b_1, and b_2 will be determined from the condition of minimizing the deviation

$$\delta_x = \sqrt{(1 - n^2)x + n^2} - \left(\frac{b_0 + b_1x + b_2x^2}{2}\right) \tag{5.132}$$

within the range $x = 0...1$. For example, given condition $\delta_x = 0$ in fixed points $x = 0, 0.5$, and 1 we will end up with

$$b_0 = n, \quad b_1 = 2N - 3n - 1, \quad b_2 = 4(1 + n - N), \quad N = \sqrt{2(1 + n^2)} \tag{5.133}$$

(at $n = 0.4319$; $b_0 = 0.4319$; $b_1 = 0.785256$; $b_2 = -0.434316$).

If the fixed points are only the extremities of an interval, then

$$b_0 = n, \quad \frac{b_2}{2} = 1 - n - b_1, \tag{5.134}$$

while the coefficient b_1 will be found (using the least-square method) from the equation

$$\frac{\partial}{\partial b_1} \int_0^1 \delta_x^2 dx = 0.$$

The result is

$$b_1 = \frac{-n^7 - 21n^6 + 35n^5 + 63n^4 - 147n^3 + 49n(n+1) - 27}{14(n-1)^3(n+1)^2} \tag{5.135}$$

(at $n = 0.4319$; $b_0 = 0.4319$; $b_1 = 0.790957$; $b_2/2 = -0.222857$). As the example illustrates, the coefficients exhibit only a slight difference in their magnitude. The simpler relation Equation 5.133 will be used from now on.

It will be demonstrated later that the ejective capacity of this putative flow is virtually the same as that of a real flow described with combined Equations 5.20 through 5.26.

Consider the case of an impermeable pipe ($E = 0$) as an example.

Recognizing, that for a uniformly accelerated flow,

$$\bar{v}_q = \int_0^1 v dx = \frac{2}{1-n^2} \int_n^1 v^2 dv = \frac{2}{1-n^2} \frac{1-n^3}{3}, \tag{5.136}$$

and, for a putative flow in view of Equations 5.131 and 5.133,

$$\bar{v}_y = b_0 + \frac{b_1}{2} + \frac{b_2}{6} \tag{5.137}$$

(at $n = 0.4319$, it results in $\bar{v}_q = 0.7535152$; $\bar{v}_y = 0.752142$).

For that flow at $E = 0$, the integral equation (Equation 5.82) in view of Equations 5.123 and 5.124 would appear as

$$u_0^2 = \frac{\text{Le}}{\sum \zeta} \cdot \begin{cases} \overline{v - 2u_0 + u_0^2\left(\dfrac{1}{v}\right)} & \text{at } \text{Le} \le \text{Le}_0, \\ J(u_0) & \text{at } \text{Le} > \text{Le}_0, \end{cases} \tag{5.138, 5.139}$$

where

$$J(u_0) = \int_0^1 |v - u_0|(v - u_0)\frac{dx}{v} = -\int_0^{x_0} (v - u_0)^2 \frac{dx}{v} + \int_{x_0}^1 (v - u_0)^2 \frac{dx}{v} \tag{5.140}$$

x_0 is the distance to the cross section at which the flow of particles accelerates to match the speed of ejected air

This variable is determined by solving the equation

$$u_0 = v(x_0) = \frac{b_0 + b_1 x_0 + b_2 x_0^2}{2}, \tag{5.141}$$

Le_0 is the ejection number at which the velocity of ejected air in a particular duct is $u_0 = n$.

Considering that the final results are somewhat unwieldy, we recommend opting for numerical solution using *Maple*, a universal mathematical software suite. In this case, Equations 5.138 and 5.139 in view of Equation 5.140 would be solved using the `fsolve` command:

```
>restart: Le: = a1; skms: = a2; n: = a3; N: = sqrt(2*(1+n^2)):
b0: = n; b1: = 2*N-3*n-1; b2: = 4*(1+n-N); v: = b0+b1*x+b2*x^2/2;
y: = (v-u0)/v*abs(v-u0); u1: = fsolve(u0^2*skms/Le = Int(y,x = 0...1), u0 = 0...n);
u2: = fsolve(uo^2*(1-n^2)*skms/2/Le = ((1-u0)^3-(abs(n-u0)^3)/3; u0 = 0...1)
```

Here, $u1$ is the velocity of air (in a duct with impermeable walls) ejected by the putative flow (in view of Equations 5.131 and 5.133); $u2$ is same for a constantly accelerated flow of particles (in view of Equation 5.24); $a1$, $a2$, and $a3$ are input numerical constants for Le, $\sum \zeta$, and n, respectively.

In a similar way, it is possible to determine the velocities of ejected air in this duct with linearized volumetric forces of interaction between components. For a constantly accelerated flow, based on Equation 5.123 (at $u = u_0$), we get

$$\sum \zeta u_0^2 = k \mathrm{Le}(\bar{v}_q - u_0), \qquad (5.142)$$

where $k = 1 - \dfrac{u_0}{\bar{v}_q}$ and the computational equations for determining u_0 have the form

$$u_0^2 = \frac{\mathrm{Le}}{\bar{v}_q \sum \zeta} \left(\bar{v}_q - u_0\right)^2, \qquad (5.143)$$

whence

$$u_0 = \frac{\bar{v}_q}{1+b}; \quad b = \sqrt{\frac{\bar{v}_q \sum \zeta}{\mathrm{Le}}}. \qquad (5.144)$$

Replacing $\bar{v}_q$ with $\bar{v}_y$ enables the same relationship to be reused to determine the velocity of air ejected by a putative flow: `> vqs: = 2/3*(1-n^3)/(1-n^2); vys: = b0+b1/2+b2/6; bq: = sqrt(vqs*skms/Le); by: = sqrt(vqs*skms/Le); ul3: = vys/(1+by); ul4: = vqs/(1+bq):`

Here, $ul3$ is the velocity of air ejected by a putative flow with linearized volumetric forces of interaction between components; $ul4$ is the same for a constantly accelerated flow. Computation results are summarized in Table 5.1.

Table 5.1 introduces the following conventions:

$$\Delta_1 = \frac{u_1 - u_2}{u_2} 100,\%, \quad \Delta_4 = \frac{ul4 - u_2}{u_2} 100,\%, \quad \Delta_3 = \frac{ul3 - u_2}{u_2} 100,\%.$$

As the table illustrates, replacement of a real flow with a putative flow does not register any significant change of putative velocity of ejected air (variation remains within 0.25%). Error introduced by linearization of volumetric forces is somewhat higher but that also remains within the accuracy margin of input data.

TABLE 5.1

Comparing the Velocities of Ejected Air in a Duct with Impermeable Walls
(at $n = 0.4319$; $\sum \zeta = 1.5$; $E = 0$)

Le	u_2	u_1	Δ_1	$u/4$	Δ_4	$u/3$	Δ_3
0.1	0.17306	0.17289	−0.1	0.17275	−0.18	0.17255	−0.29
0.2	0.22385	0.22364	−0.1	0.22311	−0.33	0.22285	−0.45
0.5	0.30311	0.30288	−0.07	0.30098	−0.70	0.30060	−0.83
0.6	0.32020	0.31998	−0.07	0.31760	−0.81	0.31719	−0.94
0.8	0.34783	0.34763	−0.06	0.34429	−1.02	0.34383	−1.15
1.0	0.36970	0.36954	−0.04	0.36523	−1.21	0.36473	−1.34
2	0.43911	0.43917	−0.01	0.43015	−2.04	0.42953	−2.18
4	0.50847	0.50895	0.01	0.49199	−3.24	0.49125	−3.39
5	0.53005	0.53064	0.11	0.51070	−3.65	0.50992	−3.80
6	0.54716	0.54781	0.12	0.52545	−3.97	0.52464	−4.12
8	0.57292	0.57358	0.12	0.54766	−4.41	0.54680	−4.56
10	0.59158	0.59218	0.10	0.56392	−4.67	0.56303	−4.83
15	0.62190	0.62221	0.05	0.59122	−4.93	0.59026	−5.09
20	0.64025	0.64024	−0.002	0.60879	−4.91	0.60779	−5.07
30	0.66148	0.66095	−0.08	0.63103	−4.60	0.62997	−4.76
50	0.68105	0.67990	−0.17	0.65503	−3.82	0.65392	−3.98
100	0.69758	0.69578	−0.25	0.68110	−2.36	0.67992	−2.53

5.3.4 SPECIAL CONCERNS OF A NUMERICAL STUDY

Let us now proceed with a numerical solution of the problem. The number of unknown variables can be reduced by introducing new functions

$$y_1 = \frac{u}{u_0}, \quad y_2 = \frac{(p_\omega - p)}{\zeta_k u_0^2} = \frac{\Delta p}{\zeta_k u_0^2}, \tag{5.145}$$

to simplify boundary conditions.

To introduce the function y_2, original equation (Equation 5.21) will be transformed. By deriving relation Equation 5.22 over x and substituting Equation 5.21 into the result, we can evaluate

$$\frac{d\Delta p}{dx} = \left[4u(1 - r^{-2}) + 4r^{-2}u_0 \right] \frac{du}{dx} - \text{Le} \left| 1 - \frac{u}{v} \right| (v - u). \tag{5.146}$$

Then, in view of the introduced functions, initial combined Equations 5.20 and 5.146 will become

$$y'_1 = E\sqrt{\zeta_k} \sqrt{|y_2|} \cdot \text{signum}(y_2), \quad E = \frac{\tilde{\prod}\tilde{l}\varepsilon}{\tilde{s}_t \sqrt{\zeta_0}}, \tag{5.147}$$

$$y'_2 = \frac{y'_1}{\zeta_k} \left[4y_1(1 - r^{-2}) + 4r^{-2} \right] - \frac{\text{Le}}{\zeta_k v} |v/u_0 - y_1| (v/u_0 - y_1), \tag{5.148}$$

and boundary conditions, due to the fact that

$$\Delta p(0) = \zeta_n u_0^2 + p_a, \quad \Delta p(1) = -\zeta_k u_0^2 + p_a, \tag{5.149}$$

would appear as

$$y_1(0) = 1, \quad y_2(0) = \frac{\zeta_n}{\zeta_k} + A_1, \tag{5.150}$$

$$y_1(1) = 1, \quad y_2(1) = A_1 + 1, \tag{5.151}$$

where

$$A_1 = \frac{p_a}{\zeta_k u_0^2}. \tag{5.152}$$

The greatest difficulty in the numerical solution of the Cauchy problem is posed by the fact that the air velocity at the inlet/outlet of the perforated pipe u_0 and excess pressure in the upstream/downstream sections of the bypass duct p_a are, at the same time, unknown variables. These quantities are the primary sought parameters of the problem, while, the boundary value problem will be approached using the false position method. The objective is to solve the *artillery* problem of hitting a *target* at points u_0 and p_a given the unknown initial position of the cannon and the inclination angle of its barrel. Preliminary computations have shown the trajectories of the *projectile* to be steeply descending, further complicating the search for the initial position of the cannon—computations have to be carried out with a much greater precision (widening the mantissa to 50 and more positions). This makes it even more important to evaluate the initial approximation of the u_0 and p_a parameters. Preliminary computations, given $\zeta_n = 0.5$ and $\zeta_k = 1$, indicate that

$$A_1 \approx 0.25, \tag{5.153}$$

therefore, the following can be assumed for p_a at the outset:

$$p_a \approx -0.25 u_0^2, \tag{5.154}$$

while $u_0 \equiv u_n$ is equal to the velocity of ejected air in a duct with impermeable walls u_2.

These parameters serve as the departing points. What follows is the dull traditional false position method. In the beginning, with specified values of Le, n, ζ_H, ζ_k, formulas given in Equations 5.88 and 5.89 are used to determine u_2 that is then used to define u_n and, consequently, P_a. In addition, parameters E and r are introduced. The Cauchy problem is solved in view of initial conditions given as Equation 5.150. Equations 5.147 and 5.148 are partially integrated. The first condition (Equation 5.151) is verified: if it happens that $y_1(1) > 1$, then a new value is set for $u_0 < u_n$ (if $y_1(1) < 1$, then $u_0 > u_n$). The value of p_a remains unchanged. The system is integrated for another time. The condition, Equation 5.151, is checked again, unless

$$|y_1(1) - 1| < \varepsilon_1 \approx 0.001, \tag{5.155}$$

where ε_1 is a predetermined margin of error. After that the second condition (Equation 5.151) is checked: if it is not met, p_a shall be adjusted in either direction. u_n is repeatedly adjusted at a new value unless the condition (Equation 5.155) is met again. The second condition (Equation 5.151) is checked, p_a is adjusted anew, and the process repeats until the second function comes within permissible deviation limits:

$$|y_2(1) - 1 + A_1| < \varepsilon_2 \approx 0.001. \tag{5.156}$$

It is possible to change the order of adjustments, that is, choose u_n to be fixed while adjusting p_a.

The results of numerical integration are compared in Tables 5.2 through 5.4. In addition, numerical computation results for a perforated duct with a large-size bypass chamber at $r = \infty$ are provided.

As these tables illustrate, the velocity of ejected air for the putative flow (u_p) deviates from the velocity of ejected air with a constantly accelerated flow (u_q) by a negligibly small margin (Table 5.2). Like the previously examined case of a duct with impermeable walls (at $E = 0$), deviations of these velocities become more noticeable if volumetric forces are linearized (Tables 5.3 and 5.4).

TABLE 5.2

Comparison of Numerical Study Findings for Changes in the Velocity of Ejected Air in a Perforated Duct in the Cases of a Constantly Accelerated Flow (u_q) and a Putative Flow with Velocities Following the Parabolic Law (u_p)

	At Le = 1; r = 1; E = 1			At Le = 5; r = 1; E = 1		
x	u_q	u_p	Δ, %	u_q	u_p	Δ, %
0	0.3520	0.3670	4.09	0.4800	0.5100	5.88
0.1	0.3566	0.3720	4.14	0.4808	0.5102	5.76
0.2	0.3644	0.3808	4.31	0.4884	0.5181	5.73
0.3	0.3745	0.3923	4.54	0.5015	0.5321	5.75
0.4	0.3860	0.4054	4.78	0.5182	0.5500	5.78
0.5	0.3983	0.4190	4.94	0.5370	0.5692	5.66
0.6	0.4103	0.4320	5.02	0.5562	0.5880	5.41
0.7	0.4199	0.4425	5.11	0.5733	0.6033	4.97
0.8	0.4132	0.4325	4.46	0.5759	0.5969	3.52
0.9	0.3889	0.4061	4.24	0.5469	0.5630	2.86
1.0	0.3520	0.3670	4.09	0.4968	0.5100	2.59

TABLE 5.3

Comparison of Numerical Study Findings for Changes in the Velocity of Ejected Air in a Perforated Duct in the Cases of a Constantly Accelerated Flow (u_q) and with Volumetric Component Interaction Forces Replaced with a Linear Relationship (Le $(v - u)$) for a Constantly Accelerated Flow (u_{lq}) at $E = 1$; $r = 1$; $n = 0.4319$

	At Le = 1.0			At Le = 5.0		
x	u_q	u_{lq}	Δ_{lq}, %	u_q	u_{lq}	Δ_{lq}, %
0	0.3520	0.4300	18.10	0.4800	0.6080	21.05
0.1	0.3566	0.4364	18.29	0.4808	0.5943	19.10
0.2	0.3644	0.4468	18.44	0.4884	0.5940	17.78
0.3	0.3745	0.4599	18.57	0.5015	0.6091	17.67
0.4	0.3860	0.4746	18.67	0.5182	0.6304	17.80
0.5	0.3983	0.4900	18.71	0.5370	0.6548	18.00
0.6	0.4103	0.5048	18.72	0.5562	0.6799	18.19
0.7	0.4199	0.5163	18.67	0.5733	0.7023	18.37
0.8	0.4132	0.5062	18.37	0.5759	0.7074	18.59
0.9	0.3889	0.4756	18.23	0.5469	0.6712	18.52
1.0	0.3520	0.4300	18.10	0.4968	0.6080	18.29

TABLE 5.4

Comparison of Numerical Study Findings for Changes in the Velocity of Ejected Air in a Perforated Duct in the Cases of a Constantly Accelerated Flow (u_q) (at $E = 1$; $r = 1$; $n = 0.4319$) and with Volumetric Component Interaction Forces Replaced with a Linear Relationship (Le ($v - u$)) for an Accelerated (Putative) Flow (u_{lu}))

x	At Le = 1.0			At Le = 5.0		
	u_q	u_{lu}	Δ_{lu}, %	u_q	u_{lu}	Δ_{lu},%
0	0.3520	0.4310	18.33	0.4800	0.6060	20.79
0.1	0.3566	0.4367	18.34	0.4808	0.5916	18.73
0.2	0.3644	0.4467	18.42	0.4884	0.5902	17.25
0.3	0.3745	0.4596	18.52	0.5015	0.6054	17.16
0.4	0.3860	0.4743	18.62	0.5182	0.6274	17.31
0.5	0.3983	0.4898	18.68	0.5370	0.6526	17.71
0.6	0.4103	0.5048	18.72	0.5562	0.6785	17.94
0.7	0.4199	0.5165	18.70	0.5733	0.7016	18.29
0.8	0.4132	0.5070	18.50	0.5759	0.7067	18.51
0.9	0.3889	0.4766	18.40	0.5469	0.6700	18.37
1.0	0.3520	0.4310	18.33	0.4968	0.6060	18.02

However, in this case as well, they remain within the accuracy of industrial experiment. A sustained positive deviation can be noticed. This is due to the fact that, in absolute values,

$$\text{Le} \frac{|v-u|}{v}(v-u) < \text{Le}(v-u), \tag{5.157}$$

and thus, the resulting ejection volumes are somewhat overstated (i.e., there is a certain margin of safety).

Thus, numerical computations indicate that the solution of a boundary-value problem with unknown parameters u_0 and P_a using the false position method with steeply declining functions $u(x)$ requires more machine time and greater precision of numerical values.

The problem can be simplified by linearizing the volumetric forces of interaction between components so that full-scale analytical studies would become possible. The solution (Equation 5.54) could be used for that purpose. However, we will resort to a different linearization technique.

5.3.5 LINEARIZING THE EQUATION FOR ABSOLUTE MOTION OF EJECTED AIR

In order to reveal regularities in the variation of the function of Equation 5.83 and, thereby, optimize the ultimate reduction in ejection volume, let us linearize combined Equations 5.20 and 5.21. In addition to introducing a putative flow (relation, Equation 5.131) and linearizing mass forces (relation, Equation 5.123), we posit that

$$p_\omega - p \approx \zeta_0 |\overline{w}| w, \tag{5.158}$$

$$u du \approx \overline{u} du, \tag{5.159}$$

$$\omega d\omega \approx \overline{\omega} d\omega. \tag{5.160}$$

Then, as a consequence of Equation 5.160, the *exact* Equation 5.13 is replaced with

$$p_\omega = p_a - 4\bar{\omega}\omega \tag{5.161}$$

or, in view of Equation 5.16,

$$p_\omega = \frac{p_a - 4\bar{\omega}\left(u - u_0\right)}{r}. \tag{5.162}$$

Let us rewrite the condition for cross flow of air through holes in walls for Equation 5.158 in view of Equation 5.15:

$$p_\omega - p = \frac{\zeta_0}{\hat{S}_t}\,\overline{|w|}\frac{du}{dx} = \frac{\sqrt{\zeta_0}}{E}\,\overline{|w|}\frac{du}{dx} \tag{5.163}$$

or, taking account of the relation Equation 5.162,

$$p = p_a - 4\bar{\omega}\frac{u - u_0}{r} - \frac{\sqrt{\zeta_0}}{E}\,\overline{|w|}\frac{du}{dx}. \tag{5.164}$$

Now, we can rewrite the initial equation in view of the condition for Equation 5.159 while linearizing the mass forces of interaction between components

$$\frac{dp}{dx} = -4\bar{u}\frac{du}{dx} + k\mathrm{Le}\cdot(v - u), \tag{5.165}$$

so that, in view of Equation 5.164, this equation can be reduced to a linear nonuniform second-order differential equation with constant coefficients:

$$u'' - 2Au' - \mathrm{Bu} = -\mathrm{Bv}, \tag{5.166}$$

where

$$A = 2\frac{\bar{u} - \bar{\omega}/r}{\sqrt{\zeta_0|w|}}E, \quad B = \frac{k\mathrm{Le}\cdot E}{\sqrt{\zeta_0|w|}}. \tag{5.167}$$

It would not be difficult to integrate this equation with the condition applied for Equation 5.131. The general solution appears as

$$u = C_1 e^{a_1 x} + C_2 e^{a_2 x} + v - k_1 - k_2 x, \tag{5.168}$$

where

$$k_2 = 2b_2\frac{A}{B}, \quad k_1 = 2\frac{A}{B}(b_1 - k_2) - b_2/B \tag{5.169}$$

a_1, a_2 are the roots of a characteristic equation corresponding to homogeneous differential equation (Equation 5.166), equal to

$$a_1 = A + \sqrt{A^2 + B}, \quad a_2 = A - \sqrt{A^2 + B} \tag{5.170}$$

What is left is to determine the constants C_1 and C_2 for Equation 5.168. To that end, we will use boundary conditions (Equations 5.25 and 5.26). Let us differentiate Equation 5.168:

$$u' = C_1 a_1 e^{a_1 x} + C_2 a_2 e^{a_2 x} + b_1 + b_2 x - k_2. \tag{5.171}$$

In view of Equations 5.164, 5.168, and 5.171, boundary conditions can be rewritten as

$$\left. \begin{aligned} u_0 &= C_1 + C_2 + n - k_1, \\ u_0 &= C_1 e^{a_1} + C_2 e^{a_2} + 1 - k_1 - k_2, \end{aligned} \right\} \tag{5.172}$$

$$\left. \begin{aligned} -\zeta_n u_0^2 &= p_a - \frac{\zeta_0}{E} \overline{|w|}(C_1 a_1 + C_2 a_2 + b_1 - k_2), \\ \zeta_k u_0^2 &= p_a - \frac{\zeta_0}{E} \overline{|w|}(C_1 a_1 e^{a_1} + C_2 a_2 e^{a_2} + b_1 + b_2 - k_2). \end{aligned} \right\} \tag{5.173}$$

Solving combined Equation 5.172 enables us to determine the constants for Equation 5.168, C_1 and C_2 as functions of the sought parameter u_0:

$$C_1 = u_0 \frac{1 - e^{a_2}}{e^{a_1} - e^{a_2}} - \frac{b_3 - a_3 e^{a_2}}{e^{a_1} - e^{a_2}}, \tag{5.174}$$

$$C_2 = u_0 \frac{e^{a_1} - 1}{e^{a_1} - e^{a_2}} - \frac{a_3 e^{a_1} - b_3}{e^{a_1} - e^{a_2}}, \tag{5.175}$$

with the following notational simplification:

$$a_3 = n - k_1, \quad b_3 = 1 - k_1 - k_2. \tag{5.176}$$

At the same time, combined Equation 5.173 leads to an equation for determining the parameters u_0 and P_a for the problem. So, by subtracting the first equation of the system from the second one, we end up with

$$\sum \zeta u_0^2 = -\frac{\sqrt{\zeta_0}}{E} \overline{|w|} \left[C_1 a_1 \left(e^{a_1} - 1 \right) + C_2 a_2 \left(e^{a_2} - 1 \right) + b_2 \right], \tag{5.177}$$

and summing these equations together, we obtain the formula

$$p_a = \frac{1}{2}(\zeta_k - \zeta_n) u_0^2 + \frac{\sqrt{\zeta_0}}{2E} \overline{|w|} \left[C_1 a_1 \left(e^{a_1} + 1 \right) + C_2 a_2 \left(e^{a_2} + 1 \right) + 2(b_1 - k_2) + b_2 \right], \tag{5.178}$$

which, with the first parameter of the problem u_0 found using Equation 5.177, can be used to determine the second parameter p_a.

Velocities $\bar{u}, \bar{\omega}$, and $\bar{w}$ averaged over the interval $x = 0...1$ must be known in order to solve Equations 5.177 and 5.178 likewise. For $\bar{u}$, we can use the relation given in Equation 5.130 following from the integral relation of Equation 5.82 at linearization of mass forces (by virtue of Equations 5.123 and 5.124):

$$\bar{u} = \bar{v} - u_0 b, \quad b = \sqrt{\frac{\nu \Sigma \zeta}{\text{Le}}}.$$ (5.179)

For the average velocity in the bypass duct, it follows due to Equation 5.16 that

$$\bar{\omega} = \frac{\bar{u} - u_0}{r}.$$ (5.180)

A greater difficulty is posed by averaging the velocity of cross flow of air through a perforated hole in the pipe wall, because this velocity varies both in its magnitude and its direction. In addition, in a certain point of the interval $x = 0...1$, the velocity $w = 0$. Therefore, the absolute value of averaged $\overline{|w|}$ could be evaluated using the relation

$$\overline{|w|} = \frac{|w(0)| + 0 + |w(1)|}{3},$$ (5.181)

while the value of the velocity w at the pipe inlet and outlet will be found from the condition of Equation 5.149:

$$|w(0)| = \frac{|p_a - p(0)|}{\zeta_0 \overline{|w|}}, \quad w(1) = \frac{|p(1) - p_a|}{\zeta_0 \overline{|w|}}$$ (5.182)

or, accounting for the boundary conditions of Equations 5.25 and 5.26, as well as Equation 5.154,

$$|w(0)| = \frac{0.25 + \zeta_n}{\zeta_0 \overline{|w|}} u_0^2, \quad |w(1)| = \frac{|0.25 - \zeta_k|}{\zeta_0 \overline{|w|}} u_0^2.$$ (5.183)

Then,

$$\overline{|w|} = \sqrt{\frac{0.25 + \zeta_n + |0.25 - \zeta_k|}{3\zeta_0}} u_0.$$ (5.184)

Consequently, the values A, B, a_1, and a_2 similarly depend on u_0 in the general case. Therefore, Equation 5.177 can only be solved numerically.

5.3.6 COMPARING INTEGRATION RESULTS

Table 5.5 and Figure 5.4 present a comparison of the results of integrating the linearized problems described by differential equation (Equation 5.166) with a numerical solution of *exact* combined Equations 5.147 and 5.148. These data show that the functions $u(x)$ are virtually identical both in the variation behavior and in the absolute values of ejected air velocity.

Now let us move on to examine the function $u_0 = f_0(E, r, \text{Le}, \zeta_n, \zeta_k, \zeta_0)$, which determines the velocity of ejected air as a functional pipe wall perforation degree, bypass chamber dimensions, particle flow rate, aerodynamic drag of inlet/outlet pipe sections, as well as perforation holes (in numerical examples the latter were quantitatively assumed as $\zeta_n = 0.5$, $\zeta_k = 1$, and $\zeta_0 = 1.5$; if different values were accepted, that was pointed out in captions under illustrations).

Increased ejection number Le, similar to the case of a flow of particles in a pipe with impermeable walls ($E = 0$), contributes to increasing u_0. The asymptotic nature of this increase (Figure 5.5) is notable along with a drop in ejection volumes compared to a flow of particles in a nonperforated pipe ($u_0 < u_2$, dash–dot curve) in the range of Le surveyed. An even greater effect of minimizing u_0 is observed at higher degrees of perforation (Figure 5.6). This case is similarly asymptotic: with $E \geq 2.5$, the decline in u_0 virtually ceases.

Bypass duct size (Figure 5.7) plays a somewhat different role. A minimum area is clearly observed (at $r \approx 0.5 - 0.6$) for u_0. Within the range $0 \leq r < 0.5$, the magnitude of u_0 rises sharply until at $r = 0$ it matches the value of u_2—the velocity of ejected air in a pipe with impermeable walls (the case of $E = 0$). As r increases up to 2.5, the value of u_0 stabilizes virtually completely.

For convenience, further comparisons will involve the coefficient of reduction in ejection volumes

$$\text{Ke} = \frac{u_{l3}}{u_0}, \tag{5.185}$$

where u_{l3} is the velocity of air ejected by a putative flow in a pipe with impermeable walls at a linearized mass force of interaction between components (the value of u_{l3} is computed using formula Equation 5.144). As the findings from computation (Table 5.6) indicate, the drop in ejection volumes

---

TABLE 5.5

Comparison of Numerical Solution u_q for the *Exact* Equation (Perforated Duct at $E = 1$; $r = 1$; $n = 0.4319$; $\zeta_n = 0.5$; $\zeta_k = 1$; $\zeta_0 = 1.5$) with the Numerical Solution u_l for a Linearized Equation (Perforated Duct, Same Constants $v = n + b_1 x + b_2 x^2/2$, $\text{Le}(v - u)\left(1 - \dfrac{\bar{u}}{\bar{v}}\right)$)

	At Le = 1.0			At Le = 5.0		
x	u_q	u_l	$\Delta u_l, \%$	u_q	u_l	$\Delta u_l, \%$
0	0.3520	0.3463	−1.6	0.4800	0.4767	−0.7
0.1	0.3566	0.3539	−0.8	0.4808	0.4834	0.5
0.2	0.3644	0.3631	−0.4	0.4884	0.4940	1.1
0.3	0.3745	0.3736	−0.2	0.5015	0.5075	1.2
0.4	0.3860	0.3849	−0.3	0.5188	0.5231	0.8
0.5	0.3983	0.3961	−0.6	0.5370	0.5394	0.4
0.6	0.4103	0.4058	−1.0	0.5562	0.5543	−0.3
0.7	0.4199	0.4116	−2.0	0.5733	0.5645	−1.5
0.8	0.4132	0.4096	−0.9	0.5759	0.5639	−2.1
0.9	0.3889	0.3922	0.8	0.5469	0.5415	−1.0
1.0	0.3520	0.3463	−1.6	0.4968	0.4767	−4.0

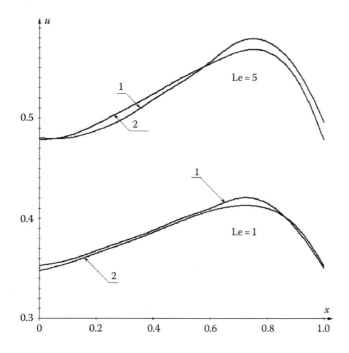

FIGURE 5.4 Variation in the velocity of ejected air within a perforated duct with a bypass chamber (at $E = 1$; $r = 1$; $n = 0.4319$): 1, the numerical solution of *exact* equation (Equations 5.147 and 5.148 combined); 2, the analytical solution of the linear equation, Equation 5.166 (at $v = n + b_1x + b_2x^2/2$).

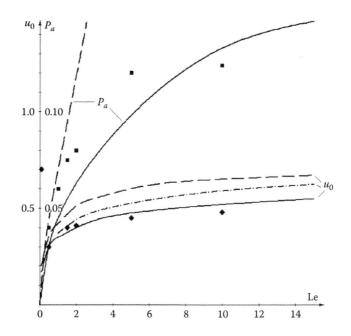

FIGURE 5.5 Ejected air velocity u_0 and bypass chamber pressure p_a as functions of the number Le (at $E = 1$; $r = 1$; $n = 0.4319$; $\zeta_n = 0.5$; $\zeta_k = 1$; $\zeta_0 = 1.5$); solid curves were plotted using formulas for the linearized problem; dotted curve plotted for maximum forces $k = 1$, Le $(v - u)$; dash–dot curve plotted for a pipe with impermeable walls ($E = 0$) using the formula given by Equation 5.139; diamonds ◆ correspond to u_0 while squares ■ correspond to p_a based on findings from the numerical solution of the *exact* equation.

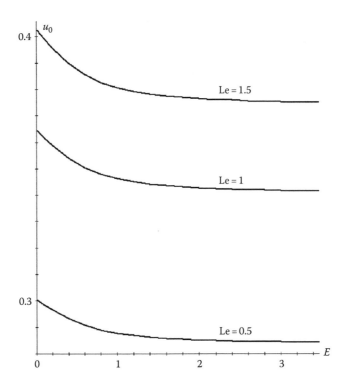

FIGURE 5.6 Variation of ejected air velocity u_0 with increased perforation of pipe walls (at $r = 1$; $n = 0.431964$; $\zeta_n = 0.5$; $\zeta_k = 1$; $\zeta_0 = 1.5$).

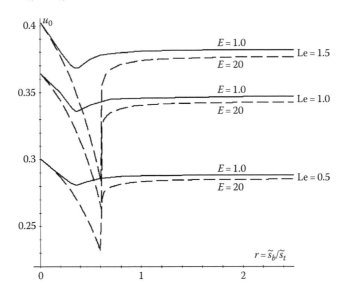

FIGURE 5.7 Variation of the velocity u_0 as a function of bypass chamber dimensions (at $E = 2$; $n = 0.4319$; $\zeta_n = 0.5$; $\zeta_k = 1$; $\zeta_0 = 1.5$).

due to the recirculation of air in the bypass duct increases with greater Le numbers and declines with smaller n (i.e., greater heights of perforated pipe). Even more notable is the reduction in ejection volumes with increased coefficients of resistance to ejected air at the inlet/outlet of the perforated pipe (Tables 5.7 and 5.8). This fact confirms that the air recirculation is to a significant extent promoted by the upper cowl sealing and a buffer capacity within the lower cowl (see Chapter 8), as

TABLE 5.6
Coefficient of Reduction in Ejection Volumes (Ke) at $E = 2.5$; $r = 2$; $\zeta_n = 0.5$; $\zeta_k = 1$; $\zeta_0 = 1.5$

	At n Equal To						
Le	0.3	0.4	0.4319	0.5	0.6	0.7	0.8
0.1	1.0324	1.0328	1.0329	1.0331	1.0333	1.0334	1.0334
0.2	1.0411	1.0419	1.0421	1.0424	1.0429	1.0432	1.0434
0.3	1.0467	1.0477	1.0479	1.0485	1.0491	1.0496	1.0500
0.5	1.0538	1.0552	1.0557	1.0565	1.0575	1.0583	1.0590
0.7	1.0584	1.0602	1.0608	1.0618	1.0632	1.0643	1.0652
1.0	1.0631	1.0654	1.0661	1.0675	1.0692	1.0707	1.0720
1.2	1.0653	1.0679	1.0687	1.0703	1.0723	1.0740	1.0755
1.4	1.0671	1.0700	1.0709	1.0726	1.0749	1.0768	1.0784
1.6	1.0686	1.0717	1.0727	1.0745	1.0770	1.0792	1.0810
1.8	1.0698	1.0732	1.0742	1.0762	1.0789	1.0812	1.0832
2.0	1.0709	1.0744	1.0755	1.0777	1.0806	1.0831	1.0852
2.2	1.0718	1.0755	1.0767	1.0790	1.0820	1.0847	1.0870
2.4	1.0725	1.0765	1.0777	1.0801	1.0834	1.0862	1.0886
2.6	1.0732	1.0779	1.0786	1.0811	1.0845	1.0875	1.0901
2.8	1.0738	1.0781	1.0794	1.0820	1.0856	1.0888	1.0915
3.0	1.0743	1.0787	1.0801	1.0829	1.0866	1.0899	1.0927
3.5	1.0753	1.0801	1.0816	1.0846	1.0887	1.0923	1.0955
4.0	1.0760	1.0812	1.0828	1.0861	1.0905	1.0944	1.0978
4.5	1.0765	1.0821	1.0838	1.0872	1.0919	1.0961	1.0998
5.0	1.0769	1.0827	1.0845	1.0882	1.0932	1.0977	1.1016
7.5	1.0775	1.0846	1.0868	1.0913	1.0974	1.1030	1.1080
10	1.0776	1.0851	1.0876	1.0927	1.0998	1.1063	1.1121

both increase the aerodynamic drag countering ejected air and create greater negative pressures in the upper part of the pipe, while providing a head in its lower part. Conditions conducive to the more intense cross flow of air through perforated holes thus arise.

Both the differential pressure between the duct inlet ($\Delta p(0)$) and outlet ($\Delta p(1)$) and the excess absolute pressure increase with increasing ejection parameter (Le) as well as with higher coefficients ζ_n^* and ζ_k^* (Figure 5.8). If one of these coefficients increases, excess pressures p_a are shifted toward greater pressures if the upper cowl is sealed or toward lower pressures (down to negative values, i.e., relative vacuum) if a buffer chamber is set up in the lower cowl promoting greater ζ_k^* (Figure 5.9).

5.4 SPECIFIC BEHAVIOR OF AIR RECIRCULATION IN A TRANSFER CHUTE WITH A COMBINED BYPASS CHAMBER

5.4.1 ORIGINAL EQUATIONS

Consider a case of a bypass chamber with a transit air interchange between the cowls of the handling facility with aspirated lower cowl (Figure 5.10). In this case, the motion of ejected and recycled air owes itself not only to the aerodynamic forces of particles of handled material falling through the chute but also to negative pressure in the lower cowl created by the fan of the aspiration system. In addition, upward flowing air in the bypass chamber is recycled both through the holes of uniformly perforated chute walls and through frontal holes at both the ends of the bypass chamber that evacuate air from the inner chamber of the lower cowl into the hollow area of the bypass chamber

TABLE 5.7

Coefficient of Reduction in Ejection Volumes (Ke) at Increasing LRC on Air Outlet from Perforated Pipe (at $E = 2.5$; $r = 2$; $n = 0.4319$; $\zeta_n = 0.5$; $\zeta_0 = 1.5$)

Le	With ζ_k Equal To					
	1	2	4	8	16	32
0.1	1.0302	1.0385	1.0485	1.0599	1.0694	1.0780
0.2	1.0387	1.0502	1.0645	1.0800	1.0946	1.1073
0.3	1.0441	1.0580	1.0754	1.0944	1.1128	1.1290
0.5	1.0512	1.0686	1.0908	1.1155	1.1396	1.1609
0.7	1.0560	1.0760	1.1019	1.1310	1.1599	1.1856
1.0	1.0609	1.0839	1.1142	1.1488	1.1835	1.2149
1.2	1.0633	1.0880	1.1207	1.1584	1.1965	1.2312
1.4	1.0653	1.0914	1.1263	1.1667	1.2079	1.2457
1.6	1.0670	1.0943	1.1311	1.1741	1.2181	1.2587
1.8	1.0684	1.0969	1.1354	1.1807	1.2273	1.2706
2.0	1.0696	1.0992	1.1393	1.1866	1.2357	1.2815
2.2	1.0707	1.1012	1.1428	1.1921	1.2435	1.2916
2.4	1.0716	1.1030	1.1459	1.1971	1.2506	1.3010
2.6	1.0725	1.1046	1.1488	1.2017	1.2573	1.3098
2.8	1.0732	1.1061	1.1515	1.2060	1.2635	1.3181
3.0	1.0739	1.1075	1.1540	1.2100	1.2694	1.3260
3.5	1.0753	1.1105	1.1595	1.2190	1.2827	1.3438
4.0	1.0764	1.1130	1.1642	1.2268	1.2943	1.3596
4.5	1.0773	1.1151	1.1683	1.2337	1.3046	1.3738
5.0	1.0781	1.1169	1.1718	1.2398	1.3139	1.3867
7.5	1.0801	1.1230	1.1848	1.2628	1.3498	1.4375
10	1.0809	1.1266	1.1932	1.2783	1.3749	1.4740
20	1.0802	1.1322	1.2096	1.3113	1.4311	1.5597

and is let out to the hollow area of the upper unaspirated cowl, decreasing the vacuum in the latter and thereby bringing down the transit airflow rate Q_1. As the upflow of air proceeds further, its flow rate is increased in the lower part of the bypass chamber by continuous ingress of ejected air through perforated chute walls. This increase gives way to a decrease in the upper part of the chamber. Thus, two loops of recirculated air arise: the lesser inner loop where ejected air circulates and the greater outer loop, which is responsible for the so-called transit air interchange between cowls of the handling facility.

Quantitative estimates of the air recycling volume and of the decrease in the flow rate of ejected air entering the lower aspirated cowl are different from those discussed earlier in that the velocities and static pressures are different both at the ends of the bypass chamber and at the ends of the transfer chute. Undoubtedly, this complicates the analysis involved in minimizing ejected airflow rate. As will be pointed out later, it would be necessary to solve three combined nonlinear algebraic (transcendental) equations that would take a significant time to compute even with numerical methods. As for the solution of differential equations for cross flows of air through perforation holes in the walls of a chute, we use the same methods for linearization of a nonuniform second-order equation.

We will proceed with selecting three characteristic cross sections to establish boundary conditions: initial, N–N (set at an infinitesimal distance ε apart from the center of coordinates, i.e., at $\tilde{x} = 0 + \tilde{\varepsilon}$); final, K–K (cross-section at $\tilde{x} = \tilde{l} - \tilde{\varepsilon}$); and the transitional maximum, M–M (a cross section set apart by $\tilde{x}_m$ from the center of coordinates) where extreme velocities/flow rates of ejected and recirculated air are observed.

TABLE 5.8

Coefficient of Reduction in Ejection Volumes (Ke) at Increasing LRC on Air Inlet into Perforated Pipe (at $E = 2.5$; $r = 2$; $n = 0.4319$; $\zeta_k = 16$; $\zeta_0 = 2.5$)

Le	With ζ_n Equal To					
	0.5	1.0	2.0	4.0	8.0	16
0.1	1.0694	1.0698	1.0706	1.0720	1.0743	1.0778
0.2	1.0946	1.0952	1.0964	1.0985	1.1019	1.1070
0.3	1.1128	1.1135	1.1150	1.1176	1.1220	1.1284
0.5	1.1396	1.1406	1.1426	1.1461	1.1519	1.1605
0.7	1.1600	1.1611	1.1634	1.1676	1.1746	1.1850
1.0	1.1835	1.1850	1.1878	1.1930	1.2015	1.2143
1.2	1.1965	1.1981	1.2013	1.2069	1.2146	1.2305
1.4	1.2079	1.2097	1.2131	1.2192	1.2295	1.2449
1.6	1.2181	1.2200	1.2236	1.2302	1.2413	1.2578
1.8	1.2273	1.2293	1.2332	1.2402	1.2520	1.2697
2.0	1.2357	1.2379	1.2419	1.2493	1.2618	1.2805
2.2	1.2435	1.2457	1.2500	1.2577	1.2708	1.2906
2.4	1.2506	1.2530	1.2574	1.2655	1.2792	1.2999
2.6	1.2573	1.2597	1.2644	1.2728	1.2871	1.3087
2.8	1.2635	1.2661	1.2709	1.2796	1.2945	1.3170
3.0	1.2694	1.2720	1.2770	1.2861	1.3014	1.3247
3.5	1.2827	1.2855	1.2908	1.3006	1.3172	1.3425
4.0	1.2943	1.2973	1.3030	1.3134	1.3312	1.3582
4.5	1.3046	1.3078	1.3138	1.3248	1.3436	1.3723
5.0	1.3139	1.3173	1.3236	1.3351	1.3549	1.3851
7.5	1.3498	1.3538	1.3613	1.3751	1.3989	1.4355
10	1.3749	1.3793	1.3878	1.4033	1.4301	1.4718
20	1.4311	1.4366	1.4474	1.4673	1.5019	1.5568

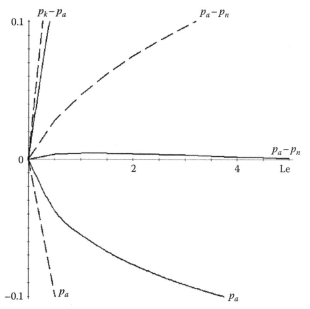

FIGURE 5.8 Variation of excess pressure in the bypass chamber (p_a) and differential pressure between the outlet ($\Delta p_k = p_a - p_k$) and inlet ($\Delta p_n = p_n - p_a$) of the perforated pipe as functions of the ejection number (at $E = 2.5$; $r = 2$; $n = 0.4319$; $\zeta_0 = 1.5$; solid curves $\zeta_n = 0.5$; $\zeta_k = 1$; dotted curves $\zeta_n^* = 22.1$; $\zeta_k^* = 16$).

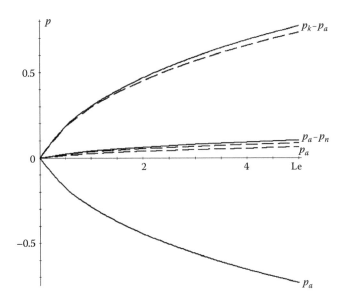

FIGURE 5.9 Variation of excess pressure in the bypass chamber and differential pressure between the pipe inlet and outlet as functions of the ejection number Le (at $E = 2.5$; $r = 2$; $n = 0.4319$; $\zeta_0 = 1.5$; $\zeta_n^* = 22.1$; solid curves $\zeta_k = 1$; dotted curves at $\zeta_n = 0.5$; $\zeta_k^* = 16$).

We will adhere to the previously established convention by denoting dimensional values with a tilde (~) above. From now on, values averaged over length l will be denoted with a subscript s (instead of a simple bar above letter) and dimensionless values will be denoted with the same letters but without a tilde. Some equations will be repeated for the benefit of readability. Thus, due to cross-sectional sizes of the chute and bypass chamber being constant and equal to $\tilde{S}_u$ (m²) and $\tilde{S}_\omega$ (m²), respectively, and perforation being uniform along chute walls, contiguity equations would appear as

$$\frac{du}{dx} = \frac{\tilde{S}_0}{\tilde{S}_u} w; \quad \frac{d\omega}{dx} = \frac{\tilde{S}_0}{\tilde{S}_\omega} w, \tag{5.186}$$

where $\tilde{S}_0 = \tilde{\Pi}\tilde{l}\varepsilon_0$ is the total area of perforation holes in chute walls ($\tilde{\Pi}$ is the perimeter of cross section of the chute (m), $\tilde{l}$ is the chute length (m), and ε_0 is the dimensionless perforation degree) (m²).

Considering that the ratio $r = \tilde{S}_\omega/\tilde{S}_u$ remains constant over the entire length, Equation 5.186 can be rewritten as

$$du = rd\omega \Rightarrow u - r\omega = z - const = u_n - r\omega_n = u_k - r\omega_k. \tag{5.187}$$

In order to identify the physical character of the constant z, let us write the obvious airflow rate balance for cowls:

$$\left.\begin{aligned}\tilde{Q}_1 + \tilde{\omega}_n\tilde{S}_\omega = \tilde{u}_n\tilde{S}_u,\\ \tilde{u}_k\tilde{S}_u = \tilde{Q}_1 + \omega_k\tilde{S}_\omega,\end{aligned}\right\} \Rightarrow \begin{aligned}\tilde{u}_n = r\tilde{\omega}_n + \tilde{Q}_1/\tilde{S}_u,\\ \tilde{u}_k = r\tilde{\omega}_k + \tilde{Q}_1/\tilde{S}_u,\end{aligned} \tag{5.188}$$

where $\tilde{Q}_1$ is the flow rate of transit air (m³/s). A comparison of these combined equations with Equation 5.187 yields

$$z = \frac{\tilde{Q}_1}{\tilde{v}_k\tilde{S}_u}, \tag{5.189}$$

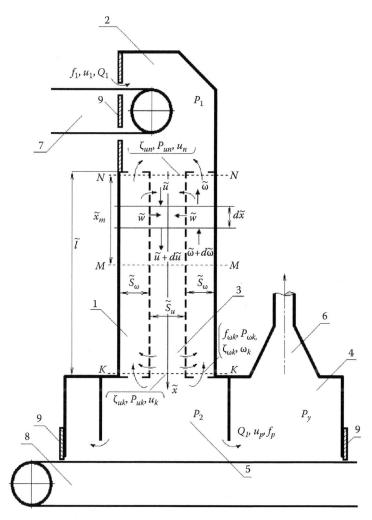

FIGURE 5.10 Diagram of ejected and recycled airflows in a chute with a combined bypass chamber: 1, bypass chamber with transit flow of recycled air; 2, upper cowl; 3, chute with perforated walls; 4, lower cowl with inner chamber 5 for taking in the material being handled; 6, aspiration connection; 7 and 8, upper and lower conveyors; 9, sealing aprons.

where the constant z is the ratio of *global transit* airflow rate to the maximum possible flow rate of ejected air. By virtue of $\tilde{Q}_1$ being a part of aspirated air (total volumetric flow rate of air aspirated from the lower cowl, $\tilde{Q}_a = \tilde{Q}_1 + \tilde{Q}_{nn}$, where $\tilde{Q}_{nn}$ is the flow rate of air coming in through leaky areas in this cowl), the constant z would become the main unknown variable in our problem.

Another important relation, easily obtainable from Equation 5.188, should be pointed out:

$$\frac{u_k - u_n}{\omega_k - \omega_n} = \frac{\tilde{S}_\omega}{\tilde{S}_u} \Rightarrow \tilde{S}_u(u_k - u_n)\tilde{v}_k = \tilde{S}_\omega(\omega_k - \omega_n)\tilde{v}_k = \tilde{Q}_R, \tag{5.190}$$

where $\tilde{Q}_R$ is the volumetric flow rate of air interchange between the flow of ejected air in the chute and the flow of air recirculated in the bypass chamber (m^3/s).

The recycling coefficient is determined with an obvious relation

$$R_z = \frac{\tilde{Q}_\omega}{\tilde{Q}_u} = \frac{r\omega_k}{u_k} = \frac{u_k - z}{u_k} = 1 - \frac{z}{u_k}. \tag{5.191}$$

Let us now rewrite the dimensionless dynamical equations for upward airflow in the bypass chamber (see Equation 5.12):

$$dp_\omega + 4\omega d\omega = 0, \quad p_\omega = 2\tilde{p}_\omega/(\tilde{\rho}\tilde{v}_k^2), \quad \omega = \tilde{\omega}/\tilde{v}_k, \tag{5.192}$$

and for the ejected air in the perforated chute (see Equation 5.7)

$$dp + 4udu = \frac{\mathrm{Le}(v-u)|v-u|}{v} \cdot dx, \tag{5.193}$$

as well as the condition for cross flow of air through perforation holes of chute wall:

$$p_\omega - p_w = \zeta_0 |w| w. \tag{5.194}$$

Here, w is a dimensionless air cross-flow velocity that can be expressed through a change in the dimensionless velocity of ejected air in view of the former of Equation 5.186:

$$w = \frac{1}{E\sqrt{\zeta_0}} \frac{du}{dx}, \tag{5.195}$$

where E is a dimensionless parameter for wall perforation degree and LRC of hole that is equal (see Equation 5.23), using the convention established by us, to

$$E = \frac{\tilde{S}_0}{\tilde{S}_u \sqrt{\zeta_0}}. \tag{5.196}$$

Before moving on to integrate Equations 5.192 and 5.193, we will formulate boundary conditions.

Without detriment to the generality of the problem at hand, we will assume that cross-sectional areas at the inlet and outlet of the chute and bypass chamber are equal to the respective cross-sectional areas of the chute and the chamber.* With that in mind, boundary conditions for air velocity could be written as follows:

at inlet section $N-N$ (at $x = 0$)

$$u(0) = u_n, \quad \omega(0) = \omega_n, \tag{5.192a}$$

$$w(0) = \frac{u'(0)}{E\sqrt{\zeta_0}} = \gamma_w(0)\sqrt{\frac{|p_\omega(0) - p_u(0)|}{\zeta_0}}, \quad \gamma_w(0) = \mathrm{signum}\left(p_\omega(0) - p_u(0)\right), \tag{5.193a}$$

at outlet section $K-K$ (at $x = 1$)

$$u(1) = u_k, \quad \omega(1) = \omega_k, \tag{5.194b}$$

---

* If the design calls for tapering cross section, this factor can be accounted for by increasing the LRC of a hole as a result of recalculating its value adjusted for velocity in the cross section of the chute or chamber.

$$w(1) = \frac{u'(1)}{E\sqrt{\zeta_0}} = \gamma_w(1)\sqrt{\frac{|p_\omega(1) - p_u(1)|}{\zeta_0}}, \quad \gamma_w(1) = \mathrm{signum}\left(p_\omega(1) - p_u(1)\right). \qquad (5.194a)$$

Boundary conditions for static pressures will be expressed using LRC and excess pressure values in the cowl of the handling facility:

at the inlet section *N–N*

$$\left.\begin{array}{l} p_u(0) = p_1 - \zeta_{un}u_n^2, \\ p_\omega(0) = p_1 - \zeta_{\omega n}\omega_n^2, \end{array}\right\}, \qquad (5.195a)$$

and at the outlet section *K–K*

$$\left.\begin{array}{l} p_u(1) = p_2 - \zeta_{uk}u_k^2, \\ p_\omega(1) = p_2 - \zeta_{\omega k}\omega_k^2, \end{array}\right\}, \qquad (5.196a)$$

where

ζ_{un}, ζ_{uk} are the respective LRCs for ejected air entering and exiting the chute

$\zeta_{\omega n}$, $\zeta_{\omega k}$ are the respective LRCs for the upward flow of air leaving and entering the bypass chamber

p_1, p_2 are the respective excess static pressures in the upper (unaspirated) cowl and in the receiving chamber of the lower (aspirated) cowl

The latter can be quantified using the LRCs for leaky areas in the upper cowl (ζ_1), receiver chamber partition wall (ζ_p), dimensionless negative pressure in the lower cowl (p_y), and the sought parameter z:

$$p_1 = -\zeta_1\left(\frac{\tilde{S}_u}{\tilde{f}_1}\right)^2 z^2 = -\zeta_1^* z^2, \quad \zeta_1^* = \zeta_1\left(\frac{\tilde{S}_u}{\tilde{f}_1}\right)^2, \qquad (5.197)$$

$$p_2 = p_y + \zeta_p\left(\frac{\tilde{S}_u}{\tilde{f}_p}\right)^2 z^2 = p_y + \zeta_p^* z^2, \quad \zeta_p^* = \zeta_p\left(\frac{\tilde{S}_u}{\tilde{f}_p}\right)^2, \qquad (5.198)$$

where

$\tilde{f}_1$ is the total leakage area of the upper cowl (m²)

$\tilde{f}_p$ is the total area of a gap between the walls of the lower cowl receiving chamber and the laden conveyor belt (m²)

$\tilde{S}_u$ is the cross-sectional area of the chute (m²)

5.4.2 Special Case with Bypassing a Nonperforated Chute

Before moving on with solving the proposed problem to determine the parameter z for a design with a combined bypass chamber, let us consider the case of loose material transfer over a chute with impermeable walls (at $\zeta_0 \to \infty$, $E = 0$). The problem could be simplified due to the possibility of positing, with a modest error margin, that

$$u_n = u_k = u \equiv u_c - \text{const,} \tag{5.199}$$

$$\omega_n = \omega_k = \omega \equiv \omega_c - \text{const.} \tag{5.200}$$

This makes it possible to integrate the equation of ejected air dynamics (Equation 5.193) in a finite form, for example, with an equally accelerated flow of falling particles with their velocities (see Equation 5.24) defined as

$$v = \sqrt{(1-n^2)x + n^2}, \quad n = \frac{\tilde{v}_0}{\tilde{v}_k} \tag{5.201}$$

could serve as the independent variable for integration. In this case, owing to

$$dx = \frac{2v\,dv}{1-n^2},$$

the initial equation would appear in a simple form:

$$dp_y = \frac{2\text{Le}}{1-n^2}\left(v - u_c\right)\left|v - u_c\right|dv, \tag{5.202}$$

which, given the boundary conditions of Equations 5.195a and 5.199 and after applying a few trivial transformations, can be written as the following criterial equation*:

$$\text{Eu} + u_c\left|u_c\right| = \text{Bu}\,\frac{\left|1-u_c\right|^3 - \left|n-u_c\right|^3}{3}, \tag{5.203}$$

where Eu is Euler's criterion equal to

$$Eu = \frac{p_2 - p_1}{\zeta_u} \tag{5.204}$$

Bu is the Butakov–Neykov criterion calculated using the formula

$$\text{Bu} = \frac{2\text{Le}}{\zeta_u(1-n^2)} \tag{5.205}$$

---

* Hereinafter, the squared velocity will be replaced with a notation $u|u|$ that extends the applicability of equations into negative velocity values.

ζ_u is the sum total of LRCs for the chute

$$\zeta_u = \zeta_{un} + \zeta_{uk} \qquad (5.206)$$

It should be noted that, by virtue of Equations 5.197 and 5.198, Euler's number can be expressed through LRCs of leaky areas in the upper cowl (ζ_1^*) and receiving chamber partition (ζ_p^*) referred to the velocity of ejected air u_c:

$$\mathrm{Eu} = \frac{\zeta_n^* z_c \left| z_c \right| + p_y}{\zeta_u}, \qquad (5.207)$$

where

$$z_c = u_c - r\omega_c \qquad (5.208)$$

$$\zeta_n^* = \zeta_1^* + \zeta_p^* \qquad (5.209)$$

On the other hand, Equation 5.192 at $\omega = \omega_c$, $dp\omega = 0$, whence, in view of the same boundary conditions for pressures, we can deduce that

$$p_2 - p_1 = \zeta_\omega \omega_c \left| \omega_c \right|, \qquad (5.210)$$

ζ_ω is the sum total of LRCs for the bypass chamber

$$\zeta_\omega = \zeta_{\omega n} + \zeta_{\omega k} \qquad (5.211)$$

and, consequently, Euler's criterion (Equation 5.204) can be expressed using the velocity of the upward flow ω_c:

$$\mathrm{Eu} = \frac{\zeta_\omega}{\zeta_u} \omega_c \left| \omega_c \right|. \qquad (5.212)$$

In view of this result, Equation 5.203 can be rewritten as follows:

$$\frac{\zeta_\omega}{\zeta_u} \omega_c \left| \omega_c \right| + u_c \left| u_c \right| = \mathrm{Bu} \frac{\left| 1 - u_c \right|^3 - \left| n - u_c \right|^3}{3}, \qquad (5.213)$$

that, solved with the aid of Equation 5.208 and the equation

$$\frac{\zeta_\omega}{\zeta_n} \omega_c \left| \omega_c \right| = \zeta_n^* z_c \left| z_c \right| + p_y, \qquad (5.214)$$

obtained by juxtaposing the right parts of the relations given in Equations 5.207 and 5.212, will result in the following equations for determining the parameter z_c, the velocity of ejected air in chute u_c, and the velocity of the ascending flow of recirculating air in the bypass chamber ω_c:

$$\frac{f}{\zeta_u} + F|F| = \mathrm{Bu}\frac{|1-F|^3 - |n-F|^3}{3},\tag{5.215}$$

$$u_c = F, \quad \omega_c = \frac{f}{\sqrt{\zeta_\omega|f|}},\tag{5.216}$$

with functions of the parameter z_c introduced to simplify notation

$$f = \zeta_n^* z_c |z_c| + p_y,\tag{5.217}$$

$$F = z_c + \frac{rf}{\sqrt{\zeta_\omega|f|}}.\tag{5.218}$$

By solving Equation 5.215 to determine z_c (that can be easily done in *Maple* using the `fsolve` command), it is possible to determine the flow rate of air $\tilde{Q}_1$ (m³/s) injected into the lower cowl as a result of dynamic interaction between the flow of loose material and due to a negative pressure inside the cowl

$$\tilde{Q}_1 = z_c \tilde{v}_k \tilde{S}_u,\tag{5.219}$$

and the flow rate of recirculated air in the bypass chamber

$$\tilde{Q}_R = \omega_c \tilde{v}_k \tilde{S}_\omega,\tag{5.220}$$

and the flow rate of air $\tilde{Q}_u$ (m³/h) entering the inner chamber from the chute

$$\tilde{Q}_u = u_c \tilde{v}_k \tilde{S}_u.\tag{5.221}$$

It goes without saying that an obvious balance of these flow rates must be observed:

$$Q_R + Q_1 = Q_u.\tag{5.222}$$

The flow rate of air coming from the chute into the lower cowl with closed end openings of the bypass chamber (i.e., at $\tilde{Q}_R = 0$) can be determined with a similar ease. For that Equation 5.203 should be solved in view of $u_c = u_0 = z_c$ and, therefore,

$$\mathrm{Eu}_0 + u_0|u_0| = \mathrm{Bu}\frac{|1-u_0|^3 - |n-u_0|^3}{3},\tag{5.223}$$

where in view of Equation 5.207

$$\mathrm{Eu}_0 = \frac{\zeta_n^* u_0^2 + p_y}{\zeta_u}.\tag{5.224}$$

By determining the dimensionless velocity of ejected air u_0 from Equation 5.223, we can determine the flow rate

$$Q_0 = u_0 \tilde{v}_k \tilde{S}_u, \tag{5.225}$$

of air coming from the chute into the lower cowl in the absence of a bypass chamber.

As calculations for the *case-study* example of handling unit (Figures 5.11 and 5.12) indicate, the flow rate of this area has noticeably increased compared to the case of transfer in a chute in presence of a low-drag bypass chamber ($\zeta_\omega < 2$) and given a minor leakage area in the upper cowl ($f_1 < 0.2$ m^2), especially at great ejection numbers (Le > 5). Increasing values of Le ejection parameter are accompanied with an increase in recycle coefficient R_z (Figure 5.12) from -1 to its positive *asymptotic* values at Le ≥ 3. Negative values of R_z at lower values of Le can be explained by the fact that the greater negative pressures in the lower cowl ($p_3 \leq -10$)) and lower ejection capacities of the loose matter flow make the bypass chamber act like a parallel duct for the air from the upper

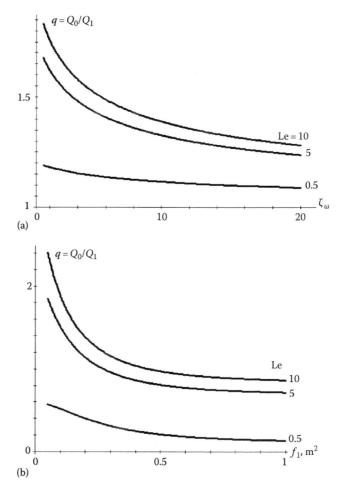

FIGURE 5.11 Variation in the relative flow rate of air ejected in the lower cowl by a flow of loose material as a function of the LRC of the bypass chamber (ζ_ω) and of the leakage area of the upper cowl f_1 at $v_0 = 3$ m/s; $v_k = 10$ m/s; $P_3 = -10$ Pa; Q_1—with a bypass chamber; Q_0—at $\zeta_\omega \to \infty$ (absent a bypass chamber): (a) at $S_u = S_\omega = 0.3$ m^2; $f_1 = f_p = 0.3$ m^2; $\zeta_u = 1.5$; $\zeta_1 = \zeta_p = 2.4$ and (b) at $S_u = S_\omega = 0.3$ m^2; $f_p = 0.3$ m^2; $\zeta_u = 1.5$; $\zeta_\omega = 4$; $\zeta_1 = \zeta_p = 2.4$.

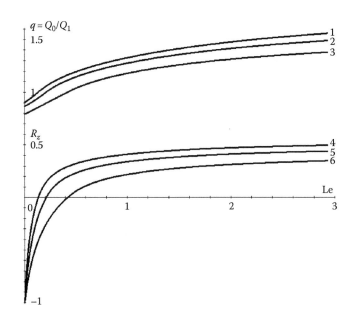

FIGURE 5.12 Variation in the relative flow rate $q = Q_0/Q_1$ of air ejected into the lower cowl as a result of ejecting the flow of loose matter and the change in ejected air recycling rate as a function of the Le number (at $S_u = S_\omega = 0.3$ m^2; $f_1 = f_p = 0.3$ m^2; $v_k = 10$ m/s; $\zeta_1 = \zeta_p = 2.4$; $n = 0.3$; negative pressure $P_3 = -10$ Pa created in the cowl by the aspiration unit fan): 1, q at $\zeta_u = \zeta_\omega = 1.5$; 2, q at $\zeta_u = \zeta_\omega = 2.4$; 3, q at $\zeta_u = \zeta_\omega = 4.8$; 4, R_z at $\zeta_u = \zeta_\omega = 1.5$; 5, R_z at $\zeta_u = \zeta_\omega = 2.4$; 6, R_z at $\zeta_u = \zeta_\omega = 4.8$.

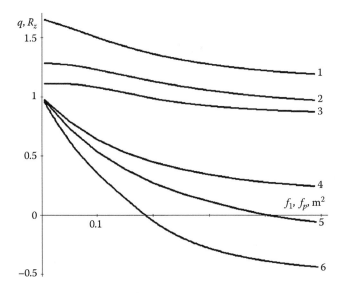

FIGURE 5.13 Variations in the relative airflow rate $q = Q_0/Q_1$ and ejected air recirculation air R_z as a function of leakage area in the upper cowl f_1 (at $f_p = 0.3$ m^2) or gap area in the receiving chamber f_p (with $f_1 = 0.3$ m^2) at $S_u = S_\omega = 0.3$ m^2; $\zeta_1 = \zeta_p = \zeta_u = \zeta_\omega = 2.4$; $v_k = 10$ m/s; $n = 0.3$; $P_3 = -10$ Pa: 1, q at Le = 1.0; 2, q at Le = 0.3; 3, q at Le = 0.1; 4, R_z at Le = 1.0; 5, R_z at Le = 0.3; 6, R_z at Le = 0.1.

cowl that reaches the receiving chamber—a kind of negative recycling is observed ($\omega_c < 0$, $q < 1$). This can be avoided by increasing aerodynamic resistance ζ_n^* and decreasing the leakage areas f_1 and f_p (Figure 5.13).

Note the importance of the upper cowl sealing and of designing the receiving chamber with small gaps for pass-through material. Of course, the absence of this chamber will make *positive* recirculation possible only as long as the negative pressure in the upper chamber is greater than the negative pressure created by the lower cowl fan due to adequate sealing.

5.4.3 Efficiency of Combined Bypass

Let us return to the general formulation of the problem. We will estimate the behavior of air recycling with circulation loops. To that end we will linearize input Equations 5.192 and 5.193, having posited that

$$udu \approx \bar{u}du, \tag{5.226}$$

$$\omega d\omega \approx \bar{\omega}d\omega. \tag{5.227}$$

Although such a simplification could be considered as rough from the mathematical point of view, the findings, as has been shown earlier, are not significantly different from *exact* solutions of the initial system of nonlinear equations. An explanation is provided by the fact that the pressure losses in the examined ducts are mainly determined by local losses rather than changes in the velocity pressure.

In addition, we will continue our assumption that the mass forces of interaction between components in a flow of loose matter change according to a linear law

$$\text{Le} \cdot \frac{|v-u|(v-u)}{v} \approx \text{Le} \cdot k(v-u), \tag{5.228}$$

$$k = \overline{\left|1 - \frac{u}{v}\right|} \approx 1 - \frac{\bar{u}}{\bar{v}}. \tag{5.229}$$

A bar above a letter (expression) will continue to denote averaging the magnitude over the length of the chute (or bypass chamber).

In order to simplify the right-hand side of the linear equation, we will replace a constantly accelerated flow of material with an accelerated (putative) with velocity defined by trinomial Equation 5.32. Then the averaged flow velocity of air would become

$$\bar{v} = b_0 + \frac{b_1}{2} + \frac{b_2}{6}. \tag{5.230}$$

As for pressure losses in the perforation holes of chute walls, we will posit that they are linearly (rather than quadratically) dependent on velocity

$$p_\omega - p_u = \zeta_0 \overline{|w|}w, \tag{5.231}$$

where
$\overline{|w|}$ is the averaged absolute magnitude of the velocity of pass-through air in holes
ζ_0 is the LRC of a hole (assuming $\zeta = 2.4$ as a rule, in a similar way to a hole in an infinitely thick wall)

Due to Equation 5.195, the last relation could be written as

$$p_\omega - p_u = \frac{|w|\sqrt{\zeta_0}}{E} \frac{du}{dx}.$$

(5.232)

In view of the accepted simplifications, we will write original Equations 5.192 and 5.193, accounting for a relationship between velocities (Equation 5.187),

$$\omega = \frac{u - z}{r}, \quad \bar{\omega} = \frac{\bar{u} - z}{r},$$

(5.233)

as the following combined equations:

$$\left.\begin{array}{l} \dfrac{dp_\omega}{dx} + 4\dfrac{\bar{u} - z}{r^2}\dfrac{du}{dx} = 0, \\[3mm] \dfrac{dp_u}{dx} + 4\bar{u}\dfrac{du}{dx} = k\,\mathrm{Le}(v - u). \end{array}\right\}$$

(5.234)

By subtracting the latter equation from the former, we can arrive at the following equation:

$$\frac{d(p_\omega - p_u)}{dx} - 4\left(\bar{u} - \frac{\bar{u} - z}{r^2}\right)\frac{du}{dx} + k\,\mathrm{Le}\cdot u = k\mathrm{Le}\cdot v,$$

(5.235)

so that in view of Equation 5.232, this equation can be reduced to a linear second-order equation with constant coefficients:

$$A = 2\left(\bar{u} - \frac{\bar{u} - z}{r^2}\right)\frac{E}{|w|\sqrt{\zeta_0}},$$

(5.236)

$$B = k\frac{\mathrm{Le}\cdot E}{|w|\sqrt{\zeta_0}}.$$

(5.237)

Thus, the general problem with a combined bypassing of a perforated chute of a handling facility can be reduced to solving the following system of differential equations:

$$\left.\begin{array}{l} u'' - 2Au' - Bu = -Bv, \\[3mm] dp_\omega + 4\dfrac{\bar{u} - z}{r^2}du = 0 \end{array}\right\}$$

(5.238)

with boundary conditions given by Equations 5.192a through 5.196a.

The first equation of these can be solved as (see Equations 5.168 and 5.171)

$$u = C_1 e^{a_1 x} + C_2 e^{a_2 x} + v - k_1 - k_2 x, \tag{5.239}$$

$$u' = C_1 a_1 e^{a_1 x} + C_2 a_2 e^{a_2 x} + b_1 + b_2 x - k_2, \tag{5.240}$$

where
$$k_2 = 2b_2 \frac{A}{B}$$
$$k_1 = 2\frac{A}{B}(b_1 - k_2) - b_2/B$$
$$a_1 = A + \sqrt{A^2 + B}$$
$$a_2 = A - \sqrt{A^2 + B}$$

In order to define integration constants C_1 and C_2, we will use boundary conditions for velocity given in Equations 5.192a and 5.194b:

$$\left.\begin{array}{l} u_n = C_1 + C_2 + n - k_1, \\ u_k = C_1 e^{a_1} + C_2 e^{a_2} + 1 - k_1 - k_2. \end{array}\right\} \tag{5.240a}$$

We end up with

$$C_1 = \frac{m_2 - m_1 e^{a_2}}{e^{a_1} - e^{a_2}}, \quad C_2 = \frac{m_1 e^{a_1} - m_2}{e^{a_1} - e^{a_2}}, \tag{5.241}$$

where

$$m_1 = u_n - n + k_1, \quad m_2 = u_k - 1 + k_1 + k_2. \tag{5.242}$$

Using Equation 5.232 in view of Equation 5.239, we get the following relation:

$$p_\omega - p_u = \frac{\overline{|w|}\sqrt{\zeta_0}}{E}\left(C_1 a_1 e^{a_1 x} + C_2 a_2 e^{a_2 x} + b_1 + b_2 x - k_2\right), \tag{5.243}$$

that, based on the first equations of boundary conditions for pressures given in Equations 5.195a and 5.196a, enable the following combined equations to be constructed:

$$\left.\begin{array}{l} \zeta_{\omega n}\dfrac{(u_n - z)^2}{r^2} + \zeta_{un}u_n^2 = \dfrac{\overline{|w|}\sqrt{\zeta_0}}{E}\left(C_1 a_1 + C a_2 + b_1 - k_2\right), \\[4mm] \zeta_{\omega k}\dfrac{(u_k - z)^2}{r^2} + \zeta_{uk}u_k^2 = -\dfrac{\overline{|w|}\sqrt{\zeta_0}}{E}\left(C_1 a_1 e^{a_1} + C a_2 e^{a_2} + b_1 - k_2 + b_2\right). \end{array}\right\} \tag{5.244}$$

To close the resulting system of two equations with three unknowns u_n, u_k, and z we will use the latter of combined differential equation (Equation 5.238). Integrating this equation over the entire length of ducts yields the following relation:

$$p_\omega(1) - p_\omega(0) + 4\frac{\bar{u} - z}{r^2}(u_k - u_n) = 0. \tag{5.245}$$

In view of the latter equations of boundary conditions for pressures given in Equations 5.195a and 5.196a, as well as the ratios Equations 5.197 and 5.198, this relation can be used to derive the third equation

$$p_y + \zeta_n^* z^2 + 4\frac{\bar{u} - z}{r^2}(u_k - u_n) = \zeta_{\omega k}\frac{(u_k - z)^2}{r^2} + \zeta_{\omega n}\frac{(u_n - z)^2}{r^2} = 0, \tag{5.246}$$

$$\zeta_n^* = \zeta_1^* + \zeta_p^* = \zeta_1\left(\frac{\tilde{S}_u}{\tilde{f}_1}\right)^2 + \zeta_p\left(\frac{\tilde{S}_u}{\tilde{f}_p}\right)^2, \tag{5.247}$$

for a closure of combined Equations 5.244.

Solving combined Equations 5.246 and 5.244 requires identifying the averaged velocities of air ejected ($\bar{u}$), recycled ($\bar{\omega}$), and flowing through perforation holes $|w|$. There is a number of averaging options. For example, in Section 5.3.3, integral relation (Equation 5.92) was used in view of the approximation of the type Equation 5.228. If this way is opted for, then determining an averaged velocity of ejected air $\bar{u}$ would call for integrating the second equation of the *simplified* system Equation 5.234 over the entire length of the chute. That gives

$$p_u(1) - p_u(0) + 4\bar{u}(u_k - u_n) = \frac{\text{Le}(\bar{v} - \bar{u})^2}{\bar{v}} \tag{5.248}$$

or, in view of the boundary requirements for pressures

$$\zeta_{uk}u_k^2 + \zeta_{un}u_n^2 + \zeta_n^* z^2 + p_y + 4\bar{u}(u_k - u_n) = \text{Le}\frac{(\bar{v} - \bar{u})^2}{\bar{v}},$$

whence the following relation can be found for z:

$$z = \sqrt{\frac{k\text{Le}(\bar{v} - \bar{u}) - F_y}{\zeta_n^*}}, \tag{5.249}$$

where it is assumed for notational convenience that

$$F_y = \zeta_{uk}u_k^2 + \zeta_{un}u_n^2 + p_y + 4\bar{u}(u_k - u_n). \tag{5.250}$$

This is different from a similar form (Equation 5.130) by the inclusion of unknown variables u_n, u_k into the averaged quantity. The problem of determining those variables, that is, solving combined transcendental equations (Equations 5.244) becomes somewhat complicated.* It would be tempting to use findings from the solution of the *exact* problem with a single circulation loop, that is, to posit for the first approximation

$$\bar{u} = u_c, \quad \bar{\omega} = \omega_c, \tag{5.251}$$

where u_c and ω_c are determined by Equation 5.216 after solving Equation 5.215 relative to parameter z_c.

Absolute cross-flow velocity $\overline{|w|}$ will be averaged by the value of this quantity in three points along the chute. Considering that the velocity w may reverse its sign, it can be written as follows (see Equation 5.181):

$$\overline{|w|} \approx \frac{|w(0)| + |w(1)|}{3}, \tag{5.252}$$

where, owing to boundary conditions given by Equations 5.193a and 5.194a; Equations 5.195a and 5.196a; and also making allowance for Equation 5.187

$$|w(0)| = \sqrt{\frac{\zeta_{\omega n}}{\zeta_0}\left(\frac{u_n - z}{r}\right)^2 + \frac{\zeta_{un}}{\zeta_0} u_n^2}, \tag{5.253}$$

$$|w(1)| = \sqrt{\frac{\zeta_{\omega k}}{\zeta_0}\left(\frac{u_k - z}{r}\right)^2 + \frac{\zeta_{uk}}{\zeta_0} u_k^2}. \tag{5.254}$$

Let us dwell on a simpler option—solving the system of Equation 5.244 with substitution Equation 5.249 by approximating averaged velocity $\bar{u}$ initially with the arithmetic mean of ejected air velocity in three points (at $x = 0$; $x = x_m$, and $x = 1$):

$$\bar{u} = \frac{u_n + u_m + u_k}{3}, \tag{5.255}$$

where

$$u_m \approx u_c. \tag{5.256}$$

---

* On the other hand, the solution of the problem is simplified as it becomes possible to reduce three combined equations to two (Equation 5.244) using Equation 5.249. Simplification can also be performed by solving the Equation 5.236 for z.

After solving combined Equation 5.244, having determined u_n and u_k, we will determine the design averaged velocity using the formula

$$\bar{u}_r = \bar{v} + \frac{C_1}{a_1}(e^{a_1} - 1) + \frac{C_2}{a_2}(e^{a_2} - 1) - k_1 - 0.5k_2, \tag{5.257}$$

obtained by integrating the function Equation 5.239 in the interval $0 \le x \le 1$. We can compare these quantities and perform another approximation, positing

$$\bar{u} \approx \bar{u}_r \tag{5.258}$$

or determine the maximum value of u_m by solving Equations 5.239 and 5.240 combined (positing $u = u_m$; $x = x_m$; $u' = 0$) and then determine the value of $\bar{u}$ with the formula Equation 5.255. We have opted for a successive approximation method using the relations given by Equations 5.257 and 5.258.

Studies of the specific case of a handling assembly with combined bypassing indicate that, given two recirculation loops (#1: outer transit loop relative to the bypass chamber of the upward flow; #2: inner upward flow of air in the bypass chamber passing through holes in the walls of this chamber into the chute in its upper part and exiting from the chute at the bottom), the flow rate of recycled air would grow whereas the flow rate of air injected (transit part of airflow in the chute relative to the system comprised by the upper chute and lower cowl) from the receiving chamber into the lower aspirated cowl diminishes (Figure 5.14). This can be easily noted by comparing the plots of

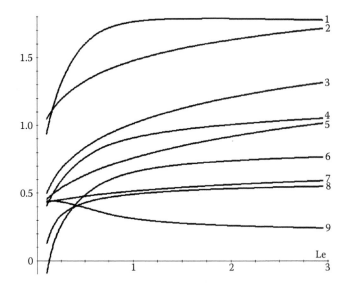

FIGURE 5.14 Variation of flow rates of ejected air that is injected into the lower aspirated cowl from the receiving chamber with a single recirculation loop Q_1, with two recirculation loops Q_2, and in the absence of these loops Q_0; variation of the flow rates of ejected air Q_{u1} with a single recirculation loop Q_{u2}, two loops Q_{u0}, and no recirculation loops; the ratios of these flow rates $q_0 = Q_{u0}/Q_1$, $q_{02} = Q_{u2}/Q_1$, as well as the air recirculation coefficients $R_{z1} = (Q_{u1} - Q_1)/Q_{u1}$; $R_{z2} = (Q_{u2} - Q_2)/Q_{u2}$ as a function of ejection number Le (at $\rho = 1.2$ kg/m³; $\varepsilon = 0.25$; $\zeta_0 = \zeta_1 = \zeta_p = 2.4$; $\zeta_{un} = 0.5$; $\zeta_{yk} = 1.0$; $\zeta_u = \zeta_\omega = 1.5$; $\zeta_{\omega n} = \zeta_{\omega k} = 0.75$; $f_1 = f_p = S_u = S_\omega = 0.3$ m²; $n = 0.3$; $v_k = 10$ m/s; $P_3 = -6$ Pa; $P_y = -0.1$; $\zeta_1^* = \zeta_p^* = 2.4$; $\zeta_n^* = 4.8$): 1, $q_2 = Q_{u2}/Q_1$; 2, $q_0 = Q_{u0}/Q_1$; 3, Q_{u1}; 4, Q_{u2}; 5, Q_{u0}; 6, R_{z2}; 7, Q_1; 8, R_{z1}; 9, Q_2.

variation in recirculation coefficients R_{z1}, R_{z2}, and injected airflow rates Q_1 and Q_2. As the ejecting capacity of the flow of handled material increases, that is, with greater values of Le, this difference becomes even more noticeable. In the case of Le = 0.6, absent bypassing, the flow rate of ejected air in absolute numbers was $Q_0 = 0.66$ m³/s. With only one recirculation loop present, this flow rate fell to $Q_1 = 0.49$ m³/s (by 26%) while with two recirculation loops it decreased to $Q_2 = 0.37$ m³/s (or 1.78 times).

These values at Le = 3 will be, respectively, equal to $Q_0 = 1.03$ m³/s, $Q_1 = 0.6$ m³/s (42% lower, $Q_2 = 0.24$ m³/s (4.29 times lower than Q_0).

Such a significant drop in the flow rate of injected air (and, as a consequence, a significant reduction in the amount of evacuated air Q_a) can be explained with growing counterpressure in cowls (Figure 5.15) with increasing Le. First of all, negative pressure $p_{1(1)}$ in the upper cowl decreases noticeably while the negative pressure in the receiving chamber of the lower cowl $p_{2(1)}$ falls, prompting increased flow rate of recirculated air with growing $\Delta p_{21(1)}$.

For a case of two recirculation loops, air recycling increases due to intensified cross flow of air through perforation holes with increasing ejection numbers Le. Despite growing negative pressure in the receiving chamber $p_{2(2)}$ and the related reduction of counterpressure $\Delta p_{21(2)}$, recycling increases in magnitude. This is explained by an increase in the ejection capacity of a flow of handled material with increasing parameter Le. Excess pressures mount (Figure 5.16) in the chute $p_u(0)$ and in the bypass chamber $p\omega(0)$ while differential pressures $\delta p(0)$ and $\delta p(1)$ increase in absolute sense, promoting increased flow rate of recycled air.

Increasing Le boosts longitudinal velocities of both ejected and recycled air (Figure 5.17) as well as the velocity of cross-flowing air (Figure 5.18).

This case similarly does not invalidate the assertion that a significant role in reducing the amount of injected air Q_1 is played by cowl sealing (p_1 increases) and the reduction of the gap area between conveyor belt and the walls of the receiving chamber (p_2 increases).

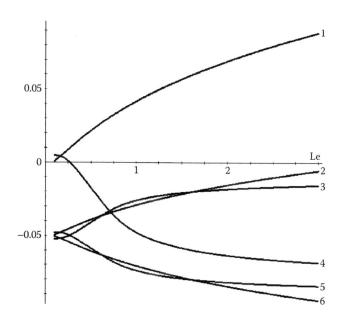

FIGURE 5.15 Variation of excess static pressures of air in the upper (unaspirated) cowl $p_{1(1)}$—with a single recirculation loop and $p_{1(2)}$—with two loops, as well as in the receiving chamber of the lower aspirated cowl $p_{2(1)}$ with a single loop and $p_{2(2)}$ with two loops and variation of the difference between these pressures $\Delta p_{21(1)}$ $= p_{2(1)} - p_{1(1)}$ and $\Delta p_{21(2)} = p_{2(2)} - p_{1(2)}$ as a function of Le (given the same input data as in Figure 5.14): 1, $\Delta p_{21(1)}$; 2, $p_{2(1)}$; 3, $p_{1(2)}$; 4, $\Delta p_{21(2)}$; 5, $p_{2(2)}$; 6, $p_{1(1)}$.

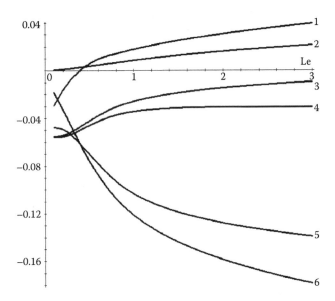

FIGURE 5.16   Variation in excess static pressures of air at chute ends: $p_u(0)$ (curve 4)—at inlet of ejected air into the chute, $p_u(1)$ (curve 1)—at its outlet from the chute; at ends of bypass chamber, $p_\omega(0)$ (curve 3)—at exit of recycled air into the upper cowl, and $p_\omega(1)$ (curve 5)—at inlet into the chamber. Also plotted are differences of these pressures $\delta p(0) = p_\omega(0) - p_u(0)$ (curve 2) and $\delta p(1) = p_\omega(1) - p_u(1)$ (a curve 6) as functions of Le (given the same input data as in Figure 5.14).

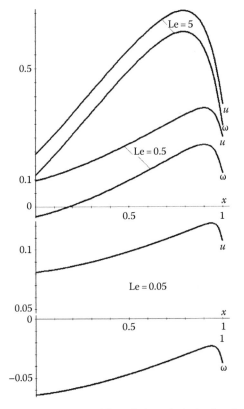

FIGURE 5.17   Variation in dimensionless velocities of ejected air in the chute (u) and recycled upward airflow in the bypass chamber (ω) over the fall height of particles of loose solid material being handled (given the same input data as in Figure 5.14).

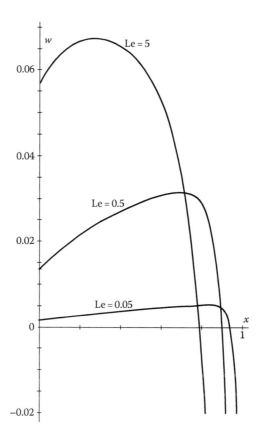

FIGURE 5.18 Variation in dimensionless velocity of cross flow of recycled air from the bypass chamber through perforation holes over chute height (given the same input data as in Figure 5.14).

5.5 CONCLUSIONS

1. Basic dynamics equations have been proposed for 1D flow of ejected air in the hollow area of a perforated vertical pipe with falling solid particles, as well as the flow of recycled air in a coaxial bypass duct (combined nonlinear (*exact*) differential equations, Equations 5.20, 5.21, 5.25, and 5.26 in view of volumetric forces of interaction between components (Equation 5.3) and cross flows of air through perforation holes (Equation 5.19)). These combined equations would be difficult to solve numerically both due to nonlinearity of equations coupled with unknown boundary conditions u_0 and p_a and due to steep-sloping velocity functions calling for computations with a much greater precision (widening the mantissa to 50 and more positions).

   In order to arrive at approximate values and facilitate numerical examination of *exact* equations, the input combined equations have been linearized by substituting linear laws for quadratic laws with respect to variations of aerodynamic drag in pipe wall holes and volumetric forces of interaction between the components given as Equations 5.35 and 5.34, as well as averaging air velocities over the height of the duct/pipe (Equations 5.38 and 5.39).

2. A nonuniform linear second-order differential equation was obtained for the difference between the velocity of ejected air and the velocity of falling particles (Equation 5.47), along with a nonuniform second-order differential equation for the velocity of ejected air (Equation 5.166). To relax the inhomogeneity of these equations, constantly accelerated flow of particles was replaced with an accelerated flow (Equation 5.32). This enabled

the discovery of an analytical solution to the problem in the form (Equation 5.54) and combined equations for boundary conditions given by Equations 5.58 through 5.61 for the difference in velocities and in a simpler form, Equations 5.168, 5.177, and 5.178 for the velocity of ejected air.

3. A comparison of the resulting solutions with findings from integration of the *exact* system, Equations 5.147 and 5.148 has shown that

    a. The function $u(x)$ behaves virtually identically both in character and in absolute magnitude along the height of a perforated pipe (Figure 5.4, Table 5.5, relative error remains within -1.6% at Le $= 1$, and -4.0% at Le $= 5$)

    b. Ejected air velocities/volumetric flows in a perforated pipe (u_0) as well as pressures in bypass chamber (p_a) expressed as functions of Le agree adequately with the findings from solving the *exact* equation (Figure 5.5) numerically.

4. Increased ejection number Le, similar to the case of a flow of particles in a pipe with impermeable walls, contributes to greater values of u_0. The asymptotic nature of this increase (Figure 5.5) is notable along with a drop in ejection volumes compared to a flow of particles in a nonperforated pipe ($u_0 < u_2$, dash–dot curve) in the range of Le surveyed. An even greater effect of minimizing u_0 is observed at higher degrees of perforation (Figure 5.6). This case is similarly asymptotic: with $E \geq 2.5$, the decline in u_0 virtually ceases.

5. Effects of cross-sectional size of the bypass duct on the character and value of velocity u_0 are manifold. An *avalanche* downslope occurring at velocities u_0 within $0.5 \leq r \leq 0.6$ defies explanation. As r decreases to zero, the value of u_0 grows sharply, reaching u_2—the velocity of ejected air in a pipe with impermeable walls. At greater $r > 0.6$, the velocity u_0 first grows fast, then increases slowly (remaining within u_2), and virtually stabilizes at $r \geq 2.5$.

6. Reduction in ejected air volume (velocity u_0) owing to its recirculation in the bypass duct is more intense at greater Le numbers and smaller values of n. Even more notable is the reduction in ejection volumes with increased drag at inlet/outlet of the perforated pipe (Tables 5.7 and 5.8). This fact confirms that air recirculation volume is to a significant extent promoted by the upper cowl sealing and a buffer capacity (flow-shaping chamber) within the lower cowl (see Chapter 8), as both lead to greater negative pressures in the upper part of the pipe while providing a head in its lower part. Conditions conducive to more intense cross flow of air through perforated holes therefore arise.

    Volume of ejected air can be increased 1.3–1.5 times by increasing ζ_k to between 8 and 32, or 1.45 to 1.55 times if ζ_n is increased to between 4 and 16 (at Le ≥ 10; $r = 2$; $E = 2.5$).

7. Installation of a combined chamber with transit interchange of air between the upper unaspirated cowl and the receiving chamber of the lower aspirated cowl (Figure 5.10) improves the energy-saving effect. The flow rate of ejected air forced into the aspirated cowl Q_1 can be reduced greatly by providing for double recirculation of air through an outer loop (using end openings in the bypass chamber) and an inner loop ensuring cross flow of air through perforation holes along the entire chute.

    So, in a specific example reviewed with a small ejection capacity of loose matter flow (Le $= 0.6$), even with a single inner recirculation loop, the flow rate of injected air was brought down 1.35 times compared to handling the same flow in a chute with impermeable walls and no bypass chamber installed. With a double recirculation loop, the effect reaches 1.78 times (Figure 5.14). It is possible to reduce the flow rate Q_1 even more with large ejection numbers. So, with Le $= 3$, the same flow rate in the same comparison (with Q_{u0}) was brought down 1.72 times with a single outer recirculation loop and 4.29 times with two loops.

6 Air Ejection in a Porous Pipe with Linear Cross-Flow Pattern

We will consider the dynamics of air ejected by an accelerated flow of particulate matter in a vertical pipe with porous walls. Dynamics equations for airflow are not different from air motion equations in a perforated pipe (see Chapter 5):

$$\frac{dp}{dx} = -4u\frac{du}{dx} + \text{Le}\,|v-u|\frac{v-u}{v},$$ (6.1)

$$\frac{du}{dx} = Ew,$$ (6.2)

$$v = b_0 + b_1 x + b_2 x^2/2,$$ (6.3)

where E is the ratio of the total area of pores in the pipe walls to the cross-sectional area of the pipe itself:

$$E = \frac{\tilde{\Pi}\tilde{l}\varepsilon}{\tilde{s}_t}.$$

What sets apart the present case is the linear dependence between the velocity of air cross flow through pipe walls and differential pressure. This makes it significantly simpler to solve the problem (as equations become naturally linearized). Assuming that pores in the wall are small enough and the wall itself is an ideal filtering layer, that is, a layer comprising cylindrically shaped parallel capillary channels, we can use the now-classical linear law of filtration in homogeneous porous fine-grained media [211].

Using a porous-tube model to study the velocities of cross flows of air through porous partitions, the following relation [212] is put forward:

$$\tilde{w} = \frac{\tilde{r}^2\Delta\tilde{p}}{8\tilde{\delta}\tilde{\mu}},$$ (6.4)

where
$\tilde{r}$ is the radius of a porous tube (m)
$\tilde{\delta}$ is the partition thickness (m)
$\Delta\tilde{p}$ is the differential pressure (Pa)
$\tilde{\mu}$ is the dynamic viscosity of air (Pa·s)

The ratio (Equation 6.4) can be rewritten in a dimensionless form as follows:

$$w = \frac{\Delta p}{\xi_0},$$ (6.5)

where

$$\xi_0 = \frac{16\tilde{\delta}\tilde{\mu}}{\tilde{r}^2\tilde{\rho}\tilde{v}_k}. \tag{6.6}$$

(For example, with $\tilde{\delta} = 0.01\,\text{m}$, $\tilde{r} = 0.0005\,\text{m}$, $\tilde{\rho} = 1.2\,\text{kg/m}^3$, $v_k = 7\,\text{m/s}$, $\tilde{\mu} = 18 \cdot 10^{-6}\,\text{Pa}\cdot\text{s}$, this results in $\xi_0 = 1.37$.)

We will continue here with defining the velocity $w(x)$ as positive when air flows through porous walls into the pipe (this is usually the case within the upstream section of the pipe with an excess pressure $p(x) < 0$). Air escaping outside from the pipe corresponds to $w(x) < 0$; this is characteristic of the downstream section of a porous pipe with a positive excess pressure $p(x)$. Therefore, the differential pressure in formulas, Equations 6.4 and 6.5, will be understood as follows:

$$\Delta p = p_\omega - p, \tag{6.7}$$

where
    p_ω is the excess pressure outside porous pipe (e.g., in the bypass chamber)
    p is the excess pressure in the pipe

6.1 POROUS PIPE WITHOUT BYPASS

6.1.1 ONE-DIMENSIONAL FLOW EQUATIONS FOR AIR EJECTED IN A POROUS PIPE

Absence of a bypass chamber in this case results in $p_\omega = 0$ and, thus,

$$\Delta p = -p, \tag{6.8}$$

so Equation 6.2, in view of Equations 6.5 and 6.8, becomes

$$\frac{du}{dx} = -\frac{E}{\xi_0}p, \tag{6.9}$$

which, when derived over x in view of Equation 6.1, produces a second-order differential equation

$$u'' = \frac{E}{\xi_0}\left[4uu' - Le\frac{|v-u|}{v}(v-u)\right], \tag{6.10}$$

which differs from a similar equation for a perforated pipe by its linearity with respect to the highest derivative. This way, Equation 6.10 can be reduced to a linear equation with minimal effort. It would be just enough to linearize the mass forces of interaction between components by accepting the condition

$$Le\frac{|v-u|}{v}(v-u) \approx Le\left(\overline{\frac{|v-u|}{v}}\right)(v-u) \approx Le \cdot k(v-u), \tag{6.11}$$

$$k \approx \frac{\overline{v}-\overline{u}}{\overline{v}} \tag{6.12}$$

and averaging the function in the expression before the first derivative u', having posited that

$$uu' \approx \overline{u}u'. \tag{6.13}$$

The quantity $\bar{u}$ will be determined from an integral condition that can be easily obtained by integrating Equation 6.1 over the entire length of the duct ($x = 0...1$) in view of Equation 6.11:

$$p(1) - p(0) = -2(u^2(1) - u^2(0)) + \text{Le}\frac{(\bar{v} - \bar{u})^2}{\bar{v}}. \tag{6.14}$$

Taking into account the boundary conditions,

$$u(0) = u_0, \quad p(0) = -\zeta_n u_0^2, \tag{6.15}$$

$$u(1) = u_k, \quad p(1) = \zeta_k u_k^2, \tag{6.16}$$

we can use Equation 6.14 to determine

$$\bar{u} = \bar{v} - \sqrt{\frac{\bar{v}}{\text{Le}}\left[(\zeta_k + 2)u_k^2 - (2 - \zeta_H)u_0^2\right]}. \tag{6.17}$$

The introduction of these conditions leads us to a linear nonuniform second-order differential equation

$$u'' - 2Nu' - Mu = -Mv \tag{6.18}$$

with constant coefficients

$$N = 2\frac{E}{\xi_0}\bar{u}, \quad M = \text{Le}\left|\frac{\bar{v} - \bar{u}}{\bar{v}}\right|\frac{E}{\xi_0}. \tag{6.19}$$

Equation 6.18 is solved as follows (for an accelerated particle flow with the fall velocity determined by the trinomial given by Equation 6.3):

$$u = C_1 e^{n_1 x} + C_2 e^{n_2 x} + v - k_1 - k_2 x, \tag{6.20}$$

where

$$k_2 = 2b_2\frac{N}{M}, \quad k_1 = 2\frac{N}{M}(b_1 - k_2) - b_2/M. \tag{6.21}$$

n_1, n_2 are the roots of a characteristic equation corresponding to the homogeneous differential equation, Equation 6.18

$$n_1 = N + \sqrt{N^2 + M}, \quad n_2 = N - \sqrt{N^2 + M}. \tag{6.22}$$

Derivation of Equation 6.20 results in

$$u' = C_1 n_1 e^{n_1 x} + C_2 n_2 e^{n_2 x} + b_1 + b_2 x - k_2. \tag{6.23}$$

In order to determine the parameters u_0 and u_k of the problem as well as constants C_1 and C_2, the following four equations will be constructed using boundary conditions from Equations 6.15 and 6.16:

$$u_0 = C_1 + C_2 + n - k_1, \tag{6.24}$$

$$u_k = C_1 e^{n_1} + C_2 e^{n_2} + 1 - k_1 - k_2, \tag{6.25}$$

$$-\zeta_n u_0^2 = -\frac{\xi_0}{E} \left(C_1 n_1 + C_2 n_2 + b_1 - k_2 \right), \tag{6.26}$$

$$\zeta_k u_k^2 = -\frac{\xi_0}{E} \left(C_1 n_1 e^{n_1} + C_2 n_2 e^{n_2} + b_1 - k_2 + b_2 \right). \tag{6.27}$$

C_1 and C_2 become known once the first two equations are solved:

$$C_1 = \frac{b_3 - a_3 e^{n_2}}{e^{n_1} - e^{n_2}}, \quad C_2 = \frac{a_3 e^{n_2} - b_3}{e^{n_1} - e^{n_2}}, \tag{6.28}$$

where the following assignments were made to simplify the notation:

$$a_3 = u_0 + k_1 - n, \quad b_3 = u_k + k_1 - 1 + k_2. \tag{6.29}$$

Sought parameters of the problems u_0 and u_k are determined numerically by solving Equations 6.26 and 6.27 together, where the known values of C_1 and C_2 are given.

With C_1, C_2, u_0, and u_k determined, the formula (Equation 6.20) can be used to determine the function $u(x)$, and then, taking account of Equations 6.9 and 6.23, the following function can be defined:

$$p(x) = -\frac{\xi_0}{E} \left(C_1 n_1 e^{n_1 x} + C_2 n_2 e^{n_2 x} + b_1 - k_2 + b_2 x \right). \tag{6.30}$$

The resulting expression is used to determine the velocity of airflow through porous tubes:

$$w(x) = \frac{-p(x)}{\xi_0}. \tag{6.31}$$

In order to compare computation results for Equation 6.20, let us perform the numeric integration of Equations 6.1, 6.9, and 5.24 in view of the boundary conditions given by Equations 6.15 and 6.16. In order to simplify the latter we will move on to new functions:

$$y_1(x) = \frac{u}{u_k}, \quad y_2(x) = \frac{p}{\zeta_k u_k^2}. \tag{6.32}$$

In this case, the system of *exact* equations and boundary conditions will appear as follows:

$$y_1' = -\frac{\zeta_k}{\xi_0} u_k E y_2, \tag{6.33}$$

$$y_2' = 4 \frac{E}{\xi_0} u_k y_1 y_2 + \frac{\text{Le}}{\zeta_k} \cdot \frac{1}{v} \left| \frac{v}{u_k} - y_1 \right| \left(\frac{v}{u_k} - y_1 \right), \tag{6.34}$$

$$y_1(0) = \alpha, \quad y_2(0) = -\frac{\zeta_n}{\zeta_k} \alpha^2, \quad y_1(1) = 1, \quad y_2(1) = 1, \tag{6.35}$$

where

$$\alpha = \frac{u_0}{u_k}. \tag{6.36}$$

The following approximate assignment will be used for the initial assignment of the velocity u_0:

$$u_0 \approx u_2, \tag{6.37}$$

where u_2 is the velocity of ejected air in a porous duct with plugged porous tubes (i.e., at $E = 0$) where a constantly accelerated flow of particles are given. This velocity is quantified with relations given by Equations 5.85 and 5.86 that can be written in a more compact form:

$$\frac{3u_2^2}{(1-u_2)^3 - |u_2 - n|^3} = \frac{2Le}{(1-n^2)\sum \zeta}. \tag{6.38}$$

Another option is to use the data from Table 5.1.

Combined Equations 6.33 through 6.35 are solved numerically.

Another possible approach would be to use the following initial values:

$$u_0 = u_0^{l_i}, \quad u_k = u_k^{l_i}, \tag{6.39}$$

where $u_0^{l_i}$ and $u_k^{l_i}$ are previously defined velocities of ejected air, respectively, at inlet and outlet of the porous pipe.

As the comparison indicates (Table 6.1), assumptions introduced for linearization of the problem have virtually no bearing on calculated ejection volumes. The discrepancy between air velocities at the outlet of the porous pipe u_q and u_L was less than 3%.

TABLE 6.1

Comparison of Ejected Air Velocities in a Porous Duct without Bypass (at $E = 1$, $n = 0.4319$) Determined by Numerical Integration of *Exact* Equations 6.33 through 6.35 in the Case of a Constantly Accelerated Flow u_q with Their Calculated Values for a Linearized Equation Obtained Using (Formula) Equation 6.20 in the Case of an Accelerated Flow u_L

	At Le = 1.0			At Le = 5.0		
x	u_q	u_L	$\Delta u_L, \%$	u_q	u_L	$\Delta u_L, \%$
0	0.3508	0.3523	0.4	0.4677	0.4650	−0.6
0.1	0.3554	0.3569	0.2	0.4763	0.4736	−0.6
0.2	0.3602	0.3614	0.3	0.4859	0.4831	−0.6
0.3	0.3648	0.3656	0.2	0.4964	0.4928	−0.7
0.4	0.3689	0.3691	0.1	0.5069	0.5021	−0.9
0.5	0.3721	0.3717	−0.1	0.5167	0.5101	−1.3
0.6	0.3742	0.3729	0.5	0.5248	0.5159	−1.7
0.7	0.3747	0.3726	−0.3	0.5301	0.5187	−2.2
0.8	0.3730	0.3702	−0.8	0.5310	0.5173	−2.5
0.9	0.3686	0.3654	−0.9	0.5257	0.5103	−2.9
1.0	0.3610	0.3577	−0.9	0.5118	0.4964	−3.0

6.1.2 FINDINGS FROM STUDIES

The degree of the duct wall porosity (the number E) affects ejection volumes in an insignificant way (u_k). Ejection volume reduction coefficient

$$\text{Ke} = \frac{u_{l3}}{u_k} \tag{6.40}$$

at smaller values of Le ≤ 5 is noticeably greater than one (Figure 6.1), falling to a bit below one at Le ≥ 20.

Let us expand on the regularities of variation in ejected air volume over the length of a porous channel as a function of key input parameters of the problem (E and Le) and aerodynamic resistance (drag) of the duct (ζ_n and ζ_k). We will determine how these parameters will affect the change in

1. Relative air consumption at the pipe inlet (Figure 6.2)

$$\alpha = \frac{Q_0}{Q_k} = \frac{u_0}{u_k}. \tag{6.41}$$

2. Relative maximum flow rate of ejected air

$$\alpha_m = \frac{Q_m}{Q_k} = \frac{u_m}{u_k}, \tag{6.42}$$

where u_m is the maximum velocity of ejected air in the duct at a point $\tilde{x}_m$ away from its inlet.

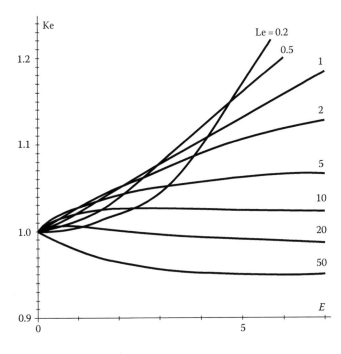

FIGURE 6.1 Change in the ejection volume reduction coefficient with increasing porosity of duct walls (at $\xi_0 = 1.37$; $\zeta_n = 0.5$; $\zeta_n = 0.5$; $\zeta_k = 1$; $n = 0.4319$).

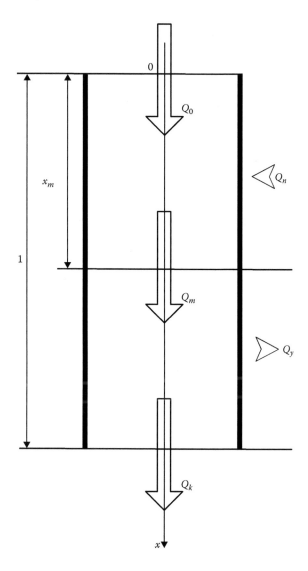

FIGURE 6.2   Illustration of airflow balance in a pipe with porous walls.

Note that these coefficients enable us to determine the relative flow rate of air entering through pores in the duct walls within the range $x = 0...x_m$

$$R_n = \frac{Q_m - Q_0}{Q_k} = \alpha_m - \alpha \tag{6.43}$$

and of air escaping through pores along the rest of the duct (within the range $x = x_m...1$)

$$R_y = \frac{Q_m - Q_k}{Q_k} = \alpha_m - 1. \tag{6.44}$$

Then the relative flow rate of recirculated air would be

$$R_z = \begin{cases} Q_y/Q_k & at \ Q_n \ge Q_y, \\ Q_n/Q_k & at \ Q_n \le Q. \end{cases} = \begin{cases} \alpha_m - 1 & at \ \alpha \le 1, \\ \alpha_m - a & at \ \alpha \ge 1. \end{cases} \tag{6.45}$$

Let us find the extremum of the function $u(x)$. Based on Equations 6.20 and 6.23,

$$u_m(x) = C_1 n_1 e^{n_1 x_m} + C_2 n_2 e^{n_2 x_m} + v(x_m) - k_1 - k_2 x_m, \tag{6.46}$$

where x_m is determined from Equation 6.23:

$$C_1 n_1 e^{n_1 x_m} + C_2 n_2 e^{n_2 x_m} + b_1 + b_2 x_m - k_2 = 0. \tag{6.47}$$

Higher values of the ejection parameter Le result in an increase of ejected air volumes at the duct inlet/outlet (Table 6.2) as well as maximum volume Q_m. Ke reaches a notable maximum with the Le $\approx$ 3.5 area (at $E = 1$). Increased porosity of the duct walls shifts the maximum toward greater injection numbers (Figure 6.3). The recycle rate (R_z) of air injected through porous ducts of a vertical pipe increases with larger ejection number (Le) and wall porosity E (Figure 6.4). Recycling in

TABLE 6.2
Key Parameters of Air Ejection in a Porous Duct at $E = 1$; $n = 0.4319$; $\zeta_k = 1$; $\xi_0 = 1.37$

Le	u_{l3} (at $E = 0$)	u_0	u_k	x_m	u_m	Ke	α	α_m	R_z
0.1	0.1726	0.1741	0.1728	0.5214	0.1776	0.9986	1.0076	1.0276	0.0200
0.2	0.2229	0.2233	0.2222	0.5497	0.2295	1.0032	1.0050	1.0331	0.0281
0.5	0.3006	0.2960	0.2970	0.5939	0.3087	1.0121	0.9966	1.0395	0.0395
0.8	0.3438	0.3344	0.3380	0.6187	0.3522	1.0173	0.9894	1.0420	0.0420
1.0	0.3647	0.3523	0.3577	0.6309	0.3730	1.0197	0.9851	1.0429	0.0429
1.2	0.3819	0,3668	0.3739	0.6410	0.3901	1.0217	0.9813	1.0436	0.0436
1.5	0.4028	0.3841	0.3935	0.6534	0.4109	1.0236	0.9760	1.0443	0.0443
1.7	0.4145	0.3935	0.4045	0.6604	0.4225	1.0247	0.9728	1.0445	0.0445
2.0	0.4295	0.4055	0.4187	0.6695	0.4375	1.0257	0.9683	1.0449	0.0449
2.5	0.4499	0.4213	0.4381	0.6819	0.4579	1.0270	0.9617	1.0452	0.0452
3.0	0.4662	0.4336	0.4537	0.6921	0.4743	1.0275	0.9558	1.0453	0.0453
3.5	0.4798	0.4436	0.4668	0.7006	0.4879	1.0279	0.9504	1.0453	0.0453
4.0	0.4912	0.4519	0.4780	0.7080	0.4996	1.0277	0.9455	1.0453	0.0453
4.5	0.5012	0.4590	0.4878	0.7144	0.5098	1.0276	0.9410	1.0452	0.0452
5.0	0.5100	0.4650	0.4964	0.7201	0.5188	1.0273	0.9368	1.0451	0.0451
5.5	0.5177	0.4703	0.5042	0.7252	0.5269	1.0268	0.9328	1.0450	0.0450
6.0	0.5246	0.4750	0.5112	0.7299	0.5342	1.0261	0.9291	1.0448	0.0448
6.5	0.5309	0.4792	0.5177	0.7341	0.5408	1.0255	0.9256	1.0447	0.0447
7.0	0.5367	0.4829	0.5236	0.7380	0.5469	1.0250	0.9222	1.0446	0.0446
8.0	0.5468	0.4893	0.5342	0.7449	0.5578	1.0236	0.9160	1.0443	0.0443
10.0	0.5630	0.4992	0.5516	0.7563	0.5757	1.0207	0.9050	1.0437	0.0437
15.0	0.5903	0.5142	0.5823	0.7759	0.6071	1.0137	0.8831	1.0426	0.0426
20.0	0.6078	0.5227	0.6036	0.7889	0.6287	1.0070	0.8660	1.0416	0.0416
25.0	0.6204	0.5281	0.6198	0.7985	0.6451	1.0010	0.8521	1.0408	0.0408
30.0	0.6300	0.5317	0.6328	0.8060	0.6583	0.9955	0.8402	1.0402	0.0402
35.0	0.6377	0.5342	0.6437	0.8120	0.6693	0.9906	0.8298	1.0397	0.0397
40.0	0.6440	0.5360	0.6531	0.8170	0.6787	0.9861	0.8207	1.0392	0.0392
50.0	0.6539	0.5383	0.6686	0.8249	0.6942	0.9781	0.8051	1.0384	0.0384
100.0	0.6800	0.5400	0.7150	0.8463	0.7408	0.9511	0.7552	1.0361	0.0361

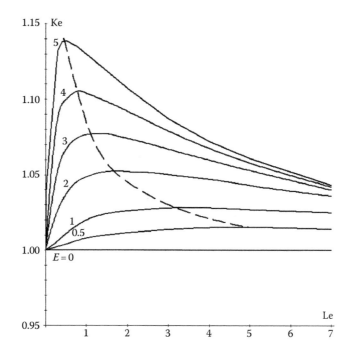

FIGURE 6.3 Change in the ejection volume reduction coefficient with increasing Le numbers (at $\xi_0 = 1.37$; $\zeta_n = 0.5$; $\zeta_n = 0.5$; $\zeta_k = 1$; $n = 0.4319$).

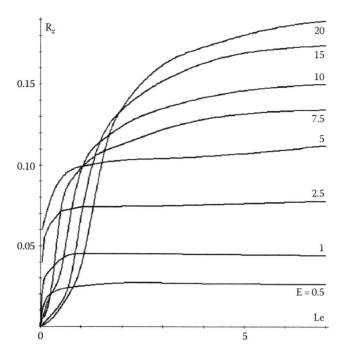

FIGURE 6.4 Change in volume of injected air recycled through porous walls of a vertical duct with increasing Le numbers (at $\xi_0 = 1.37$; $\zeta_n = 0.5$; $\zeta_n = 0.5$; $\zeta_k = 1$; $n = 0.4319$).

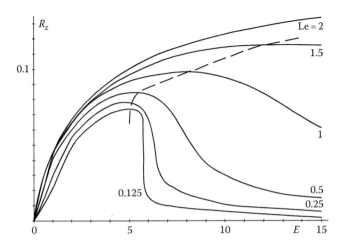

FIGURE 6.5   Change in air recycling rate in a porous duct without bypass with increased porosity of walls (at $\xi_0 = 1.37$; $\zeta_n = 0.5$; $\zeta_n = 0.5$; $\zeta_k = 1$; $n = 0.4319$).

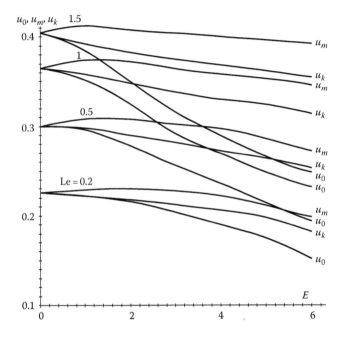

FIGURE 6.6   Change in velocities of ejected air in a duct with increasing porosity of duct walls (at $\xi_0 = 1.37$; $\zeta_0 = 0.5$; $\zeta_k = 1$; $n = 0.4319$).

this case reaches a prominent maximum within $E \approx 5$ (Figure 6.5). Increased Le numbers cause this maximum to shift toward $E > 5$. The presence of these maxima can be explained by varying the effect of changes in ejected air velocity at increased Le and E (Figure 6.6). A common tendency is the reduction in velocities with increasing porosity of the duct walls as well as increase in velocities at higher ejection numbers.

Yet, these changes have a different nature. For instance, the velocity of ejected air entering the duct u_0 is the one exhibiting the most intense decline. This is because increased porosity of walls promotes increased injection, conversely decreasing u_0. Air velocity at duct outlet u_k, as well as maximum velocity u_m, responds more conservatively to changes in porosity. In addition, the extreme

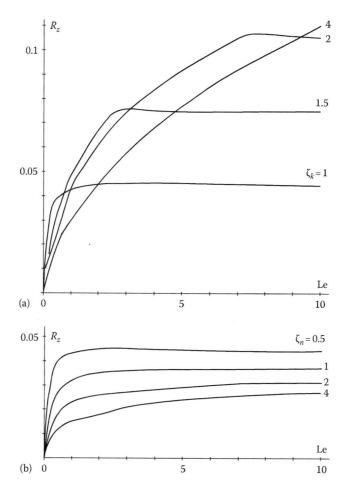

FIGURE 6.7 Change in ejected air recycling rate in a duct as a function of Le number at various resistance coefficients (at $E = 1$; $\xi_0 = 1.37$; $n = 0.4319$): (a) at $\zeta_n = 0.5$ and (b) at $\zeta_k = 1$.

value of u_m with corresponding combinations of Le and E produces a maximum in recirculated air volume. Recycling amounts significantly change with increased LRCs of the inlet and outlet duct sections as well as increased Le (Figure 6.7).

Moreover, the latter case exhibits certain asymptotics inherent in ejecting blowers due to limited volumetric forces of interaction between components: volumes of recirculated air first increase rapidly, then, at Le > 1, R_z stabilizes.

6.2 POROUS PIPE WITH A BYPASS CHAMBER

6.2.1 ANALYTICAL SOLUTIONS

When a bypass chamber is used, the pressure drop determining the cross flow of air through porous tubes in a duct wall will be determined not only by the pressure in the porous pipe but also by the pressure in the chamber p_ω:

$$\Delta p = p_\omega - p. \tag{6.48}$$

The value of p_ω will be determined using linearized Equation 5.41. Then, original Equation 6.2 in view of Equations 6.5 and 6.48 would appear as

$$\frac{du}{dx} = -\frac{E}{\xi_0}\left[p - p_a + 4\bar{\omega}\frac{(u - u_0)}{r}\right], \tag{6.49}$$

which, derived over x in view of Equations 6.1, 6.11, and 6.13, after certain trivial transformations, would be reduced to a linear nonuniform second-order equation:

$$u'' - 2N_r u' - M_r u = -M_r v, \tag{6.50}$$

with constant coefficients

$$N_r = 2\left(\bar{u} - \frac{\bar{\omega}}{r}\right)\frac{E}{\xi_0}, \quad M_r = M. \tag{6.51}$$

The roots of the characteristic equation corresponding to homogeneous Equation 6.50 appear as

$$n_1 = N_r + \sqrt{N_r^2 + M}, \quad n_2 = N_r - \sqrt{N_r^2 + M}, \tag{6.52}$$

and Equation 6.50 is resolved into the following relations:

$$u = C_1 e^{n_1 x} + C_2 e^{n_2 x} + v - k_1 - k_2 x, \tag{6.53}$$

$$u' = n_1 C_1 e^{n_1 x} + n_2 C_2 e^{n_2 x} + b_1 + b_2 x - k_2, \tag{6.54}$$

where

$$k_2 = 2b_2 \frac{N_r}{M}, \quad k_1 = 2\frac{N_r}{M}(b_1 - k_2) - \frac{b_2}{M}. \tag{6.55}$$

Boundary conditions in this case appear as

$$u(0) = u(1) = u_0, \tag{6.56}$$

$$\Delta p(0) = p_a - p(0) = p_a + \zeta_n u_0^2, \tag{6.57}$$

$$\Delta p(1) = p_a - p(1) = p_a - \zeta_k u_0^2 \tag{6.58}$$

or, in view of Equation 6.49, the latter could be written as

$$u'(0) = \frac{E}{\xi_0}\left(\zeta_n u_0^2 + p_a\right), \tag{6.59}$$

$$u'(1) = -\frac{E}{\xi_0}\left(\zeta_k u_0^2 - p_a\right). \tag{6.60}$$

Then in view of Equations 6.53 and 6.54, we arrive at the following combined algebraic equations for determining the parameters u_0 and p_a of the problem as well as the constants C_1 and C_2:

$$u_0 = C_1 + C_2 + n - k_1,$$ (6.61)

$$u_0 = C_1 e^{n_1} + C_2 e^{n_2} + 1 - k_1 - k_2,$$ (6.62)

$$p_a + \zeta_n u_0^2 = \frac{\xi_0}{E}\left[C_1 n_1 + C_2 n_2 + b_1 - k_2\right],$$ (6.63)

$$p_a - \zeta_k u_0^2 = \frac{\xi_0}{E}\left[C_1 n_1 e^{n_1} + C_2 n_2 e^{n_2} + b_1 + b_2 - k_2\right].$$ (6.64)

A solution of the first two equations relative to C_1 and C_2 yields

$$C_1 = \frac{\left[u_0\left(1 - e^{n_2}\right) - \left(b_3 - a_3 e^{n_2}\right)\right]}{z_1},$$ (6.65)

$$C_2 = \frac{\left[u_0\left(e^{n_1} - 1\right) - \left(a_3 e^{n_1} - b_3\right)\right]}{z_1},$$ (6.66)

where the following assignments were made to simplify the notation:

$$z_1 = e^{n_1} - e^{n_2},$$ (6.67)

$$a_3 = n - k_1, \quad b_3 = 1 - k_1 - k_2.$$ (6.68)

If Equation 6.64 is subtracted from Equation 6.63, the resulting equation enables us to determine the parameter u_0 of the problem:

$$\left(\zeta_n + \zeta_k\right)u_0^2 \frac{E}{\xi_0} = C_1 n_1\left(1 - e^{n_1}\right) + C_2 n_2\left(1 - e^{n_2}\right) - b_2,$$ (6.69)

while adding the two together results in the computational relation for the parameter p_a:

$$p_a = \frac{1}{2}\left(\zeta_k - \zeta_n\right)u_0^2 + \frac{\xi_0}{2E}\left[C_1 n_1\left(1 + e^{n_1}\right) + C_2 n_2\left(1 + e^{n_2}\right) + 2\left(b_1 - k_2\right) + b_2\right].$$ (6.70)

In order to obtain computational relations for determining n_1, n_2, C_1, C_2, velocities $\bar{u}$ and $\bar{\omega}$ have to be averaged within $x = 0\ldots1$. Of these, $\bar{u}$ can be determined using the integral relation Equation 6.14, which, in view of boundary conditions in Equations 6.56, 6.15, and 6.16, will become

$$\left(\zeta_n + \zeta_k\right)u_0^2 = Le\frac{\left(\bar{v} - \bar{u}\right)^2}{\bar{v}},$$ (6.71)

whence we can find

$$\bar{u} = \bar{v} - u_0 \sqrt{\frac{\bar{v}}{Le} \left(\zeta_n + \zeta_k \right)}. \tag{6.72}$$

We will use Equation 5.180 to determine the average velocity in the bypass duct.

Air cross-flow velocity in porous tubes, due to Equation 6.2, is

$$w = \frac{u'}{E}, \tag{6.73}$$

where the function u' is determined using Equation 6.54.

6.2.2 NUMERICAL EXAMINATIONS

In order to determine the error introduced by assumptions made in the previous section to linearize Equation 6.49 into the problem of air ejection in a porous duct, we will solve the original equation numerically:

$$\frac{du}{dx} = \frac{E}{\xi_0} \Delta p, \quad \Delta p = p_\omega - p, \tag{6.74}$$

$$\frac{d\Delta p}{dx} = 4 \frac{E}{\xi_0} \Delta p \left[\left(1 - \frac{1}{r^2} \right) u + \frac{u_0}{r^2} \right] - Le |v - u| \frac{v - u}{v}, \tag{6.75}$$

$$v = \sqrt{(1 - n^2)x + n^2}, \tag{6.76}$$

with boundary conditions given in Equations 6.56 through 6.58.

Like the previous case, we will introduce the following functions to simplify these conditions:

$$y_1 = \frac{u}{u_0}, \quad y_2 = \frac{\Delta p}{\zeta_k u_0^2}. \tag{6.77}$$

Then, Equations 6.74 and 6.75, as well as boundary conditions, would appear as

$$y_1' = \frac{E}{\xi_0} \zeta_k u_0 y_2, \tag{6.78}$$

$$y_2' = 4 \frac{E}{\xi_0} u_0 y_2 \left[\left(1 - \frac{1}{r^2} \right) y_1 + \frac{1}{r^2} \right] - \frac{Le}{\zeta_k v} \left| \frac{v}{u_0} - y_1 \right| \left(\frac{v}{u_0} - y_1 \right), \tag{6.79}$$

$$y_1(0) = y_1(1) = 1, \tag{6.80}$$

$$y_2(0) = \zeta_n / \zeta_k + \beta, \quad \beta = P_a / \left(\zeta_k u_0^2 \right), \tag{6.81}$$

$$y_2(1) = \beta - 1. \tag{6.82}$$

TABLE 6.3

Comparison of Ejected Air Velocities in a Porous Duct with a Bypass Chamber (at $E = 1.0$; $r = 2$; $n = 0.4319$; $\xi_0 = 1.37$; $\zeta_n = 0.5$; $\zeta_k = 1.0$) Determined by a Numerical Integration of *Exact* Equations 6.78 through 6.82 in the Case of a Constantly Accelerated Flow u_q with Their Calculated Values for a Linearized Equation Obtained for an Accelerated (Putative) Flow u_L Using Formula (Equation 6.53)

x	At Le = 1.0 ($u_2 = 0.3697$)			At Le = 10.0 ($u_2 = 0.5916$)		
	u_q^*	u_L	Δu_L^{***}, %	u_q^{**}	u_L	Δu_L, %
0	0.3552	0.3592	1.1	0.5446	0.5449	0.05
0.1	0.3592	0.3634	1.2	0.5506	0.5503	−0.05
0.2	0.3632	0.3674	1.2	0.5579	0.5575	−0.03
0.3	0.3671	0.3711	1.1	0.5664	0.5656	−0.15
0.4	0.3704	0.3741	1.0	0.5753	0.5735	−0.3
0.5	0.3728	0.3762	0.9	0.5837	0.5802	−0.6
0.6	0.3738	0.3769	0.8	0.5901	0.5845	−1.0
0.7	0.3732	0.3760	0.7	0.5924	0.5850	−1.3
0.8	0.3702	0.3730	0.8	0.5882	0.5799	−1.4
0.9	0.3645	0.3676	0.9	0.5739	0.5673	−1.2
1.0	0.3552	0.3592	1.1	0.5446	0.5449	0.05

* At $p_a = -0.01$.

** At $p_a = -0.081$.

*** At $\Delta u_L = \dfrac{u_L - u_q}{u_q} \cdot 100$.

A comparison of the results shows (Table 6.3) that simplifications made for linearizing the problem have caused significant changes in the velocity of ejected air. Its value at greater Le numbers becomes somewhat lower than the value determined by the numerical integration of the *exact* equation. However, the relative deviation stays within 1.4%, that is, it is less than input data error that may rise to or exceed 10% due to instability of current mechanical processing technology and loose matter handling.

6.2.3 FINDINGS FROM STUDIES

Studies performed so far point out a significant difference in the kinematics of ejected airflow owing itself to the structure of the bypass chamber surrounding the porous duct (not unlike the one observed in the case of a perforated pipe). First of all, a downflow of recycled air may arise in the upper section of this chamber. In addition, deviations of velocities u_0 and u_m from process (Le) and structural design parameters (r, E) of the handling facility in question are monotonous with the exception of areas $E = 16 \div 20$ and $r = 0.56$ where a certain decline in velocity u_0 occurs (Figure 6.8).

The chamber, as a rule, operates under negative pressure that ensures environmental safety of the assembly by preventing the escape of dust-laden air through accidental leaks in the bypass chamber.

6.2.3.1 Specific Features of Recirculation

We will formulate the condition for the emergence of recirculation loops (at the top of the handling facility in question) using the original equations for conservation of momentum and volumetric flow rate of air stream in the bypass chamber.

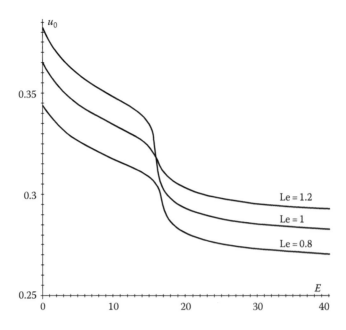

FIGURE 6.8   Change in the velocity of ejected air u_0 in a porous pipe with a bypass chamber with increasing porosity of its walls (at $r = 0.56$; $\xi_0 = 1.37$; $\zeta_n = 0.5$; $\zeta_n = 0.5$; $\zeta_k = 1$; $n = 0.4319$).

First of all, we will treat air velocities inside the pipe and in the chamber as positive if they are directed downward (along the positive axis OX). In this case, dimensionless conservation equations for volumetric airflows in the pipe and in the bypass chamber will appear in the same form:

$$dp_\omega + 4\omega d\omega = 0, \tag{6.83}$$

$$\frac{d\omega}{dx} = \frac{E}{r}w, \quad \frac{du}{dx} = Ew, \tag{6.84}$$

where w is the velocity of air in porous channels of tube walls (this velocity is similarly considered to be positive when air flows from the bypass channel into the pipe).

Figure 6.9 shows a possible air circulation pattern with two circulation loops. Due to symmetry, only right-hand halves of the pipe and the bypass chamber are shown here.

A downward flow of recirculated air is shaped inside the upper part of the duct of a length x_H. Its flow rate is

$$Q_{zH} = \omega(x_{\min})s_b = (u_0 - u_{\min})s_t, \tag{6.85}$$

where $x_{\min}$ is the distance from duct inlet to the cross section with the minimum velocity of ejected air ($u_{\min}$).

A countervailing (upward) airflow arises in the lower part of the bypass duct of a length x_b. Its flow rate is

$$Q_{zb} = \omega(x_{\max})s_b = (u_{\max} - u_0)s_t. \tag{6.86}$$

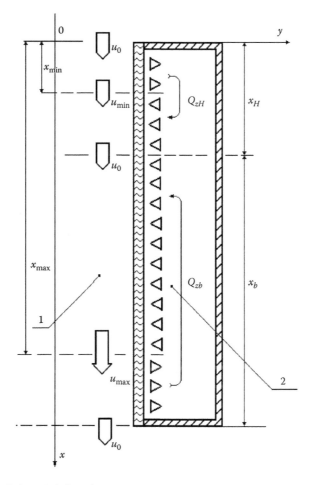

FIGURE 6.9   Chart of ejected airflows in porous duct (1) equipped with bypass chamber (2).

Two recirculation loops in this case, with a total relative flow rate of

$$R_z = \frac{Q_{zH} + Q_{zb}}{u_0 s_t} = \frac{u_{max} - u_{min}}{u_0}. \tag{6.87}$$

It is obvious that, in the absence of upper circulation loop ($Q_{zn} = 0$),

$$R_z = \frac{u_{max} - u_0}{u_0}. \tag{6.88}$$

Solving combined Equations 6.84 results in the differential equation

$$\frac{d\omega}{du} = \frac{1}{r}, \tag{6.89}$$

which describes the relationship between the velocity of recirculated air in the bypass duct and the velocity of ejected air in the pipe. Integration of this equation over the length of the upper circulation duct (within the range $x = 0\ldots x_H$) produces

$$\omega = \frac{u - u_0}{r}. \tag{6.90}$$

A similar relation is obtained as well by integrating Equation 6.89 over the length of the lower circulation loop (within $x = x_H\ldots 1$; this accounts for an obvious condition $\omega(x_H) = 0$; $u(x_H) = u_0$; moreover, the accepted direction of velocity $\omega < 0$ should be observed, that is a minus sign is required before the right-hand side of Equation 6.90).

It can be seen from Equation 6.90 that downflow of air is possible at $u < u_0$ while an upflow is possible at $u > u_0$. Or, accounting for the integral relation obtained from Equation 6.74

$$u = u_0 + \frac{E}{\xi_0} \int_0^x \left(p_\omega - p \right) dx, \quad 0 < x < x_H, \tag{6.91}$$

that determines the continuity condition for the flow of ejected air in a pipe (due to Equation 6.74). The condition for the existence of a downward stream would accept the form

$$p > p_\omega \quad at \ 0 \le x \le x_H. \tag{6.92}$$

In particular, at the duct inlet

$$p(0) > p_a \tag{6.93}$$

or, in view of the boundary condition from Equation 6.15,

$$p_a < -\zeta_n u_0^2. \tag{6.94}$$

Therefore, downward flow in the bypass duct is only possible with the emergence of a strongly vacuumed zone near its inlet provided that absolute pressure within this zone is below that of the absolute pressure at pipe inlet. This development is possible when the parameter Le is large enough (see the following).

It should be noted that the changes in the value of available pressure at the ends of porous tubes in duct walls, owing to Equations 6.74, 6.75, and 6.48, are determined by the equation

$$\frac{d\left(p_\omega - p \right)}{dx} = \text{Le} \frac{|v - u|}{v}(v - u) - 4\left[u - \frac{u - u_0}{r^2} \right] \frac{du}{dx}, \tag{6.95}$$

that, when integrated within the range $x = 0\ldots x_H$, gives

$$p_\omega - p - \left(-\zeta_n u_0^2 - P_a \right) = P_{ex} + J_v, \tag{6.96}$$

where P_{ex} is the ejection pressure force created by a flow of particles in the inlet section of a pipe of length x:

$$P_{ex} = \text{Le} \int\limits_0^x \frac{|v-u|}{v}(v-u)dx, \tag{6.97}$$

$$J_v = -2(u-u_0)\left[u+u_0-\frac{u-u_0}{r^2}\right]. \tag{6.98}$$

The first term of the right-hand side of Equation 6.95, as a rule, is positive at smaller Le numbers when $v > u$ holds for the entire length of the pipe. However, for a pipe with greater Le numbers (of length $x_0 < x_H$), it reverses to $u > v$ and the mass forces of interaction between components in this area fall below zero. The second summand on the right-hand side of this equation can only be positive within $0 \le x \le x_H$ as both $\dfrac{du}{dx} < 0$ and $u < u_0$ hold here.

As Le increases, leading to $u_0 > n$, the right-hand side of Equation 6.95 can turn negative (due to the second summand being small enough) and the available pressure would drop, leading to a decline in the velocity w of air escaping into the bypass chamber through porous channels in pipe walls, declining to zero in a point located x_{min} away from pipe inlet. Velocity of ejected air reaches its minimum u_{min} in this section. Both the magnitude u_{min} and the distance x_{min} are determined by solving equations

$$u(x_{min}) = C_1 e^{n_1 x_{min}} + C_2 e^{n_2 x_{min}} + v(x_{min}) - k_1 - k_2 x_{min}, \tag{6.99}$$

$$C_1 n_1 e^{n_1 x_{min}} + C_2 n_2 e^{n_2 x_{min}} + b_1 - b_2 x_{min} - k_2 = 0. \tag{6.100}$$

The cross section $x = x_{min}$ may be named the center of the upper circulation loop. Similarly, the section $x = x_{max}$ can be designated as the center of the lower circulation loop. Maximum velocity of ejected air u_{max} and the distance x_{max} are determined by the same combined equations.

The circumference of the upper loop is determined from the condition $u(x_H) = u_0$, that is, by solving the equation

$$C_1 e^{n_1 x_H} + C_2 e^{n_2 x_H} + v(x_H) - k_1 - k_2 x_H - u_0 = 0. \tag{6.101}$$

Let us now refer to numerical examples. Figure 6.10 shows plots of changes in air velocities along the pipe and the bypass duct at Le = 30, $E = 20$, and the cross-sectional area of the bypass chamber equal to the cross-sectional area of the pipe ($r = 1$). In this case, the ejected air velocity would have two extrema: the section $x = x_{min} = 0.14$ will have the minimum velocity $u_{min} = 0.5641$ while $x = x_{max} = 0.9148$ will have the maximum velocity $u_{max} = 0.7609$.

The circumference of the upper recirculation loop is $x_H = 0.32$, available head at pipe inlet is $\Delta p(0) = -0.01594$, $p_a = -0.1844$, and $p(0) = -0.1684$ at $u_0 = 0.5804$. Distances from the centers of recirculation loops are dependent not only on Le numbers but also on r (Figure 6.11). The higher these parameters are, the greater the distances x_{min} and $x_{max} \cdot x_{min}$ decreases with decreasing Le.

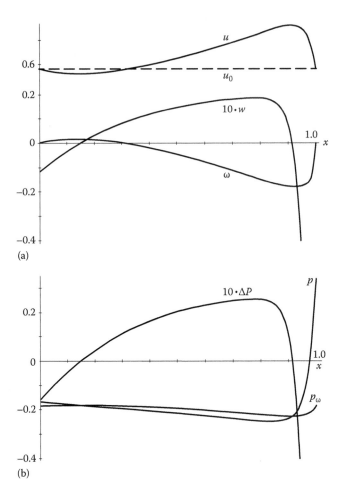

FIGURE 6.10 Change in (a) air velocities u, ω, and w and (b) pressures p, p_ω, $\Delta p = p_\omega - p$ in a porous duct with a bypass chamber (with Le = 30; $r = 1$; $E = 20$; $\xi_0 = 1.37$; $\zeta_n = 0.5$; $\zeta_k = 1$).

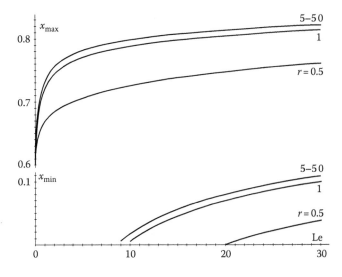

FIGURE 6.11 Shifting centers of downward (x_{min}) and upward (x_{max}) airflows in the bypass duct with various ejection numbers and duct cross-sectional sizes (at $E = 5$; $\xi_0 = 1.37$; $n = 0.4319$; $\zeta_n = 0.5$; $\zeta_k = 1$).

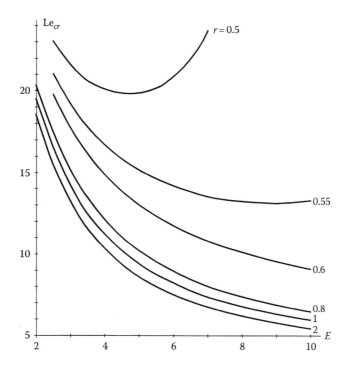

FIGURE 6.12 Effect of the design parameters of a bypassed porous duct on its critical ejection numbers (at $\xi_0 = 1.37$; $n = 0.4319$; $\zeta_n = 0.5$; $\zeta_k = 1$).

At Le ≈ 9 ($r = 1$), it follows that $x_{min} \approx 0$; $x_H \approx 0$, that is, the downward flow of recirculated air disappears so that only upward airflow is shaped along the entire duct. Ejection number giving rise to a second circulation loop will be referred to as critical (Le_{cr}). Then the condition for the existence of a second recirculation loop of ejected air would appear as

$$\mathrm{Le} \leq \mathrm{Le}_{cr}. \tag{6.102}$$

The critical number depends on handling facility design (Figure 6.12).

6.2.3.2 Change in Ejected Air Velocity

In contrast to a porous pipe without a bypass chamber, velocity of ejected air at the pipe outlet (u_0) changes monotonously (Figure 6.13) with the exception of the area $0.16 < r < 0.24$ where the velocity u_0 is declining both in the solution of a linearized equation (solid curves in Figure 6.13) and in the numeric solution of the exact combined Equations 6.78 through 6.82 (dotted curve in Figure 6.13).

Change in the bypass duct size within the area $r > 1$ does not affect the value of u_0 in any substantial way. A much greater effect is made by the ejection number Le, especially in the lower range (Le < 1) when the limitation of mass forces of component interaction is yet to become evident. Connecting a bypass duct as a parallel run of air ducts to a volumetric blower helps to improve its efficiency.

Maximum velocity u_m rises with greater ejection numbers and increased porosity of the duct walls (Figure 6.14). However, the magnitude u_0 is lower than the velocity of ejected air in a duct with impermeable walls, other conditions being equal. Thus, an important energy-saving

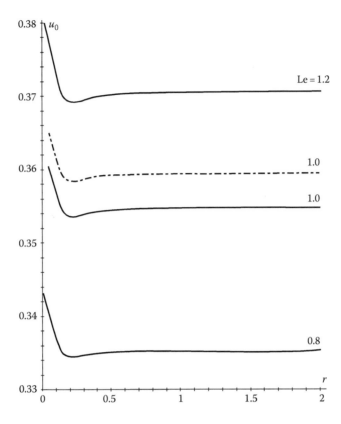

FIGURE 6.13   Change in the velocity u_0 as a function of bypass chamber dimensions r (at $E = 2$; $\xi_0 = 1.37$; $\zeta_n = 0.5$; $\zeta_k = 1$): solid curves—solution of linearized equation, dotted line—numerical solution of the *exact* equation.

effect of bypassing is revealed. The coefficient of reduction in ejection volumes increases with increasing parameters Le and E (Figure 6.14). Recycling volumes also increase considerably (Figure 6.15).

Both the Le number and bypass channel dimensions play a considerable role here. The effect of reduction in ejection and recycling volumes is especially noticeable with increasing aerodynamic drag at the inlet (Figure 6.15), at outlet (Figure 6.15), or at both openings (Figure 6.16).

Thus, at greater aerodynamic drags ($\zeta_n = \zeta_k = 20$) and significant values of the parameter Le > 50, it is possible to reduce volumes of ejected air in a porous pipe 1.5–1.7 times and more, given a bypass duct with a cross-sectional area that equals that of the pipe. A significant decline in recirculated air volume is observed as well in *narrow* areas of the minimum u_0 at $r = r_{min}$. The value u_0^{min}, as well as the function r_{min}, can be deduced using the rules for determining extrema of an implicit function, that is, by considering Equation 6.69 and solving combined equations

$$f(r, u_0) = -(\zeta_n + \zeta_k)u_0^2 \frac{E}{\xi_0} + C_1 n_1 \left(1 - e^{n_1}\right) + C_2 n_2 \left(1 - e^{n_2}\right) - b_2 = 0,$$

$$\frac{\partial f}{\partial r} = 0,$$

where the parameters C_1, C_2, n_1, n_2, owing to Equations 6.65 through 6.68 and 6.51 and 6.52, are in turn the functions of r and u_0.

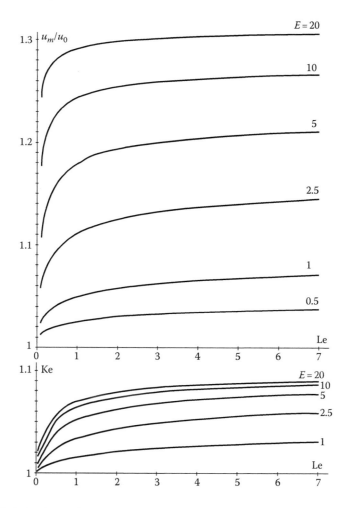

FIGURE 6.14 Variation in maximum velocity u_m/u_0 of air ejected by an accelerated flow of particles and ejection volume reduction coefficient Ke with increased wall porosity in a duct equipped with a bypass chamber (at $r = 1$; $\xi_0 = 1.37$; $n = 0.4319$; $\zeta_n = 0.5$; $\zeta_k = 1$).

Table 6.4 summarizes findings from the numerical solution of these equations. In addition to values of u_0^{min} and r_{min}, the table also lists values of the coefficient of reduction in ejected air volumes, computed using the formula

$$K_2 = \frac{u_2}{u_0^{min}},$$

where u_2 is the velocity of air ejected by a constantly accelerated flow of particles inside a pipe with impermeable walls (at $E = 0$), given identical parameters Le, ζ_n, ζ_k, n.

6.2.3.3 Pressure Variations

Higher drag values at openings produce observable maxima not only in velocities $u(x)$ and $\omega(x)$ but also in pressure $p_\omega(x)$ (Figure 6.17). It is known that excess pressures in ducts increase the likeliness of dust-laden air escaping from them. Even when it is possible to seal the walls, risks for handling facilities include accidental loss of tightness, off-design installation, and operation leakages. Therefore, it is important that the bypass chamber be under negative pressure. As diagrams

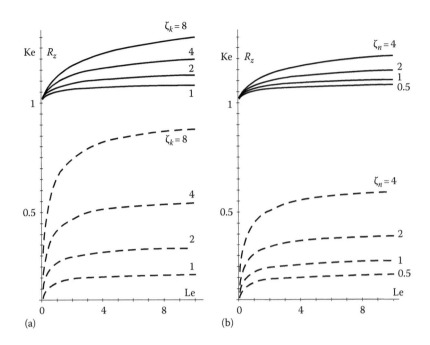

FIGURE 6.15   Change in recycling volumes R_z (dotted curves) and ejection volume reduction coefficients Ke (solid curves) with increasing Le number and increasing resistance of the outlet (a) and inlet (b) duct sections (at $r = 1$; $E = 5$; $\xi_0 = 1.37$; $n = 0.4319$). (a) $\zeta_n = 0.5$; ζ_k–var and (b) $\zeta_k = 1$; ζ_n–var.

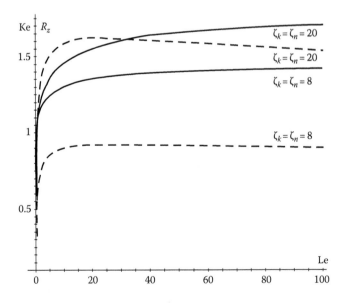

FIGURE 6.16   Change in recycling volume R_z (dotted curves) and ejection volume reduction coefficients Ke (solid curves) at greater Le numbers and intense drag in the inlet and outlet sections of the duct (at $r = 1$; $E = 5$; $\xi_0 = 1.37$; $n = 0.4319$).

TABLE 6.4
Value of the Minimum Velocity of Ejected Air in a Porous Pipe with a Bypass Chamber Measuring r_{min} (at $\zeta_n = 0.5$; $\zeta_k = 1$; $\xi_0 = 1.37$; $n = 0.4319$)

Le	u_0^{min}	r_{min}	K_2	u_0^{min}	r_{min}	K_2	u_0^{min}	r_{min}	K_2
	$E = 2.0$			$E = 4.0$			$E = 6.0$		
0.1	0.1713	0.1447	1.0102	0.1701	0.2192	1.0174	0.1689	0.2724	1.0244
0.2	0.2202	0.1675	1.0166	0.2177	0.2494	1.0284	0.2153	0.3067	1.0399
0.4	0.2755	0.1893	1.0263	0.2706	0.2776	1.0447	0.2662	0.3380	1.0622
0.6	0.3096	0.2011	1.0339	0.3029	0.2926	1.0570	0.2968	0.3543	1.0788
0.8	0.3343	0.2089	1.0403	0.3260	0.3024	1.0670	0.3185	0.3647	1.0922
1.0	0.3535	0.2145	1.0458	0.3437	0.3093	1.0756	0.3350	0.3721	1.1034
1.2	0.3690	0.2189	1.0508	0.3580	0.3146	1.0832	0.3484	0.3777	1.1132
1.4	0.3821	0.2223	1.0553	0.3699	0.3188	1.0899	0.3594	0.3820	1.1219
1.6	0.3932	0.2251	1.0594	0.3801	0.3222	1.0960	0.3688	0.3855	1.1297
1.8	0.4030	0.2275	1.0633	0.3889	0.3251	1.1016	0.3769	0.3884	1.1368
2.0	0.4116	0.2294	1.0669	0.3967	0.3274	1.1068	0.3841	0.3908	1.1433
2.2	0.4193	0.2312	1.0702	0.4037	0.3295	1.1117	0.3905	0.3929	1.1493
2.4	0.4262	0.2326	1.0734	0.4099	0.3313	1.1162	0.3962	0.3947	1.1549
2.6	0.4325	0.2339	1.0764	0.4155	0.3328	1.1204	0.4013	0.3962	1.1601
2.8	0.4383	0.2351	1.0793	0.4207	0.3342	1.1244	0.4060	0.3975	1.1650
3.0	0.4436	0.2361	1.0819	0.4254	0.3354	1.1281	0.4104	0.3987	1.1696
3.2	0.4485	0.2370	1.0845	0.4298	0.3364	1.1316	0.4143	0.3997	1.1739
	$E = 8.0$			$E = 10.0$			$E = 12.0$		
0.1	0.1678	0.3145	1.0314	0.1667	0.3493	1.0382	0.1656	0.3791	1.0449
0.2	0.2130	0.3512	1.0510	0.2108	0.3876	1.0619	0.2087	0.4183	1.0725
0.4	0.2620	0.3841	1.0791	0.2581	0.4212	1.0953	0.2545	0.4520	1.1110
0.6	0.2912	0.4009	1.0996	0.2860	0.4380	1.1194	0.2813	0.4686	1.1384
0.8	0.3117	0.4114	1.1159	0.3055	0.4484	1.1385	0.2999	0.4788	1.1600
1.0	0.3273	0.4188	1.1296	0.3203	0.4556	1.1543	0.3139	0.4858	1.1777
1.2	0.3398	0.4243	1.1413	0.3321	0.4610	1.1678	0.3251	0.4909	1.1928
1.4	0.3501	0.4286	1.1516	0.3418	0.4651	1.1796	0.3343	0.4948	1.2059
1.6	0.3589	0.4320	1.1609	0.3501	0.4683	1.1901	0.3422	0.4979	1.2176
1.8	0.3665	0.4348	1.1692	0.3572	0.4710	1.1996	0.3489	0.5003	1.2280
2.0	0.3731	0.4371	1.1769	0.3634	0.4732	1.2082	0.3548	0.5023	1.2375
2.2	0.3790	0.4391	1.1839	0.3690	0.4750	1.2161	0.3601	0.5040	1.2462
2.4	0.3843	0.4408	1.1904	0.3740	0.4765	1.2234	0.3648	0.5054	1.2542
2.6	0.3891	0.4422	1.1965	0.3785	0.4778	1.2302	0.3690	0.5066	1.2616
2.8	0.3935	0.4434	1.2021	0.3826	0.4790	1.2365	0.3729	0.5076	1.2685
3.0	0.3975	0.4445	1.2074	0.3863	0.4799	1.2423	0.3765	0.5085	1.2749
3.2	0.4012	0.4455	1.2124	0.3898	0.4808	1.2478	0.3798	0.5092	1.2808
	$E = 14.0$			$E = 16.0$			$E = 18.0$		
0.1	0.1646	0.4051	1.5015	0.1636	0.4281	1.0580	0.1626	0.4487	1.0645
0.2	0.2067	0.4448	1.0829	0.2048	0.4679	1.0930	0.2030	0.4885	1.1030
0.4	0.2511	0.4783	1.1261	0.2478	0.5011	1.1408	0.2448	0.5211	1.1550
0.6	0.2768	0.4945	1.1567	0.2727	0.5169	1.1743	0.2688	0.5365	1.1913
0.8	0.2946	0.5044	1.1805	0.2898	0.5264	1.2003	0.2853	0.5456	1.2193
1.0	0.3081	0.5111	1.2001	0.3027	0.5328	1.2214	0.2977	0.5517	1.2419
1.2	0.3188	0.5160	1.2166	0.3129	0.5374	1.2393	0.3075	0.5560	1.2610

(Continued)

TABLE 6.4 (Continued)
Value of the Minimum Velocity of Ejected Air in a Porous Pipe with a Bypass Chamber Measuring r_{min} (at $\zeta_n = 0.5$; $\zeta_k = 1$; $\xi_0 = 1.37$; $n = 0.4319$)

Le	u_0^{min}	r_{min}	K_2	u_0^{min}	r_{min}	K_2	u_0^{min}	r_{min}	K_2
	$E = 14.0$			$E = 16.0$			$E = 18.0$		
1.4	0.3276	0.5196	1.2309	0.3213	0.5409	1.2547	0.3156	0.5593	1.2775
1.6	0.3350	0.5225	1.2436	0.3285	0.5436	1.2683	0.3225	0.5618	1.2919
1.8	0.3414	0.5248	1.2549	0.3346	0.5457	1.2805	0.3284	0.5637	1.3048
2.0	0.3471	0.5267	1.2652	0.3400	0.5474	1.2914	0.3336	0.5653	1.3164
2.2	0.3521	0.5282	1.2746	0.3448	0.5488	1.3014	0.3382	0.5667	1.3270
2.4	0.3565	0.5295	1.2832	0.3491	0.5499	1.3106	0.3423	0.5676	1.3366
2.6	0.3606	0.5305	1.2911	0.3530	0.5509	1.3190	0.3460	0.5684	1.3455
2.8	0.3643	0.5314	1.2985	0.3565	0.5517	1.3268	0.3494	0.5691	1.3537
3.0	0.3677	0.5322	1.3053	0.3598	0.5523	1.3341	0.3526	0.5697	1.3613
3.2	0.3708	0.5329	1.3117	0.3628	0.5529	1.3408	0.3555	0.5702	1.3684
	$E = 20.0$			$E = 22.0$			$E = 24.0$		
0.1	0.1616	0.4674	1.0708	0.1607	0.4843	1.0770	0.1598	0.4998	1.0832
0.2	0.2012	0.5069	1.1127	0.1995	0.5235	1.1222	0.1978	0.5387	1.1315
0.4	0.2419	0.5389	1.1689	0.2391	0.5549	1.1824	0.2365	0.5693	1.1955
0.6	0.2651	0.5538	1.2078	0.2617	0.5692	1.2237	0.2584	0.5831	1.2392
0.8	0.2811	0.5625	1.2376	0.2771	0.5776	1.2553	0.2734	0.5911	1.2724
1.0	0.2930	0.5683	1.2617	0.2887	0.5831	1.2807	0.2846	0.5964	1.2991
1.2	0.3025	0.5724	1.2819	0.2978	0.5870	1.3020	0.2935	0.6000	1.3213
1.4	0.3103	0.5755	1.2993	0.3054	0.5898	1.3202	0.3008	0.6027	1.3404
1.6	0.3169	0.5778	1.3145	0.3118	0.5920	1.3362	0.3070	0.6047	1.3571
1.8	0.3226	0.5796	1.3280	0.3173	0.5936	1.3504	0.3123	0.6062	1.3718
2.0	0.3276	0.5810	1.3403	0.3221	0.5949	1.3631	0.3170	0.6074	1.3850
2.2	0.3321	0.5821	1.3513	0.3264	0.5960	1.3746	0.3212	0.6083	1.3970
2.4	0.3361	0.5831	1.3614	0.3303	0.5968	1.3852	0.3250	0.6091	1.4079
2.6	0.3397	0.5838	1.3707	0.3338	0.5974	1.3948	0.3284	0.6096	1.4179
2.8	0.3430	0.5844	1.3793	0.3370	0.5980	1.4037	0.3315	0.6101	1.4271
3.0	0.3460	0.5849	1.3872	0.3399	0.5984	1.4120	0.3343	0.6104	1.4357
3.2	0.3488	0.5853	1.3946	0.3426	0.5987	1.4196	0.3369	0.6107	1.4436

illustrate, the negative pressure in the bypass chamber at higher Le values is maintained throughout the entire length of the duct. It should be noted that no recirculation loop arises within inlet section $p_\omega(0) > p(0)$ despite a significant process value Le. An appreciable role in this case is played by the quantities ζ_n and ζ_k that intensify air recirculation as they increase. Decreasing LRCs result in increased parameters u_0 and P_a of the problem (Figure 6.18).

However, the value P_a is lower than zero in all cases, decreasing the likeliness of outward escape of dust-laden air. Nevertheless, with increasing ζ_k a *back-up head* arises in the bypass chamber, likely to cause excess pressure at

$$\zeta_k > \zeta_k^*.$$

The limit value of ζ_k^* depends on the parameters Le and E (Table 6.5).

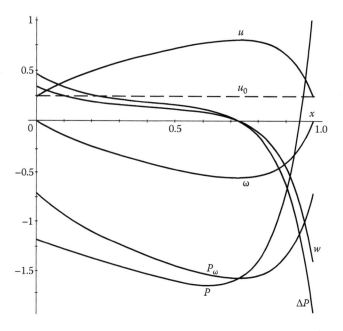

FIGURE 6.17 Variation in the parameters of ejected air along the height of the duct with porous walls and a bypass chamber (with Le = 100; $r = 1$; $E = 20$; $\xi_0 = 1.37$; $\zeta_n = \zeta_k = 20$; $n = 0.4319$).

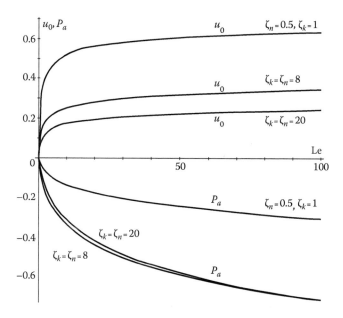

FIGURE 6.18 Variation of problem parameters u_0, p_a for a porous duct with a bypass chamber as a function of Le at $r = 1$; $E = 5$; $\xi_0 = 1.37$; $n = 0.4319$.

TABLE 6.5

Maximum Values of ζ_k^* That Give Rise to Excess Pressures in the Bypass Chamber if Exceeded (at $r = 1$; $\xi_0 = 1.37$; $\zeta_n = 0.5$)

Le	$E = 0.5$	$E = 1$	$E = 3$	$E = 5$	$E = 7$	$E = 10$
0.5	0.90345	1.04000	1.6473	2.2898	2.9225	3.8200
1	0.99874	1.17802	1.9693	2.7800	3.5482	4.5952
5	1.41303	1.73622	3.05816	4.2057	5.1570	6.3216
10	1.73312	2.16964	3.7232	4.91704	5.8398	6.6225
30	2.76065	3.37004	5.12654	6.1675	6.8918	7.7047
50	3.59082	4.28780	5.9677	6.8266	7.4024	8.0474
70	4.34353	5.07630	6.6031	7.30194	7.7633	8.2867
100	5.38303	6.11026	7.35814	7.8525	8.17885	8.5644

6.3   CONCLUSIONS

1. One-dimensional airflow in a porous pipe with a linear law for cross flow through capillary channels in the pipe walls (Equation 6.2) was introduced to put forward pseudolinear dynamics equation (Equation 6.10) for air ejected by an accelerated flow of particles prompted by the volumetric forces of component interaction. Linearization of these forces (Equation 6.11) together with kinetic energy (Equation 6.13) makes it possible to reduce the equation for dynamics of ejected air in a pipe into a linear nonuniform second-order differential equation with constant coefficients. Analytical solution of the latter is in good agreement with numerical solution of Equation 6.10. Assumptions introduced as a part of linearization of the problem have virtually no impact on the velocity of air exiting from the porous pipe and the character of change in the velocity of ejected air along pipe height. Peak deviation of these velocities is within 0.9% at small ejection numbers (Le $\leq$ 1) and within 3% at larger ejection numbers (at Le = 10).

2. By analyzing the solution of the linearized problem, it was possible to identify the effect of the process parameter (ejection number, Le) and pipe wall porosity (E) on air recycling rate within the pipe (R_z) and on the magnitude of the related coefficient of reduction in the amount of air ejected at the pipe outlet (Ke = $\tilde{u}_2 / \tilde{u}_k$). Recycling rate increases with growing Le and wall porosity E (Figure 6.4). The value Ke has a noticeable maximum (dotted curve, Figure 6.3) that shifts with increasing porosity toward lower ejection numbers. This is due to the fact that lower Le values offset the peak recycling rate R_z into the area of low pipe wall porosity (dotted line, Figure 6.5). The presence of these maxima can be explained by a different nature of changes in the velocity of ejected air u_0 as porosity of walls E becomes greater and ejection numbers Le increase (Figure 6.6). Velocity u_0 exhibits the greatest decline due to the growing ejection of air with increasing E. At the same time, both the velocity u_k at duct outlet and the maximum velocity u_m respond more conservatively to changes in E. In addition, the extreme value of u_m with corresponding combinations of Le and E produces a maximum in recirculated air volume.

3. It was shown, in theoretical terms, that an increase in porosity E of pipe walls alone confers a significant reduction in the amount of ejected air (by more than 15% at $E > 5$ and Le = 1) compared to the flow rate of air entrained by accelerated flow of falling particles in a pipe with impermeable walls.

   With Le unchanged, the recycling rate R_z exhibits a significant increase with higher LRCs at outlet ζ_k and increasing LRCs at inlet ζ_n sections of a porous pipe (Figure 6.7).

Higher Le numbers display asymptotics inherent in ejecting blowers because the volumetric forces of interaction between components are limited: volumes of recirculated air first increase rapidly, then, at Le > 1, R_z stabilizes.

4. With a bypass chamber installed around the porous pipe, kinematics of air ejected by a flow of falling particles in the pipe changes together with that of the air recirculated inside the chamber. Regularities of 1D airflows in the pipe are determined by combined *exact* Equations 6.74 through 6.76 accounting for the ejecting action of a constantly accelerated flow of loose particulate matter and cross flows of recycled air due to a difference of pressures between the bypass chamber and the porous pipe.

   Linearization of volumetric forces of component interaction in pipe, together with kinetic energy of ejected and recycled air, enable conversion of the exact equations to the linear nonuniform second-order differential equation (Equation 6.50) with the constant coefficients (Equation 6.51). The resulting solution of a boundary-value problem described by Equations 6.53, 6.54, 6.69, and 6.70, compared to the numerical solution of the *exact* equations, testifies that simplifications introduced to linearize the problem only cause insignificant changes in the velocity of ejected air—relative deviation from integration results for *exact* equations remains within 1.4% (Table 6.3).

5. By analyzing the resulting solution of a linearized equation for air ejection in a porous pipe with a bypass chamber, the following features have been identified that enable energy consumption to be reduced significantly due to decreasing velocity (volume) of ejected air.

   At greater ejection numbers (Le > Le_{cr}, Figure 6.12) a downward flow of recycled air arises in the upper part of the bypass chamber while an upward flow arises in the lower part. Thus, two circulation loops may exist in the bypass chamber. Their sizes (circumferences) depend on Le numbers, E value, as well as bypass chamber size r (Figure 6.11). In this case, a zone of negative pressure arises along the entire height (Figures 6.10 and 6.17, see diagrams for changes in $p_\omega(x)$), ensuring environmental safety of the assembly by preventing the escape of dust-laden air through off-design leaky spots in the chamber.

6. The velocity of ejected air (u_0) at an outlet of a porous pipe with a bypass chamber, other conditions being equal (at Le = idem; $\sum \zeta = idem$), is significantly less than the velocity of ejected air (u_2) in a duct with impermeable walls (although $u_m > u_2$). The coefficient of reduction in ejection volumes rises with increasing Le and E parameters (Figure 6.14) as a result of increasing volumes of recirculated air (Figure 6.15). An even greater effect of reduction in the volume of ejected air due to bypassing (up to 50%–70%) is observed at greater aerodynamic drags both at the inlet and at the outlet of the porous pipe (Figure 6.16).

7. Areas of decreasing velocity u_0 have been identified within certain narrow ranges of bypass chamber sizes r (Figure 6.13) and pipe wall porosities (E) (Figure 6.8). The coefficient of reduction in ejection volumes (K_2) in these areas reaches considerable values (up to 40% and more, Table 6.4) even with modest drag coefficients.

7 Specific Features of Air Ejection in Elevator Handling of Grain

7.1 VARIATION OF EJECTION HEADS CREATED BY FALLING PARTICLES IN CHUTES WITH VARYING FRONTAL DRAG COEFFICIENTS

Gravity flows of loose-matter particles are accompanied with aerodynamic forces that produce ejection of air in the loading and unloading chutes. A flow of loose matter acts like a blower. Head created by this blower (that we choose to call ejection head) comprises a sum total of aerodynamic forces of falling particles divided by the cross-sectional area of the flow [213]. Aerodynamic force may be expressed using the drag coefficient

$$R = \psi F_m \frac{|v-u|(v-u)}{2} \rho, \tag{7.1}$$

where
  R is the aerodynamic force of a single particle (N)
  ψ is the drag coefficient
  F_m is the midsection area of a particle (m²)
  v is the velocity of falling particles (m/s)
  u is the velocity of ejected air (m/s)
  ρ is the density of air (kg/m³)

Absolute value of relative velocity $|v-u|$ had to be isolated into a dedicated variable due to the need to vectorize the square of relative velocity for the case involving unidirectional motion of a two-component particle–air medium. This notation means that at $v > u$, R is positive and directed parallel with the leading motion of solid particles. At $v < u$, this force is negative, that is, opposite to the falling particles.

The drag coefficient is determined not only by the geometrical shape of particles, but also by their concentration. The greater the volumetric concentration of falling particles is, the more pronounced is the effect of aerodynamic shadow on a particle moving toward the *aft* side of a particle ahead of it. The following empirical relationship [214] has been identified for a flow of firm mineral particles sized between 2 and 20 mm falling in inclined chutes:

$$\psi_y = \psi_0 \exp\left(\frac{-1.8\sqrt{\beta_y \cdot 10^3}}{d_e}\right), \tag{7.2}$$

where

ψ_0 is the drag coefficient for a single particle in turbulent surface flow mode (within the self-similarity area)

d_e is the equivalent particle diameter (mm)

β_y is the volumetric concentration, averaged over the length of the chute and equal to

$$\beta_y = \frac{2G_m}{S\rho_m(v_n + v_k)} \tag{7.3}$$

where

G_m is the mass flow rate of the material (kg/s)

S is the cross-sectional area of the chute (m²)

ρ_m is the air density (kg/m³)

v_n, v_k is the fall velocity of particles at inlet/outlet of the chute (m/s)

The majority of crushed mineral particles have a roughly pointed granular shape and a coefficient $\psi_0 \approx 1.8$. In contrast, cereals are likely to produce oval-shaped fragments when crushed, therefore, $\psi_0 \approx 1.0$ can be assumed. If the geometrical shape of these particles were to be evaluated by their dynamic form coefficient, then,

$$K_d = \left.\frac{\psi_0}{\psi_b}\right|_{Re=idem} \approx 2, \tag{7.4}$$

where ψ_b is the drag coefficient of a sphere within the self-similarity area ($\psi_b \approx 0.5$).

Let us determine the ejection head p_e arising within unloading and loading chutes during the handling of cereal grain at elevators.

By definition, the ejection pressure created in chutes as a result of interaction between components,

$$p_e = \frac{1}{S}\int_0^l R\frac{\beta S dx}{V_p}, \tag{7.5}$$

where

V_p is the volume of a single particle (m³)

β is the current volumetric concentration of falling particles

$$\beta = \frac{G_m}{S\rho_m v} \tag{7.6}$$

l is the chute length (m)

With a small fall height, when the chute length is

$$l < l_y = \frac{1}{\psi_y K_m \varepsilon}, \tag{7.7}$$

the aerodynamic force is negligibly small compared to the gravitational force of a particle and the flow of material can, therefore, be considered constantly accelerated:

$$\frac{dv}{dt} = \frac{vdv}{dx} = a_\tau, \tag{7.8}$$

where

$$K_m = \frac{F_m}{V_p} \approx \frac{1.5}{d_e} \tag{7.9}$$

$$\varepsilon = \frac{\rho}{\rho_m} \tag{7.10}$$

d_e is the equivalent (volumetric) particle diameter (m)

a_τ is the acceleration that, for chutes installed at an angle α to horizontal surface, is equal to

$$a_\tau = g \sin\alpha \left(1 - f_w \mathrm{tg}\alpha\right) \tag{7.10a}$$

g is the gravitational acceleration (m/s^2)

f_w is the friction coefficient for particles rubbing against chute walls (for steel walls, $f_w \approx 0.5$)

In this case, the ejection head for prismatic chutes at $S = \text{const}$, $u = \text{const}$, $\psi = \psi_y = \text{const}$ is determined using the equation

$$P_{ey} = \frac{\psi_y K_m \varepsilon}{2} \frac{G_m}{Sa_\tau} \frac{\left|v_k - u\right|^3 - \left|v_n - u\right|^3}{3} \tag{7.11}$$

or

$$P_{ey} = \frac{\psi_y K_m \varepsilon}{2} \frac{G_m v_k^3}{Sa_\tau} \frac{\left|1 - \varphi\right|^3 - \left|n - \varphi\right|^3}{3} = \psi_y \frac{G_m v_k}{S} \Phi \frac{\left|1 - \varphi\right|^3 - \left|n - \varphi\right|^3}{3}, \tag{7.12}$$

where φ is the component slip ratio (ejection coefficient)

$$\varphi = \frac{u}{v_k}, \tag{7.13}$$

$$n = \frac{v_n}{v_k}, \tag{7.14}$$

$$\Phi = \varepsilon \frac{K_m v_k^2}{2a_\tau}. \tag{7.15}$$

Velocity of cereal grains, considering their modest mass together with a large fall height ($l > l_y$), is affected by the drag force. Motion equation for particles falling in an inclined chute in this case would appear as

$$\frac{dv}{dt} = v\frac{dv}{dx} = a_\tau - \frac{R}{V_p \rho_m}. \tag{7.16}$$

Equations 7.5 and 7.16 enable the ejection head p_e to be discovered [185,213] not only for heavier particles and lower heights ($l < l_y$), but for light particles falling from great heights with an observable drag. However, changes in the ψ coefficient over the fall height was not taken into account in this case due to the decreasing volumetric concentration of particles owing to Equation 7.6

$$\psi = \psi_0 \exp\left(\frac{-1.8\sqrt{\beta \cdot 10^3}}{d_e}\right). \tag{7.17}$$

Here, ψ is a variable quantity that changes over the fall height due to decreasing volumetric concentration β as a result of increasing velocity of particles v.

Instantaneous drag coefficient will be expressed by a putative

$$\psi = \psi_0 B^{\sqrt{\frac{v_a}{v}}}, \tag{7.18}$$

with the following notational simplification:

$$B = e^{-1.8\frac{\sqrt{\beta_y \cdot 10^3}}{d_e}} = \frac{\psi_y}{\psi_0}, \quad v_a = \frac{v_n + v_k}{2}. \tag{7.19}$$

Let us now compare this effect for the case of short chutes when

$$a_\tau \gg \frac{R}{V_p \rho_m}.$$

In view of Equations 7.18 and 7.8, the following relation determines the ejection pressure for a constantly accelerated flow of particles in a prismatic inclined chute (at $S = const$):

$$p_{eu} = \psi_0 \frac{G_m v_k}{S} \Phi \int_n^1 B^{\sqrt{\frac{1+n}{2\zeta}}} |\zeta - \varphi|(\zeta - \varphi)d\zeta. \tag{7.20}$$

This relation also accounts for change in the drag coefficient ψ along the trajectory of falling particles.

In order to enable computational comparisons, we will introduce the value of adjustment coefficient K that will determine the ratio of the true ejection head value to its putative value, Equation 7.12:

$$K = \frac{p_{eu}}{p_{ey}}. \tag{7.21}$$

Let us quantify it. In view of Equations 7.12, 7.19, and 7.20 the expression appears as

$$K = \frac{3}{(1-\varphi)^3 - |n-\varphi|^3} \frac{1}{B} \int_n^1 B^{\sqrt{\frac{1+n}{2\zeta}}} |\zeta - \varphi|(\zeta - \varphi) d\zeta.$$ (7.22)

Then the ratio of maximum ejection heads would be (at $u = \varphi = 0$)

$$K_0 = \frac{3}{1-n^3} \frac{1}{B} \int_n^1 B^{\sqrt{\frac{1+n}{2\zeta}}} \zeta^2 d\zeta = \frac{6}{1-n^3} \frac{1}{B} \int_1^{\frac{1}{\sqrt{n}}} \frac{e^{pt}}{t^7} dt,$$ (7.23)

where

$$p = \sqrt{\frac{1+n}{2}} \ln B.$$ (7.24)

Let us resolve the integral by representing the integrand function as a series:

$$\frac{e^{pt}}{t^7} = \sum_{k=0}^{\infty} \frac{p^k}{k!} \zeta^{k-7},$$ (7.25)

which results in

$$\int_1^{\frac{1}{\sqrt{n}}} \frac{e^{P_t}}{t^7} dt = F(t) \Big|_1^{\frac{1}{\sqrt{n}}} = F\left(\frac{1}{\sqrt{n}}\right) - F(1),$$ (7.26)

where the antiderivative function F is equal to

$$F(t) = -\frac{1}{6}\frac{1}{t^6} - \frac{p}{5}\frac{1}{t^5} - \frac{p^2}{8}\frac{1}{t^4} - \frac{p^3}{18}\frac{1}{t^3} - \frac{p^4}{48}\frac{1}{t^2} - \frac{p^5}{120}\frac{1}{t} - \frac{p^6}{720}\ln t + \sum_{k \to 7} \frac{p^k}{k!}\frac{1^{k-6}}{k-6},$$ (7.27)

then

$$K_0 = \frac{6}{1-n^3} \frac{1}{B} \left[F\left(\frac{1}{\sqrt{n}}\right) - F(1) \right].$$ (7.28)

Tables 7.1 and 7.2 summarize the values of the adjustment coefficient K_0 computed using the formula given by Equation 7.28 and K computed using the formula given by Equation 7.22. It is apparent from Table 7.1 that changes in the drag coefficient ψ along a trajectory travelled by a particle only cause a noticeable impact within the area of significant volumetric concentrations, when $B \leq 0.3$. For example, for grains of wheat ($d_e \approx 3$ mm), this corresponds to the averaged volume concentration $\beta_y = 0.004$. The adjustment factor is significantly higher for smaller initial

TABLE 7.1

Adjustment Coefficient Values for the Case of Still Air in the Chute

	Values of K_0 at n Equal To...								
B	0.1	0.2	0.3	0.4	0.5	0.6	0.7	0.8	0.9
0.01	2.017	1.717	1.489	1.319	1.198	1.113	1.057	1.023	1.005
0.02	1.786	1.556	1.382	1.251	1.155	1.089	1.045	1.018	1.004
0.03	1.667	1.475	1.326	1.214	1.133	1.076	1.038	1.015	1.003
0.04	1.589	1.420	1.288	1.189	1.118	1.067	1.034	1.014	1.003
0.05	1.531	1.379	1.261	1.171	1.106	1.061	1.031	1.012	1.003
0.06	1.487	1.348	1.239	1.157	1.098	1.056	1.028	1.011	1.003
0.07	1.450	1.322	1.221	1.146	1.090	1.052	1.026	1.010	1.002
0.08	1.420	1.300	1.207	1.136	1.084	1.048	1.024	1.010	1.002
0.09	1.390	1.281	1.194	1.128	1.079	1.045	1.023	1.009	1.002
0.1	1.371	1.265	1.183	1.120	1.075	1.043	1.022	1.009	1.002
0.2	1.234	1.168	1.116	1.076	1.047	1.027	1.014	1.005	1.001
0.3	1.165	1.118	1.081	1.053	1.033	1.020	1.010	1.004	1.001
0.4	1.120	1.086	1.059	1.039	1.024	1.014	1.007	1.003	1.001
0.5	1.088	1.063	1.043	1.028	1.018	1.010	1.005	1.002	1.000
0.6	1.063	1.045	1.031	1.020	1.013	1.007	1.004	1.001	1.000
0.7	1.043	1.030	1.021	1.014	1.009	1.005	1.002	1.001	1.000
0.8	1.026	1.019	1.013	1.008	1.005	1.003	1.002	1.001	1.000
0.9	1.012	1.009	1.006	1.004	1.002	1.001	1.001	1.000	1.000
1.0	1.000	1.000	1.000	1.000	1.000	1.000	1.000	1.000	1.000

velocities (at $n < 0.5$) of the flow (Figure 7.1a). Its value increases somewhat (Figure 7.1b) when air is flowing inside a chute (at $\varphi \neq 0$). However, considering that the ejection head

$$p_e = K\psi_y \frac{G_m v_k}{S} \Phi \frac{|1-\varphi|^3 - |n-\varphi|^3}{3} \tag{7.29}$$

diminishes with increasing φ as a result of the braking action of the initial section of accelerated flow, the following assumption would be justified for computations:

$$K \approx K_0. \tag{7.30}$$

Keeping in mind that grain handling facilities in real-world elevator will have a flow velocity at duct inlet of $(0.3 \div 0.5 v_k)$, ejection head can, thus, be determined using the formula given by Equation 7.12 by introducing an adjustment coefficient exclusively within the area of large volumetric concentrations, at

$$B < 0.1 \quad \text{and} \quad n < 0.3. \tag{7.31}$$

Thus, the flow of loose particulate matter can be considered as a kind of blower with the output described as

$$p_e(Q) = K_0 z \frac{1}{3} \left[\left| 1 - \frac{Q}{Sv_k} \right|^3 - \left| n - \frac{Q}{Sv_k} \right|^3 \right], \tag{7.32}$$

TABLE 7.2

Adjustment Coefficient Values for the Case of Moving Air in the Chute

B	Values of K at n Equal To...								
	0.1	**0.2**	**0.3**	**0.4**	**0.5**	**0.6**	**0.7**	**0.8**	**0.9**
At $\varphi = 0.25$									
0.02	2.026	1.746	1.526	1.349	1.216	1.122	1.061	1.024	1.005
0.04	1.778	1.571	1.405	1.270	1.167	1.095	1.047	1.019	1.004
0.06	1.650	1.479	1.341	1.227	1.141	1.080	1.040	1.016	1.003
0.08	1.565	1.417	1.298	1.199	1.123	1.070	1.035	1.014	1.003
0.1	1.502	1.371	1.266	1.178	1.110	1.063	1.031	1.012	1.003
0.2	1.326	1.242	1.174	1.116	1.072	1.041	1.020	1.008	1.002
0.4	1.173	1.129	1.093	1.062	1.038	1.022	1.011	1.004	1.001
0.6	1.093	1.069	1.050	1.033	1.021	1.012	1.006	1.002	1.000
0.8	1.039	1.029	1.021	1.014	1.009	1.005	1.002	1.001	1.000
1.0	1.000	1.000	1.000	1.000	1.000	1.000	1.000	1.000	1.000
At $\varphi = 0.5$									
0.02	4.507	2.455	1.822	1.526	1.342	1.198	1.097	1.037	1.008
0.04	3.840	2.152	1.644	1.412	1.270	1.157	1.077	1.030	1.006
0.06	3.482	1.989	1.548	1.351	1.230	1.134	1.066	1.025	1.005
0.08	3.239	1.878	1.483	1.309	1.203	1.119	1.058	1.022	1.005
0.1	3.056	1.794	1.435	1.277	1.183	1.107	1.053	1.020	1.004
0.2	2.499	1.547	1.292	1.185	1.123	1.072	1.035	1.013	1.003
0.4	1.926	1.311	1.161	1.101	1.067	1.039	1.019	1.007	1.002
0.6	1.556	1.173	1.089	1.055	1.037	1.021	1.010	1.004	1.001
0.8	1.261	1.075	1.039	1.023	1.016	1.009	1.005	1.002	1.000
0.9	1.131	1.034	1.019	1.011	1.007	1.004	1.002	1.001	1.000
1.0	1.000	1.000	1.000	1.000	1.000	1.000	1.000	1.000	1.000

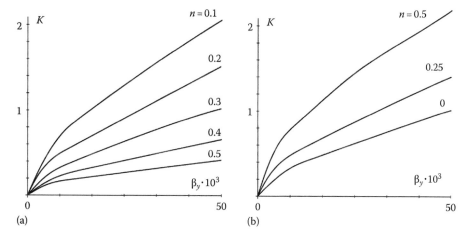

FIGURE 7.1 Variation of the adjustment coefficient as a function of volumetric concentration for a constantly accelerated flow of particles $d_e = 3$ mm: (a) at $\varphi = 0$ and (b) at $\varphi \neq 0$ and $n = 0.3$.

with the following assignment made to simplify the notation:

$$z = \psi_y \frac{G_m v_k}{S} \quad \Phi = \psi_y \frac{G_m v_k}{S} \varepsilon \frac{K_m v_K^2}{2a_\tau};$$

(7.33)

Q is the volumetric flow rate of air ejected through the loading chute of the elevator (m³/s).

7.2   EJECTING PROPERTIES OF A BUCKET ELEVATOR

Let us now consider ejection properties of a bucket-belt elevator. Air inside the enclosure of a belt elevator may be brought into motion both by moving bucket belt and by spillage flows during loading and unloading of buckets. Initial findings from studies performed to evaluate air motion in ducts with mobile partitions have been published in our earlier books [185,214]. Here we will consider the process of air ejection in bucket elevators from the standpoint of the classical laws of change in air mass and momentum.

Let us begin with considering airflow in a duct of elevator return line of length dx (Figure 7.2). For this area a momentum conservation equation can be written in projections onto axis Ox directed vertically downward. We will formulate a 1D problem assuming that the velocity of ejected air is directed downward and is equal to the cross-sectional average:

$$u = \frac{Q}{S}, \quad S = ab,$$

(7.34)

where
     Q is the flow rate of air ejected inside the duct (m³/s)
     a, b are the cross-sectional dimensions of the duct (m)

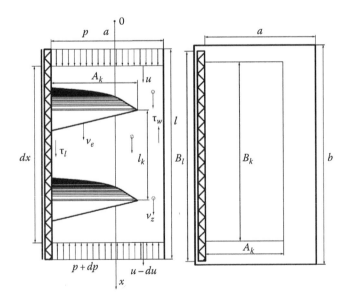

FIGURE 7.2   Diagram of forces acting on an element of elevator return run enclosure with a length dx.

The tangential frictional stress on the surface of a moving belt is equal to

$$\tau_l = c_l \frac{(v_e - u)|v_e - u|}{2} \rho, \tag{7.35}$$

where
    c_l is a dimensionless resistance coefficient
    v_e is the velocity of elevator belt (m/s)

Similarly, tangential frictional stress on the surface of enclosure walls is

$$\tau_w = c_w \frac{u|u|}{2} \rho, \tag{7.36}$$

where c_w is a dimensionless resistance coefficient of enclosure walls.
    It is known that coefficients c_l, c_w are related to friction coefficients in the Darcy–Weisbach equation for determining pressure losses in straight pipe sections

$$c_l = \frac{\lambda_l}{4}, \quad c_w = \frac{\lambda_w}{4},$$

where
    λ_l is the hydraulic friction coefficient of the belt
    λ_w is the hydraulic friction coefficient of enclosure walls

The aerodynamic force of a bucket, expressed similarly to the aerodynamic force of particles, is

$$R_k = c_k F_k \frac{(v_e - u)|v_e - u|}{2} \rho, \tag{7.37}$$

where
    F_k is the midsection area of a bucket ($F_k = A_k B_k$) (m²)
    c_k is the drag coefficient of an empty bucket

In this case, the equation for change in momentum would appear as

$$\rho u S(-u) + \rho(u + du) S(u + du)$$
$$= pS - (p + dp)S - \tau_w(b + 2a)dx + \tau_l B_l dx + R_k \frac{dx}{l_k} + R_z \frac{\beta dx S}{V_p}, \tag{7.38}$$

where
    l_k is the spacing of buckets on the belt (m)
    β is the volumetric concentration of grain spillage, equal to

$$\beta = \frac{G_p}{\rho_z S v_z} \tag{7.39}$$

G_p is the mass flow rate of grain spillage during bucket unloading (kg/s)
ρ_z is the grain density (kg/m³)
v_z is the fall velocity of grain (m/s)
V_p is the volume of an individual particle (m³)

R_z is the aerodynamic force of a single particle of spilled material,

$$R_z = \psi_z F_z \frac{|v_z - u|(v_z - u)}{2} \rho \qquad (7.40)$$

F_z is the midsection area of a particle (m²)

$$F_z = \frac{\pi d_e^2}{4} \qquad (7.41)$$

d_e is the equivalent grain diameter (m)

Considering that, in this case,

$$u = \text{const}, \qquad (7.42)$$

after a trivial transformation of Equation 7.38, ignoring infinitesimal second-order terms, we will obtain the following relation for determining differential pressure inside the enclosure of the return run of the elevator:

$$p(0) - p(l) + E_k + E_p = p_w, \qquad (7.43)$$

where
  $p(0)$, $p(l)$ are static pressures at the inlet/outlet of the enclosure (Pa)
  p_w is the aerodynamic drag of enclosure walls (Pa)

$$p_w = \int_0^l \lambda_w \frac{2a + b}{4S} \frac{u^2}{2} \rho dx, \qquad (7.44)$$

equal at $u = \text{const}$ and $S = \text{const}$

$$p_w = \lambda_w \frac{l}{D_w} \frac{u^2}{2} \rho, \qquad (7.45)$$

$$D_w = \frac{4S}{b + 2a}, \qquad (7.46)$$

l is the total length of elevator enclosure (distance between the axes of driving and return drums along the belt of the bucket elevator) (m); E_k is the ejection head created by a conveyor belt with buckets:

$$E_k = \frac{1}{S} \int_0^l \left(c_k \frac{F_k}{l_k} + \frac{\lambda_l}{4} B_l \right) \frac{|v_e - u|(v_e - u)}{2} \rho dx, \qquad (7.47)$$

which, at constant relative velocity, is equal to

$$E_k = c_{ek} \frac{|v_e - u|(v_e - u)}{2} \rho; \tag{7.48}$$

c_{ek} is the aerodynamic coefficient of the return run of elevator belt (with an account of empty buckets and conveyor belt carrying them):

$$c_{ek} = \frac{l}{S}\left(c_k \frac{F_k}{l_k} + \frac{\lambda_l}{4} B_l\right), \tag{7.49}$$

E_p is the ejection head created by a flow of spilled grain during the unloading of elevator buckets:

$$E_p = \frac{1}{S}\int_0^l \psi_z K_m \varepsilon G_p \frac{|v_z - u|(v_z - u)}{2} \frac{dx}{v_z}, \tag{7.50}$$

which, for a constantly accelerated vertical flow of particles at ψ_z = const, equals

$$E_p = \frac{\psi_z K_m \varepsilon}{2} \frac{G_p}{Sg} \frac{|v_k - u|^3 - |v_n - u|^3}{3} \tag{7.51}$$

or, for a uniformly accelerated flow of particles at $v_z = v_e$ = const, u = const,

$$E_p = \frac{\psi_z K_m \varepsilon}{2} \frac{G_p}{S} \frac{l}{v_e}\left[|v_e - u|(v_e - u)\right], \tag{7.52}$$

where
   K_m is the ratio of midsectional area of a particle to its volume (1/m)
   ε is the ratio of air density to particle density

Let us now consider a more complex case with carrying and return runs of a bucket elevator both located in a common enclosure (Figure 7.3). In this case, air may flow laterally from one part of the enclosure (for example, the one with grain-laden buckets running) to another (with empty buckets running). The velocity of air cross flow in the gap between the belt and enclosure walls will be designated as ω. Parameters of airflow in the right-hand side of the enclosure (where empty buckets run and spilled particles fall) will be denoted with a subscript u (from the designation of air velocity in the return run), while those in the left-hand side will be denoted with v (from the designation of air velocity in the carrying run of the conveyor).

The cross-flow velocity is determined by differential pressure and aerodynamic drag of gaps:

$$\Delta p = p_v - p_u = \zeta_z \frac{w|w|}{2}\rho, \tag{7.53}$$

where
   p_v, p_u are the respective excess static pressures in the left-hand and right-hand sides of the enclosure (Pa)
   ζ_z is the total local resistance coefficient (LRC) for two gaps between chute walls and the end sides part of carrying and return runs of the conveyor

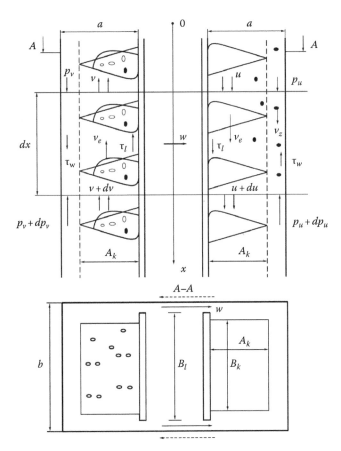

FIGURE 7.3 Longitudinal airflow diagram for return and carrying runs both located in a common elevator enclosure.

The sign of the absolute value in the right-hand side was introduced to ensure universality of the relation that in this case would be just as good for the case of a reverse flow (with $p_u > p_v$). The velocity w in such *vector* case will be negative, that is, velocity vector will be directed oppositely (from right to left). This vector is represented with a dotted line in Figure 7.3.

Due to the presence of lateral cross flow of air, velocities u and v will not be constant but rather would change along the height of the elevator enclosure.

Now we will write an airflow conservation equation while still assuming velocities u and v to be averaged throughout the cross section and the positive direction of these velocities to coincide with bucket traveling direction:

$$uab = (u + du)ab - w(b - B_l)dx, \tag{7.54}$$

$$vab = (v + dv)ab - w(b - B_l)dx. \tag{7.55}$$

It can be seen from here that change in the absolute value of dilatational velocities is equal:

$$\frac{du}{dx} = w\frac{b - B}{ab},$$

$$\frac{dv}{dx} = w\frac{b - B}{ab}. \tag{7.56}$$

And the difference of these velocities does not vary along the enclosure:

$$u - v = k = \text{const.} \tag{7.57}$$

Let us now write a motion preservation equation for the chosen element of enclosure of length dx. For the right-hand (downward) airflow the momentum conservation equation in the projection onto axis Ox does not differ in any way from Equation 7.38. Trivial transformations would yield the following differential equation for the dynamics of the air current at hand:

$$2\rho u \frac{du}{dx} = -\frac{dp_u}{dx} - \frac{\lambda_w}{D_w} \frac{u^2}{2}\rho + \zeta_u \frac{|v_e - u|(v_e - u)}{2}\rho + A_n \frac{|v_e - u|(v_e - u)}{2v}\rho, \tag{7.58}$$

with the following notational simplification:

$$\zeta_u = \frac{\lambda_l}{4}\frac{B_l}{ab} + c_k \frac{A_k B_k}{abl_k}, \tag{7.59}$$

$$A_n = \frac{1.5}{d_e}\frac{G_p}{\rho_z}\frac{\psi_e}{ab}. \tag{7.60}$$

For the left-hand (upward) airflow we will first write the momentum preservation equation in differentials:

$$-\rho vvab - \rho(v + dv)ab\left[-(v + dv)\right] = p_v ab - (p_v + dp_v)ab + \tau_w(b + 2a)dx - \tau_l B_l dx - R_k \frac{dx}{l_k}, \tag{7.61}$$

where

$$\tau_w = \frac{\lambda_w}{4}\frac{v|v|}{2}\rho \tag{7.62}$$

$$\tau_l = \frac{\lambda_l}{4}\frac{(v_e - v)|v_e - v|}{2}\rho \tag{7.63}$$

$$R_k = c_{kz}A_k B_k \frac{(v_e - v)|v_e - v|}{2}\rho \tag{7.64}$$

c_{kz} is a dimensionless coefficient of aerodynamic drag of a grain-laden bucket.

After trivial transformations, Equation 7.61 would appear as follows:

$$2\rho v \frac{dv}{dx} = -\frac{dp_v}{dx} + \frac{\lambda_w}{D_w}\frac{v^2}{2}\rho - \zeta_v \frac{|v_e - v|(v_e - v)}{2}\rho, \tag{7.65}$$

$$\zeta_v = \frac{1}{ab}\left(\frac{\lambda_l B_l}{4} + c_{kz}\frac{A_k B_k}{l_k}\right). \tag{7.66}$$

Thus, longitudinal airflow in the case of colocation of the carrying and return runs in the same enclosure can be described with combined equations:

$$2\rho u \frac{du}{dx} = -\frac{dp_u}{dx} + f_u, \tag{7.67}$$

$$2\rho v \frac{dv}{dx} = -\frac{dp_v}{dx} + f_v, \tag{7.68}$$

$$\frac{du}{dx} = w \frac{b - B_l}{ab}, \tag{7.69}$$

$$v = u - k, \tag{7.70}$$

$$\Delta p = \zeta_z \frac{w|w|}{2}\rho, \quad \Delta p = p_v - p_u, \tag{7.71}$$

with the following assignments for brevity:

$$f_u = -\frac{\lambda_w}{D_w} \frac{u|u|}{2}\rho + \zeta_u \frac{|v_e - u|(v_e - u)}{2}\rho + A_n \frac{|v_z - u|(v_z - u)}{2v_z}\rho, \tag{7.72}$$

$$f_v = \frac{\lambda_w}{D_w} \frac{v|v|}{2}\rho - \zeta_v \frac{|v_e - v|(v_e - v)}{2}\rho. \tag{7.73}$$

Newly introduced functions f_u and f_v can be written in a more convenient (symmetric) form. First of all let us assume that spillage velocity is equal to the velocity of the return run (considering that gap between buckets and enclosure walls is rather narrow, particles would first impinge on the bottom of a bucket and then accelerate gravitationally and *catch up* with uniformly moving buckets):

$$v_z \approx v_e. \tag{7.74}$$

Then,

$$f_u = -\xi_u \frac{T}{l} \frac{\rho u^2}{2} + \gamma_u \frac{M_u}{l} \frac{\rho(v_e - u)^2}{2}, \tag{7.75}$$

$$f_v = \xi_v \frac{T}{l} \frac{\rho v^2}{2} - \gamma_v \frac{M_v}{l} \frac{\rho(v_e - v)^2}{2}, \tag{7.76}$$

where T, M_u, and M_v are dimensionless parameters:

$$T = \lambda_w \frac{l}{D_w}, \tag{7.77}$$

$$M_u = \lambda_l \frac{l}{D_l} + c_k \frac{l}{l_k} \frac{A_k B_k}{S} + 1.5\psi \frac{l}{d}\beta_e, \tag{7.78}$$

$$M_v = \lambda_l \frac{l}{D_l} + c_{kz} \frac{l}{l_k} \frac{A_k B_k}{S}, \tag{7.79}$$

$$D_w = \frac{4S}{b + 2a},$$

$$D_l = \frac{4S}{B_l}, \tag{7.80}$$

$$\beta_e = \frac{G_p}{\rho_z v_e S}, \tag{7.81}$$

$$\xi_u = \text{signum}(u), \quad \xi_v = \text{signum}(v) = \text{signum}(u - k), \tag{7.82}$$

$$\gamma_u = \text{signum}(v_e - u), \quad \gamma_v = \text{signum}(v_e - v) = \text{signum}(v_e + k - u). \tag{7.83}$$

Combined Equations 7.67 through 7.71 can be simplified significantly. Cross-flow velocity of a lateral flow can be derived from Equation 7.71:

$$w = \delta \sqrt{\frac{2|\Delta p|}{\zeta_z \rho}}, \quad \Delta p = p_v - p_u, \tag{7.84}$$

where

$$\delta = \text{signum}(\Delta p). \tag{7.85}$$

Substitution of Equation 7.84 into Equation 7.69 yields

$$\frac{du}{dx} = \delta \frac{L}{l} \sqrt{\frac{2|\Delta p|}{\rho}}, \tag{7.86}$$

where

$$L = \frac{l(b - B)}{S\sqrt{\zeta_z}}. \tag{7.87}$$

By subtracting the relation Equation 7.68 from the Equation 7.67 and considering $\frac{dv}{dx} = \frac{du}{dx}$ in view of Equation 7.70 we will get

$$2\rho k \frac{du}{dx} = \frac{d\Delta p}{dx} - \xi_u \frac{T}{l} \frac{u^2}{2} \rho + \gamma_u \frac{M_u}{l} \frac{(v_e - u)^2}{2} \rho - \xi_v \frac{T}{l} \frac{v^2}{2} \rho + \gamma_v \frac{M_v}{l} \frac{(v_e - v)^2}{2} \rho \tag{7.88}$$

or, in view of Equation 7.86,

$$-\frac{d\Delta p}{dx} = -2\delta k \rho \frac{L}{l} \sqrt{\frac{2|\Delta p|}{\rho}} - \xi_x \frac{T}{l} \frac{u^2}{2} \rho + \gamma_u \frac{M_u}{l} \frac{(v_e - u)^2}{2} \rho - \xi_v \frac{T}{l} \frac{v^2}{2} \rho + \gamma_v \frac{M_v}{l} \frac{(v_e - v)^2}{2} \rho. \tag{7.89}$$

Equation 7.89 can be supplemented with Equation 7.67 in view of Equations 7.75 and 7.86:

$$\frac{dp_u}{dx} = -2\rho u\delta \frac{L}{l}\sqrt{\frac{2|\Delta p|}{\rho}} - \xi_u \frac{T}{l}\frac{u^2}{2}\rho + \gamma_u \frac{M_u}{l}\frac{(v_e - u)^2}{2}\rho. \tag{7.90}$$

This, in view of Equation 7.86, yields a familiar combined set (considering Equation 7.70) of three differential equations (Equations 7.90, 7.89, and 7.86) describing the process of averaged longitudinal airflows in an enclosure with colocation of the carrying and return runs of bucket elevator inside it.

Changes in the velocity u can be found from Equation 7.88 using Equation 7.86, which can be written as follows:

$$\Delta p = A\left(\frac{du}{dx}\right)^2 \frac{\rho}{2}\left(\frac{l}{L}\right)^2, \tag{7.91}$$

where

$$A = \mathrm{signum}\left(\frac{du}{dx}\right).$$

A substitution of Equation 7.91 into Equation 7.88 using the relation in Equation 7.70 results in a second-order nonlinear equation relative to the sought function u:

$$-A\rho\left(\frac{l}{L}\right)^2 \frac{d^2u}{dx^2}\frac{du}{dx} + 2\rho k\frac{du}{dx} = -\xi_u \frac{T}{l}\frac{u^2}{2}\rho + \gamma_u \frac{M_u}{l}\frac{(v_e - u)^2}{2}\rho$$

$$-\xi_v \frac{T}{l}\frac{(u-k)^2}{2}\rho + \gamma_v \frac{M_v}{l}\frac{(v_e + k - u)^2}{2}\rho. \tag{7.92}$$

We will use a dimensionless differential equation formula to facilitate our numerical integration of the resulting dimensional equations. It will enable us to reduce the number of constants. The following quantities will be considered as basic values:

- Elevator belt velocity v_e
- Elevator enclosure length or height l
- Dynamic pressure $\rho v_e^2/2$

Thus, let

$$p_u = p\frac{v_e^2 \rho}{2}, \quad \Delta p = R\frac{v_e^2 \rho}{2}, \quad u = u^* v_e, \quad k = m^* v_e, \quad x = zl, \tag{7.93}$$

then, after we substitute the accepted conventions into Equations 7.67, 7.89, and 7.86 and perform certain trivial transformations, the following system of dimensionless differential equations will result:

$$\frac{dp}{dz} = -\delta u^* 4L\sqrt{|R|} - \xi_u T\left(u^*\right)^2 + \gamma_u M_u\left(1 - u^*\right)^2, \tag{7.94}$$

$$\frac{dR}{dz} = 4\delta m^* L\sqrt{|R|} + \xi_u T\left(u^*\right)^2 - \gamma_u M_u\left(1-u^*\right)^2 + \xi_v T\left(u^*-m^*\right)^2 - \gamma_v M_v\left(1+m^*-u^*\right)^2, \quad (7.95)$$

$$\frac{du^*}{dz} = \delta L\sqrt{|R|}, \quad (7.96)$$

where

$$\delta = \text{signum}(R), \quad \xi_u = \text{signum}(u^*), \quad \xi_v = \text{signum}(u^*-m^*) \quad (7.97)$$

$$\gamma_u = \text{signum}(1-u^*), \quad \gamma_v = \text{signum}(1+m^*-u^*). \quad (7.98)$$

Similarly, Equation 7.92 will now appear as

$$\left(-A_3 \frac{2}{L^2} \frac{d^2 u^*}{dz^2} + 4m^*\right)\frac{du^*}{dz}$$

$$= -\xi_u T\left(u^*\right)^2 + \gamma_u M_u\left(1-u^*\right)^2 - \xi_v T\left(u^*-m^*\right)^2 + \gamma_v M_v\left(1+m^*-u^*\right)^2, \quad (7.99)$$

where

$$A_3 = \text{signum}\left(\frac{du^*}{dz}\right). \quad (7.100)$$

7.3 CROSS FLOW OF AIR THROUGH SEALED ELEVATOR ENCLOSURES

Consider the most common case of an elevator with two buckets. Let the carrying and return runs of the elevator with buckets be located in separate sealed enclosures that will not experience cross flows of air over their entire length. These enclosures are aerodynamically coupled only in their bottom (loading) and top (unloading) parts. Let static pressure p be maintained in these parts, respectively, at p_k and p_n, additionally assuming that

$$p_k < p_n. \quad (7.101)$$

In this case, air will flow from the upper into the lower zone through the return run enclosure but will only pass through the carrying run enclosure when the ejection head caused by laden buckets is lower that differential static pressure.

$$\Delta p_e \le p_n - p_k, \quad (7.102)$$

then, the limit case (at $v = 0$) owing to Equations 7.65 and 7.66 will be written as

$$p_n - p_k > \zeta_1 \frac{v_e^2}{2}\rho \quad (7.103)$$

or the following inequality can be used to describe the trigger condition for the ejection properties of the carrying run:

$$h_a = \frac{p_n - p_k}{\frac{v_e^2}{2}\rho} > \zeta_1 \tag{7.104}$$

or

$$t_1 = \frac{h_a}{M_1} \geq 1. \tag{7.105}$$

Hereinafter, a subscript *1* will continue to denote the characteristic parameters of airflow inside the enclosure of conveyor carrying run (i.e., *1* will be substituted for index *v*, e.g., longitudinal air velocity u_1 instead of *v*, ζ_1 instead of ζ_v, M_1 instead of M_v, etc.), whereas subscript *2* will denote longitudinal airflow in the return run enclosure (airflow velocity u_2 instead of *u*, parameter ζ_2 instead of ζ_u, M_2 instead of M_u, etc.)

Generally (when p_n may also be less than p_k), two patterns of air cross flows through bucket elevator enclosures are possible: a direct-flow pattern with positive velocities u_1 and u_2 and air moving in the same direction with buckets, and a combined pattern whereby airflow and bucket traveling directions are the same in one enclosure but opposite in the other (Figure 7.4).

Let us determine airflow rates Q_1 and Q_2 as well as their difference:

$$\Delta Q = Q_2 - Q_1.$$

Airflow rates can also become negative, depending on the sign and magnitude of velocity vectors u_1 and u_2.

First we will determine the flow rate Q_2. Dynamics Equation 7.43 will be used to find out the flow rate in the return run enclosure. In this case, the static pressure at inlet and outlet of the enclosure will be expressed through pressures p_n and p_k using local resistance coefficients for air entering the enclosure (ζ_{2n}) and leaving the enclosure (ζ_{2k}).

$$p_2(0) = p_n - \zeta_{2n}\frac{u_2^2}{2}\rho, \tag{7.106}$$

$$p_2(l) = p_k + \zeta_{2k}\frac{u_2^2}{2}\rho. \tag{7.107}$$

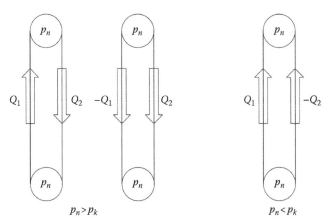

FIGURE 7.4   Aerodynamic diagrams for cross flows of air in two sealed enclosures of a bucket elevator.

In view of the accepted conditions we will rewrite Equation 7.43 and expand the values of E_k and E_p based on Equations 7.47, 7.48, and 7.52:

$$p_k - p_n + \sum \zeta_2 \frac{u_2^2}{2} \rho = M_2 \frac{|v_e - u_2|(v_e - u_2)}{2} \rho, \qquad (7.108)$$

where $\sum \zeta_2$ is the sum total of the LRCs of the enclosure.

$$\sum \zeta_2 = \zeta_{2n} + \lambda_w \frac{l}{D_w} + \zeta_{2k}, \qquad (7.109)$$

M_2 is a parameter describing the ejection capacity of the return run of the conveyor and the flow of spilled material (in accordance with the formula given in Equation 7.78).

If both sides of the equation are divided by $\rho(v_e/2)$, a dimensionless equation would result:

$$-h_a + \sum \zeta_2 \varphi_2^2 = M_2 |1 - \varphi_2|(1 - \varphi_2), \qquad (7.110)$$

where

$$\varphi_2 = \frac{u_2}{v_e}$$

$$h_a = \frac{p_n - p_k}{\rho \dfrac{v_e^2}{2}} = \frac{h_k - h_n}{\rho \dfrac{v_e^2}{2}} \qquad (7.111)$$

where

h_k is the negative pressure maintained inside an aspirated cowl of elevator boot by an aspiration system fan (Pa)

h_n is the sustained negative pressure occurring inside an unaspirated cowl of the bucket elevator head as a result of air cross flow through the unloading chute and elevator enclosures (Pa)

The sought flow rate Q_2 is determined by an obvious relation

$$Q_2 = \varphi_2 v_e S, \qquad (7.112)$$

where $S = a \cdot b$.

The value of φ_2 is determined with Equation 7.110, which, owing to random nature of h_a, can be written as the following dimensionless equation:

$$h_a + M_2 \left| 1 - \varphi_2 \right| \left(1 - \varphi_2 \right) = \sum \zeta_2 \varphi_2 \left| \varphi_2 \right|, \tag{7.113}$$

or

$$t_2 = g_2 \varphi_2 \left| \varphi_2 \right| - \left| 1 - \varphi_2 \right| \left(1 - \varphi_2 \right), \tag{7.114}$$

where dimensionless numbers have been introduced to dispense with some determinant parameters:

$$t_2 = \frac{h_a}{M_2}, \tag{7.115}$$

$$g_2 = \frac{\sum \zeta_2}{M_2}, \tag{7.116}$$

representing a ratio of the available pressure and pressure losses to the total resistance that the enclosure poses to ejection head created by the bucket elevator.*

Expanding signs of absolute values reduce Equation 7.114 to the following three combined equations:

$$t_2 = g_2 \varphi_2^2 - \left(1 - \varphi_2 \right)^2 \quad \text{at } 1 \geq \varphi_2 \geq 0, \tag{7.117}$$

$$t_2 = -g_2 \varphi_2^2 - \left(1 - \varphi_2 \right)^2 \quad \text{at } -\infty < \varphi_2 \leq 0, \tag{7.118}$$

$$t_2 = g_2 \varphi_2^2 + \left(1 - \varphi_2 \right)^2 \quad \text{at } +\infty > \varphi_2 \geq 1. \tag{7.119}$$

The single-valued function $\varphi_2 = f(t_2)$ is plotted as a joint set of three parabolic arcs (Figure 7.5):

$$y_1 \equiv t_2 = -\varphi_2 \left(g_2 + 1 \right) + 2\varphi_2 - 1 \quad \text{at } -\infty < \frac{\varphi}{2} \leq 0, \tag{7.120}$$

$$y_2 \equiv t_2 = -\varphi_2^2 \left(1 - g_2 \right) + 2\varphi_2 - 1 \quad \text{at } 1 \geq \varphi_2 \geq 0, \tag{7.121}$$

$$y_3 \equiv t_2 = -\varphi_2^2 \left(1 + g_2 \right) - 2\varphi_2 + 1 \quad \text{at } \infty > \varphi_2 \geq 1. \tag{7.122}$$

---

* One should keep in mind that the parameter g_2 may change as a result of possible changes in $\sum \zeta$ when the sign of φ_2 reverses.

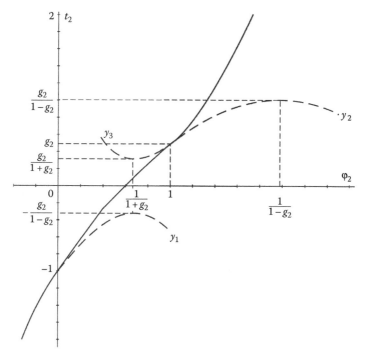

FIGURE 7.5 Variation in the relative flow rate of air transferred over the enclosure of elevator conveyor return run as a function of pressure transitions (solid curve—plot of single-valued function $\varphi_2 = f(t_2)$).

In order to obtain a single value for dimensionless flow rate φ_2 across the entire variation range of the parameter t_2, roots of the following equations have to be found:

$$-\varphi_2^2\left(1+g_2\right)+2\varphi_2-\left(1+t_2\right)=0 \quad \text{at} \ -\infty < t_2 < -1, \tag{7.123}$$

$$-\varphi_2^2\left(1-g_2\right)+2\varphi_2-\left(1+t_2\right)=0 \quad \text{at} \ g_2 > t_2 > -1, \tag{7.124}$$

$$\varphi_2^2\left(1+g_2\right)-2\varphi_2+\left(1-t_2\right)=0 \quad \text{at} \ \infty > t_2 > g_2. \tag{7.125}$$

As a result (at $g_2 < 1$),

$$\varphi_2 = \frac{1}{1+g_2}\left[1-\sqrt{1-\left(1+t_2\right)\left(1+g_2\right)}\right] \quad \text{at} \ -\infty < t_2 \le -1, \tag{7.126}$$

$$\varphi_2 = \frac{1}{1-g_2}\left[1-\sqrt{1-\left(1+t_2\right)\left(1-g_2\right)}\right] \quad \text{at} \ g_2 > t_2 \ge -1, \tag{7.127}$$

$$\varphi_2 = \frac{1}{1+g_2}\left[1+\sqrt{1-\left(1-t_2\right)\left(1+g_2\right)}\right] \quad \text{at} \ \infty > t_2 > g_2. \tag{7.128}$$

Thus, forward airflow arises in the enclosure of the return run at $t_2 > -1$, that is, at

$$M_2 > -h_a. \tag{7.129}$$

Otherwise, at

$$M_2 < -h_a, \tag{7.130}$$

a countercurrent of air is promoted by a significant negative pressure in the upper cowl at

$$h_n > h_k + M_2 \frac{v_e^2}{2} \rho. \tag{7.131}$$

Absent differential pressure ($h_n = h_k$), φ_2 reaches its limit value

$$\lim_{t_2 \to 0} \varphi_2 = \varphi_{2pr} = \frac{1}{1 + \sqrt{g_2}}, \tag{7.132}$$

which converges toward one with increasing ejection forces

$$\varphi_{2pr} \approx 1 \quad \text{at } M_2 \gg \sum \zeta_2, \tag{7.133}$$

that is, the velocity of air inside the return run conveyor belt enclosure of the elevator reaches the velocity of the belt only with significant ejection forces.

Let us now determine the airflow rate inside the enclosure of the carrying run of the bucket elevator. To that end we will put forward an equation for the dynamics of air in this enclosure with a compound effect of differential pressure $\Delta p = p_n - p_k$ and of ejection head created by a belt with laden buckets. It will be recognized that air velocity u_1 may turn negative at significant differential pressures. The dynamics equation for airflow in this enclosure will be put down as follows*:

$$M_1 |1 - \varphi_1| (1 - \varphi_1) - h_a = \sum \zeta_1 \varphi_1 |\varphi_1| \tag{7.134}$$

or

$$t_1 = |1 - \varphi_1| (1 - \varphi_1) - g_1 \varphi_1 |\varphi_1|, \tag{7.135}$$

where

$$t_1 = \frac{h_a}{M_1}, \quad g_1 = \frac{\sum \zeta_1}{M_1}. \tag{7.136}$$

---

* It should be noted that, generally, a reversal of airflow inside elevator enclosure will also change $\sum \zeta_1$ and $\sum \zeta_2$ $\left(\sum \zeta_1 \uparrow \neq \sum \zeta_1 \downarrow, \ \sum \zeta_2 \downarrow \neq \sum \zeta_2 \uparrow \text{ arrows indicate downward } (\downarrow) \text{ and upward} (\uparrow) \text{airflow direction} \right)$.

An expansion of the signs at absolute values breaks down the Equation 7.135 into three separate ones:

$$t_1 = \left(1 - \varphi_1\right)^2 + g_1 \varphi_1^2 \quad \text{at} \ -\infty < \varphi_1 \le 0, \tag{7.137}$$

$$t_1 = \left(1 - \varphi_1\right)^2 - g_1 \varphi_1^2 \quad \text{at} \ 1 \ge \varphi_1 \ge 0, \tag{7.138}$$

$$t_1 = -\left(1 - \varphi_1\right)^2 - g_1 \varphi_1^2 \quad \text{at} \ +\infty > \varphi_1 \ge 1. \tag{7.139}$$

The first of these equations describes the balance of dimensionless forces in the case of downward (from top to bottom) motion of air arising as a result of a significant difference in the available static differential pressure ($\Delta p = p_n - p_k$). The ejection head of a bucket-carrying belt reduces airflow, further hindering the downward motion. When the available differential pressure is small enough (the second and third equations), an upward airflow (from bottom to top) arises. In this case it is counteracted not only by the drag of enclosure walls but also by the differential pressure Δp.

The single-valued function $\varphi_1 = f(t_1)$ is plotted as a joint set of three parabolic arcs* (Figure 7.6):

$$y_1 \equiv t_1 = \left(1 + g_1\right)\varphi_1^2 - 2\varphi_1 + 1 \quad \text{at} \ -\infty < \varphi_1 \le 0, \tag{7.140}$$

$$y_2 \equiv t_1 = \left(1 - g_1\right)\varphi_1^2 - 2\varphi_1 + 1 \quad \text{at} \ 1 \ge \varphi_1 \ge 0, \tag{7.141}$$

$$y_3 \equiv t_1 = -\left(1 + g_1\right)\varphi_1^2 + 2\varphi_1 - 1 \quad \text{at} \ +\infty > \varphi_1 \ge 1. \tag{7.142}$$

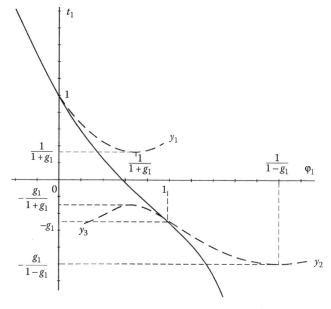

FIGURE 7.6 Variation in the relative flow rate of air flowing over the enclosure of elevator conveyor carrying run as a function of differential pressure (solid curve—plot of single-valued function $\varphi_1 = f(t_1)$).

* Parabola sections outside of the range of single-valued function $t = f(\varphi)$ are shown in Figure 3.6 as dotted lines.

Single-valued functions of dimensionless air velocity φ_1 inside the enclosure of the carrying run of the conveyor within the entire range of variations in the parameter t_1 are determined by roots of the following equations:

$$(1+g_1)\varphi_1^2 - 2\varphi_1 + (1-t_1) = 0 \quad \text{at} \ \infty > t_1 \geq 1, \tag{7.143}$$

$$(1-g_1)\varphi_1^2 - 2\varphi_1 + (1-t_1) = 0 \quad \text{at} \ -g_1 \leq t_1 \leq 1, \tag{7.144}$$

$$(1-g_1)\varphi_1^2 - 2\varphi_1 - (1+t_1) = 0 \quad \text{at} \ -\infty < t_1 \leq -g_1, \tag{7.145}$$

which gives (at $g_1 < 1$)

$$\varphi_1 = \frac{1}{1+g_1}\left[1 + \sqrt{1-(1+t_1)(1+g_1)}\right] \quad \text{at} \ -\infty \leq t_1 \leq -g_1, \tag{7.146}$$

$$\varphi_1 = \frac{1}{1-g_1}\left[1 - \sqrt{1+(t_1-1)(1-g_1)}\right] \quad \text{at} \ 1 \geq t_1 \geq -g_1, \tag{7.147}$$

$$\varphi_1 = \frac{1}{1+g_1}\left[1 - \sqrt{1+(t_1-1)(1+g_1)}\right] \quad \text{at} \ \infty > t_1 \geq 1. \tag{7.148}$$

As is evident from these results, counterflow of air ($\varphi_1 < 0$) inside the enclosure of conveyor belt carrying run may only arise at greater values of the parameter t_1, that is, at

$$t_1 > 1, \quad h_k > h_n + M_1\frac{v_e^2}{2}\rho. \tag{7.149}$$

The limit value of dimensionless air velocity (flow rate) inside the enclosure of the conveyor carrying run (at $h_k = h_n$) is

$$\lim_{t_1 \to 0} \varphi_1 = \varphi_{1pr} = \frac{1}{1+\sqrt{g_1}} \tag{7.150}$$

and air velocity u_1 reaches the velocity of buckets v_e

$$\varphi_{1pr} = 1 \quad \text{at} \ M_1 \gg \sum \zeta_1. \tag{7.151}$$

When static differential pressure is small ($t_1 < 1$), only the forward airflow pattern may arise inside enclosures. Airflow follows the traveling conveyor belt inside enclosures of the bucket elevator. Additionally, as a rule,

$$Q_2 > Q_1.$$

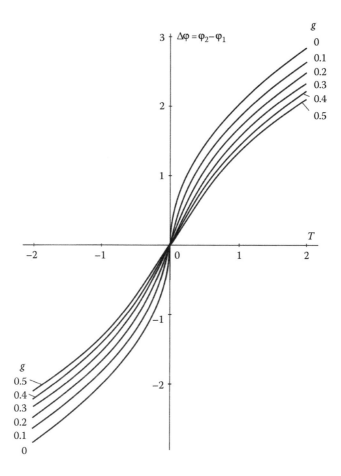

FIGURE 7.7 Changes in the relative flow rate of air transferred through bucket elevator enclosures (at $t_1 = t_2 = t$, $g_1 = g_2 = g$).

This is explained by the influence of ejecting capacity of spillage when grain is unloaded from buckets in the upper part of elevator, and by the difference between the drag of an empty bucket and a grain-laden bucket. In addition the available differential pressure promotes airflow inside return run enclosures while hindering it in the carrying run enclosure. The latter explains the fact that, given equal ejection forces ($M_1 = M_2$) and aerodynamic drag forces $\left(\sum \varsigma_1 = \sum \varsigma_2 \right)$, velocities (airflows in bucket elevator enclosures) fail to equalize (Figure 7.7).

In this case the difference between airflows

$$Q_2 - Q_1 = \left(\varphi_2 - \varphi_1 \right) v_e S \tag{7.152}$$

will be either positive (at $t > 0$) or negative (at $t < 0$) and will increase in its absolute value with increasing parameter t.

Tables 7.3 and 7.4 summarize the calculated values of relative air flow rates inside bucket elevator enclosures. These values have been determined using the formulas given in Equations 7.126 through 7.128 and Equations 7.146 through 7.148.

These findings reveal that, within the range of low available pressures (at $t < 1$), flow rate of ascending air inside the return run enclosure declines both with increasing parameter t_2 and decreasing parameter g_2. Within the range of high available pressures (at $t_2 > g_2$), air velocity $u_2 > v_e$. In this

TABLE 7.3
Relative Air Flow Rate φ_2 inside the Enclosure of Bucket Elevator Return Run

t_2 \ g_2	0	0.1	0.2	0.3	0.4	0.5
−2.0	−0.4142	−0.4083	−0.4027	−0.3974	−0.3923	−0.3874
−1.9	−0.3784	−0.3733	−0.3685	−0.3639	−0.3595	−0.3553
−1.8	−0.3416	−0.3374	−0.3333	−0.3295	−0.3257	−0.3222
−1.7	−0.3036	−0.3004	−0.3971	−0.2939	−0.2908	−0.2879
−1.6	−0.2649	−0.2622	−0.2596	−0.2571	−0.2546	−0.2523
−1.5	−0.2247	−0.2227	−0.2208	−0.2189	−0.2170	−0.2153
−1.4	−0.1832	−0.1818	−0.1805	−0.1791	−0.1779	−0.1766
−1.3	−0.1402	−0.1393	−0.1385	−0.1377	−0.1369	−0.1361
−1.2	−0.0954	−0.0950	−0.0946	−0.0942	−0.0938	−0.0935
−1.1	−0.0488	−0.0487	−0.0486	−0.0485	−0.0484	−0.0483
−1.0	0.000	0.000	0.000	0.000	0.000	0.000
−0.9	0.0513	0.0512	0.0510	0.0509	0.0508	0.0506
−0.8	0.1056	0.1050	0.1044	0.1038	0.1032	0.1026
−0.7	0.1633	0.1618	0.1603	0.1588	0.1574	0.1561
−0.6	0.2254	0.2222	0.2192	0.2164	0.2137	0.211
−0.5	0.2929	0.2871	0.2818	0.2768	0.3722	0.2679
−0.4	0.3675	0.3575	0.3486	0.3406	0.3333	0.3267
−0.3	0.4523	0.4352	0.4208	0.4084	0.3974	0.3875
−0.2	0.5528	0.5232	0.5000	0.4810	0.4648	0.4508
−0.1	0.6838	0.6268	0.5886	0.5596	0.5363	0.5168
0	1.000	0.7597	0.6910	0.6461	0.6126	0.5858
0.1	1.3162	1.000	0.8170	0.7435	0.6948	0.6584
0.2	1.4472	1.2240	1.000	0.8511	0.7847	0.7351
0.3	1.5477	1.3451	1.1667	1.000	0.8849	0.8168
0.4	1.6325	1.4392	1.2743	1.1300	1.000	0.9046
0.5	1.7071	1.5189	1.3604	1.2243	1.1055	1.000
0.6	1.7746	1.5894	1.4343	1.3022	1.1881	1.0883
0.7	1.8367	1.6535	1.5000	1.3700	1.2583	1.1611
0.8	1.8944	1.7120	1.5598	1.4309	1.3204	1.2244
0.9	1.9487	1.7667	1.6151	1.4867	1.3767	1.2813
1.0	2.0000	1.8182	1.5667	1.5385	1.4286	1.3333
1.1	2.0488	1.8669	1.7153	1.5869	1.4769	1.3816
1.2	2.0954	1.9132	1.7613	1.6357	1.5224	1.4268
1.3	2.1402	1.9575	1.8052	1.66761	1.5655	1.4694
1.4	2.1832	2.0000	1.8471	1.7176	1.6064	1.5099
1.5	2.2247	2.0409	1.8874	1.7573	1.6456	1.5486
1.6	2.2649	2.0804	1.9262	1.7955	1.6832	1.5856
1.7	2.3038	2.1186	1.9637	1.8323	1.7194	1.6212
1.8	2.3416	2.1556	2.0000	1.8679	1.7543	1.6555
1.9	2.3784	2.1915	2.0354	1.9024	1.7881	1.6886
2	2.4142	2.2265	2.0694	1.9358	1.8209	1.7208

TABLE 7.4

Relative Air Flow Rate φ_1 inside the Enclosure of Bucket Elevator Carrying Run

t_1 \ g_1	0	0.1	0.2	0.3	0.4	0.5
−2.0	2.4142	2.2265	2.0694	1.9358	1.8209	1.7208
−1.9	2.3784	2.1915	2.0354	1.9024	1.7881	1.6886
−1.8	2.3416	2.1556	2.0000	1.8679	1.7543	1.6555
−1.7	2.3038	2.1186	1.9637	1.8323	1.7194	1.6212
−1.6	2.2649	2.0804	1.9262	1.7955	1.6832	1.5856
−1.5	2.2247	2.0409	1.8874	1.7573	1.6456	1.5486
−1.4	2.1832	2.0000	1.8471	1.7176	1.6064	1.5099
−1.3	2.1402	1.9575	1.8052	1.6761	1.5655	1.4694
−1.2	2.0954	1.9132	1.7613	1.6357	1.5224	1.4268
−1.1	2.0488	1.8669	1.7153	1.5869	1.4769	1.3816
−1.0	2.0000	1.8182	1.5667	1.5385	1.4286	1.3333
−0.9	1.9487	1.7667	1.6151	1.4867	1.3767	1.2813
−0.8	1.8944	1.7120	1.5598	1.4309	1.3204	1.2244
−0.7	1.8367	1.6532	1.5000	1.3700	1.2583	1.1611
−0.6	1.7746	1.5894	1.4343	1.3022	1.1881	1.0883
−0.5	1.7071	1.5189	1.3604	1.2243	1.1055	1.000
−0.4	1.6325	1.4392	1.2743	1.1300	1.000	0.9046
−0.3	1.5477	1.3451	1.1667	1.000	0.8849	0.8168
−0.2	1.4472	1.2240	1.000	0.8511	0.7847	0.7351
−0.1	1.3162	1.000	0.8170	0.7435	0.6948	0.6584
0	1.000	0.7597	0.6910	0.6461	0.6126	0.5858
0.1	0.6838	0.6268	0.5886	0.5596	0.5363	0.5168
0.2	0.5528	0.5232	0.5000	0.4810	0.4648	0.4508
0.3	0.4523	0.4352	0.4208	0.4084	0.3974	0.3875
0.4	0.3675	0.3775	0.3486	0.3406	0.3333	0.3267
0.5	0.2929	0.2871	0.2818	0.2768	0.3722	0.2679
0.6	0.2254	0.2222	0.2192	0.2164	0.2137	0.2111
0.7	0.1633	0.1618	0.1603	0.1588	0.1574	0.1561
0.8	0.1056	0.1050	0.1044	0.1038	0.1032	0.1026
0.9	0.0513	0.0512	0.0510	0.0509	0.0508	0.0506
1.0	0.000	0.000	0.000	0.000	0.000	0.000
1.1	−0.0488	−0.0487	−0.0486	−0.0485	−0.0484	−0.0483
1.2	−0.0954	−0.0950	−0.0946	−0.0942	−0.0938	−0.0935
1.3	−0.1402	−0.1393	−0.1385	−0.1377	−0.1369	−0.1361
1.4	−0.1832	−0.1818	−0.1805	−0.1791	−0.1779	−0.1766
1.5	−0.2247	−0.2227	−0.2208	−0.2189	−0.2170	−0.2153
1.6	−0.2649	−0.2622	−0.2596	−0.2571	−0.2546	−0.2523
1.7	−0.3036	−0.3004	−0.3971	−0.2939	−0.2908	−0.2879
1.8	−0.3416	−0.3374	−0.3333	−0.3295	−0.3257	−0.3222
1.9	−0.3784	−0.3733	−0.3685	−0.3639	−0.3595	−0.3553
2	−0.4142	−0.4083	−0.4027	−0.3974	−0.3923	−0.3874

case the ejecting capacity of empty buckets poses additional resistance and, therefore, relative flow air φ_2 would decrease with increasing M_2 at unchanged Δp (and decreasing t_2).

The relative flow rate of ascending air decreases with increasing Δp inside the enclosure of the carrying run of the conveyor and a downward airflow arises at $t_1 > 1$ (φ_1 is negative). Higher values of M_1 (decreasing t_1) result in decreased flow rate φ_1.

7.4 CONCLUSIONS

1. Flows of loose material in loading and unloading chutes that arise during the operation of high-performance bucket elevators feature elevated volumetric concentrations as high as $\beta_y = 0.01$. Estimates of air ejection caused by such flows should be based on instantaneous (relation of Equation 7.18) rather than averaged aerodynamic drag coefficient ψ_y.

   Within the range of significant volumetric concentrations, varying volumetric concentration of falling particles leads to fluctuations in instantaneous values of the coefficient ψ. These fluctuations cause the ejection head, even in the case of short chutes ($l < l_y$), to significantly diverge from the head p_{ey} determined using averaged coefficient ψ_y. In order to compute ejection heads inside loading and unloading chutes, it is necessary to introduce an adjustment coefficient K, the value of which will noticeably diverge from one at small initial velocities of particle flow (at $n < 0.5$, Table 7.2, Figure 7.1).

   It is possible to view flows of particles in chutes at $B < 0.1$ as blowers with a performance curve determined with the formula given in Equation 7.32 in view of the resulting coefficient K_0. For a flow of wheat ($d_e \approx 3$ mm) within the ranges $n > 0.5$ and $\beta_y = 0.001$, increasing drag coefficient along the fall height is able to produce only negligibly small changes in the intensity of ejection head. The adjustment coefficient may be dispensed with ($K = K_0 = 1$) and head value will then be determined using the formula given by Equation 7.12.

2. Direction of airflow inside the enclosures of the carrying and return runs of a bucket elevator is determined by the drag of buckets and moving conveyor belt as well as the ejection head created by a stream of spilled particles when buckets are unloaded. As a result of these forces acting together inside an enclosure, differential pressure given Equation 7.43 arises. This differential pressure is equal to the sum total of ejection heads created by the conveyor belt with buckets E_k (Equation 7.47) and flow rate of spilled material E_p (Equation 7.50) minus the aerodynamic drag of enclosure walls (Equation 7.44).

   The ejection head E_k created by a bucket-carrying conveyor belt is determined by aerodynamic coefficient c_{ek} (Equation 7.49) (proportional to the number of buckets, their head resistances and squared midsectional dimensions) together with an absolute value and the direction of bucket velocity relative to the velocity of airflow inside the enclosure.

   Ejection head of spilled particles E_p (Equation 7.52) depends on the drag coefficient of particles, their size and flow rate, as well as the enclosure length, enclosure cross section, and relative flow velocity of particles.

3. When both the carrying and return runs of the conveyor belt are located in a common enclosure, the velocity of forward airflow varies over its length as a result of cross flows of air through gaps between the conveyor runs and enclosure walls. Cross flows are caused by a differential pressure between the carrying and return run enclosures and is dependent on the drag of the gap (Equation 7.53). Cross flow direction depends on the ratio between p_v and p_u.

   Given identical size of elevator enclosures, change in absolute values of longitudinal velocities is identical and depends on absolute values of cross-flow velocities and geometrical dimensions of the gap, as well as enclosure cross section (Equations 7.56 and 7.57). The momentum of longitudinal airflow in this case is determined by variable magnitudes of aerodynamic forces of buckets due to changes in their relative motion velocities.

The flow rate of air in enclosures may be determined by numerically integrating three dimensionless combined differential equations, Equations 7.94 through 7.96.

4. Both the direction and the flow rate of ejected air in bucket elevator enclosures that feature a separate arrangement of carrying and idle conveyor runs would depend on the ratio between ejection heads and the difference between static pressures inside the enclosures of elevator head and elevator boot. A forward motion of air (along the bucket travel direction) arises inside the enclosure of the carrying run when ejection forces prevail (at $\Delta \bar{p} < M_1$, $t_1 < 1$ and $\Delta \bar{p} < M_2$) and inside the return run enclosure at any ejection forces differential pressures (at $\Delta \bar{p} > -M_2$, $t_2 > -1$). A counterflow of air is only possible in a single enclosure: within the carrying run enclosure at $\Delta \bar{p} > M_1$ or within the return run enclosure at $\Delta \bar{p} < -M_2$. The other enclosure would experience a forward flow of air in this case.

5. Relative velocities and flow rates of air inside the elevator enclosures depend on two parameters, t and g (Equations 7.114 and 7.135), representing the ratio of differential pressures and resistances of enclosures to ejection forces. Single-valued variables φ_1 and φ_2 within a wide range of differential pressures ($-\infty < t_1 < \infty$; $-\infty < t_2 < \infty$) can be determined using formulas (given by Equations 7.146 through 7.148 and Equations 7.126 through 7.128).

When pressures inside the upper and lower elevator enclosures are equal, relative velocities reach their maxima determined by relations given by Equations 7.150 and 7.132. With ejection forces, large enough $\left(M_1 \gg \sum \zeta_1 \text{ and } M_2 \gg \sum \zeta_2 \right)$ air velocities become equal to the velocity of traveling elevator buckets.

Absolute velocities of airflows inside enclosures are dependent not only on the velocity of moving buckets but also on the differential pressure, the head resistance of elevator buckets, and the aerodynamic drag of enclosures, as well as spillage of particles.

6. In the case of a forward flow pattern, airflow rate inside the return run enclosure is greater than the one inside the carrying run enclosure of the elevator conveyor. The explanation is that ejection forces arise in an opposite direction to forces caused by differential pressure inside the carrying run enclosure (both forces act in the same direction inside the return run, thus intensifying the air ejection process and boosting additional ejection forces that occur when buckets are unloaded, producing streams of spilled particles), as well as different values of the drag coefficient for empty and laden buckets.

When air moves in a counterflow pattern, ejection forces of buckets create additional drag and, therefore, the absolute flow rate of ascending air inside the return run enclosure, as well as descending air inside the carrying run enclosure, increase less markedly than in the forward flow case (Tables 7.3 and 7.4).

8 Measures for Reducing Aspiration Volumes in Elevator Handling of Grain

8.1 AERODYNAMIC PERFORMANCE OF ELEVATOR BUCKETS

Obtaining data on the coefficient c_b is a prerequisite for determining the parameters M_1 and M_2 describing the behavior of air cross-flow patterns in elevator enclosures. Using a definition based on Equation 7.37, this coefficient represents the ratio of frontal drag force to the dynamic pressure of air multiplied by midsection area of a bucket traveling at a relative velocity w:

$$c_b = \frac{R}{F_b \dfrac{w^2 \rho}{2}}, \quad w = v_e - u. \tag{8.1}$$

In case of fixed buckets, airflow with the same velocity w within elevator enclosure with a cross-sectional size of S produces drag forces equal in their absolute value to the force R:

$$R = \Delta p_b \cdot S, \tag{8.2}$$

where Δp_b are pressure losses caused by the drag of the fixed end. In the design practice these are determined using the local resistance coefficient

$$\Delta p_b = \zeta_b \frac{w^2 \rho}{2}. \tag{8.3}$$

Substitution of Equations 8.2 and 8.3 into Equation 8.1 results in the ratio

$$c_b = \frac{S}{F_b} \zeta_b, \tag{8.4}$$

which can be used to determine c_b given local resistance coefficient (LRC) of an enclosure member with a fixed bucket is known. We will use experimental data on orifice drag [215] determining LRC as a function of orifice area and cross-sectional size of the duct S. In our case, the orifice area would be

$$S_o = S - A_b \cdot B_b = S - F_b. \tag{8.5}$$

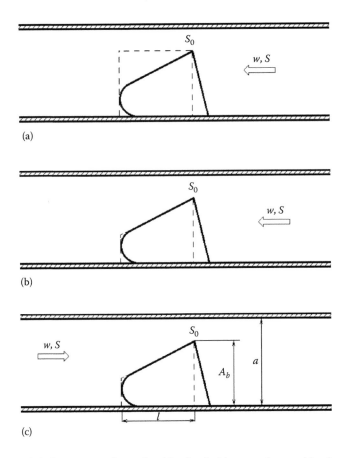

FIGURE 8.1   Chart of airflow surrounding a fixed bucket inside an enclosure: (a) scheme 1, (b) scheme 2, and (c) scheme 3.

Bucket dimensions in lengthwise cross section will be compared either with a rectangle (Figure 8.1a), a sharp-edged trapezoid (Figure 8.1b) if the surrounding airflow passes along the head side of the bucket, or a dull-edged shape if the surrounding airflow passes along the back of the bucket (Figure 8.1c). In the first case we will use the results for an orifice with beaded edges in a direct pipe [215, p. 138]:

$$\zeta_b = \left(\zeta_o + \lambda \frac{l}{D_h} \right) \left(\frac{S}{S_0} \right)^2, \tag{8.6}$$

where

$$\zeta_o = 0.5 \left(1 - \frac{S}{S_o} \right) + \left(1 - \frac{S_o}{S} \right)^2 + \tau \sqrt{1 - \frac{S_o}{S}} \left(1 - \frac{S_o}{S} \right) \tag{8.7}$$

D_h is hydraulic diameter of the orifice, equal in our case to

$$D_h = \frac{4S_o}{\Pi_o} = \frac{2S_o}{a + b + A_b} \tag{8.8}$$

λ is the coefficient of air friction against bucket walls ($\lambda \approx 0.02$)
τ is a correction coefficient that depends on bucket depth (Table 8.1)

TABLE 8.1

Correction Coefficient τ

l/D_h	0	0.2	0.4	0.6	0.8	1.0	1.4	2	≥ 3
τ	1.35	1.22	1.10	0.84	0.42	0.24	0.1	0.02	0

It is possible to use a simpler model of pressure losses inside an orifice with sharp edges [215, p. 139]:

$$\zeta_b = \bar{\zeta}_b = \left(1 + 0.707\sqrt{1 - \frac{S_o}{S}} - \frac{S_o}{S}\right)^2 \left(\frac{S}{S_o}\right)^2. \tag{8.9}$$

Let $A_k = 250$ mm, $B_b = 500$ mm, $l = 300$ mm, $a = 400$ mm, and $b = 700$ mm, then $F_b = 0.125$ m^2, $S = 0.28$ m^2, $S_0 = 0.28 - 0.125 = 0.155$ m^2, $D_h = \dfrac{2 \cdot 0.155}{0.4 + 0.7 + 0.25} = 0.23$ m, and $\dfrac{l}{D_h} = 0.3/0.23 = 1.3$; $\tau \approx 0.08$.

If these values were substituted into the formulas (Equations 8.7, 8.6, and 8.9), they would resolve into

$$\zeta_o = 0.947; \quad \zeta_b = 1.56; \quad \bar{\zeta}_b = 2.75,$$

that is, the formula given in Equation 8.9 would produce somewhat higher LRC values.

For a flow surrounding the back of a bucket we will use experimental data for an orifice with edges chamfered downstream [215, p. 141]:

$$\zeta_b = \vec{\zeta}_b = \left(1 + \sqrt{\xi\left(1 - \frac{S_o}{S}\right)} - \frac{S_o}{S}\right)^2 \left(\frac{S_o}{S}\right)^2, \tag{8.10}$$

where $\xi = f(l/D_h)$ is the adjustment coefficient (Table 8.2).

The following ratio will result if formulas from Equations 8.9 and 8.10 are compared:

$$\frac{\bar{\zeta}_b}{\vec{\zeta}} = \left(\frac{1 + 0.707\sqrt{\left(1 - \frac{S_o}{S}\right)} - \frac{S_o}{S}}{1 + \sqrt{\xi}\sqrt{1 - \frac{S_o}{S}} - \frac{S_o}{S}}\right)^2. \tag{8.11}$$

As charts (Figure 8.2) illustrate, the aerodynamic drag of a bucket increases noticeably when airflow impinges on the receiving orifice. The deeper the bucket, the more noticeable is the difference. It is unfortunate that the formula in Equation 8.10 is only applicable at $l/D_h \leq 1.6$ whereas l/D_h is much greater in grain elevators ($l/D_h > 0.5$).

TABLE 8.2

Correction Coefficient ξ

l/D_b	0.01	0.02	0.03	0.04	0.06	0.08	0.12	0.16
ξ	0.46	0.42	0.38	0.35	0.29	0.23	0.16	0.13

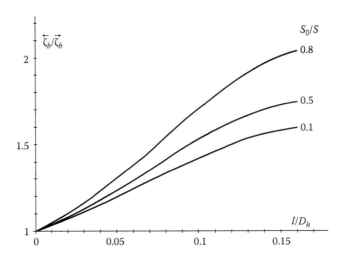

FIGURE 8.2 Variation in the LRC of elevator bucket as a function of airflow and depth (according to the data provided by Prof. Dr. I.E. Idelchik).

Let us now compare the results in the particular case of thin partitions with the resistance of a gate (damper) in a straight pipe [215]. For a rectangular duct ($a/b = 0.5$), the LRC of a single-side gate depends on the relative gap height h/a. Listed in Table 8.3 (assuming identical $h/a \equiv S_o/S$) are LRCs of gate (ζ_g), LRCs of a sharp-edged orifice ($\overleftarrow{\zeta}_b$) computed using the formula given in Equation 8.9, LRCs of an orifice with beaded edges (ζ_b) computed using the formula given in Equation 8.6 at $l/D_h = 0$ ($\tau = 1.35$), and LRCs of an orifice chamfered downstream ($\overrightarrow{\zeta}_b$) computed using the formula given in Equation 8.10 at $l/D_h = 0.01$ ($\xi = 0.46$).

It can be seen from these data that LRCs of thin orifices are, for all practical purposes, in an adequate agreement within a wide range of relative orifice areas (within $0.1 \leq S_o/S \leq 0.9$). However, the LRC of an orifice significantly exceeds the LRC of a gate, especially within the range of small orifices (at $S_o/S < 0.2$), despite the two being quite similar structurally.

To examine the effect that elongation of a body has on its frontal drag coefficient and resistance to airflow, it would be instructive to consider the aerodynamic resistance of classic bodies: the hemisphere and the cylinder. At $Re = wd_p/v = 4 \cdot 10^5$, boundless airflow around a convex hemisphere of a cup (so that the cup opens leeward) would encounter a drag coefficient of $\vec{c}_p = 0.36$ [215, p. 402]. With the same Reynolds number but with oppositely directed airflow (the cup opens windward), this coefficient quadruples ($\overleftarrow{c}_p = 1.44$).

For an airflow around a smooth circular cylinder parallel to its generatrix (i.e., perpendicular to the impermeable cylinder base), the drag coefficient first declines with increasing cylinder length but then increases. A cylinder with a length equal to double its base diameter would have a minimal drag coefficient $c_p = 0.85$ (for a thin disk-like cylinder, $c_d = 1$, same in the case of cylinder length equal to seven times its diameter).

TABLE 8.3

LRCs

h/a	0.1	0.15	0.20	0.30	0.40	0.50	0.60	0.70	0.80	0.90
ζ_g	105	51.5	30.6	13.5	6.85	3.34	1.73	0.83	0.32	0.09
$\overleftarrow{\zeta}_b$	246.7	100	51.3	18.5	8.23	4.0	2.0	0.96	0.42	0.13
ζ_b	241.3	98	50	18.1	8.04	3.91	1.95	0.94	0.41	0.13
$\overrightarrow{\zeta}_b$	238.2	96.7	49.5	17.8	7.92	3.84	1.91	0.92	0.40	0.12

So, despite the apparent structural similarity of partitions surveyed, LRC values differ sharply in the case of beaded (lengthy) partitions. Therefore, great care is needed when using these findings to quantitatively estimate the drag of various bucket designs as they may be filled up with grain to a different extent. It would be much more reliable to investigate individual cases experimentally.

8.2 FEATURES OF THE PROPOSED DESIGN FOR ASPIRATION LAYOUT OF GRAIN ELEVATORS

Dust releases during handling of unheated grain are usually contained by sucking air from bottom cowls (Figure 8.3):

- From the cowl of elevator loading location (from the elevator *boot* enclosure).
- From the cowl of conveyor loading location (from the cowl of the boot of elevator discharge chute).

Let us assume that, in the case of unaspirated cowls, a sufficient negative pressure (preventing escape of dust-laden air through poorly sealed locations in cowls) is maintained inside the cowl of the bottom conveyor driving drum (feeding mechanism) and inside the enclosure of the elevator driving drum (the *head* of the elevator) by cross flows of air through adjacent chutes and elevator enclosures.

The air cross-flow pattern through these ducts is determined by the following combined equations for the dynamics of air in ducts:

$$P_s - P_b - P_2(Q_2) = R_2 Q_2^2, \tag{8.12}$$

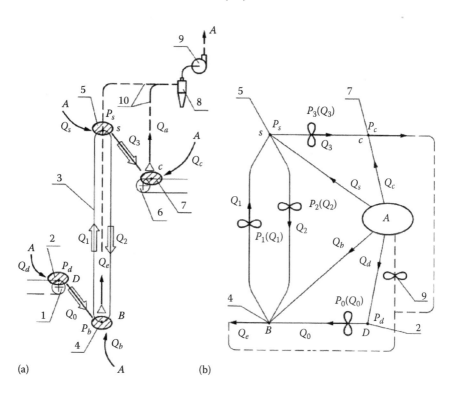

FIGURE 8.3 (a) Aspiration design and (b) its aerodynamic equivalent: 1, belt-conveyor feeder; 2, cowl of the driving drum of belt-conveyor feeder; 3, (bucket) elevator; 4, cowl of elevator boot; 5, upper cowl of bucket elevator (cowl of the drive sprocket of elevator); 6, belt conveyor; 7, cowl of the filling location; 8, dust trap (cyclone); 9, aspiration system fan; 10, air ducts.

$$P_b - P_s + P_1(Q_1) = R_1 Q_1^2, \tag{8.13}$$

$$P_s - P_c + P_3(Q_3) = R_3 Q_3^2, \tag{8.14}$$

$$P_d - P_b + P_0(Q_0) = R_0 Q_0^2, \tag{8.15}$$

$$P_A - P_c = R_c Q_c^2, \tag{8.16}$$

$$P_A - P_s = R_s Q_s^2, \tag{8.17}$$

$$P_A - P_b = R_b Q_b^2, \tag{8.18}$$

$$P_A - P_d = R_d Q_d^2 \tag{8.19}$$

and balance of air in junction points

$$Q_e + Q_1 - Q_2 - Q_o - Q_b = 0, \tag{8.20}$$

$$Q_3 + Q_2 - Q_1 - Q_s = 0, \tag{8.21}$$

$$Q_a - Q_3 - Q_c = 0, \tag{8.22}$$

$$Q_0 - Q_d = 0, \tag{8.23}$$

$$Q_s + Q_c + Q_d + Q_b - Q_a - Q_e = 0, \tag{8.24}$$

where

P_s, P_b, P_d, P_c are the respective absolute pressures within the cowls of elevator head and boot and upper and lower conveyor feeder driving drums (Pa)

Q_0, Q_1, Q_2, Q_3 are the respective flow rates of air arriving through the loading chute, enclosures of carrying and return runs of elevator conveyor, and the discharge chute (m³/s)

Q_d, Q_b, Q_s, Q_c are flow rates of air coming in through leaky areas in cowls of the feeder drive drum, elevator boot and head, and upper conveyor loading location (m³/s)

Q_e, Q_a are the respective flow rates of air evacuated by the suction system from the elevator boot cowl and the location of grain dumping onto upper conveyor (m³/s)

R_0, R_1, R_2, R_3 are the respective aerodynamic properties of the loading chute, enclosures of the carrying and return runs of elevator conveyor, and the discharge chute (Pa/(m³/s)²) that determine resistance posed by ducts to cross flows of air:

$$R_0 = \zeta_0 \frac{\rho}{2 S_0^2}, \tag{8.25}$$

$$R_1 = \sum \zeta_1 \frac{\rho}{2 S_1^2}, \tag{8.26}$$

$$R_2 = \sum \zeta_2 \frac{\rho}{2S_2^2},$$ (8.27)

$$R_3 = \zeta_3 \frac{\rho}{2S_3^2},$$ (8.28)

where

$\zeta_0, \sum \zeta_1, \sum \zeta_2, \zeta_3$ are the respective sum totals of LRCs for the loading chute, enclosures of carrying and return runs of elevator conveyor, and discharge chute

S_0, S_1, S_2, S_3 are the respective cross-sectional areas of the loading chute, enclosures of carrying and return runs of elevator conveyor, and discharge chute (m²)

ρ is the air density (kg/m³)

P_A is the atmospheric pressure (absolute pressure in the room) (Pa)

R_d, R_b, R_s, R_c are the respective aerodynamic properties of leaky areas in the feeder drive drum cowl, elevator boot and head cowls, and upper conveyor loading location cowl (Pa/(m³/s)²), which determine the resistance to ingress of air into cowls through leaky areas

$$R_d = \zeta_d \frac{\rho}{2F_d^2},$$ (8.29)

$$R_b = \zeta_b \frac{\rho}{2F_b^2},$$ (8.30)

$$R_s = \zeta_s \frac{\rho}{2F_s^2},$$ (8.31)

$$R_c = \zeta_c \frac{\rho}{2F_c^2}.$$ (8.32)

where

$\zeta_d, \zeta_b, \zeta_s, \zeta_c$ are the LRCs of leakage areas in the respective cowls, accepted equal to the LRC of a hole in a thin wall [215]:

$$\zeta_d = \zeta_b = \zeta_s = \zeta_c = 2.4$$ (8.33)

F_d, F_b, F_s, F_c are the respective leakage areas of the feeder cowl, elevator boot and head cowls, and upper conveyor cowl (m²)

$P_0(Q_0), P_1(Q_1), P_2(Q_2), P_3(Q_3)$ are the ejection pressures as functions of airflows in the loading chute, inside enclosures of the carrying and return runs of elevator conveyor, and in the discharge chute, respectively (Pa), determined with the relations of Equations 7.32 and 7.48

While combined Equations 8.12 through 8.15 describe the cross flow of air through ducts, the second set of equations that include the equalities (Equations 8.16 through 8.19) determine the flow rates of air entering through leakage areas of the respective cowls due to differential pressure. Combined, relations of Equations 8.20 through 8.24 comprise balance equations of the respective airflow rates: in boot (Equation 8.20) and head (Equation 8.21) cowls of the elevator, in the conveyor cowl (Equation 8.22), in the feeder cowl (Equation 8.23) and, finally, inside an imagined junction point (Equation 8.24) for airflows in the atmosphere.

The latter equation can be transformed into the relation

$$Q_e + Q_a = Q_d + Q_s + Q_c, \tag{8.34}$$

describing an obvious but, nevertheless, a rather important fact: the total performance of an aspiration unit is determined by the sum total of flow rates of air entering the cowl system through leaky areas. In addition, in order to reduce power consumption of the aspiration system and to bring down dust releases, it is necessary to seal not only cowls provided with local suction units for evacuating air but also to adequately seal aspirated cowls.

For the purpose of arriving at computational relations for determining flow rates of air in ducts, we will put forth, based on Equations 8.21 and 8.23, that

$$Q_s = Q_3 + Q_2 - Q_1 = Q_3 + \Delta Q; \quad \Delta Q = Q_2 - Q_1, \tag{8.35}$$

$$Q_d = Q_0. \tag{8.36}$$

Keeping in mind that differential pressures in left-hand sides of Equations 8.16 through 8.19 represent negative pressures in the respective cowls, we will rewrite the second set of combined equations as follows:

$$h_d = R_d Q_d^2 = R_d Q_0^2, \tag{8.37}$$

$$h_b = R_b Q_b^2, \tag{8.38}$$

$$h_s = R_s Q_s^2 = R_s \left(Q_3 + \Delta Q \right)^2, \tag{8.39}$$

$$h_c = R_c Q_c^2, \tag{8.40}$$

where
  h_d, h_s are negative pressures established within the unaspirated cowls of elevator feeder due to cross flows of air through adjacent ducts (Pa)
  h_b, h_c are negative pressures maintained in the aspirated cowls of elevator boot and upper conveyor loading location by a running fan (Pa)

The first set of combined Equations 8.12 through 8.15 will be rewritten in view of these relations as follows:

$$h_b + P_0 \left(Q_0 \right) = \left(R_d + R_0 \right) Q_0^2, \tag{8.41}$$

$$P_1 \left(Q_1 \right) + P_s Q_s^2 - h_b = R_1 Q_1^2, \tag{8.42}$$

$$P_2 \left(Q_2 \right) - R_s Q_s^2 + h_b = R_2 Q_2^2, \tag{8.43}$$

$$P_3 \left(Q_3 \right) + h_c - R_s Q_s^2 = R_3 Q_3^2. \tag{8.44}$$

By assuming h_b and h_c to be predefined and solving these combined equations in view of Equation 8.35, we will determine the sought flow rates of air in ducts Q_0, Q_1, Q_2, and Q_3.

In order to combine equations for computation purposes, we will explicitly present the formulas for determining ejection pressures in ducts. According to the formulas from Equations 7.48 and 7.49,* (in view of the accepted conventions of Equations 7.78 and 7.79), the result for elevator enclosures is

$$P_1(Q_1) = E_1(L_1 - Q_1)^2, \tag{8.45}$$

$$P_2(Q_2) = E_2(L_2 - Q_2)^2, \tag{8.46}$$

with the following notational simplification:

$$L_1 = v_e S_1, \tag{8.47}$$

$$L_2 = v_e S_2, \tag{8.48}$$

$$E_1 = M_1 \frac{\rho}{2S_1^2}, \tag{8.49}$$

$$E_2 = M_2 \frac{\rho}{2S_2^2}. \tag{8.50}$$

For loading and discharge chutes, formulas of Equations 7.32 and 7.33 result in

$$P_0(Q_0) = E_0 \left[|L_{k0} - Q_0|^3 - |L_{n0} - Q_0|^3 \right], \tag{8.51}$$

$$P_3(Q_3) = E_3 \left[|L_{k3} - Q_3|^3 - |L_{n3} - Q_3|^3 \right], \tag{8.52}$$

$$L_{ki} = v_{ki} S_i; \quad L_{ni} = v_{ni} S_i, \tag{8.53}$$

$$E_i = K_0 \psi_{yi} \beta_{ki} \frac{v_{ki} \rho}{4 d_e a_{Ti} S_i^3}, \tag{8.54}$$

$$\beta_{ki} = \frac{G_m}{\rho_m S_i v_{ki}}. \tag{8.55}$$

Here a subscript $i = 0$ is used to refer to the parameters of a loose-matter flow in the loading chute, and $i = 3$ refers to the flow in the discharge chute.

After simple algebraic transformations in view of the relations of Equations 8.45, 8.46, 8.51, and 8.52, combined Equations 8.41 through 8.44 can be presented as the following computational set:

$$E_0 \left[|L_{k0} - Q_0|^3 - |L_{n0} - Q_0|^3 \right] + h_b = (R_d + R_0) Q_0^2, \tag{8.56}$$

---

* Two-stage numbering will be used from now on to refer to formulas listed in other sections.

$$E_1\left(L_1-Q_1\right)^2+E_2\left(L_2-Q_2\right)^2=R_1Q_1^2+R_2Q_2^2, \tag{8.57}$$

$$E_2\left(L_2-Q_2\right)^2+h_b-R_s\left(Q_3+Q_2-Q_1\right)^2=R_2Q_2^2, \tag{8.58}$$

$$E_3\left[\left|L_{k3}-Q_3\right|^3-\left|L_{n3}-Q_3\right|^3\right]+E_1\left(L_1-Q_1\right)^2+h_c-h_b=R_1Q_1^2+R_3Q_3^2. \tag{8.59}$$

However, the first equation of this set is independent of others and can be formally solved as follows:

$$Q_0=\sqrt{\frac{E_0\left[\left|L_{k0}-Q_0\right|^3-\left|L_{n0}-Q_0\right|^3\right]+h_b}{R_d+R_0}}.$$

The last three equations are dependent and due to their nonlinearity it would be difficult to solve combined Equations 8.57 through 8.59 in a general form.

Formally, Equation 8.57 can be used to determine

$$Q_2=f_1\left(Q_1\right), \tag{8.60}$$

a substitution into Equation 8.58 enables Q_3 to be expressed using a new function

$$Q_3=f_2\left(Q_1,Q_2\right)=f_2\left(Q_1,f_1\left(Q_1\right)\right)=f_2\left(Q_1\right), \tag{8.61}$$

that, after being substituted into Equation 8.59, would result in an equation for determining Q_1:

$$E_3\left[\left|L_{k3}-f_2\left(Q_1\right)\right|^3-\left|L_{n3}-f_2\left(Q_1\right)\right|^3\right]+E_1\left(L_1-Q_1\right)^2+h_k-h_b=R_1Q_1^2+R_3\left(f_2\left(Q_1\right)\right)^2. \tag{8.62}$$

Equation 8.62 can be solved numerically, for example, using bisection (dichotomy). Nevertheless, the well-known difficulties of choosing an unambiguous branch of dependencies Q_1 and Q_2 from the difference in negative pressures h_s-h_b should not be dismissed. In particular, square values in input Equations 8.56 through 8.59 should be written as

$$R_1Q_1^2=R_1Q_1\left|Q_1\right|; \quad R_2Q_2^2=R_2Q_2\left|Q_2\right|, \tag{8.63}$$

$$E_1\left(L_1-Q_1\right)^2=E_1\left(L_1-Q_1\right)\left|L_1-Q_1\right|, \tag{8.64}$$

etc.

In order to avoid losing necessary real roots, combined Equations 8.57 through 8.59 will be solved by choosing a negative pressure inside the enclosure of elevator *head h* ($h\equiv h_s$) and comparing this value with the true negative pressure s:

$$s=R_s\left(Q_3+Q_2-Q_1\right)\left|Q_3+Q_2-Q_1\right|. \tag{8.65}$$

An iterative approach was used to solve the combined equations in shorter time:

$$h_i = 0.5(h_{i-1} + s_{i-1}), \tag{8.66}$$

where s_{i-1} is the real negative pressure at determined values of $Q_1 = f_1(h_{i-1})$, $Q_2 = f_2(h_{i-1})$, and $Q_3 = f_3(h_{i-1})$.

Values of Q_1 and Q_2 have been determined using the relations in Equations 7.148 through 7.150 and Equations 7.128 through 7.130 at

$$g_1 = \frac{R_1}{E_1}; \quad g_2 = \frac{R_2}{E_2}, \tag{8.67}$$

$$t_1 = \frac{h_b - h}{E_1 L_1^2}; \quad t_2 = \frac{h_b - h}{E_2 L_2^2}. \tag{8.68}$$

At the same time, the unknown flow rates were determined using the formulas

$$Q_1 = \varphi_1 v_e S_1 = \varphi_1 L_1, \tag{8.69}$$

$$Q_2 = \varphi_2 v_e S_2 = \varphi_2 L_2. \tag{8.70}$$

The values Q_0 and Q_3 were determined by solving equations using the bisection method:

$$f_0 = E_0 \left[|L_{k0} - Q_0|^3 - |L_{n0} - Q_0|^3 \right] + h_b - R_0 Q_0 |Q_0| = 0, \tag{8.71}$$

$$f_3 = E_3 \left[|L_{k3} - Q_3|^3 - |L_{n3} - Q_3|^3 \right] + h_c - h - R_3 Q_3 |Q_3| = 0. \tag{8.72}$$

8.3 ASPIRATION VOLUMES

Flows rate of air evacuated from cowls by local suction units of the aspiration system are determined using the air balance equation. So, for aspirated cowls for elevator handing of grain, the following equations result:

For the cowl of the elevator *boot*, due to Equation 8.20:

$$Q_e = Q_0 + Q_2 + Q_b - Q_1 = Q_0 + \Delta Q + Q_b, \tag{8.73}$$

For the cowl of the location of grain loading from the elevator onto the upper conveyor, due to Equation 8.22:

$$Q_a = Q_3 + Q_c. \tag{8.74}$$

Flow rates of air entering the cowl through leaky locations are determined using the value of negative pressure maintained by a local suction unit and the total leakage area.

For the cowl of the elevator *boot*, the recommended target negative pressure is $h_b = 10 \div 30$ Pa, then, due to Equations 8.18, 8.30, and 8.33, the flow rate of air arriving through leakage areas would be

$$Q_b = \sqrt{h_b / R_b} = 0.913 F_b \sqrt{h_b / \rho}. \tag{8.75}$$

Negative pressure for the upper cowl is determined by the structural design of the cowl. For a grain measuring $d_e > 3$ mm, it equals to $h_b = 6$ for a double-walled cowl or $h_b = 11$ for a single-walled cowl.

The flow rate of air entering the cowl through leakage areas is equal, in view of Equations 8.16, 8.32, and 8.33, to

$$Q_c = \sqrt{h_c/R_c} = 0.913 F_c \sqrt{h_c/\rho}. \tag{8.76}$$

The flow rate of air passing through the loading chute is determined by Equation 8.56. If both sides of this equation are divided by L_{k0}, the following dimensionless equation for the cross flow of air will result:

$$\varphi_0^2 = \frac{\mathrm{Bu}}{3}\left[\left|1-\varphi_0\right|^3 - \left|n-\varphi_0\right|^3\right] + \mathrm{Eu}_0, \tag{8.77}$$

where

$$\varphi_0 = \frac{Q_0}{L_{k0}} = \frac{u_0}{v_{k0}}; \quad n = \frac{v_{n0}}{v_{k0}} \tag{8.78}$$

u_0 is the velocity of air in loading chute (m/s)
v_{n0}, v_{k0} are the velocities of a grain stream at the inlet/outlet of the loading chute (m/s)

Bu is the Butakov–Neykov number [213] that in this case is equal to

$$\mathrm{Bu}_0 = 3 \cdot E_0 \frac{L_{k0}}{R_d + R_0} = K_0 \psi_{y0} \frac{1.5}{d_e} \cdot \frac{G_m v_{k0}}{\rho_m S_0 a_{t0}} \sum \zeta_0 \tag{8.79}$$

$$\sum \zeta_0 = \zeta_0 + \zeta_n \left(\frac{S_0}{F_d}\right)^2 \tag{8.80}$$

Eu_0 is the Euler's number, in this case equal to

$$\mathrm{Eu}_0 = \frac{h_b}{(R_d + R_0)L_{k0}^2} = \frac{2h_b}{\sum \zeta_0 v_{k0}^2 \rho} \tag{8.81}$$

A similar criterial equation can be proposed as well for the discharge chute:

$$\varphi_g^2 = \frac{\mathrm{Bu}_g}{3}\left[\left|1-\varphi_g\right|^3 - \left|n-\varphi_g\right|^3\right] + \mathrm{Eu}_g - \frac{h_s}{0.5\zeta_g v_{kg}^2 \rho}, \tag{8.82}$$

$$\mathrm{Bu}_g = K_0 \psi_{yg} \frac{1.5}{d_e} \cdot \frac{G_m v_{kg}}{\rho_m S_g a_{tg} \zeta_g}, \tag{8.83}$$

$$\mathrm{Eu}_g = \frac{h_c}{0.5\zeta_g v_{kg}^2 \rho}. \tag{8.84}$$

However, it would be impossible to determine the respective flow rate as the negative pressure inside the unaspirated cowl of the elevator *head* is unknown.

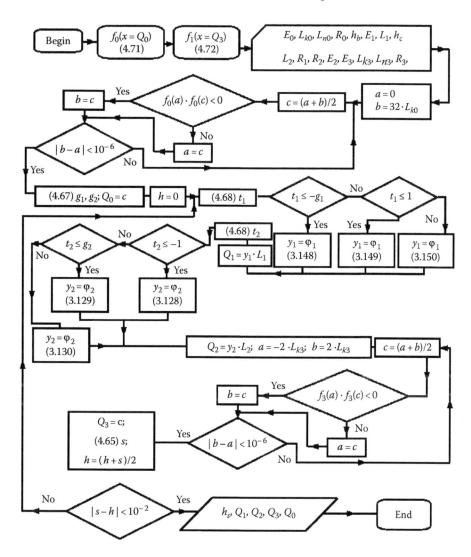

FIGURE 8.4 Flowchart of the algorithm for computing airflow rates (Q_0,Q_1,Q_2,Q_3) in the ducts of grain-handling elevators without bypass connections.

Therefore, air cross-flow rates will be determined iteratively using the procedure given by Equation 8.67. The algorithm for the aspiration volume computation program (Figure 8.4) includes determining Q_0 and Q_g using the bisection method and computing flow rates Q_1 and Q_2 using the formulas given in Equations 8.69 and 8.70.

As an example, consider elevator handling of wheat with the following parameters for the loading chute: $E_0 = 4$, $L_{k0} = 1.6$, $L_{n0} = 1$, h_b–var ($h_b = 16.8$ for charts plotted in Figure 8.6), $R_0 = 50$; for enclosures of the carrying and return runs of elevator conveyor: $E_1 = 200$, $R_1 = 21$, $L_1 = L_2 = 1$, $E_2 = 400$, h_s–var, $R_2 = 22$; for the discharge chute: $E_g = 5$, $L_{kg} = 2$, $L_{ng} = 0.5$, R_s–var ($R_s = 40$ for charts plotted on Figure 8.5), $R_3 = 10$, $h_k = 10$.

Findings from the computations of air cross-flow rates are summarized in Figures 8.5 and 8.6. As this data indicates, increasing negative pressures inside the aspirated cowl of elevator boot (Figure 8.5) lead to increased flow rates of ejected air in the loading chute Q_0 and a greater flow rate of air in the return run enclosure of the conveyor. At the same time, negative pressure within elevator head cowl h_s increases. Only the flow rates of air in discharge chute Q_3 and the enclosure of

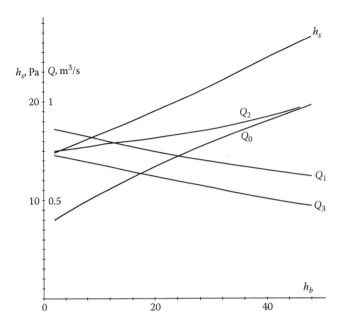

FIGURE 8.5   Variation in airflow rates (Q_0, Q_1, Q_2, Q_3) in ducts and negative pressure (h_s) within the cowl of elevator head as a function of negative pressure in the elevator boot cowl.

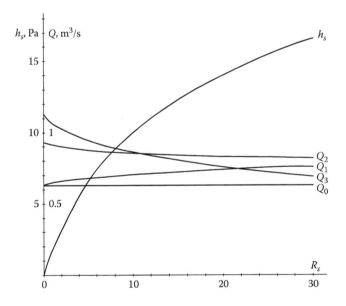

FIGURE 8.6   Variation of airflow rate (Q_0, Q_1, Q_2, Q_3) in ducts and negative pressure in the elevator head enclosure as a function of sealing degree.

conveyor carrying run Q_1 decrease. This happens because increasing negative pressure h_b reduces the ejecting capacity of the grain stream inside the discharge chute (as a direct consequence of increasing h_s) and of laden buckets inside the carrying run enclosure of the bucket elevator.

The value of h_s increases especially strongly with tighter sealing of the elevator head cowl (Figure 8.6). This reduces the flow rates Q_3, Q_2 and increases the flow rate of ejected air inside the enclosure of the elevator carrying run Q_1.

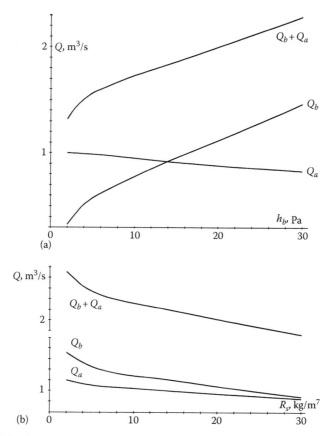

FIGURE 8.7 Variation in aspiration volumes (a) as a function of negative pressure inside the elevator boot cowl at $R_s = 40$, $h_c = 10$ Pa and (b) as a function of the elevator head cowl sealing degree at $h_c = 16.8$ Pa.

As for aspiration volumes (Figure 8.7), they naturally reflect the regularities of air cross-flow patterns in elevator ducts noted earlier. Thus, aspiration volumes for the cowl of elevator boot Q_b rise with increasing negative pressure h_b.

This occurs not only due to increasing Q_0 but also as a result of greater ejecting capacity of buckets within the return run enclosure of bucket elevator as well as an increased flow rate of air arriving into the cowl in question through leakage areas as negative pressure h_b increases (Figure 8.7a). With increasing negative pressure h_s, as a result of decreasing leakage area of the elevator cowl (at higher R_s), aspiration volumes Q_b and Q_a decline somewhat (Figure 8.7b). Aspiration volume Q_a decreases as well with increasing negative pressure in the lower cowl h_b.

The total aspiration volume in the first case (with increasing h_b) grows due to increasing volumes of air sucked through the leakage areas of elevator boot cowl. If the upper cowl is sealed, the total performance of local suction decreases (Figure 8.7b). Despite that, the required aspiration volume remains high enough even at small negative pressures maintained inside cowls (at $h_b = 16.8$ Pa, $h_c = 10$ Pa, and $F_s = 0.1$ m^2 ($R_s = 14.4$ Pa/(m^3/s)^2). The total volume in the example case at hand is 2.06 m^3/s).

8.4 REDUCING REQUIRED ASPIRATION VOLUMES IN BYPASSED ELEVATOR HANDLING SYSTEMS

The performance of an aspiration system is determined by the total flow rate of air entering through all aspirated cowls as well as unaspirated cowls that are aerodynamically coupled with the former (due to Equation 2.4). Reducing aspiration volumes requires not only sealing the cowl but also decreasing the negative pressures h_b and h_c.

It is well known [216] that the use of double-walled cowls enables a significant reduction in the optimum negative pressure (at $d_3 \geq 3\text{mm}$ the optimum negative pressure inside a single cowl of conveyor loading is $h_c = 11$ Pa, decreasing to $h_c = 6$ Pa in the case of a rigidly partitioned double-walled cowl).

Unaspirated cowls are purposefully joined by flow-around (bypass) ducts with a positive-pressure zone to inhibit negative pressures. The resulting internal circulation of air would reduce the total volume of air evacuated by suction.

Let us now study the role of these bypass connections by choosing the aspiration of elevator loading and unloading units as an example. Let the bottom part of the loading chute be equipped with a special (buffer) chamber that has its internal space connected by flow-around ducts with the cowl of feeder drive drum and the enclosure of elevator head. Grain being handled proceeds from the chamber into the elevator boot enclosure via an orifice of a flapper-type dust trap. In this case aerodynamic drag of the valve produces an excess pressure inside the chamber that forces a part of ejected air to proceed into the flow-around duct and to return into the upper cowls, creating internal circulation flows with flow rates Q_4 and Q_6.

The cowl of elevator head enclosure can also be connected using a bypass duct with the inner chamber of the doubled-walled cowl at upper conveyor loading location (a dust trap in this case is provided by a rigid partition).

Thus, three circulation flows may exist with flow rates of Q_4, Q_6, and Q_7 (see Figure 8.6).

We will begin with considering a classic grain-handling facility transferring grain from a belt feeder onto a belt conveyor (Figure 8.8).

When a double-walled cowl is used (to contain dust releases when loading grain onto belt conveyor), a bypass duct connects the inner chamber of this cowl with an unaspirated cowl of the feeder driving drum. A positive effect is provided with the airflow direction indicated in the chart. In this case

$$Q_y = Q_c - Q_{bc} = Q_f, \tag{8.85}$$

that is, the flow rate of air coming from the inner chamber into unaspirated (outer) chamber Q_y is less than the flow rate of air Q_c flowing over through a chute due to ejection head $P_e(Q_c)$ and negative pressure in the outer chamber h_y caused by a running fan of the aspiration system.

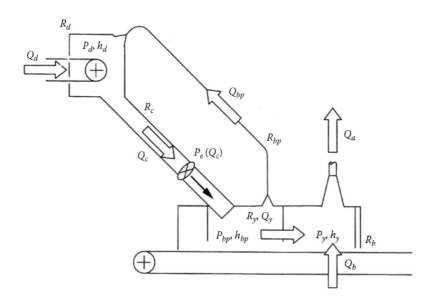

FIGURE 8.8   Aspiration layout of a bypassed handling facility.

Let us write down the obvious equations of pressure losses in ducts,

$$P_d + P_e(Q_c) - P_{bc} = R_c Q_c^2 = h_{bc} + P_e(Q_c) - h_d, \tag{8.86}$$

$$P_{bc} - P_d = P_{bc} Q_{bp} |Q_{bp}| = h_d - h_{bc}, \tag{8.87}$$

$$P_{bc} - P_y = R_y Q_y^2 = h_y - h_{bc}, \tag{8.88}$$

$$P_a - P_y = R_d Q_d^2 = h_y, \tag{8.89}$$

$$P_a - P_d = R_d Q_d^2 = R_d Q_y^2 = h_d \tag{8.90}$$

as well as air balance equations,

$$Q_d = Q_c - Q_{bp}, \tag{8.91}$$

$$Q_c = Q_{bp} + Q_y, \tag{8.92}$$

$$Q_a = Q_y + Q_b. \tag{8.93}$$

From Equation 8.86 it follows that

$$P_e(Q_c) - R_c Q_c^2 = h_d - h_{bc} \tag{8.94}$$

where it is evident (considering that the ejection pressure in the case at hand is greater than the pressure losses in the chute) that the negative pressure in the feeder cowl must be greater in magnitude than the negative pressure in the buffer (inner) chamber:

$$h_d \geq h_{bc}. \tag{8.95}$$

This inequality, the driver of cost savings, can be written in an expanded form accounting for Equations 8.90 and 8.88:

$$R_d Q_y^2 \geq h_y - R_y Q_y^2. \tag{8.96}$$

This enables us to determine the minimum flow rate Q_y:

$$Q_y^{min} = \sqrt{\frac{h_y}{R_d + R_y}}. \tag{8.97}$$

The expanded condition of efficient bypass operation Equation 8.96 can be presented in another form:

$$R_y \geq \left(\frac{h_y}{h_d} - 1\right) R_d, \tag{8.98}$$

where the role of the buffer chamber is evident. So, absent $(R_y = 0)$, the condition (Equation 8.98) will appear as

$$h_d \geq h_y, \tag{8.99}$$

which would be troublesome to accomplish in actual conditions (considering the difficulties of sealing the feeder cowl). Thus, the buffer chamber increases the likelihood of efficient bypass operation.

How precisely is the flow rate Q_y determined by process and structural design parameters of the chute, cowls, and the bypass duct?

Let us designate

$$P_e\left(Q_c\right) - R_c Q_c \left| Q_c \right| = L\left(Q_c\right). \tag{8.100}$$

Then, a combined solution of Equations 8.86 and 8.87 gives

$$Q_{bp} = \frac{L\left(Q_c\right)}{\sqrt{\left|L\left(Q_c\right)\right| R_{bp}}}, \tag{8.101}$$

while Equations 8.87, 8.88, and 8.90 combine into

$$\left(R_d + R_y\right) Q_y^2 = h_y + R_{bp} Q_{bp} \left| Q_{bp} \right|, \tag{8.102}$$

whence

$$Q_y = \frac{h_y + R_{bp} Q_{bp} \left| Q_{bp} \right|}{\sqrt{\left| h_y + R_{bp} Q_{bp} \left| Q_{bp} \right| \right| \left(R_d + R_y\right)}}. \tag{8.103}$$

Substitution the resulting expressions for $Q_f(Q_c)$ and $Q_y\left(Q_f\left(Q_c\right)\right)$ into Equation 8.92 produces the following functional equation for determining Q_c:

$$F\left(Q_c\right) = Q_c - Q_{bp}\left(Q_c\right) - Q_y\left(Q_c\right) = 0. \tag{8.104}$$

Deriving the root of the equation and substituting its value into Equations 8.100, 8.101, and 8.103 enables us to determine the unknown variables Q_c, Q_{bp}, and Q_y.

Consider a handling assembly with the following parameters as an example. Ejection head defined in the form of Equation 8.51 has the following parameters: $E = 4$, $L_c = 1.6$, $L_d = 1$. The aerodynamic performance of the chute is $R_e = 25$.

We will consider two cases: good sealing of the upper cowl $R_d = 25$ and mediocre sealing $R_d = 5$. We will be measuring the negative pressure maintained in the lower cowl ($h_y = 5$, 10 and 20 Pa) as well as aerodynamic properties of the buffer chamber/partition within the range $R_y = 0$–300 and of the bypass duct within the range $R_{bp} = 0$–150 (a case of no bypass, i.e., $R_{bp} = 15 \cdot 10^9$ will be included for comparison). A flowchart of the computation algorithm is shown in Figure 8.9.

Without bypass, the flow rate of air arriving from the inner (buffer) chamber into the outer chamber ($Q_{y\infty}$) is significantly higher than with a bypass duct installed (Figure 8.10). Particularly, increasing the drag of cowl partition and the cross section of bypass duct ($R_{bp} < 20$) cause the flow rate Q_y to deviate even further from $Q_{y\infty}$ (Q_y is the flow rate of air entering the outer chamber with a bypass connection present).

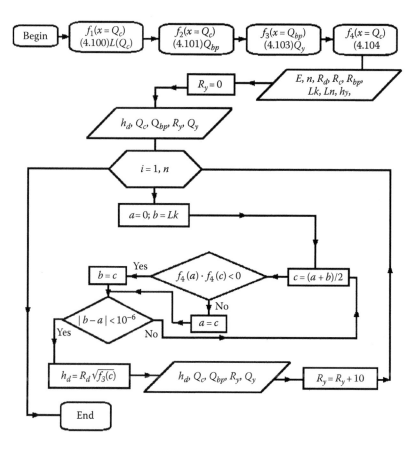

FIGURE 8.9 Flowchart of a software algorithm for computing airflow rates (Q_c, Q_{bp}, Q_y) in conveyor handling of grain.

Variation of airflow rates in ducts and negative pressure in the unaspirated cowl of the feeder driving drum is indicated in Figures 8.11 through 8.15.

Curves Q_c, Q_{bp}, Q_y have been plotted using Equations 8.101, 8.103, and 8.104, while negative pressure has been plotted according to the formula, Equation 8.90. As these plots illustrate, the flow rate of air entering through the chute (Q_c) decreases with increasing aerodynamic performance index of the buffer chamber (Figures 8.11 and 8.12) and the bypass duct (Figures 8.13 and 8.14).

With regard to airflow in the latter (Q_{bp}), it should be noted that the flow rate increases with increasing partition drag and declines with narrowing cross section of the bypass duct. In addition, in the case of mediocre sealing of the feeder cowl, airflow direction in the bypass duct reverses at a low drag of the buffer chamber ($R_y < 20$).

This zone expands to $R_y = 40$–90 with negative pressure inside the conveyer enclosure increasing to 20 Pa (Figure 8.15). This can be explained by diminishing pressure difference at the ends of the bypass duct with increasing gap between cowl partition and conveyor belt. In this case, the flow rate of air entering the outer cowl chamber rises above the flow rate of air in the chute ($Q_y > Q_c$).

The bypass duct works like a channel allowing an additional volume of air in along with the main flow of ejected air. Improved feeder sealing is needed for this situation to change. As Figure 8.12 illustrates, the airflow rate Q_y will be less than Q_c even if less drag is produced by the enclosure partition.

This flow rate will remain significantly below Q_c with increasing drag. At $R_y > 200$, the flow rate Q_y is reduced to almost a third of the airflow rate in the chute while the flow rate of recirculated air Q_{bp} reaches 60% of Q_c. One should keep in mind, however, that the flow rate of air inside the chute

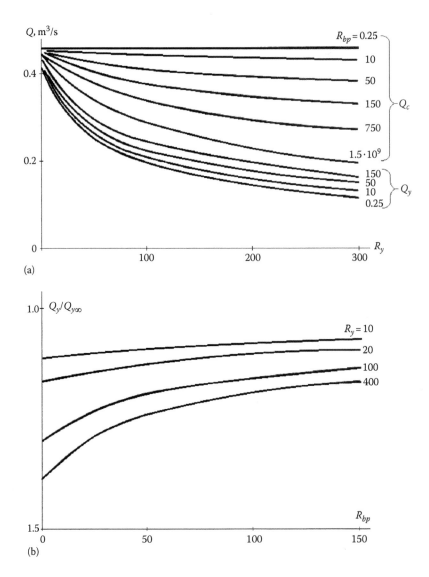

FIGURE 8.10   Variation in (a) airflow rates Q_y and Q_c as a function of resistances of the buffer chamber and (b) bypass duct (with a good sealing of the upper cowl ($h_y = 5$; $R_d = 25$)).

is higher with a bypass duct than when a bypass duct is absent (Figure 8.10a). This happens because the bypass duct is connected in parallel and the chute-with-bypass system has a lower overall drag than any individual component of this system. As a result, when air is forced by ejection, its flow rate increases similarly to the flow rate of air blown by a fan with air ducts connected in parallel. Therefore, it would be more precise to compare the flow rate Q_y with the flow rate Q_c in the absence of a bypass connection (Figure 8.10b). Thus, at $R_y = 100$ and a low drag of the bypass duct ($R_{bp} = 25$), the flow rate of air entering from the bypass chamber in the case at hand will measure 70% of air velocity in the duct $Q_{y\infty}$ (in the absence of a bypass connection $Q_{y\infty} = Q_{c\infty}$); and the flow rate of recirculated air is 30% of $Q_{c\infty}$.

Negative pressure inside the feeder cowl decreases with growing pressure losses in the buffer chamber and may become low enough for untight locations with depressed velocities to begin leaking dust into the working room due to diffusion transfer. Velocity within openings must be

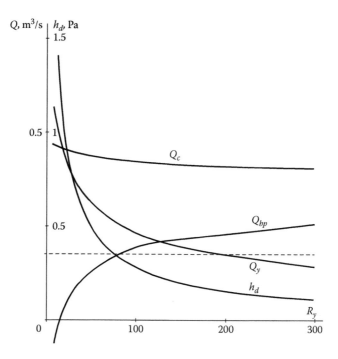

FIGURE 8.11 Variation in airflow rates and negative pressures with increasing drag of the buffer chamber of the handling facility in the case of mediocre sealing of the upper cowl: ($R_d = 5$; $h_y = 5$; $R_{bp} = 25$).

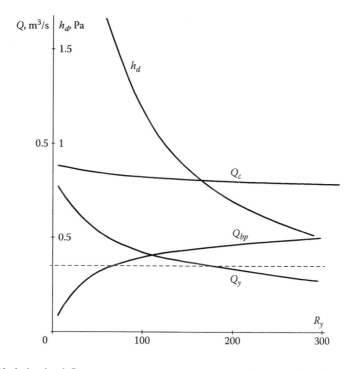

FIGURE 8.12 Variation in airflow rates and negative pressures as a function of buffer chamber drag with a properly sealed upper cowl ($R_f = 25$; $h_y = 5$; $R_{bp} = 25$).

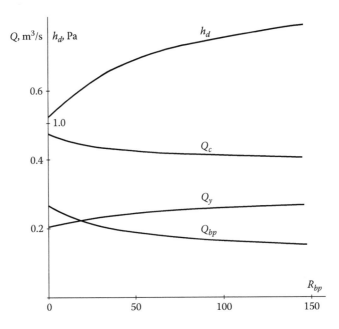

FIGURE 8.13   Variation in airflow rate and negative pressure as a function of bypass duct drag assuming moderate pressure losses in the buffer chamber ($R_y = 100$; $h_y = 5$; $R_d = 25$).

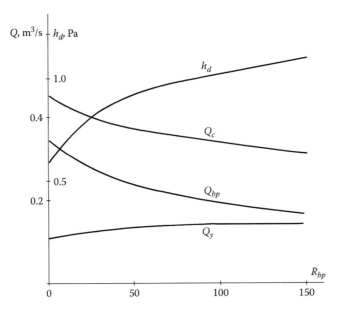

FIGURE 8.14   Variation in airflow rate and negative pressure as a function of bypass duct drag assuming elevated pressure losses in the buffer chamber ($R_y = 400$; $h_y = 5$; $R_d = 25$).

maintained at or above the safety margin of 0.5 m/s [217]. In this case (given an LRC 2.4 of leakage areas) the negative pressure in leakage areas must not become lower than 0.36 Pa (Figures 8.11, 8.12, and 8.15 show the ultimate boundary of negative pressure with a dotted horizontal line). With a good seal of the feeder cowl (at $R_d = 20$) (Figures 8.12 through 8.14), as well as with negative pressure inside the cowl increased to 20 Pa (Figure 8.15), negative pressure within leakage areas will never fall below 0.36 Pa. Thus, it would be wise from the cost standpoint to use a bypass duct with a buffer chamber and properly sealed cowls.

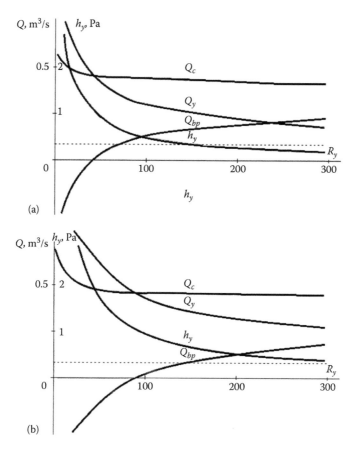

FIGURE 8.15 Variation in airflow rates as a function of buffer chamber drag for the case of mediocre sealing of the upper cowl with increased negative pressure in the lower cowl ($R_d = 5$; $R_{bp} = 25$): (a) $h_y = 10$ and (b) $h_y = 20$.

Let us now move on to examine regularities in cross flows of air given a more complicated case, using three bypass ducts for an aspirated elevator handling facility (Figure 8.16). In particular we will be studying the behavior of flow rates Q_5, Q_8, and $\Delta Q = Q_2 - Q_1$.

For this purpose we will put together the obvious equations of air dynamics and balance for ducts and flow junction points:

$$P_s - P_{bc} + P_2(Q_2) = R_2 Q_2^2, \tag{8.105}$$

$$P_{bc} - P_s + P_1(Q_1) = R_1 Q_1^2, \tag{8.106}$$

$$P_s - P_y + P_3(Q_3) = R_3 Q_3^2, \tag{8.107}$$

$$P_y - P_c = R_8 Q_8^2, \tag{8.108}$$

$$P_y - P_s = R_7 Q_7^2, \tag{8.109}$$

$$P_d - P_{bc} + P_0(Q_0) = R_0 Q_0^2, \tag{8.110}$$

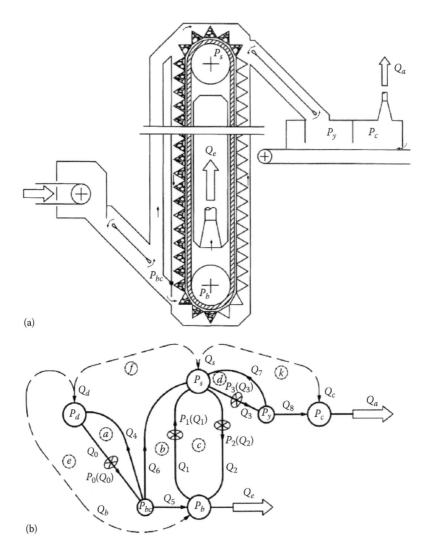

FIGURE 8.16 Aspiration layout for elevator handling with a bypass connection (a) and its aerodynamic equivalent (b).

$$P_{bc} - P_d = R_4 Q_4^2, \tag{8.111}$$

$$P_{bc} - P_b = R_5 Q_5^2, \tag{8.112}$$

$$P_{bc} - P_s = R_6 Q_6^2, \tag{8.113}$$

$$Q_0 = Q_5 + Q_4 + Q_6, \tag{8.114}$$

$$Q_e = Q_5 + Q_2 - Q_1 + Q_e, \tag{8.115}$$

$$Q_3 = Q_s + Q_6 + Q_7 + Q_1 - Q_2, \tag{8.116}$$

$$Q_f = Q_0 - Q_4, \tag{8.117}$$

$$Q_8 = Q_3 - Q_7, \tag{8.118}$$

$$Q_a = Q_8 + Q_c. \tag{8.119}$$

In addition to conventions used for combined Equations 8.12 through 8.24, the following conventions are introduced:

P_c is the absolute pressure in the outer chamber of the upper (receiving) conveyor (Pa)
P_{bc}, P_y are the absolute pressures in the buffer chamber of the loading chute and in the inner chamber of the upper conveyor cowl (Pa)
Q_4, Q_6, Q_7 are the flow rates of air circulating in bypass ducts (m³/s)
Q_5, Q_8 are the flow rates of air coming in, respectively, from the buffer chamber into the bucket elevator boot enclosure and from the inner chamber into the outer chamber of upper conveyor cowl (m³/s)
R_4, R_6, R_7 are the aerodynamic properties of the respective bypass ducts (Pa/$(m^3/s)^2$) determined by their LRCs and cross-sectional areas:

$$R_4 = \rho\zeta_4/\left(2S_4^2\right); \quad R_6 = \rho\zeta_6/\left(2S_6^2\right); \quad R_7 = \rho\zeta_7/\left(2S_7^2\right) \tag{8.120}$$

ζ_4, ζ_6, ζ_7 are the LRCs of the respective bypass ducts referenced to dynamic heads in their cross sections
S_4, S_6, S_7 are the cross-sectional areas of the respective ducts (m²)

The foregoing translates into a total of 19 unknown combined Equations 8.105 through 8.119, Equations 8.17, 8.18, 8.38, and 8.40 with 19 unknown variables in the general case. These include airflow rates in ducts Q_0, Q_1,..., Q_8, airflow rates in the leakage areas of cowls Q_s, Q_d, Q_b, Q_c, airflow rates in unaspirated cowls P_s and P_d, pressures within recirculating flow separators P_{bc} and P_y, and flow rates of aspirated air Q_e and Q_a.

The number of equations may be reduced by applying the law for pressure losses in loops of complex ventilating systems as known from the aerology of mining. Let us write down the corresponding equations:

For the loop a:

$$P_0\left(Q_0\right) = R_0 Q_0^2 + R_4 Q_4^2, \tag{8.121}$$

For the loop b:

$$P_1\left(Q_1\right) = R_1 Q_1^2 + R_5 Q_5^2 - R_6 Q_6^2, \tag{8.122}$$

For the loop c:

$$P_1\left(Q_1\right) + P_2\left(Q_2\right) = R_1 Q_1^2 + R_2 Q_2^2, \tag{8.123}$$

For the loop d:

$$P_3\left(Q_3\right) = R_3 Q_3^2 + R_7 Q_7^2, \tag{8.124}$$

For the loop e^*:

$$P_0\left(Q_0\right) = R_0 Q_0^2 + R_5 Q_5^2 - R_s Q_s^2 + R_d\left(Q_0 - Q_4\right)^2,$$ (8.125)

For the loop f:

$$-R_s Q_s^2 + R_6 Q_6^2 + R_d\left(Q_0 - Q_4\right)^2 - R_4 Q_4^2 = 0,$$ (8.126)

For the loop k:

$$R_c Q_c^2 = R_8 Q_8^2 + R_s Q_s^2 - R_7 Q_7^2.$$ (8.127)

The 10 unknown variables $(Q_0, Q_1,...,Q_8, Q_s)$ are represented with 10 equations, Equations 8.121 through 8.127, 8.114, 8.118, and 8.116.

These combined nonlinear equations will be solved iteratively. Specifically, we will proceed from a set negative pressure in the unaspirated cowl of elevator head (which would be equivalent to setting the initial value of Q_s due to Equation 8.39). For example, let this negative pressure be h ($h_1 = 0$ or $h_1 = h_c$ may be posited for the first approximation). Proceeding from these premises we will determine airflow rates in ducts $Q_0(h_i), Q_1(h_i), Q_2(h_i),...,Q_8(h_i)$ as functions of h_i. The next step is refining the negative pressure value:

$$s_i = R_s \left|Q_s\left(h_i\right)\right| Q_s\left(h_i\right),$$ (8.128)

where

$$Q_s\left(h_i\right) = Q_2\left(h_i\right) - Q_1\left(h_i\right) + Q_3\left(h_i\right) - Q_6\left(h_i\right) - Q_7\left(h_i\right).$$ (8.129)

The computation process repeats with a new value of h_{i+1}.

Computation can be facilitated if the following is posited for $i + 1$-th approximation:

$$h_{i+1} = \frac{h_i + s_i}{2}.$$ (8.130)

Computation stops as soon as the inequality is met:

$$\left|s_i - h_i\right| \le E,$$ (8.131)

where E is the accepted precision ($E \approx 10^{-2}$).

Now let us derive computation formulas for ith approximations of airflow rates in ducts:

$$Q_0\left(h_i\right), Q_1\left(h_i\right),...,Q_8\left(h_i\right).$$

---

* Loops e, f, k are fictitious. They connect airflows in ducts by terminating them in the atmosphere of the room through leakage areas in cowls (these links are shown with dotted *ducts* in Figure 8.16).

We will begin with considering upper ducts adjacent to the cowl of bucket elevator *head*. Equation 8.124 gives

$$f_7 = Q_7(h_i) = \frac{L_3(Q_3(h_i))}{\sqrt{R_7 |L_3(Q_3(h_i))|}},$$
(8.132)

where

$$L_3(Q_3(h_i)) = P_3(Q_3(h_i)) - R_3 |Q_3(h_i)|(Q_3(h_i)).$$
(8.133)

In view of the connection between negative pressure in the cowl and airflow rates in leakage area of this cowl, Equation 8.127 will appear as

$$-h_i + h_b = R_8 [Q_8(h_i)]^2 - R_7 |Q_7(h_i)|(Q_7(h_i)),$$
(8.134)

whence we can find

$$Q_8(h_i) = \frac{L_9}{\sqrt{R_8 |L_9|}},$$
(8.135)

where

$$f_9 = L_9 = h_k - h_i + R_7 f_7(Q_3(h_i)) |f_7(Q_3(h_i))|.$$
(8.136)

Equation 8.118 enables us to put forth the following equation with a single unknown variable $Q_3(h_i)$:

$$f_{10} = Q_3(h_i) - Q_8(Q_3(h_i)) - Q_7(Q_3(h_i)) = 0,$$
(8.137)

which can be solved using bisection. As soon as $(Q_3(h_i))$ is found, one formula (Equation 8.132) can be used to determine $Q_7(h_i)$ and another formula (Equation 8.135) can be used for $Q_8(h_i)$.

Let us now move on to the lower assembly (bucket elevator boot cowl). For the sake of clarity we will omit the argument of airflow rate functions. Equation 8.125 in view of the connection between negative pressure and airflow rate in the leakage areas of the elevator boot cowl will appear as

$$h_b + L_0(Q_0) = R_5 Q_5^2 + R_d(Q_0 - Q_4)^2,$$
(8.138)

whence

$$f_5 = Q_5 = f_{15}/\sqrt{|f_{15}|},$$
(8.139)

with the following functions introduced for a more convenient notation:

$$f_{15} = L_{15} = \frac{h_b + L_0(Q_0)}{R_d} - R_d(Q_0 - Q_4)^2,$$
(8.140)

$$f_0 = L_0(Q_0) = P_0(Q_0) - R_0 Q_0 |Q_0|.$$
(8.141)

Equation 8.121 results in

$$f_4 = Q_4 = \frac{L_0(Q_0)}{\sqrt{R_4|L_0(Q_0)|}}. \tag{8.142}$$

In order to resolve Q_6 as a function Q_0, we will use Equation 8.126, which will be written as

$$h_i - R_d(Q_0 - Q_4)|Q_0 - Q_4| + R_4Q_4|Q_4| = R_6Q_6|Q_6| \tag{8.143}$$

or

$$L_{11} = R_6Q_6|Q_6|, \tag{8.144}$$

where

$$L_{11} = f_{11} = h_i - R_4(Q_0 - Q_4)|Q_0 - Q_4| + R_4Q_4|Q_4|, \tag{8.145}$$

whence

$$f_6 = Q_6 = \frac{L_{11}}{\sqrt{R_6|L_{11}|}}. \tag{8.146}$$

Substitution of Equations 8.139, 8.142, and 8.146 into the air balance Equation 8.114 results in the following functional equation for determining Q_0:

$$f_{12} = Q_0 - Q_4(Q_0) - Q_5(Q_0) - Q_6(Q_0) = 0. \tag{8.147}$$

After determining $Q_0(h_i)$ and substituting it into Equations 8.139, 8.142, and 8.146, flow rates $Q_4(h_i)$, $Q_5(h_i)$, and $Q_6(h_i)$ will become known.

In order to determine the remaining unknown variables Q_1 and Q_2, Equations 7.146 through 7.148 and 7.126 through 7.128 can be used.

The resolved values of $Q_1(h_i)$, $Q_2(h_i)$, $Q_3(h_i)$, $Q_6(h_i)$, and $Q_7(h_i)$ are substituted into Equation 8.128 in order to determine the next iterative value of negative pressure.

The iterative process completes as soon as the condition of Equation 8.131 is met. A flowchart of the algorithm described is shown in Figure 8.17.

Computed flow rates of $Q_0 \ldots Q_8$, in addition to determining negative pressures in the unaspirated cowls of elevator head (h) and the feeder drive drum cowl,

$$h_d = R_d(Q_0 - Q_4)|Q_0 - Q_4|, \tag{8.148}$$

enable the required aspiration volumes to be computed as well:

$$Q_a = Q_8 + \sqrt{\frac{h_c}{R_c}}, \tag{8.149}$$

$$Q_e = Q_5 + Q_2 - Q_1 + \sqrt{\frac{h_b}{R_b}}. \tag{8.150}$$

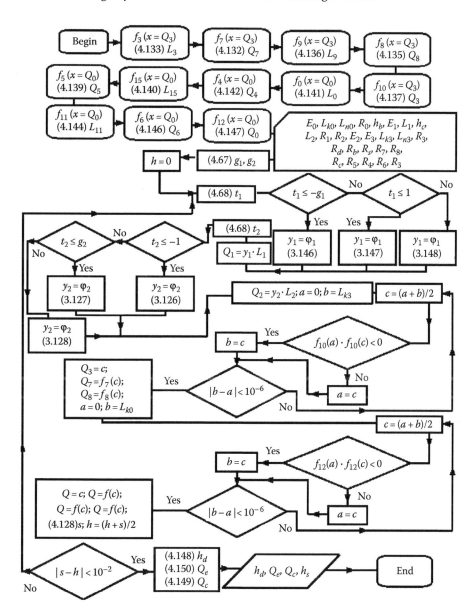

FIGURE 8.17 Flowchart of the algorithm for computing airflow rates in ducts $Q_0, Q_1, \ldots, Q_8$ and aspiration volumes Q_b and Q_c of a grain-handling elevator unit equipped with bypass connections.

The role of bypass connections can be evaluated by comparing the current values of Q_5, Q_8, and ΔQ with respective values of $Q_{5\infty}$, $Q_{8\infty}$, and ΔQ_∞ for an aspiration system without bypass connections. It would be best to choose a *standard* case of elevator facility design for this study. We will be using the previously considered example of elevator handling facility (see Section 8.3) that will be supplemented generally with three bypass channels by analogy with Figure 8.16a. The following values will serve as constant parameters describing this handling facility:

$$E_0 = 4\frac{\text{Pa}}{\left(\text{m}^3/\text{s}\right)^3}, \quad L_{k0} = 1.6\frac{\text{m}^3}{\text{s}}, \quad L_{n0} = 1\frac{\text{m}^3}{\text{s}}, \quad h_b = 5\,\text{Pa}, \quad R_d = 10\frac{\text{Pa}}{\left(\text{m}^3/\text{s}\right)^2},$$

$$E_1 = 200\frac{\text{Pa}}{\left(\text{m}^3/\text{s}\right)^2}, \quad E_2 = 400\frac{\text{Pa}}{\left(\text{m}^3/\text{s}\right)^2}, \quad R_1 = 21\frac{\text{Pa}}{\left(\text{m}^3/\text{s}\right)^2}, \quad R_2 = 22\frac{\text{Pa}}{\left(\text{m}^3/\text{s}\right)^2},$$

$$L_1 = L_2 = 1\frac{\text{m}^3}{\text{s}}, \quad R_0 = 10\frac{\text{Pa}}{\left(\text{m}^3/\text{s}\right)^2}, \quad R_5 = 50\frac{\text{Pa}}{\left(\text{m}^3/\text{s}\right)^2}, \quad R_s = 25\frac{\text{Pa}}{\left(\text{m}^3/\text{s}\right)^2},$$

$$R_3 = 50\frac{\text{Pa}}{\left(\text{m}^3/\text{s}\right)^2}, \quad R_8 = 50\frac{\text{Pa}}{\left(\text{m}^3/\text{s}\right)^2}, \quad R_b = 36\frac{\text{Pa}}{\left(\text{m}^3/\text{s}\right)^2}, \quad E_3 = 5\frac{\text{Pa}}{\left(\text{m}^3/\text{s}\right)^3},$$

$$L_{k3} = 2\frac{\text{m}^3}{\text{s}}, \quad L_{n3} = 0.5\frac{\text{m}^3}{\text{s}}, \quad h_c = 5\ Pa, \quad R_c = 40\frac{\text{Pa}}{\left(\text{m}^3/\text{s}\right)^2}.$$

Aerodynamic performance indices of bypass ducts R_4, R_6, and R_7 within the range of 0.5...150 Pa/$\left(\text{m}^3/\text{s}\right)^2$ will serve as variables. These values become fixed at 10^9 Pa/$\left(\text{m}^3/\text{s}\right)^2$ when the respective bypass duct is closed or disconnected. For example, if $R_4 = R_6 = R_7 = 10^9$, then airflow rates in bypass ducts will be $Q_4 = Q_6 = Q_7 = 0$, that is, bypass ducts are disconnected. Calculated values are presented in Figures 8.18 through 8.20. On these plots, airflow rates corresponding to the unbypassed case are designated with a subscript ∞ and their changes are plotted using straight horizontal dotted lines. Another instance of a fixed aerodynamic performance index of bypass ducts occurs at

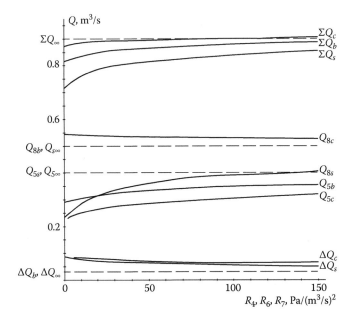

FIGURE 8.18 Airflow rates Q_5, Q_8, ΔQ, and $\sum Q$ as functions of aerodynamic drag for an elevator equipped with a single bypass duct (H—lower, M—middle, B—upper, ∞—no bypass ducts).

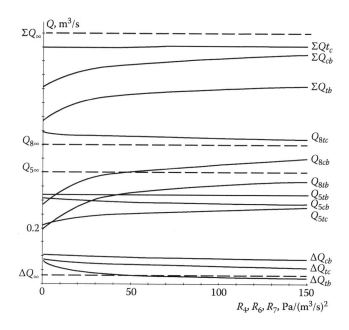

FIGURE 8.19 Airflow rates Q_5, Q_8, ΔQ, and $\sum Q$ as functions of aerodynamic resistance for an elevator equipped with two bypass ducts (HC—lower and medium, CB—medium and upper, HB—lower and upper, and ∞—no bypass ducts).

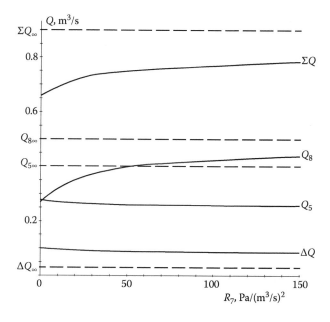

FIGURE 8.20 Airflow rates Q_5, Q_8, ΔQ, and $\sum Q$ as functions of aerodynamic drag R_7 for an elevator equipped with three bypass ducts.

25 Pa/$(m^3/s)^2$. For example, Figure 8.18 indicates variations in airflow rates Q_5, Q_8, ΔQ and their sum total $\sum Q$ for three possible bypass layouts:

   Case 1: Only a bottom bypass duct is provided (on the loading chute); its aerodynamic performance varies $0.5 \leq R_4 \leq 150$ while other bypass ducts are absent (their aerodynamic performance converges toward ∞ or, more precisely, is $R_6 = R_7 = 10^9$);
   Case 2: Only the middle bypass duct (located on elevator enclosure) is arranged, that is, $0.5 \leq R_s \leq 150$ and $R_4 = R_7 = 10^9$;
   Case 3: Only the upper bypass duct is provided (on the discharge chute), that is, $0.5 \leq R_7 \leq 150$ and $R_4 = R_6 = 10^9$ (this case is illustrated in Figure 8.18).

Figure 8.19 plots airflow rates for three possible design layouts of bypass ducts: case 1—lower and middle bypass ducts are installed so that $R_4 = 25$ Pa/$(m^3/s)^2$, 0.5 Pa/$(m^3/s)^2 \leq R_6 \leq 150$ Pa/$(m^3/s)^2$, $R_7 = 10^9$ Pa/$(m^3/s)^2$; case 2—middle and upper bypass ducts are installed so that $R_4 = 10^9$ Pa/$(m^3/s)^2$, $R_6 = 10^9$ Pa/$(m^3/s)^2$, 0.5 Pa/$(m^3/s)^2 \leq R_7 \leq 150$ Pa/$(m^3/s)^2$; case 3—lower and upper bypass ducts are installed, so that $R_4 = R_6 = 25$ Pa/$(m^3/s)^2$ and 0.5 Pa/$(m^3/s)^2 \leq R_7 \leq 150$ Pa/$(m^3/s)^2$.

Figure 8.20 contains plots of airflow rates for the only possible configuration with three bypass ducts: lower, medium, and upper, so that $R_4 = 25$ Pa/$(m^3/s)^2$, $R_6 = 25$ Pa/$(m^3/s)^2$, and 0.5 Pa/$(m^3/s)^2 \leq R_7 \leq 150$ Pa/$(m^3/s)^2$.

As the plots illustrate, installation of bypass ducts noticeably decreases the total amount of air entering through leakage areas of unaspirated cowls in elevator feeder and head:

$$\sum Q = Q_5 + Q_8 + \Delta Q,$$

where

$$\Delta Q = Q_2 - Q_1.$$

So, with a single bypass duct installed, installation of the upper bypass duct (Figure 8.18) provides the greatest effect: at $R_7 = 25$ Pa/$(m^3/s)^2$ $(R_6 = R_7 = 10^9)$ the total volume of air

$$\sum Q_s = \frac{0.79}{0.9} \sum Q_\infty = 0.88 \sum Q_\infty,$$

that is, 1.14 times smaller than without bypass ducts installed. An even greater effect is achieved when the elevator is equipped with two bypass connections at the top and at the bottom (Figure 8.18): at $R_4 = 25$ Pa/$(m^3/s)^2$, $R_6 = 10^9$ Pa/$(m^3/s)^2$, and $R_6 = 25$ Pa/$(m^3/s)^2$ the total volume of air is

$$\sum Q_{tb} = \frac{0.66}{0.9} \sum Q_\infty = 0.73 \sum Q_\infty,$$

that is, 1.37 times less than for the case of aspiration without bypass connections installed.
   It would be unwise to install three bypass connections in this case, as with

$$R_4 = R_6 = R_7 = 25 \text{ Pa}/(m^3/s)^2,$$

$$\sum Q_{tbm} = \frac{0.73}{0.9} \sum Q_\infty = 0.81 \sum Q_\infty,$$

that is, the efficiency would be lower than with two (upper and lower) bypass connections. This happens because cross flows of air from the upper cowl (at the elevator head) into the lower cowl (elevator boot cowl) increases as negative pressure in the upper cowl is reduced to 2.1 Pa ($h_s = 5.6$ Pa for the case of upper and lower bypass ducts). Meanwhile, ΔQ increases from 0.046 m³/s with two bypass ducts to 0.093 m³/s with three bypass ducts.

With regard to aspiration volumes, the effect of bypassing appears less prominent. For example, in the case of upper and lower bypass ducts in the example considered,

$$Q_a = \sum Q_{tb} + Q_b + Q_c = 0.66 + \sqrt{\frac{5}{36}} + \sqrt{\frac{5}{40}} = 0.66 + 0.73 = 1.39 \text{ m}^3/\text{s},$$

while, if any bypass ducts are absent,

$$Q_{a\infty} = \sum Q_\infty + Q_b + Q_c = 0.9 + 0.73 = 1.63 \text{ m}^3/\text{s},$$

that is, 17% higher.

It should be noted that values indicated may be higher. A specially developed computation methodology and software enable to evaluate the effect in each specific case on the design stage, making it possible to devise efficient layouts of aspiration and bypassing for elevator handling of grain.

8.5 CONCLUSIONS

1. Analysis of the known aerodynamic properties of structurally similar equivalents (orifices, gates, and geometric bodies) makes it possible to conclude that the drag coefficient of a bucket (ζ_b) depends not only on its geometrical shape and the width of the gap between a traveling bucket and elevator enclosure but also on the direction of relative airflow velocity (Figure 8.2). When air flows around the back of a bucket, ζ_b can be determined by analogy with a diaphragm having downstream-chamfered edges. When air flows freely into bucket opening, the drag of the bucket increases several times. However, the coefficient ζ_b is dominated by spatial constraints to airflow—the width of the gap between the elevator enclosure walls and buckets. The combined effect of bucket shape, filling degree, and flow regimes around them should be determined experimentally.
2. Aspiration layouts proposed for elevator handling of grain must take into consideration the predominant effect of ejection forces in ducts. Downward-directed action of ejection head in chutes together with predominant ejecting properties of the return run with empty buckets, predetermine the use of a classical aspiration layout (Figure 8.3): designing for local suction units to evacuate air from the elevator boot cowl and from the cowl at the loading location of the upper (receiving) conveyor or from the internal space of the receiving hopper.
3. Fundamental relations for balance equations of air exchange in aspirated cowls (with the purpose of determining necessary aspiration volumes) may be provided by combined equations of air dynamics in four ducts, Equations 8.56 through 8.59 (or Equations 8.69 through 8.72): inside loading and discharge chutes and in the enclosure of carrying and return runs of the elevator conveyor.
4. Aspiration volumes in the classical aspiration layout for elevator handling of grain are determined by balance equations (Equations 8.73 and 8.74). Basic components of both

equations comprise volumetric flow rates of air entrained into aspirated cowls through leaking joints and process openings by the negative pressure (maintained inside cowls by an aspiration unit) and flow rates of air transferred in ducts.

5. Flow rates of transferred air Q_0, Q_1, Q_2, Q_3 may be determined using algorithms (Figure 8.4) devised by solving combined nonlinear equations, Equations 8.69 through 8.72, with a joint use of iteration and bisection methods.

6. An analysis of computation results for an exemplary (*standard*) assembly shows that increasing negative pressure inside the boot cowl and decreasing leakage area in the cowl of elevator head both produce increasing negative pressure in the head cowl (h_s), causing flow rates of ejected air to increase in the loading chute ((Q_0), Figure 8.5) and to decrease in the discharge chute ((Q_3), Figure 8.6). Total airflow rate in the elevator enclosures ($\Delta Q = Q_2 - Q_1$) varies similarly, rising with increasing h_b and declining with increasing R_s. Specific regularities of air cross flow in ducts also affect aspiration volumes: the required airflow rate in the local suction unit of elevator boot cowl (Q_e) increases while the volume of aspiration from the upper conveyor cowl (Q_a) decreases with increasing h_b (Figure 8.7a), whereas, both flow rates decrease with increasing sealing degree (Figure 8.7b). However, the total flow rate of aspirated air remains sufficiently high: even with a small negative pressure inside the boot cowl ($h_b = 16.8$ Pa) and adequate sealing of the elevator head cowl $\left(R_s = 14.4 \text{ Pa/} \left(\text{m}^3\text{/s} \right)^2 \right)$, the total aspiration volume for the exemplary assembly would be about 2.06 m³/s.

7. The necessary design performance of aspiration system may be reduced not only by reducing the optimum negative pressure in cowls and their sealing but also by using bypass channels to enable closed-loop circulation of ejected air.

8. Regularities of air cross flow and circulation patterns in conveyor-to-conveyor grain handling with a bypass duct connecting the inner chamber of a double-walled cowl with the cowl of the driving drum (Figure 8.8) can be studied using equations for pressure losses in ducts, Equations 8.86 through 8.90, and airflow rate balance equations for cowls and junction points, Equations 8.91 through 8.93. Analysis of these regularities has shown that the efficient operation of the bypass duct requires negative pressure inside the cowl of the upper unaspirated conveyor to exceed the negative pressure in the buffer (inner) chamber, Equations 8.96 or 8.98. This chamber improves the reliability of bypass operation with a circulation layout as opposed to a parallel cross-flow layout whereby the total flow rate of incoming air from the upper cowl into the lower cowl would increase, leading to higher required aspiration volumes.

9. An analysis of the numerical solution for Equation 8.104 in view of Equation 8.103 for the exemplary conveyor handling facility indicates that, in the absence of bypass, the flow rate of air coming in from the inner chamber into the outer chamber of a double-walled cowl Q_{yoo} is significantly higher than when the assembly is furnished with a bypass duct (Q_y). With increasing drag of inner cowl walls as well as increasing cross-sectional area of the bypass duct ($R_{bp} < 20$), flow rates Q_y and Q_{yoo} deviate further apart (Figure 8.10).

The flow rate Q_{bp} of air entering through the bypass duct of a handling facility increases with increasing drag of the buffer (inner) chamber and decreasing drag of the bypass duct (Figures 8.11 and 8.13). When the upper (unaspirated) cowl is poorly sealed $\left(R_d \leq 5 \text{ Pa/} \left(\text{m}^3\text{/s} \right)^2 \right)$ and partitions have a low drag $\left(R_y < 20 \text{ Pa/} \left(\text{m}^3\text{/s} \right)^2 \right)$, airflow in the bypass duct may reverse its direction ($Q_{bp} < 0$). The zone of negative airflow in the duct extends to $R_y = 40$–90 Pa/(m³/s)² when negative pressure inside the conveyor cowl increases to 20 Pa (Figure 8.15). To prevent the occurrence of such a zone, sealing of the upper cowl must be improved (Figure 8.12). As the drag of the buffer chamber increases (with $R_y > 200 \text{ Pa/} \left(\text{m}^3\text{/s} \right)^2$), flow rate of recirculated air Q_{bp} reaches 60% of Q_c.

The flow rate of air Q_c coming in through the chute in the presence of a bypass connection is greater than the flow rate $Q_{c\infty}$ of air in the same chute where the bypass is absent (Figure 8.10a). This is because ejection head building up inside the chute acts like a blower (fan) for a parallel network of two ducts—the bypass duct and the gap between the inner cowl walls and the conveyor belt of the upper conveyor, with a combined drag lower than they would be in the case for either duct individually. Therefore the relation $Q_{bp}/Q_{c\infty}$ would be more fitting to describe the degree of recirculation in the bypass duct.

10. The effect of reducing aspiration volumes for a handling facility equipped with a bypass duct is maximized by ensuring proper sealing of ducts and designing for a buffer channel with increased drag in the inner cowl. In the example considered earlier, at $R_y = 100$ Pa/$(m^3/s)^2$ and a bypass duct with a low drag $R_{bp} = 25$ Pa/$(m^3/s)^2$, airflow rate Q_y would make up 70% of $Q_{c\infty}$, that is, decrease 1.428 times compared to the case of an unbypassed assembly.

11. Three bypass channels (Figure 8.16a) can be used to reduce aspiration volumes in elevator handling facilities by maintaining a smaller negative pressure inside unaspirated cowls of the elevator feeder and head. Airflows can be quantitatively decomposed using combined Equations 8.121 through 8.127 describing pressure losses in loops of a proposed aerodynamic equivalent for the assembly (Figure 8.16b). These combined nonlinear equations have been solved numerically using a specially developed algorithm (Figure 8.17) based on joint application of iterative and bisection procedures.

Analysis of the findings for a *standard* elevator (Figures 8.18 through 8.20) indicates that reduction in the flow rates of air $\left(\sum Q\right)$ entering through leakage areas of unaspirated cowls is maximized when two bypass connections are installed: the lower one, connecting feeder cowl with a buffer chamber upstream of elevator boot cowl, and the upper one, connecting elevator head cowl with the inner chamber of a double-walled cowl of the upper (receiving) conveyor. The total flow rate $\sum Q$ in the case reviewed was 1.37 times lower than $\sum Q_\infty$ (a corresponding aspiration case with identical process and design parameters of the assembly, but with no bypass ducts).

The effect cited may differ from the actual effect in every particular case. The computer program developed for the purpose can be used to provide a design-stage estimate.

References

1. I.M. Della Valle. *Exhaust Hoods*. New York: Industrial Press, 1952.
2. L.-H. Engels und G. Willert. Kriterien und Möglichkeiten zur Erfassung des Staubes in Industriebetrieben. *Staub-Reinhlat*. 1973. No. 3. pp. 140–141.
3. H.A. Koop. Staubdüsen an Schleifmaschinen. *Zeitschrift des Vereins Deutscher Ingenieure für Maschinenbau und Metallbearbeitung*. 1944. pp. 21–44.
4. A.S. Pruzner. Flow pattern in the effective area of suction inlets. *Heating and Ventilation*. 1939. No. 10. pp. 13–21.
5. M.F. Bromley. Flow pattern in the effective area of suction inlets. *Heating and Ventilation*. 1934. No. 3 pp. 2–8.
6. V.V. Baturin. *The Fundamentals of Industrial Ventilation*. Moscow, Russia: Profizdat, 1990. 448pp.
7. G.A. Maksimov and V.V. Deryugin. *Air Motion in Operation of Ventilation Systems*. Leningrad, Russia: Construction Literature Publishing, 1972. 97pp.
8. E.M. Ivanus, S.S. Zhukovsky, and Yu.S. Yurkevich. *The Study of Local Ventilation from Brazing Joints of Radio Equipment Units*. Lvov, Ukraine: Lvov Polytechnic Institute, 1985. 5pp. Ukraine NIINTI Department, December 22, 1986. No. 2797.
9. L.F. Shevchenko. The study of suction spectrum of local exhausts of rock-cutting machines. Researches in the field of air dedusting: Interuniversity collection of scientific papers. Perm, Russia, 1986. pp. 81–85.
10. L.F. Shevchenko. *Design Principles for Local Exhausts of Rock-Cutting Machines. Dust Control at Bulk Handling Enterprises*. Belgorod, Russia: BTISM Publishing, 1990. pp. 112–114.
11. I.N. Logachev, V.G. Stetsenko, and L.K. Saplinov. Resolving some problems of industrial ventilation aerodynamics using the electro-hydrodynamic analogy (EHDA) method. *Ventilation and Air Cleaning*. Moscow, Russia: Nedra, 1969. 5th edn. pp. 144–149.
12. V.N. Taliev. *Ventilation Aerodynamics*. Moscow, Russia: Stroyizdat, 1979. 296pp.
13. V.N. Taliev. Plane wall slot suction flare. *Izvestiya vuzov/Construction and Architecture*. 1979. No. 3. pp. 124–127.
14. V.N. Taliev and L.R. Aleksandrov. Suction flare of a longitudinal slot of constant width through a round cross-section pipe. *Izvestiya vuzov/Fabric Industry Technology*. 1990. No. 4. pp. 76–78.
15. V.N. Posokhin. Regarding the flow analysis at a flat askew inlet pipe. *Izvestiya vuzov/Fabric Industry Technology*. 1982. No. 3. pp. 78–81.
16. V.N. Posokhin. *Design of Local Exhausts from Heat- and Gas-Generating Equipment*. Moscow, Russia: Mashinostroeniye, 1984. 160pp.
17. V.N. Posokhin and I.L. Gurevich. Regarding the flow analysis at a bell-shaped suction slot. *Izvestiya vuzov/Fabric Industry Technology*. 1981. No. 3. pp. 84–88.
18. G.M. Pozin and V.N. Posokhin. Calculation methods for fields of velocities, which are formed by slotted hoods in a confined space. *Occupational Safety and Health*. Moscow, Russia. 1980. pp. 52–57.
19. V.N. Posokhin. Calculating velocities of flow along the walls of a parallelepipedal confined volume wherefrom air is exhausted through a slotted opening. *Izvestiya vuzov/Construction and Architecture*. 1983. No. 11. pp. 97–103.
20. V.N. Posokhin. Exhaust inflow in air motion interference conditions. *Flow Mechanics and Heat Exchange in Heating and Ventilation Devices*. Kazan, Russia, 1981. pp. 9–11.
21. I.N. Logachev. Regarding design of double-flange exhausts. *Ventilation in Iron and Steel Industry*. Moscow, Russia: Metallurgiya, 1968. pp. 88–92.
22. I.N. Logachev. Irrotational air motion at a suction slot. *Ventilation and Air Cleaning*. Moscow, Russia: Nedra, 1969. pp. 143–150.
23. V.L. Makhover, L.S. Khalezov, and A.G. Chesnokov. Suction flare of slotted openings. *Izvestiya vuzov/Fabric Industry Technology*. 1969. No. 1. pp. 143–147.
24. E.I. Shulekina. Determining velocity in a confined suction flare. *Flow Mechanics and Heat Exchange in Heating and Ventilation Devices*. Kazan, Russia, 1981. pp. 17–19.
25. A.V. Kryuchkov. *Suction Inlet Slot Inflow Velocity Profiles with Regard to Boundary Surfaces*. Dust control of the constructions materials industry processes: Collected scientific papers. Moscow, Russia: MISI Publishing; Belgorod, Russia: BTISM Publishing, 1992. pp. 72–76.
26. M.S. Kuzmin and P.A. Ovchinnikov. Exhaust and supply air outlets. Moscow, Russia: Stroyizdat, 1987. 168pp.

27. I.I. Konyshev, A.G. Chesnokov, and S.N. Schadrova. Design of some spatial suction flares. *Izvestiya vuzov/Fabric Industry Technology.* 1976. No. 4. pp. 103–116.

28. I.A. Shepelev. Air flows at suction inlets. Plumbing Research Institute Papers. Moscow, Russia. 1967. No. 24. pp. 180–209.

29. I.A. Shepelev. Aerodynamics of indoor air flows. Moscow, Russia: Stroyizdat, 1978. 145pp.

30. A.L. Altynova. Variation in axial velocity in a plane wall elliptical hole suction flare. *Water Supply and Sanitary Engineering.* 1974. No. 5. pp. 26–28.

31. A.L. Altynova. Variation in axial velocity on the right angle edge with a round intake port located inside the angle. *Heating and Ventilation.* Irkutsk, Russia, 1976. pp. 53–57.

32. I.G. Tyaglo and I.A. Shepelev. Regarding air flow parameters at a rectangular exhaust outlet. Plumbing Research Institute papers. Moscow, Russia, 1969. No. 30. pp. 169–171.

33. Yu.A. Korostelev and G.D. Livshits. Regarding the study of suction flares. *Izvestiya vuzov/Construction and Architecture.* 1974. No. 12. pp. 132–136.

34. G.D. Livshits. The study of velocity variation regularities on the round sharp-edged suction inlet flow axis. *Izvestiya vuzov/Construction and Architecture.* 1973. No. 7. pp. 153–158.

35. G.D. Livshits. The study of the field of velocities in a round semi-infinite pipe suction flare. *Izvestiya vuzov/Construction and Architecture.* 1974. No. 10. pp. 115–119.

36. G.D. Livshits. Regarding the study of suction flare regularities. *Izvestiya vuzov/Construction and Architecture.* 1975. No. 12. pp. 135–141.

37. G.D. Livshits. The study of local exhaust plumes using the method of singularities. *Izvestiya vuzov/Construction and Architecture.* 1974. No. 4. pp. 104–108.

38. G.D. Livshits. The study of local exhausts built in the mechanized gas-shielded welding equipment. *Izvestiya vuzov/Construction and Architecture.* 1978. No. 12. pp. 130–133.

39. Fr. Drkal. Theoretishe Bestimmung der Strömungsverhälthisse bei Saugshlitzen. HLH 22. 1971. No. 5. pp.167–172.

40. M. Crawford. *Air Pollution Control Theory. Design of Industrial Ventilation Systems.* New York: McGraw-Hill, 1976. pp. 165–187.

41. W. Pfeiffer. Absaugluftmengen von Erfassungseinrichtungen offener Bauart. Staub—Reinhalt. Luft 42. 1982. No. 8. pp. 303–308.

42. B. Fletcher and A.E. Johnson. Velocity profiles around hoods and slots the effects of om adjacent plane. *Annals of Occupational Hygiene.* 1982. No. 4. pp. 365–367.

43. S. Gomula and W. Kosczynialski. Wpływ ukstaltowania paszczyznu wlotowej w ssacym urzadzeniem. *Zeszyty Naukowe AGH.* 1981. No. 819. pp. 5–9.

44. A.G. Ceag Dominit. Überine Näherung zur Bestimmung von Strömungsfeldem von Absaugehauben. Staub-Reinhalt. Luft 33. 1973. No. 3. pp. 142–146.

45. D. Haase. Die Wirkung des Windes bei der Absaugung. Staub-Reinhalt. Luft 27. 1967. pp. 131–133.

46. V. Filek, S. Gomula, B. Nowak, and W. Roszczynomalki. Opis pola predkosci przed wentylacyjnym irzadzeiniem odciayowym za pomoca rozwiazon numerycznych rownania potencjalei Laplacia. *Zeszyty Naukowe AGH.* 1979. No. 761. pp. 337–342.

47. S. Gomula and W. Roszczynomalki. Teoretyczne okreslanie rozkladu predkosci przed zasusajacum urzadzeniem wentylacyinum. *Zeszyty Naukowe AGH.* 1980. No. 804. pp. 261–266.

48. Heating and ventilation: In 2 parts, V.N. Bogoslavskiy (eds.). Moscow, Russia, 1976. Part 2: Ventilation. 435pp.

49. M.I. Grimitlin. Indoor air distribution. Moscow, Russia: Stroyizdat, 1983. 164pp.

50. E.V. Sazonov. Design of one-sided canopy hoods of electrical furnace work orifices. *Izvestiya vuzov/Construction and Architecture.* 1986. No. 9. pp. 92–95.

51. E.V. Sazonov. *Theoretical Ventilation Design Basis.* Voronezh, Russia: VSU Publishing, 1990. 208pp.

52. J. Batchelor. *Introduction to Fluid Dynamics.* Moscow, Russia: Mir, 1973. 758pp.

53. N.Ya. Fabrikant. *Aerodynamics.* Moscow, Russia: Nauka, 1964. 816pp.

54. P. Banerjee and R. Butterfield. *Boundary Element Method in Applied Sciences.* Moscow, Russia: Mir, 1984. 486pp.

55. Boundary integral equations method. Computational aspects and applications in mechanics. Novelties of foreign science, T. Cruz, F. Rizzo (eds.). Moscow, Russia: Mir, 1979. 210pp.

56. K. Brebbia, G. Telles, and L. Vroubel. *Boundary Element Methods.* Moscow, Russia: Mir, 1987. 525pp.

57. K. Brebbia and S. Walker. *Boundary Element Method Application in Engineering.* Moscow, Russia: Mir, 1982. 248pp.

58. S.G. Mikhlin. *Integral Equations Applications.* Moscow, Russia: OGIZ, 1947. 304pp.

59. J.C. Wu and M.M. Wahbah. *Numerical Solution of Viscous Flow Equation Using Integral Representations: Lecture Notes in Physics.* New York: Springer-Verlag, 1976. Vol. 59. pp. 448–453.

60. R.Kh. Akhmadeev, I.L. Gurevich, and V.N. Posokhin. Regarding design of slotted hoods from axiosymmetric diffusion sources. *Izvestiya vuzov/Construction and Architecture*. 1990. No. 6. pp. 78–83.

61. M.R. Flynn and M.J. Ellenbecker. The potential flow solution for air flow into a flanget cirailar hood. *American Industrial Hygiene Association*. 1985. No. 6. pp. 318–322.

62. I.N. Logachev, K.I. Logachev, and O.D. Neykov. Containment of particulate emissions in powder compression. *Metal Powder Industry*. 1995. No. 3,4. pp. 100–103.

63. V.A. Minko, I.N. Logachev, and K.I. Logachev. Dynamics of air flows in suction flares of local exhausts of industrial building dust extract ventilation. *Izvestiya vuzov/Construction*. 1996. No. 10. pp. 110–113.

64. V.A. Minko, I.N. Logachev, K.I. Logachev, and V.A. Titov. The study of dust particles dynamics in cavities of silo bins. Scientific-and-technological advance in the area of industrial ecology and environmental monitoring: Collected reports. *International Conference on Constructions Materials Industry and Building Industry, Efficient Use of Energy and Resources in the Context of Market Relations*. Belgorod, Russia: BelGTASM Publishing, 1997. Part 9. pp. 3–7.

65. K.I. Logachev. Environmental industry: Mathematical modeling of industrial ventilation systems. *Engineering Ecology*. 1999. No. 1. pp. 8–18.

66. K.I. Logachev. Environmental industry: Numerical modeling of shielded exhausts of industrial ventilation systems. *Engineering Ecology*. 1999. No. 5. pp. 30–40.

67. K.I. Logachev. On increasing an suction flare range by means of straight jets. Advanced technologies in industry and construction on the threshold of the 21th century: Collected reports. *International Conference School Workshop of Young Scientists and Post-Graduates*. Belgorod, Russia: BelGTASM Publishing, 1998. Part 2. pp. 694–701.

68. K.I. Logachev and L.V. Balukhtina. Regarding the suction inlet geometrical shape effect on the suction inlet range capability. Scientific-and-technological advance in the area of industrial ecology and environmental monitoring: Collected reports. *International Conference on Constructions Materials Industry and Building Industry, Efficient Use of Energy and Resources in the Context of Market Relations*. Belgorod, Russia: BelGTASM Publishing, 1997. Part 9. pp. 15–19.

69. M.A. Morev. Simulation and optimization of operation of compensation air-jet hoods: Synopsis of thesis for the degree of Cand.Sc. (Engineering). Voronezh, Russia, 2000. 17pp.

70. V.G. Shaptala and G.L. Okuneva. *Numerical Modeling of Workshop Air Exchange Based on Navier-Stokes Equations*. Mathematical modeling in construction materials industry: Collected scientific papers. Belgorod, Russia: BTISM Publishing, 1992. pp. 49–54.

71. G.L. Okuneva. *Numerical Study of Separated Flows in Ventilation Premises*. Mathematical modeling in construction materials industry: Collected scientific papers. Belgorod, Russia: BTISM Publishing, 1992. pp. 54–59.

72. V.G. Shaptala and G.L. Okuneva. Numerical modeling of impurity concentration and temperature distribution in a plane air flow. Topical issues of fundamental sciences: Papers of the *Second International Scientific-and-Technical Conference*. Moscow, Russia: Technosfera-Inform, 1994. Vol. 1, Part 1. pp. 129–131.

73. V.V. Shaptala. *Computational Modeling of Heat- and Gas-Generating Equipment Aspiration*. Mathematical modeling of processes in production of construction materials and structures: Collected scientific papers. Belgorod, Russia: BelGTASM Publishing, 1998. pp. 74–79.

74. V.V. Shaptala. *Computational Experiment in Study of Local Exhaust Ventilation Efficiency*. Mathematical modeling of processes in production of construction materials and structures: Collected scientific papers. Belgorod, Russia: BelGTASM Publishing, 1998. pp. 80–85.

75. V.G. Shaptala, I.N. Logachev, G.L. Okuneva, and V.G. Shaptala. Mathematical modeling of impurity concentration and temperature distribution in burning departments: Collected reports. *International Conference on Constructions Materials Industry and Building Industry, Efficient Use of Energy and Resources in the Context of Market Relations*. Belgorod, Russia: BelGTASM Publishing, 1997. pp. 60–63.

76. V.G. Shaptala, V.A. Minko, I.N. Logachev et al. *Mathematical Support of CAD Installations of Ventilation Systems*. Tutorial. Belgorod, Russia: BelGTASM Publishing, 1998. 77pp.

77. V.K. Khrushch. Mathematical modeling of pollutant dispersion in natural spheres. Visnik Dnipropetrovskogo universitetu. Mekhanika. Vol. 1. Issue 1. 1998. pp. 53–65.

78. N.N. Belyaev and V.K. Khrushch. Numerical computation of aerosol pollution propagation. Dnepropetrovsk: Dnepropetrovsk State University, Ukraine, 1990. 80pp.

79. O.G. Goman. Discrete vortex method in hydrodynamics: Background and application experience. Visnik Dnipropetrovskogo universitetu. Mekhanika. Vol. 1. Issue 1. 1998. pp. 21–29.

80. V.G. Ivenskiy. Development of suction systems based on the swirling effect: Synopsis of thesis for the degree of Cand. Sci. (Engineering). Rostov-on-Don, Russia, 1991. 18pp.

81. O.D. Neykov, I.N. Logachev, and R.N. Shumilov. Aspiration of vapor and dust mixtures in dedusting of process equipment. Kiev, Ukraine: Naukova Dumka, 1974. 127pp.

82. I.N. Logachev. Air injection in a free-discharging jet of bombarding particles. Reducing harmful occupational factors at ore mining enterprises. Moscow, Russia: Nedra, 1985. pp. 56–63.

83. I.N. Logachev. On air circulation in chutes in transfer of unheated bulk materials. In *Occupational Safety in Metal Mining Industry*. Moscow, Russia: Nedra, 1985. pp. 56–63.

84. I.N. Logachev and L.M. Chernenko. Peculiarities of air injection in a free-discharging jet of coalescent powder. In *Occupational Safety in Metal Mining Industry*. Moscow, Russia: Nedra, 1985. pp. 65–69.

85. I.N. Logachev. Basis for calculation of technical means of air containment and dedusting in order to reduce the intensity of airborne dust emissions in transfer of unheated bulk materials at preparation factories: Thesis for the degree of Doctor of Engineering. Krivoy Rog, Ukraine; Belgorod, Russia, 1996. 688pp.

86. O.D. Neykov and I.N. Logachev. *Suction in Production of Powders*. Moscow, Russia: Metallurgiya, 1973. 224pp.

87. V.V. Nedin and O.D. Neykov. *Dust Control at Mines*. Moscow, Russia: Nedra, 1965. 200pp.

88. I.N. Logachev. Flow strain in ventilated hood openings and strain impact on suction volumes. In *Control of Hazardous and Harmful Occupational Factors at Ore Mining Enterprises*. Moscow, Russia: Nedra, 1988. pp. 65–71.

89. M.I. Gurevich. *Theory of Ideal Fluid Jets*. Moscow, Russia: Physmathgiz, 1961. 496pp.

90. N.V. Polyakov. *Numerical Analytic Methods of Resolving Non-Linear Boundary Problems*. Dnepropetrovsk, Ukraine: DSU Publishing, 1991. 144pp.

91. N.D. Vorobyev, V.S. Bogdanov, and M. Yu. Eltsov. *Simulating a Grinding Medium Interaction with a Tube-Mill Liner. Physico-Mathematical Methods in Construction Materials Science*. Moscow, Russia: MISI Publishing; Belgorod, Russia: BTISM Publishing, 1986. pp. 168–173.

92. Sh.F. Araslanov and Sh.Kh. Zaripov. Analysis of dust-laden gas flow in an inertial air cleaner. *Proceedings of the Russian Academy of Sciences*. Fluid Mechanics. 1996. No. 6. pp. 62–68.

93. O.N. Zaitsev. The improvement of the processes re-movement of harmful things by tightened streams from nonfiction heat sources. Abstract Thesis … Cand.Sc. (Engineering). Odessa, Ukraine, 1996. 16p.

94. G.A. Kruglov. Development of resource saving mining process dust control technologies: Synopsis of thesis for the degree of Doctor of Engineering. Chelyabinsk, Russia, 1997. 43pp.

95. L.G. Loytsyanskiy. *Fluid Mechanics*. Moscow, Russia: Nauka, 1973. 848pp.

96. L.I. Ilizarova. The pattern of flow behind bluff body. Industrial aerodynamics. Moscow, Russia: Mashinostroeniye, 1966. 27th edn. pp. 96–110.

97. S.M. Tsoy. Enhancing dust suction efficiency by means of reshaping the suction flare. Dust control or air and process equipment in industry. Rostov-on-Don, Russia, 1991. pp. 5–6.

98. A.N. Tikhonov and A.A. Samarskiy. *Equations of Mathematical Physics*. Moscow, Russia: Nauka, 1977. 735pp.

99. G. Lamb. Hydrodynamics. M, 1947. 930pp.

100. B.A. Fuks and B.V. Shabat. *Functions of Complex Variable and Some Applications of the Same*. Moscow, Russia: Physmathgiz, 1959. 376pp.

101. V.I. Krylov. *Approximate Integration*. Moscow, Russia: Nauka, 1967. 500pp.

102. V.I. Krylov, V.V. Lugin, and L.A. Yanovich. *Tables for Numerical Integration of Exponential Functions*. Minsk, Belarus: BSSR AS Publishing, 1963.

103. V.M. Ryasnoy. Research into methods of dust reduction in short-hole drilling with stopper drills: Synopsis of thesis for the degree of Cand. Sci. (Engineering). L., 1981. 24pp.

104. V.A. Minko. *Dust Control of Construction Materials Production Processes*. Voronezh, Russia: VSU Publishing, 1981. 176pp.

105. I.I. Afanasiev, I.N. Logachev, V.A. Minko et al. *Air Dedusting at Mining-and-Processing Integrated Works Factories*. Moscow, Russia: Nedra, 1972. 184pp.

106. A.F. Vlasov. *Cutting Tool Clearance of Dust and Chips*. Moscow, Russia: Mashinostroeniye, 1982. 240pp.

107. V.A. Larionov and V.P. Sozinov. Controlled suction systems in woodwork and timber industry. Moscow, Russia: Forest industry, 1989. 240pp.

108. V.V. Ryabov. Pneumatic removal of metal chips. Moscow, Russia: Mashinostroeniye, 1988. 144pp.

109. A.N. Zheltkov. Calculating volumes of suction from high-production machines intended for rough grinding of steel and alloys. Local exhaust ventilation. Moscow, Russia: MDNTP n.a. F.E. Dzerzhinskiy, 1969. pp. 67–71.

110. D.V. Koptev. Dust control at electrode and electrical coal works. Moscow, Russia: Metallurgiya, 1980. 127pp.

111. L.D. Landau and E.M. Livshits. Continuum mechanics. Moscow, Russia: Gostekhizdat, 1944. 624pp.

112. S.M. Belotserkovskiy and I.K. Lifanov. *Numerical Method in Singular Integral Equation*. Moscow, Russia: Nauka, 1985. 256pp.

113. I.K. Lifanov. *The Method of Singular Integral Equations and Numerical Experiment*. Moscow, Russia: Janus, 1995. 520pp.

114. M.V. Katkov, A.G. Labutkin, N.B. Salimov, and V.N. Posokhin. Flow at a slotted lateral hood. *Izvestiya vuzov/Construction*. 1998. No. 11–12. pp. 96–100.

115. V.N. Posokhin and M.V. Katkov. Observational study of vortex zones in flow at slotted suction inlets. *Izvestiya vuzov/Aeronautical Engineering*. 2001. No. 1. pp. 61–63.

116. S.M. Belotserkovskiy and A.S. Ginevskiy. *Simulating Turbulent Jets and Wakes Based on the Discrete Vortex Method*. Moscow, Russia: Physmathlit, 1995. 368pp.

117. V.A. Sabelnikov and E.A. Smirnykh. Numerical computation of turbulent flow at the initial section of a sharp-edged flat duct using the discrete vortex method. *Proceedings of CAHI*. 1985. T.XVI. pp. 59–64.

118. V.L. Zimont, V.E. Kozlov, and A.A. Praskovskiy. The study of turbulent flow at the initial section of a sharp-edged cylindrical duct. *Proceedings of CAHI*. 1981. Vol. XII. No. 1. pp. 145–152.

119. V.M. Voloschuk. *Introduction to Hydrodynamics of Coarsely Dispersed Aerosols*. Leningrad, Russia: Hydrometeoizdat, 1975. 448pp.

120. V.G. Shaptala. *Mathematical Modeling in Applied Problems of Two-Phase Flow Mechanics*. Belgorod, Russia: BelGTASM Publishing, 1996. 102pp.

121. M.E. Berlyand. Present problems of atmospheric diffusion and atmospheric pollution. Leningrad, Russia: Hydrometeoizdat, 1975. 448pp.

122. V.N. Uzhov, A.Yu. Valdberg, B.I. Myagkov, and I.K. Reshidov. *Dedusting Dust-Laden Gases*. Moscow, Russia: Chemistry, 1981. 392pp.

123. Manual of special functions, M. Abramovits and I. Stigan (eds.). Moscow, Russia: Nauka, 1979. 832pp.

124. V.N. Posokhin, N.B. Salimov, K.I. Logachev, and A.M. Zhivov. Regarding analysis of flows around a slotted bell-shaped hood inlet. *Izvestiya vuzov/Construction*. 2002. Paper 1. No. 8. pp. 70–76; Paper 2. No. 9. pp. 80–85; Paper 3. No. 10. pp. 81–85.

125. M.V. Katkov. The study of slotted drain flows: Thesis for the degree of Cand. Sci. (Engineering). Kazan, Russia. KSTU n.a. A.N. Tupolev, 2000, 153pp.

126. O.G. Goman, V.I. Karplyuk, M.I. Nisht, and A.G. Sudakov. Numerical modeling of axiosymmetrical separated flows of incompressible liquid. M.I. Nisht (ed.). Moscow, Russia: Mashinostroeniye, 1993. 288pp.

127. J.L. Alden. *Design of Industrial Exhaust Systems*. New York: The Industrial Press, 1959.

128. G.D. Livshits and B.L. Gil. Mathematical computer-aided modeling of suction flares of local exhausts built in equipment. *Izvestiya vuzov/Construction and Architecture*. 1986. No.7, pp. 90–93.

129. I.E. Idelchik. *Hydraulic Resistance Guide Book*. Moscow, Russia: Mashinostroeniye, 1977. 559pp.

130. K.I. Logachev, A.I. Puzanok, and V.N. Posokhin. Analysis of flows around a slotted bell-shaped hood inlets using the discrete vortex method. *Izvestiya vuzov/Problems of Energetics*. 2004. No. 7–8. pp. 61–69.

131. K.I. Logachev. Regarding design of slotted hoods from rotating cylindrical parts. *Izvestiya vuzov/Construction*. 2002. No. 11. pp. 67–73.

132. I.N. Logachev and K.I. Logachev. On forecasting dispersed composition and concentration of coarsely dispersed aerosols in local hoods of suction systems. *Izvestiya vuzov/Construction*. 2002. No. 9. pp. 85–90.

133. K.I. Logachev and N.M. Anzheurov. On simulation of air flows around slotted suction inlets surrounded with thin canopies. *Izvestiya vuzov/Construction*. 2003. No. 1. pp. 58–62.

134. K.I. Logachev and R.V. Prokopenko. On numerical modeling of dimensional air flows around suction inlets of local hoods from rotating cylindrical parts. *Izvestiya vuzov/Construction*. 2003. No. 8. pp. 74–82.

135. K.I. Logachev and R.V. Prokopenko. Regarding modeling of air flows around slotted hoods using the vortex method. *Izvestiya vuzov/Construction*. 2003. No. 9. pp. 100–105.

136. K.I. Logachev and A.I. Puzanok. "Spectrum" software complex intended for modeling air-and-coal flows next to slotted suction inlets. *Izvestiya vuzov/Construction*. 2004. No. 1. pp. 59–64.

137. K.I. Logachev and V.N. Posokhin. Analysis of flows around intake pipes of round cross-section. *Izvestiya vuzov/Aeronautical Engineering*. 2004. No. 1. pp. 29–32.

138. K.I. Logachev, I.N. Logachev, and A.I. Puzanok. Computational modeling of air-and-coal flows next to suction holes. *CD-Proceedings of European Congress on Computational Methods in Applied Sciences and Engineering ECCOMAS 2004*, Jyväskylä, Finlane, July 24–28, 2004, 19pp.

139. K.I. Logachev, A.I. Puzanok, and V.N. Posokhin. Design of vortex flows around slotted lateral hood. *Izvestiya vuzov/Construction*. 2004. No. 6. pp. 64–69.

140. K.I. Logachev and A.I. Puzanok. Numerical modeling of air-and-coal flows next to rotating suction cylinder. *Izvestiya vuzov/Construction*. 2005. No. 2. pp. 63–70.

141. K.I. Logachev, A.I. Puzanok, and E.V. Selivanova. Numerical computation of flows next to shielded bell-shaped suction hood. *Izvestiya vuzov/Construction*. 2005. No. 6. pp. 53–58.

142. W.A. Sirignano. *Fluid Dynamics and Transport of Droplets and Sprays*. Cambridge, U.K.: Cambridge University Press, 1999. 311pp.

143. A.D. Altshul and P.G. Kiselyov. *Hydraulics and Aerodynamics*. Moscow, Russia: Stroyizdat, 1975. 323pp.

144. V.N. Khanzhonokov. Reduction in aerodynamic drag of openings using ring shaped ribs and grooves. *Industrial Aerodynamics*, No. 12. Moscow, Russia: Oborongiz, 1959. pp. 181–196.

145. I.E. Idelchik. Determining outflow resistance coefficients. *Hydraulic Engineering*. 1958. No. 5. pp. 31–36.

146. M.M. Nosova. Resistance of shielded inlet and outlet pipes. *Industrial Aerodynamics*. No. 12. Moscow, Russia: Oborongiz, 1959. pp. 197–215.

147. I.N. Logachev and K.I. Logachev. *Aerodynamic Basis of Suctions*. S-Pb: Khimizdat, 2005. 659pp.

148. I.N. Logachev, K.I. Logachev, and O.A. Averkova. Mathematical modeling of separated flows at inlet of a shielded flat duct. *Computational Methods and Programming*. 2010. Vol. 11. No. 1. pp. 68–77.

149. I.N. Logachev, K.I. Logachev, V.Yu. Zorya, and O.A. Averkova. Modeling of separated flows next to suction slots. *Computational Methods and Programming*. 2010. Vol. 11, No. 1. pp. 43–52.

150. F. Vorheimer. Hydraulics. USSR NKTP Main Office of Energetic Literature. Moscow, Russia. 1935. 615pp.

151. V.I. Khanzhonkov. Resistance of supply and exhaust shafts. *Industrial Aerodynamics*. 1947. No. 3. pp. 210–214.

152. I.E. Idelchik. Flow resistances in flow upstream of channels and at ports. *Industrial Aerodynamics Digest*, No. 2. BNT, NKAP, 1944. pp. 27–57.

153. V. Dyke. *Sketchbook of Liquid and Gas Flows*. Moscow, Russia: Mir, 1986. 182pp.

154. V.V. Baturin and V.I. Khanzhonkov. Indoor air circulation based on arrangement of supply and exhaust ports. *Heating and Ventilation*. 1939. No. 4–5. pp. 29–33.

155. V.G. Shaptala. *Mathematical Modeling in Applied Problems of Two-Phase Flow Mechanics*. Tutorial. Belgorod, Russia: BelGTASM Publishing. 1996. 103pp.

156. A.V. Khoperskov. Forming instable conditions when simulating suction flows: Kelvin-Helmholtz instability, A.A. Khoperskov, V.N. Azarov, S.A. Khoperskov, E.A. Korotkov, A.G. Zhurmaliev (eds.), *Volgograd State University Bulletin*. Series 1. Mathematics. Physics. 2011. No. 1. pp. 141–155.

157. I.N. Logachev, A.M. Golishev et al., Local exhausts and extract hoods of preparation complexes. Krivoy Rog, Ukraine: VNIIBTG, 1985. 87pp.

158. P.A. Kouzov. *The Basis for Analysis of Dispersed Composition of Industrial Dust and Crushed Materials*. Leningrad, Russia: "Chemistry" Publishing, 1974. 280pp.

159. Fluent 6.1 Users' Guide, http://202.185.100.7/homepage/fluent/html/ug/main_pre.htm. 2003.

160. J.M. McDonough. Introductory lectures on turbulence: Physics, mathematics and modeling. http://www.engr.uky.edu/~acfd/lctr-notes634.pdf. 2007.

161. J.M. McDonough and J. Endean. Parallel simulation of type IIa supernovae explosions, using a simplified physical model. *Lecture Notes in Computational Science and Engineering*, 2009. Vol. 67. pp. 355–362.

162. Industry Standard OST 14-17-98-83. Preparation of metallurgical source materials. Aspiration. Method for computing the efficiency of local suction units in cowls at loose-matter handling locations. Effective March 1, 1984. Moscow, Russia: Ministry of Ferrous Metallurgy of the USSR. 1983. 32pp. (Occupational Safety Standards System, USSR).

163. O.D. Neykov and I.N. Logachev. *Suction and Air Dedusting in Production of Powders*. Moscow, Russia: Metallurgiya, 1981. 192pp.

164. A.V. Kalmykov and D.F. Zhurbinsky. *Dust and Noise Control at Beneficiation Plants*. Moscow, Russia: Nedra, 1984. 222pp.

165. V.D. Afanasiev, I.N. Logachev, V.I. Stukanov et al., Procedure manual for comprehensive improvement of labor conditions at metallurgical ore beneficiation plants. Leningrad, Russia: Institute for Mechanical Processing of Mineral Products, 1984. 169pp.

166. A.M. Golyshev. A survey of local exhaust ventilation for firing and screening of iron-ore pellets. PhD thesis in Technology, May 23, 2003. Krivyy Rih, Ukraine, 1980. 242pp.

167. Survey and development of means for optimizing aspiration cowls in loose-matter handling facilities, Final Research Study Report (All-Union Institute for Occupational Safety in Ore Mining [VNIIBTG]), Study directors: I.N. Logachev, A.M. Golyshev, No. GR 81010690; Inv. No. 02830003679. Krivyy Rih, Ukraine, 1982. 90pp.

168. I.N. Logachev, A.M. Golyshev, and L.M. Chernenko. Reducing aspiration losses of dustlike materials in conditions specific to iron-ore pelletization plants. Gornyi Zhurnal. Issue 3, 1985, pp. 57–59.

169. USSR Certificate of Authorship (CoA) #931601. IPC B65 G11/00. Pneumatic classifier sealing point, G.K. Suldimirov, Z.I. Khomenko, A.S. Liventsev et al., No. 2765531/27-11; filed May 3, 1979. Otkrytiya. Izobreteniya. 1982. No. 20. pp. 92–93.

170. CoA #662462. PIC B65 G 65/32. Loose matter loading device, I.P. Polyakov, A.V. Snagorsky. No. 2441103/22-11; filed January 4, 1977. Otkrytiya. Izobreteniya. 1979. No. 18. 100pp.

171. CoA #745817. IPC B65 G 47/78. A device for unloading materials from conveyor, G.A. Tsimbal, E.D. Gorbatko. No. 2072953; filed November 4, 1974. Otkrytiya. Izobreteniya. 1979. No. 25. 100pp.

172. CoA #615001. IPC B65 G 11/02. A device for gravity-flow transportation of loose matter, N.F. Grashchenkov, B. Tsay, 2398123/29–11; filed September 2, 1976. Otkrytiya. Izobreteniya. 1978. No. 26. 61pp.

173. CoA #644679. IPC G11/02. A device for gravity-flow transportation of loose matter, E.A. Dmitruk, V.P. Sukhenko, V.P. Chobotov. No. 2473005/29–11; filed April 13, 1977. Otkrytiya. Izobreteniya. 1979. No. 4. 64pp.

174. CoA #998269. IPC B65 G69/18. An aspiration device for loose-matter handling facilities, N.F. Grashchenkov, V.S. Kharkovsky, A.F. Kirsik et al. No. 3300808/27–11; filed March 20, 1981. Otkrytiya. Izobreteniya. 1983. No. 7. 126pp.

175. CoA #103063. IPC E21 F5/00. Cowl design for loose-matter handling locations, V.D. Olifer, S.A. Kozinets. No. 3381533/22–03; filed November 25, 1981. Otkrytiya. Izobreteniya. 1983. No. 27. 143pp.

176. CoA #445599. IPC B65 G3/18. Cowl design for loose-matter pouring locations/N.S. Tsitsorin. 1703602/23–26; filed October 6, 1971. Otkrytiya. Izobreteniya. 1974. No. 37. 51pp.

177. CoA #1105406. IPC B65 G21/00. Cowl design for belt conveyor loading location, I.N. Logachev, A.M. Golyshev, S.I. Zadorozhny et al. No. 3589109/27–03; filed May 10, 1983. Otkrytiya. Izobreteniya. 1984. No. 28. 52pp.

178. CoA #485928. IPC B65 G19/28. A gravity-flow chute for granular materials, O.D. Neykov, Ya.I. Zilberberg, E.D. Nesterov et al. No. 1688112/27-11; field September 08, 1971. Otkrytiya. Izobreteniya. 1975. No. 36. 45pp.

179. CoA #777238. IPC E21 F5/00. Cowl design for loose-matter transfer locations in belt conveyor systems, M.I. Feskov, N.N. Dmitriyenko. No. 2728240/22-03; filed February 12, 1979. Otkrytiya. Izobreteniya. 1979. No. 41, 131pp.

180. M.I. Fekov. Using plumes of sprayed water for dust extraction. Bezopasnost truda v promyshlennosti, 1982. Issue 9, pp. 44–46.

181. CoA #939347. IPC B65 G21/08. A dust control device for material handling locations in belt conveyor systems. S.I. Sergeyev, V.G. Shlyapin, I.D. Galin et al. No. 3232898/27-03; filed December 31, 1980, Otkrytiya. Izobreteniya. 1982. No. 24. 107pp.

182. CoA #962127. IPC B65 G21/00. Aspiration cowl design for transfer facilities in belt conveyor systems, B.I. Berezhnoy, A.M. Kirichenko, L.V. Kaplenko et al. No. 3265798/27-03; filed March 24, 1981. Otkrytiya. Izobreteniya. 1982. No. 36. 91pp.

183. CoA #1190064. IPC E21 F5/00. A dust control device, R.S. Sharafutdinov, A.N. Byov, G.R.Antonyants. No. 3471887/22-03; filed July 19, 1982. Otkrytiya. Izobreteniya. 1985. No. 41. 146pp.

184. CoA #950925. IPC E21 F5/00. Cowl design for loose-matter handling locations, V.D. Olifer, V.B. Rabinovich, S.A. Kozinets. No. 3233960/22-03; filed September 1, 1981. Otkrytiya. Izobreteniya. 1979. No. 30. 124pp.

185. O.D. Neykov and I.N. Logachev. Suction and air dedusting in production of powders. Moscow, Russia: Metallurgiya, 1981. 124pp.

186. CoA #921994. IPC B65 G21/00. Aspiration cowl design for belt conveyor loading location, S.A. Kozinets, V.D. Olifer, G.Yu. Khvostov, No. 2972381/27-03; filed August 15, 1980. Otkrytiya. Izobreteniya. 1982. No. 15. 98pp.

187. CoA #962130. IPC B65 G21/20. A device for sealing the sides of belt conveyor aspiration cowl, N.I. Kuz'minok, A.V. Kolpakov, V.M. Marinchenko et al. No. 3271544/27-03; filed March 18, 1981. Otkrytiya. Izobreteniya. 1982. No. 36. 91pp.

188. CoA #882854. IPC B65 G21/00. Belt conveyor loading location cowl. I.F. Yufit, No. 2897490/27-03; filed March 19, 1980. Otkrytiya. Izobreteniya. 1981. No. 43. 75pp.

189. CoA #589431. IPC B65 F5/00. Aspiration cowl design for belt conveyor transfer location, K.V. Kuzminov, A.P. Mikulevich, Yu.V. Vdovin, No. 2187663/22-03; filed October 29, 1975. Otkrytiya. Izobreteniya. 1978. No. 3. 114pp.

190. CoA #1148814. IPC B65 G21/08. Cowl design for belt conveyor transfer location, S.I. Sergeyev, V.G. Shlyapin, V.I. Popov et al. No. 3670778/27-03; filed December 12, 1983. Otkrytiya. Izobreteniya. 1985. No. 13. 54pp.

191. CoA #947015. IPC B65 G21/00. Cowl design for belt conveyor loading location, N.F. Graschenkov, V.S. Kharkovsky, B. Tsay et al. No. 2874795/27-03; filed January 24, 1980. Otkrytiya. Izobreteniya. 1982. No. 28. 97pp.

192. A survey of dust and air dynamics in aspiration networks of ore beneficiation plant shops. Part 1. Final Research Study Report. All-Union Institute for Occupational Safety in Ore Mining (VNIIBTG). Study directors: I.N. Logachev, A.M. Golyshev, No. GR 01530008202; Inv. No. 02850048954. Krivyy Rih, Ukraine, 1985. 82pp.

193. N.A. Bobrovnikov. *Ambient Dust Control in Transportation Industry*. Moscow, Russia: Transport, 1984. 72pp.

194. A.F. Germoni. Centralized exhaust system for aspiration air/gas with high concentrations of dust. Gornyi Zhurnal, 1983, No. 5. pp. 51–53.

195. CoA #973447. IPC B65 G21/00. Belt conveyor aspiration cowl. V.M. Kolesnichenko, L.K. Saplinov. No. 328705/27-03; filed May 18, 1981. Otkrytiya. Izobreteniya. 1982. No. 42. pp. 70–71.

196. Yu.Ya. Tikhonov. The effect of ionization on increasing efficiency of cowls at crushstone plants. Innovations in rail transport occupational safety: Works of the All-Union Railway Transport Research Institute. Moscow, Russia: Transport, 1973. Issue 493, pp. 31–43.

197. An inquiry into airborne dust control and occupational safety improvement in iron ore agglomeration shops. Final Research Study Report, All-Union Institute for Occupational Safety in Ore Mining (VNIIBTG); Director: I.N. Logachev, No. GR 73052312; Inv. No. B452597. Krivyy Rih, Ukraine, 1975. 158pp.

198. Survey and basic design studies for dust control and removal in iron-ore and ore concentrate processing. Final Research Study Report, All-Union Institute for Occupational Safety in Ore Mining (VNIIBTG); Director: I.N. Logachev, No. GR 78016036; Inv. No. B920758. Krivyy Rih, Ukraine, 1975. 242pp.

199. Development of guidance for identification and computation of fugitive dust emissions in the case of outdoor stockpiling of dust-releasing materials at ore beneficiation plant shops. Final Research Study Report, All-Union Institute for Occupational Safety in Ore Mining (VNIIBTG). Study directors: I.N. Logachev, A.M. Golyshev, No. GR 81063744; Inv. No. 02830004287. Krivyy Rih, Ukraine, 1985. 103pp.

200. I.N. Logachev. Ejection of air in a free jet of falling particles. Mitigating harmful production factors at ore processing enterprises. Works All-Union Institute for Occupational Safety in Ore Mining. Moscow, Russia: Nedra, 1985. pp. 56–63.

201. I.N. Logachev. The aerodynamics of one-dimensional flow of loose matter in inclined chutes. *Proceedings of the All-Union Conference on Processes in Disperse Through-Flows*. Odessa, Ukraine, 1967 (Engineering and Research Information Department, Institute for Thermal Physics Engineering), 23pp.

202. A.S. Serenko. *Airborne Dust Control in the Refractories Industry*. Kharkiv, Ukraine: Metallurgizdat, 1953. 146pp.

203. O.D. Neykov and Ya.I. Zilberberg. A survey of aerodynamics of closed chutes with gravitation flows of powdered materials. Prevention of spontaneous inflammation of powders and explosions in gas-solid dispersion systems. Works Institute of Material Sciences Problems under the Academy of Sciences of the Ukrainian SSR. Kyiv, Ukraine: Nauka Dumka, 1975. pp. 196–203.

204. I.M. Khaletsky. *Ventilation and Heating in Ferrous Metallurgy Plants*. Moscow, Russia: Metallurgiya, 1981. 240pp.

205. A.I. Sukhareva. *Reducing Airborne Dust Content at Glass Works*. Leningrad, Russia: Stroyizdat, 1976. 136pp.

206. I.I. Afanasyev, V.I. Vashenko, I.N. Logachev et al. *Airborne Dust Control in Ore Beneficiation Plant Shops*. Moscow, Russia: Nedra, 1972. 184pp.

207. A.V. Kalmykov. *Dust Control in Crushing Shops*. Moscow, Russia: Nedra, 1976. 207pp.

208. M.A. Lavrentyev and B.V. Shabat. *Methods of the Theory of Functions of a Complex Variable*. Moscow, Russia: Fizmatgid, 1958. 678pp.

209. V.M. Belyakov, R.I. Kravtsova, and M.G. Rappoport. *Tables of Ellyptical Integrals*. Moscow, Russia: Publishing House of the Academy of Sciences of the Soviet Union, 1963. Vol. 1. 783pp.

210. I.N. Logachev. *One-Dimensional Flow of Loose Matter in Inclined Chutes*. Ventilyatsiya i ochistka vozdukha. Issue 6. Moscow, Russia: Nedra, 1970 (Research Institute for Ventilation and Air Cleaning in Mining). pp. 121–128.

211. P.Ya. Polubarinova-Kochina. *Groundwater Motion Theory*. Moscow, Russia: Nauka, 1977. 664pp.

212. A.D. Altshul and P.G. Kiselyov. *Hydraulics and Aerodynamics*. Moscow, Russia: Stroyizdat, 1975, 327pp.

213. I.N. Logachev and K.I. Logachev. *Industrial Air Quality and Ventilation: Controlling Dust Emissions*. Boca Raton: CRC press, 2014. 417pp.

214. I.N. Logachev. *Aspiration in Loose-Matter Handling Systems of Agglomeration Shops*. Local exhaust ventilation. Moscow, Russia: Moscow House for Research and Engineering Information (MDNTI), 1969. pp. 93–106.

215. I.E. Idelchik. *Handbook of Hydraulic Resistance*. New York: Begell House, 2008. 861pp.

216. V.A. Minko, I.N. Logachev, K.I. Logachev et al. *Dedusting Ventilation*. Belgorod, Russia: Printing Office of the V.G. Shukhov State Technology University of Belgorod, 2010. Vol. 2. 565pp.

217. V.V. Nedin and O.D. Neykov. *Dust Control at Mines*. Moscow, Russia: Nedra, 1965. 200pp.

Index

A

Aerodynamic drag value of suction slot
  auxiliary variables, 62
  complex potential, 53
  computational solution, 65–66
  flow function, 63
  flow lines of suction flare, 63–64
  geometrical dimensions of physical flow range, 58, 62
  hydrodynamic picture, 65
  integrals, 54–55, 59
  jet height, 65–68
  jet thickness, 64–65, 67
  Joukowski function, 53–54
  maximum rate of kinetism, 68
  nature of infinity, 60–61
  parameters determination, plane slot with screen, 53–54
  parametric equation of free flow line, 57
  positive real semiaxis, 59
  relative coefficient of kinetism, 66, 68
  role of inner screen, 64–65
  sum of partial fractions, 55
Air ejection
  bypass ducts (*see* Bypass ducts)
  porous pipe (*see* Porous pipe)
Air exhaust duct, *see* Local exhaust ventilation
Aspiration system
  aerodynamic properties
    airflow and depth, 499–500
    bucket dimensions, 498–499
    drag coefficient, 497, 500
    LRC, 497, 500–501
  air flow rates
    air balance equation, 507
    bisection method, 509
    Butakov–Neykov number, 508
    cross-flow patterns, 511
    Euler's number, 508
    leaky locations, 507–508
    loading chute, 507–508
    negative pressure, 509–510
    sealing degree, 509–510
  aspirated cowls
    aerodynamic drag, 526–528
    aerology of mining, 521–522
    with bypass, 519–521
    computation algorithm, 514–515
    cost savings, 513
    design parameters, 514
    double-walled cowls, 511–512
    elevator head, 522–524, 528–529
    feeder drive drum cowl, 524–526
    feeder sealing, 515, 517
    flow rates, 510, 512–513
    loading and unloading units, 512

    negative pressure, leakages, 516, 518–519
    unaspirated cowls, 512, 515, 517–519
    unknown variables, 524
    without bypass, 514, 516
  ejection pressures, 503, 505
  iterative approach, 507
  junction points, 502
  LCR, 502–503
  negative pressure, 501–502, 504
  suction units, 504

B

BIEM, *see* Boundary integral equations method (BIEM)
Biot–Savart law, 5–6
Borda–Carnot effect, 98–99
Boundary integral equations method (BIEM)
  flow superposition method, 6
  multiply connected regions
    flat flows (*see* Flat flows in multiply connected regions)
    three-dimensional flows (*see* Three-dimensional flows in multiply connected regions)
Bucket elevators
  cross-flow velocity, 477, 481–483
  cross-sectional average, 474
  elevator enclosure, 476–477
  longitudinal airflow diagram, 478–481
  sealed enclosures
    differential pressure, 488–489
    dimensionless equation, 485–486
    direct-flow pattern, 484
    ejection capacity, 485
    relative flow rate, 490–493
    single-valued function, 486–490
    static pressure, 483
    trigger condition, 483–484
  tangential frictional stress, 475–476
Bypass ducts
  air recycling
    aerodynamic resistance forces, 386
    aspirated lower cowl, 415, 419
    averaged velocity, 430–432
    boundary-value problem, 389
    circulation loops, 427
    contiguity equations, 418
    cross-flow velocity, 431, 433–435
    design, 385–386
    differential equations, 428–430
    dimensionless pressure, 387–388
    elementary volume, 386–388
    excess pressures, 433–434
    global transit, 418–419
    immobile volume, 385
    inlet section, 420–421
    linear law, 427